Fourier and Laplace Transforms

This book presents in a unified manner the fundamentals of both continuous and discrete versions of the Fourier and Laplace transforms. These transforms play an important role in the analysis of all kinds of physical phenomena. As a link between the various applications of these transforms the authors use the theory of signals and systems, as well as the theory of ordinary and partial differential equations.

The book is divided into four major parts: periodic functions and Fourier series, non-periodic functions and the Fourier integral, switched-on signals and the Laplace transform, and finally the discrete versions of these transforms, in particular the Discrete Fourier Transform together with its fast implementation, and the z-transform. Each part closes with a separate chapter on the applications of the specific transform to signals, systems, and differential equations. The book includes a preliminary part which develops the relevant concepts in signal and systems theory and also contains a review of mathematical prerequisites.

This textbook is designed for self-study. It includes many worked examples, together with more than 450 exercises, and will be of great value to undergraduates and graduate students in applied mathematics, electrical engineering, physics and computer science.

Fourier and Laplace Transforms

R. J. Beerends, H. G. ter Morsche,
J. C. van den Berg and E. M. van de Vrie

Translated from Dutch by
R. J. Beerends

CAMBRIDGE
UNIVERSITY PRESS

CAMBRIDGE UNIVERSITY PRESS
Cambridge, New York, Melbourne, Madrid, Cape Town,
Singapore, São Paulo, Delhi, Tokyo, Mexico City

Cambridge University Press
The Edinburgh Building, Cambridge CB2 8RU, UK

Published in the United States of America by Cambridge University Press, New York

www.cambridge.org
Information on this title: www.cambridge.org/9780521534413

© Cambridge University Press 2003

This publication is in copyright. Subject to statutory exception
and to the provisions of relevant collective licensing agreements,
no reproduction of any part may take place without the written
permission of Cambridge University Press.

First published 2003
Reprinted 2006

A catalogue record for this publication is available from the British Library

ISBN 978-0-521-80689-3 Hardback
ISBN 978-0-521-53441-3 Paperback

Cambridge University Press has no responsibility for the persistence or
accuracy of URLs for external or third-party internet websites referred to in
this publication, and does not guarantee that any content on such websites is,
or will remain, accurate or appropriate. Information regarding prices, travel
timetables, and other factual information given in this work is correct at
the time of first printing but Cambridge University Press does not guarantee
the accuracy of such information thereafter.

Contents

Preface

This book arose from the development of a course on Fourier and Laplace transforms for the Open University of the Netherlands. Originally it was the intention to get a suitable course by revising part of the book *Analysis and numerical analysis*, part 3 in the series *Mathematics for higher education* by R. van Asselt et al. (in Dutch). However, the revision turned out to be so thorough that in fact a completely new book was created. We are grateful that Educaboek was willing to publish the original Dutch edition of the book besides the existing series.

In writing this book, the authors were led by a twofold objective:
- the 'didactical structure' should be such that the book is suitable for those who want to learn this material through self-study or distance teaching, without damaging its usefulness for classroom use;
- the material should be of interest to those who want to apply the Fourier and Laplace transforms as well as to those who appreciate a mathematically sound treatment of the theory.

We assume that the reader has a mathematical background comparable to an undergraduate student in one of the technical sciences. In particular we assume a basic understanding and skill in differential and integral calculus. Some familiarity with complex numbers and series is also presumed, although chapter 2 provides an opportunity to refresh this subject.

The material in this book is subdivided into parts. Each part consists of a number of coherent chapters covering a specific part of the field of Fourier and Laplace transforms. In each chapter we accurately state all the *learning objectives*, so that the reader will know what we expect from him or her when studying that particular chapter. Besides this, we start each chapter with an introduction and we close each chapter with a summary and a selftest. The *selftest* consists of a series of exercises that readers can use to test their own knowledge and skills. For selected exercises, answers and extensive hints will be available on the CUP website.

Sections contain such items as definitions, theorems, examples, and so on. These are clearly marked in the left margin, often with a number attached to them. In the remainder of the text we then refer to these numbered items.

For almost all theorems proofs are given following the heading *Proof*. The end of a proof is indicated by a right-aligned black square: ■. In some cases it may be wise to skip the proof of a theorem in a first reading, in order not to lose the main line of argument. The proof can be studied later on.

Examples are sometimes included in the running text, but often they are presented separately. In the latter case they are again clearly marked in the left margin (with possibly a number, if this is needed as a reference later on). The end of an example is indicated by a right-aligned black triangle: ◄.

Mathematical formulas that are displayed on a separate line may or may not be numbered. Only formulas referred to later on in the text have a number (right-aligned and in brackets).

Some parts of the book have been marked with an asterisk: *. This concerns elements such as sections, parts of sections, or exercises which are considerably more difficult than the rest of the text. In those parts we go deeper into the material or we present more detailed background material. The book is written in such a way that these parts can be omitted.

The major part of this book has been written by Dr R.J. Beerends and Dr H.G. ter Morsche. Smaller parts have been written by Drs J.C. van den Berg and Ir E.M. van de Vrie. In writing this book we gratefully used the comments made by Prof. Dr J. Boersma and the valuable remarks of Ir G. Verkroost, Ir R. de Roo and Ir F.J. Oosterhof.

Finally we would like to thank Drs A.H.D.M. van Gijsel, E.D.S. van den Heuvel, H.M. Welte and P.N. Truijen for their unremitting efforts to get this book to the highest editorial level possible.

Introduction

Fourier and Laplace transforms are examples of mathematical operations which can play an important role in the analysis of mathematical models for problems originating from a broad spectrum of fields. These transforms are certainly not new, but the strong development of digital computers has given a new impulse to both the applications and the theory. The first applications actually appeared in astronomy, prior to the publication in 1822 of the famous book *Théorie analytique de la chaleur* by Joseph Fourier (1768 – 1830). In astronomy, sums of sine and cosine functions were already used as a tool to describe *periodic* phenomena. However, in Fourier's time one came to the surprising conclusion that the Fourier theory could also be applied to *non-periodic* phenomena, such as the important physical problem of heat conduction. Fundamental for this was the discovery that an arbitrary function could be represented as a superposition of sine and cosine functions, hence, of simple periodic functions. This also reflects the essential starting point of the various Fourier and Laplace transforms: to represent functions or signals as a sum or an integral of simple functions or signals. The information thus obtained turns out to be of great importance for several applications. In electrical networks, for example, the sinusoidal voltages or currents are important, since these can be used to describe the operation of such a network in a convenient way. If one now knows how to express the voltage of a voltage source in terms of these sinusoidal signals, then this information often enables one to calculate the resulting currents and voltages in the network.

Applications of Fourier and Laplace transforms occur, for example, in physical problems, such as heat conduction, and when analyzing the transfer of signals in various systems. Some examples are electrical networks, communication systems, and analogue and digital filters. Mechanical networks consisting of springs, masses and dampers, for the production of shock absorbers for example, processes to analyze chemical components, optical systems, and computer programs to process digitized sounds or images, can all be considered as systems for which one can use Fourier and Laplace transforms as well. The specific Fourier and Laplace transform being used may differ from application to application. For electrical networks the Fourier and Laplace transforms are applied to functions describing a current or voltage as function of *time*. In heat conduction problems, transforms occur that are applied to, for example, a temperature distribution as a function of *position*. In the modern theory of digital signal processing, *discrete* versions of the Fourier and Laplace transforms are used to analyze and process a sequence of measurements or data, originating for example from an audio signal or a digitized photo.

In this book the various transforms are all treated in detail. They are introduced in a mathematically sound way, and many mutually related properties are derived, so that the reader may experience not only the differences, but above all the great coherence between the various transforms.

As a link between the various applications of the Fourier and Laplace transforms, we use the theory of signals and systems as well as the theory of ordinary and partial

FIGURE 0.1
When digitizing a photo, information is lost. Conditions under which a good reconstruction can be obtained will be discussed in part 5. *Copyright*: Archives de l'Académie des Sciences de Paris, Paris

differential equations. We do not assume that the reader is familiar with systems theory. It is, however, an advantage to have some prior knowledge of some of the elementary properties of linear differential equations.

Considering the importance of the applications, our first chapter deals with *signals* and *systems*. It is also meant to incite interest in the theory of Fourier and Laplace transforms. Besides this, part 1 also contains a chapter with mathematical preparations for the parts to follow. Readers with a limited mathematical background are offered an opportunity here to supplement their knowledge.

In part 2 we meet our first transform, specifically meant for *periodic* functions or signals. This is the theory of *Fourier series*. The central issue in this part is to investigate the information on a periodic function that is contained in the so-called Fourier coefficients, and especially if and how a periodic function can be described by these Fourier coefficients. The final chapter of this part examines some of the applications of Fourier series in continuous-time systems and in solving ordinary and partial differential equations. Differential equations often originate from a physical problem, such as heat conduction, or from electrical networks.

Part 3 treats the *Fourier integral* as a transform that is applied to functions which are *no longer periodic*. In order to construct a sound theory for the Fourier integral – keeping the applications in mind – we can no longer content ourselves with the classical notion of a function. In this part we therefore pay special attention in chapters 8 and 9 to *distributions*, among which is the well-known delta function. Usually, a consistent treatment of the theory of distributions is only found in advanced textbooks on mathematics. This book shows that a satisfactory treatment is also feasible for readers without a background in theoretical mathematics. In the final chapter of this part, the use of the Fourier integral in systems theory and in solving partial differential equations is explained in detail.

The *Laplace transform* is the subject of part 4. This transform is particularly relevant when we are dealing with phenomena that are *switched on*. In the first chapter an introduction is given to the theory of complex functions. It is then easier for the reader to conceive of a Laplace transform as a function defined on the complex numbers. The treatment in part 4 proceeds more or less along the same lines as in parts 2 and 3, with a focus on the applications in systems theory and in solving differential equations in the closing chapter.

In parts 2, 3 and 4, transforms were considered for functions defined on the real numbers or on a part of these real numbers. Part 5 is dedicated to the *discrete* transforms, which are intended for functions or signals defined on the integers.

These functions or signals may arise by *sampling* a continuous-time signal, as in the digitization of an audiosignal (or a photo, as in figure 0.1). In the first chapter of this part, we discuss how this can be achieved without loss of information. This results in the important sampling theorem. The second chapter in this part starts with the treatment of the first discrete transform in this book, which is the so-called *discrete Fourier transform*, abbreviated as DFT. The *Fast Fourier Transform*, abbreviated as FFT, is the general term for several fast algorithms to calculate the DFT numerically. In the third chapter of part 5 an FFT, based on the popular situation where the 'length of the DFT' is a power of two, is treated extensively. In part 5 we also consider the z-transform, which plays an important role in the analysis of discrete systems. The final chapter is again dedicated to the applications. This time, the use of discrete transforms in the study of discrete systems is explained.

Part 1
Applications and foundations

Contents of Chapter 1

Signals and systems

CHAPTER 1

Signals and systems

INTRODUCTION

Fourier and Laplace transforms provide a technique to solve differential equations which frequently occur when translating a physical problem into a mathematical model. Examples are the vibrating string and the problem of heat conduction. These will be discussed in chapters 5, 10 and 14.

Besides solving differential equations, Fourier and Laplace transforms are important tools in analyzing signals and the transfer of signals by systems. Hence, the Fourier and Laplace transforms play a predominant role in the theory of signals and systems. In the present chapter we will introduce those parts of the theory of signals and systems that are crucial to the application of the Fourier and Laplace transforms. In chapters 5, 10, 14 and 19 we will then show how the Fourier and Laplace transforms are utilized.

Signals and systems are introduced in section 1.1 and then classified in sections 1.2 and 1.3, which means that on the basis of a number of properties they will be divided into certain classes that are relevant to applications. The fundamental signals are the sinusoidal signals (i.e. sine-shaped signals) and the time-harmonic signals. Time-harmonic signals are complex-valued functions (the values of these functions are complex numbers) which contain only one frequency. These constitute the fundamental building blocks of the Fourier and Laplace transforms.

The most important properties of systems, treated in section 1.3, are linearity and time-invariance. It is these two properties that turn Fourier and Laplace transforms into an attractive tool. When a linear time-invariant system receives a time-harmonic signal as input, the resulting signal is again a time-harmonic signal with the same frequency. The way in which a linear time-invariant system transforms a time-harmonic signal is expressed by the so-called frequency response, which will also be considered in section 1.3.

The presentation of the theory of signals and systems, and of the Fourier and Laplace transforms as well, turns out to be much more convenient and much simpler if we allow the signals to have complex numbers as values, even though in practice the values of signals will usually be real numbers. This chapter will therefore assume that the reader has some familiarity with the complex numbers; if necessary one can first consult part of chapter 2, where the complex numbers are treated in more detail.

LEARNING OBJECTIVES
After studying this chapter it is expected that you
- know what is meant by a signal and a system
- can distinguish between continuous-time, discrete-time, real, complex, periodic, power, energy and causal signals
- know what a sinusoidal and a time-harmonic signal are
- are familiar with the terms amplitude, frequency and initial phase of a sinusoidal and a time-harmonic signal
- know what is meant by the power- and energy-content of a signal and in particular know what the power of a periodic signal is
- can distinguish between continuous-time, discrete-time, time-invariant, linear, real, stable and causal systems
- know what is meant by the frequency response, amplitude response and phase response for a linear time-invariant system
- know the significance of a sinusoidal signal for a real linear time-invariant system
- know the significance of causal signals for linear time-invariant causal systems.

1.1   Signals and systems

To clarify what will be meant by signals and systems in this book, we will first consider an example.

In figure 1.1 a simple electric network is shown in which we have a series connection of a resistor R, a coil L and a voltage generator. The generator in the network

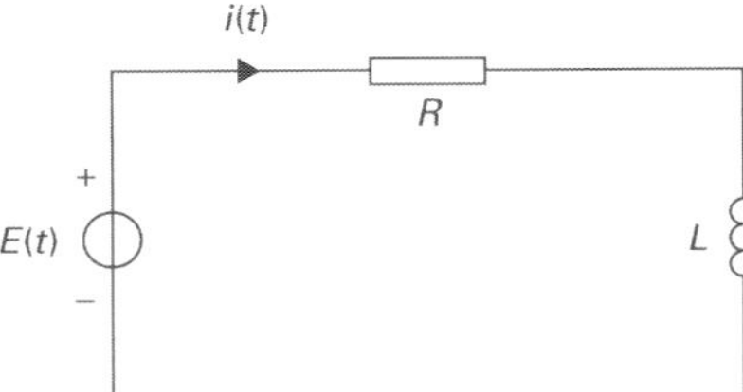

FIGURE 1.1
Electric network with resistor, coil and voltage generator.

supplies a voltage $E(t)$ and as a consequence a current $i(t)$ will flow in the network. From the theory of electrical networks it follows that the current $i(t)$ is determined unambiguously by the voltage $E(t)$, assuming that before we switch on the voltage generator, the network is at rest and hence there is no current flowing through the coil and resistor. We say that the current $i(t)$ is uniquely determined by the voltage $E(t)$. Using the Kirchhoff voltage-law and the current–voltage relationship for the resistor R and coil L, one can derive an equation from which the current $i(t)$ can be calculated explicitly as a function of time. Here we shall not be concerned with this derivation and merely state the result:

$$i(t) = \frac{1}{L} \int_{-\infty}^{t} e^{-(t-\tau)R/L} E(\tau)\, d\tau. \tag{1.1}$$

This is an integral relationship of a type that we shall encounter quite frequently in this book. The causal relation between $E(t)$ and $i(t)$ can be represented by the diagram of figure 1.2. The way in which $i(t)$ follows from $E(t)$ is thus given by the

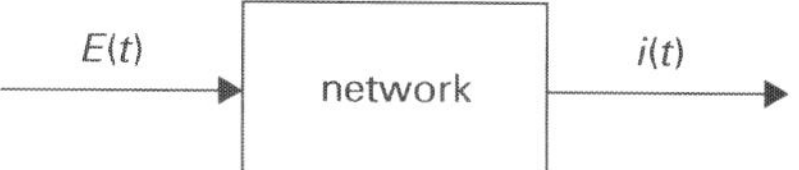

FIGURE 1.2
The relation between $E(t)$ and $i(t)$.

relation (1.1). Mathematically, this can be viewed as a mapping which assigns to a function $E(t)$ the function $i(t)$. In systems theory this mapping is called a system. The functions $E(t)$ and $i(t)$ are called the input and output respectively.

So a system is determined once the relationship is known between input and corresponding output. It is of no importance how this relationship can be realized physically (in our example by the electrical network). Often a system can even be realized in several ways. To this end we consider the mechanical system in figure 1.3, where a point-mass P with mass m is connected by an arm to a damper D. The point-mass P is acted upon by a force $F(t)$. As a result of the force the point P

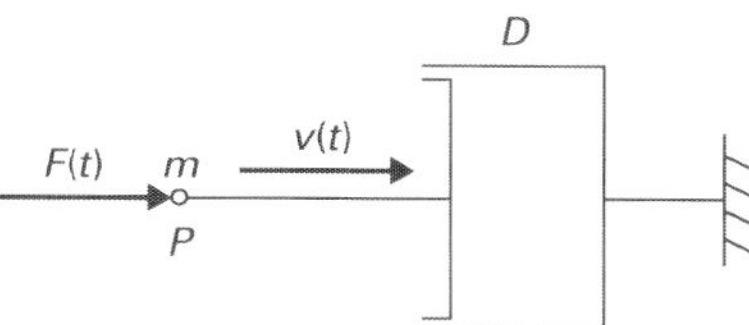

FIGURE 1.3
Mechanical system.

moves with velocity $v(t)$. The movement causes a frictional force K in the damper which is proportional to the velocity $v(t)$, but in direction opposite to the direction of $v(t)$. Let k be the proportionality constant (the damping constant of the damper), then $K = -kv(t)$. Using Newton's law one can derive an equation of motion for the velocity $v(t)$. Given $F(t)$ one can then obtain from this equation a unique solution for the velocity $v(t)$, assuming that when the force $F(t)$ starts acting, the mechanical system is at rest. Again we shall not be concerned with the derivation and only state the result:

$$v(t) = \frac{1}{m} \int_{-\infty}^{t} e^{-(t-\tau)k/m} F(\tau) \, d\tau. \tag{1.2}$$

Relation (1.2) defines, in the same way as relation (1.1), a system which assigns to an input $F(t)$ the output $v(t)$. But when $R = k$ and $L = m$ then, apart from the dimensions of the physical quantities involved, relations (1.1) and (1.2) are identical and hence the systems are equal as well. The realizations however, are different!

This way of looking at systems has the advantage that the properties which can be deduced from a system apply to all realizations. This will in particular be the case for the applications of the Fourier and Laplace transforms.

It is now the right moment to introduce the concept of a signal. The previous examples give rise to the following description of the notion of a signal.

Signal   A *signal* is a function.

Thus, in the example of the electrical network, the voltage $E(t)$ is a signal, which is defined as a function of time. The preceding description of the concept of a signal is very general and has thus a broad application. It merely states that it is a function. Even the domain, the set on which the function is defined, and the range, the set of function-values, are not prescribed. For instance, the yearly energy consumption in the Netherlands can be considered as a signal. See figure 1.4.

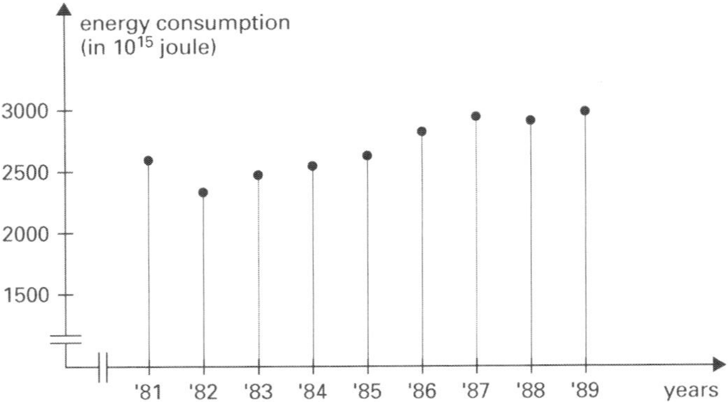

FIGURE 1.4
Energy consumption in the Netherlands.

Now that we have introduced the notion of a signal, it will also be clear from the foregoing what the concept of a system will mean in this book.

System

A *system* is a mapping L assigning to an *input u* a unique *output y*.

It is customary to represent a system as a 'black box' with an input and an output (see figure 1.5). The output *y* corresponding to the input *u* is uniquely determined by *u* and is called the *response* of the system to the input *u*.

Response

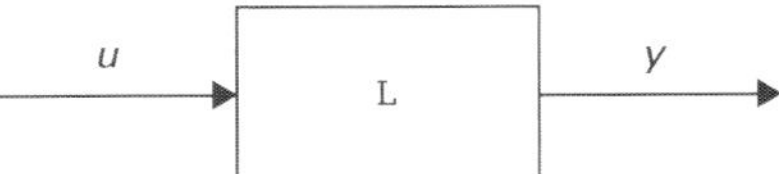

FIGURE 1.5
System.

When *y* is the response of a system L to the input *u*, then, depending on the context, we use either of the two notations

$$y = Lu,$$
$$u \mapsto y.$$

Our description of the concept of a system allows only one input and one output. In general more inputs and outputs are possible. In this book we only consider systems with one input and one output.

In the next section, signals will be classified on the basis of a number of properties.

1.2 Classification of signals

Real signal

The values that a signal can attain will in general be real numbers. This has been the case in all previous examples. Such signals are called *real* or *real-valued signals*. However, in the treatment of Fourier and Laplace transforms it is a great advantage to work with signals that have complex numbers as values. This means that we will suppose that a signal f has the form

$$f = f_1 + if_2,$$

Complex signal

where i is the imaginary unit for which $i^2 = -1$, and f_1 and f_2 are two real-valued signals. The signal f_1 is called the real part of the *complex signal* f (notation $\mathrm{Re}\, f$) and f_2 the imaginary part (notation $\mathrm{Im}\, f$). If necessary, one can first consult chapter 2, where a review of the theory of complex numbers can be found. In section 1.2.2 we will encounter an important example of a complex signal, the so-called time-harmonic signal.

Note that two complex signals are equal if the real parts and the imaginary parts of the complex signals agree. When for a signal f one has that $f_2 = \mathrm{Im}\, f = 0$, then the signal is real. When $f_1 = \mathrm{Re}\, f = 0$ and $f_2 = \mathrm{Im}\, f = 0$, then the signal f

Null-signal

is equal to zero. This signal is called the *null-signal*.

Usually, the signals occurring in practice are real. Hence, when dealing with results obtained from the application of Fourier and Laplace transforms, it will be important to consider specifically the consequences for real signals.

1.2.1 Continuous-time and discrete-time signals

In electrical networks and mechanical systems, the signals are a function of the time-variable t, a real variable which may assume all real values. Such signals are called

Continuous-time signal

continuous-time signals. However, it is not necessary that the adjective continuous-time has any relation with time as a variable. It only expresses the fact that the function is defined on $\mathbb{R}$ or a subinterval of $\mathbb{R}$. Hence, a continuous-time signal is a function defined on $\mathbb{R}$ or a subinterval of $\mathbb{R}$. One should not confuse the concept of a continuous-time signal with the concept of a continuous function as it is used in mathematics.

In the example of the yearly energy consumption in the Netherlands, the signal is not defined on $\mathbb{R}$, but only defined for discrete moments of time. Such a signal can be considered as a function defined on a part of $\mathbb{Z}$, which is the set of integers. In our example the value at $n \in \mathbb{Z}$ is the energy consumption in year n. A signal

Discrete-time signal

defined on $\mathbb{Z}$, or on a part of $\mathbb{Z}$, will be called a *discrete-time signal*.

As a matter of fact we assume in this book, unless explicitly stated otherwise, that continuous-time signals are defined on the whole of $\mathbb{R}$ and discrete-time signals on the whole of $\mathbb{Z}$. In theory, a signal can always be extended to, respectively, the whole of $\mathbb{R}$ or the whole of $\mathbb{Z}$.

We denote continuous-time signals by $f(t)$, $g(t)$, etc. and discrete-time signals by $f[n]$, $g[n]$, etc., hence using square brackets surrounding the argument n.

The introduction of continuous-time and discrete-time signals that we have given above excludes functions of more than one variable. In this book we thus confine ourselves to signals depending on one variable only. As a consequence we also confine ourselves to systems where the occurring signals depend on one variable only.

1.2.2   Periodic signals

An important class of signals consists of the periodic signals.

Periodic continuous-
time signal

A continuous-time signal $f(t)$ is called *periodic* with period $T > 0$ if

$$f(t + T) = f(t) \quad \text{for } t \in \mathbb{R}.$$

Periodic discrete-
time signal

A discrete-time signal $f[n]$ is called *periodic* with period $N \in \mathbb{N}$ if

$$f[n + N] = f[n] \quad \text{for } n \in \mathbb{Z}.$$

Sinusoidal signal

In the class of periodic signals the so-called *sinusoidal signals* play an important role. These are real signals which, in the continuous-time case, can be written as:

$$f(t) = A \cos(\omega t + \phi_0) \quad \text{for } t \in \mathbb{R}.$$

Amplitude
Frequency
Initial phase

Here A is the *amplitude*, ω the (radial)*frequency* and ϕ_0 the *initial phase* of the signal. The period T equals $T = 2\pi/\omega$.

In the discrete-time case the sinusoidal signals have the form:

$$f[n] = A \cos(\omega n + \phi_0) \quad \text{for } n \in \mathbb{N}.$$

Again A is the amplitude and ϕ_0 the initial phase. The period N equals $N = 2\pi/\omega$. From this it follows that ω cannot be arbitrary since N is a natural number!

We now introduce an important complex periodic signal which we will repeatedly come across in Fourier transforms. In order to do so, we will use Euler's formula for complex numbers, which is treated extensively in chapter 2, but will also be introduced here in a nutshell.

A complex number $z = x + iy$ with real part x and imaginary part y is represented in the complex plane by a point with coordinates (x, y). Then the distance $r = \sqrt{x^2 + y^2}$ to the origin is called the *modulus* $|z|$ of z, while the angle ϕ of the radius vector with the positive real axis is called the *argument* of z; notation $\phi = \arg z$. The argument is thus determined up to an integral multiple of 2π. See figure 1.6. Using

Modulus
Argument

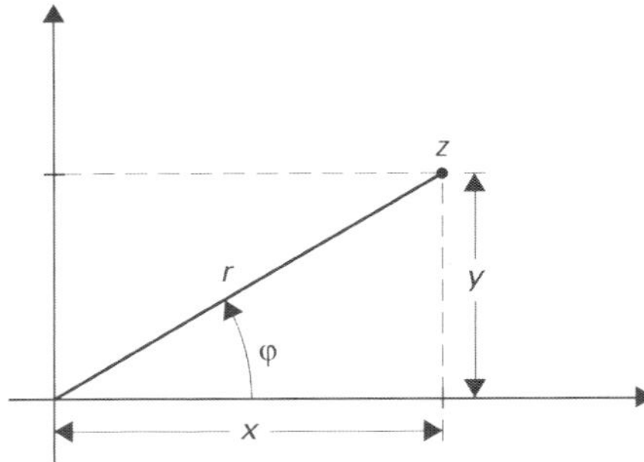

FIGURE 1.6
Representation of a complex number in the complex plane.

Euler's formula

polar coordinates in the complex plane, the complex number z can also be written as $z = r(\cos\phi + i \sin\phi)$. For the complex number $z = \cos\phi + i \sin\phi$ *Euler's formula* gives the following representation as a complex exponential:

$$e^{i\phi} = \cos\phi + i \sin\phi.$$

Hence, a complex number z with modulus r and argument ϕ can be written as

$$z = re^{i\phi}.$$

This formula has major advantages. One can compute with this complex exponential as if it were a real exponential. The most important rule is the product formula. When $z_1 = r_1 e^{i\phi_1}$ and $z_2 = r_2 e^{i\phi_2}$, then one has, as expected,

$$z_1 z_2 = r_1 r_2 e^{i(\phi_1 + \phi_2)}.$$

This means that $|z_1 z_2| = |z_1||z_2|$ and $\arg(z_1 z_2) = \arg z_1 + \arg z_2$. The latter relationship obviously only holds up to an integral multiple of 2π.

There are, however, also differences with the real exponential. For the real exponential it follows from $e^x = e^y$ that $x = y$. This does not hold for complex exponentials because $e^{i(\phi + 2\pi)} = e^{i\phi}$ for all real ϕ, since $e^{2\pi i} = 1$.

In this chapter it is not our intention to go into the theory of complex numbers any further. We will return to this in chapter 2. The preceding part of the theory of complex numbers is only intended to allow the introduction of the following complex periodic signal.

Let $\omega \in \mathbb{R}$ and c be a complex constant. The complex signal $f(t)$ is called a *time-harmonic* continuous-time signal when it is given by

Time-harmonic
continuous-time signal

$$f(t) = ce^{i\omega t} \quad \text{for } t \in \mathbb{R}.$$

If we write the complex number c as $c = Ae^{i\phi_0}$, where A is the modulus of c and ϕ_0 the argument, then the time-harmonic signal can also be written as follows:

$$f(t) = Ae^{i\phi_0} e^{i\omega t} = Ae^{i(\omega t + \phi_0)}.$$

For a given value of t, one can represent $f(t)$ in the complex plane by a point on the circle having the origin as centre and A as radius. At time $t = 0$ the argument is equal to ϕ_0, the initial phase. In the complex plane the signal $f(t)$ corresponds to a circular movement with constant angular velocity $|\omega|$. See figure 1.7. The movement is in the clockwise direction if $\omega < 0$ and counter-clockwise if $\omega > 0$. Note that the time-harmonic signal $f(t)$ is periodic with period $2\pi/|\omega|$. The real number ω is called the *frequency* of the time-harmonic signal, A the *amplitude* and ϕ_0 the *initial phase*. Hence, the frequency can be negative and it then loses its physical meaning. In the complex plane the sign of ω does indicate the *direction* of the circular movement and then $|\omega|$ is the frequency.

Amplitude, frequency and
initial phase of a
time-harmonic signal

Above we introduced a time-harmonic signal in the continuous-time case. Similarly we define in the discrete-time case:

Let $\omega \in \mathbb{R}$ and c be a complex constant. The discrete-time signal $f[n]$ is called a *time-harmonic* discrete-time signal when it is given by

Time-harmonic discrete-time
signal

$$f[n] = ce^{i\omega n} \quad \text{for } n \in \mathbb{Z}.$$

In contrast to time-harmonic continuous-time signals, a time-harmonic discrete-time signal will in general not be periodic. Only when $|\omega| = 2\pi/N$ for some positive integer N will the time-harmonic discrete-time signal be periodic with period N.

A final remark we wish to make concerns the relationship between a time-harmonic signal and a sinusoidal signal. We only consider the continuous-time case; the discrete-time case follows from this by replacing t with the variable n. From Euler's formula it follows that

$$f(t) = Ae^{i(\omega t + \phi_0)} = A(\cos(\omega t + \phi_0) + i\sin(\omega t + \phi_0)).$$

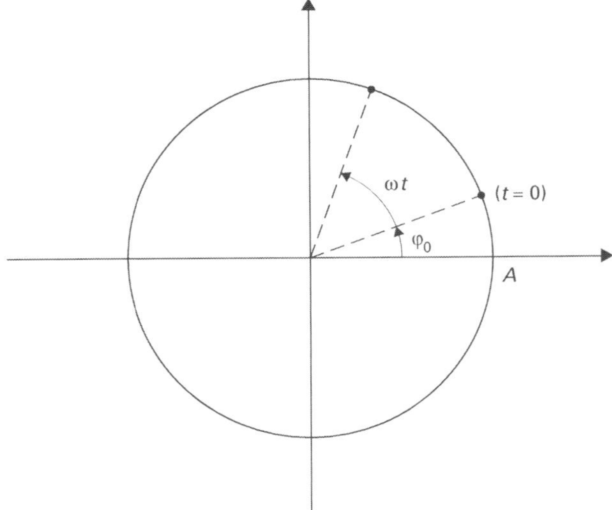

FIGURE 1.7
Time-harmonic signal.

We conclude that a sinusoidal signal is the real part of a time-harmonic signal. A
sinusoidal signal can also be written as a combination of time-harmonic signals. For
this we use the complex conjugate $\bar{z} = x - iy$ of a complex number $z = x + iy$.
Then the complex conjugate of $e^{i\phi}$ equals $e^{-i\phi}$. Since $\cos\phi = (e^{i\phi} + e^{-i\phi})/2$,
one has (verify this):

$$A\cos(\omega t + \phi_0) = \frac{ce^{i\omega t} + \bar{c}e^{-i\omega t}}{2} \quad \text{with } c = Ae^{i\phi_0}.$$

1.2.3   *Power and energy signals*

In electrical engineering it is customary to define the power of an element in an
electric network, through which a current $i(t)$ flows and which has a drop in voltage
$v(t)$, as the product $i(t)v(t)$. The average power over the time-interval $[t_0, t_1]$ then
equals

$$\frac{1}{t_1 - t_0} \int_{t_0}^{t_1} i(t)v(t)\,dt.$$

For a resistor of 1 ohm one has, ignoring the dimensions of the quantities involved,
that $v(t) = i(t)$, so that in this case the average power equals

$$\frac{1}{t_1 - t_0} \int_{t_0}^{t_1} i^2(t)\,dt.$$

In signal theory this expression is called the average power of the signal $i(t)$ over the
time-interval $[t_0, t_1]$. The limiting-case, in which the average power is taken over
ever increasing time-intervals, leads to the definition of the power of a signal.

Power (continuous-time)   The *power P* of a continuous-time signal $f(t)$ is defined by

$$P = \lim_{A \to \infty} \frac{1}{2A} \int_{-A}^{A} |f(t)|^2 \, dt. \tag{1.3}$$

Power-signal (continuous-time)   If the power of a signal is finite, then the signal is called a *power-signal*. Notice that in equation (1.3) we have not used $f^2(t)$, but $|f(t)|^2$. The reason is that the power of a signal should not be negative, while for complex numbers the square could indeed become negative, since $i^2 = -1$, and could even result in a complex value.

An example of a power-signal is a periodic signal. One can calculate the limit in (1.3) explicitly for a periodic signal $f(t)$ with period T. We formulate the result, without proof, as follows.

Let $f(t)$ be a periodic continuous-time signal with period T. Then $f(t)$ is a power-signal with power P equal to

Power of a periodic continuous-time signal

$$P = \frac{1}{T} \int_{-T/2}^{T/2} |f(t)|^2 \, dt.$$

Besides the power-signals, one also has the so-called energy-signals. In the preceding example of the resistor of 1 ohm, the amount of energy absorbed by the resistor during the time-interval $[t_0, t_1]$ equals

$$\int_{t_0}^{t_1} i^2(t) \, dt.$$

The definition of the energy-content of a signal concerns the time-interval $-\infty$ to ∞. Hence, the *energy-content* of a continuous-time signal $f(t)$ is defined by

Energy-content (continuous-time)

$$E = \int_{-\infty}^{\infty} |f(t)|^2 \, dt.$$

Energy-signal (continuous-time)   A continuous-time signal with a finite energy-content is called an *energy-signal*.

For discrete-time signals one uses analogous concepts. Integrals turn into sums. We will contend ourselves here with stating the definitions.

The *power P* of a discrete-time signal $f[n]$ is defined by

Power (discrete-time)

$$P = \lim_{M \to \infty} \frac{1}{2M} \sum_{n=-M}^{M} |f[n]|^2.$$

Power-signal (discrete-time)   If the power of a discrete-time signal is finite, then the signal is called a *power-signal*. For a periodic discrete-time signal $f[n]$ with period N one has, as for the continuous-time case, that

Power of a periodic discrete-time signal

$$P = \frac{1}{N} \sum_{n=0}^{N-1} |f[n]|^2.$$

The *energy-content E* of a discrete-time signal $f[n]$ is defined by

Energy-content (discrete-time)

$$E = \sum_{n=-\infty}^{\infty} |f[n]|^2.$$

Energy-signal (discrete-time)   If the energy-content of a signal is finite, then the signal is called an *energy-signal*.

1.2.4   Causal signals

A final characteristic of a signal that we want to mention in this section has to do
with causality. In the next section the concept of causality is also introduced for
systems. In that context it will be easier to understand why the following definition
of a causal signal is used.

A continuous-time signal $f(t)$, or a discrete-time signal $f[n]$ respectively, is called
Causal   *causal* if

$$f(t) = 0 \qquad \text{for } t < 0,$$
$$f[n] = 0 \qquad \text{for } n < 0.$$

Periodic signals are thus not causal, with the exception of the null-signal. If a signal
has the property that $f(t) = 0$ for $t < t_0$ and some t_0, then we call t_0 the *switch-on*
Switch-on time   *time* of the signal $f(t)$. Note that this definition does not fix the switch-on time
uniquely. If a signal has the property that $f(t) = 0$ for all $t < 1$, then besides $t = 1$
as switch-on time, one can also use $t = 0$. Similar definitions apply to discrete-time
signals. Notice that *causal* signals have switch-on time $t = 0$, or $n = 0$ in the
discrete case.

EXERCISES

1.1   Given are the two sinusoidal signals $f_1(t) = A_1 \cos(\omega t + \phi_1)$ and $f_2(t) = A_2 \cos(\omega t + \phi_2)$ with the same frequency ω. Show that the sum $f_1(t) + f_2(t)$ is also a sinusoidal signal with frequency ω and determine its amplitude and initial phase.

1.2   Show that the sum of two time-harmonic signals $f_1(t)$ and $f_2(t)$ with the same frequency ω and with amplitudes A_1 and A_2 and initial phases ϕ_1 and ϕ_2 respectively is again a time-harmonic signal with frequency ω and determine its amplitude and initial phase.

1.3   Show that the sum of two discrete-time sinusoidal signals with the same frequency is again a discrete-time sinusoidal signal and determine its amplitude and initial phase.

1.4   Show that the sum of two discrete-time time-harmonic signals with the same frequency is again a discrete-time time-harmonic signal and determine its amplitude and initial phase.

1.5   Calculate the power of the sinusoidal signal $f(t) = A \cos(\omega t + \phi_0)$.

1.6   Calculate the energy-content of the signal $f(t)$ given by

$$f(t) = \begin{cases} e^{-t} & \text{for } t > 0, \\ 0 & \text{for } t \leq 0. \end{cases}$$

1.7   Calculate the power of the periodic discrete-time signal $f[n] = \cos(n\pi/2)$.

1.8   Calculate the energy-content of the causal discrete-time signal $f[n]$ given by

$$f[n] = \begin{cases} e^{-n} & \text{for } n \geq 0, \\ 0 & \text{for } n < 0. \end{cases}$$

1.3   Classification of systems

Besides signals one can also classify systems. It is customary to do this on the basis
of the type of signals being processed by that system.

1.3.1 Continuous-time and discrete-time systems

Continuous-time system

A *continuous-time system* is a system for which the input and output signals are continuous-time signals.

Discrete-time system

A *discrete-time system* is a system for which the input and output signals are discrete-time signals.

Discrete-time systems are of major importance in the modern field of digital signal processing (e.g. Van den Enden and Verhoeckx (in Dutch), 1987 – see *Literature* at the back of the book).

1.3.2 Linear time-invariant systems

We will now formulate two properties of systems that are crucial for the application of Fourier and Laplace transforms. The first property concerns the linearity of systems, while the second one concerns the time-invariance.

DEFINITION 1.1
Linear system

A system L is called linear *if for each two inputs u and v and arbitrary complex a and b one has*

$$\mathrm{L}(au + bv) = a\mathrm{L}u + b\mathrm{L}v. \tag{1.4}$$

For continuous-time systems this property can be denoted as

$$au(t) + bv(t) \mapsto a(\mathrm{L}u)(t) + b(\mathrm{L}v)(t)$$

and for discrete-time systems as

$$au[n] + bv[n] \mapsto a(\mathrm{L}u)[n] + b(\mathrm{L}v)[n].$$

Note that in the preceding linear combination $au + bv$ of the signals u and v, the coefficients a and b may be complex. Since in general we assume that the signals are complex, we will also allow complex numbers a and b in (1.4).

EXAMPLE 1.1

Let L be the continuous-time system described by equation (1.1). In order to show that the system is linear, we simply have to use the fact that integration is a linear operation. The proof then proceeds as follows:

$$au(t) + bv(t) \mapsto \frac{1}{L} \int_{-\infty}^{t} e^{-(t-\tau)R/L}(au(\tau) + bv(\tau))\,d\tau$$

$$= \frac{a}{L} \int_{-\infty}^{t} e^{-(t-\tau)R/L}u(\tau)\,d\tau + \frac{b}{L} \int_{-\infty}^{t} e^{-(t-\tau)R/L}v(\tau)\,d\tau$$

$$= a(\mathrm{L}u)(t) + b(\mathrm{L}v)(t).$$

$\blacktriangleleft$

EXAMPLE 1.2

For a discrete-time system L, the response $y[n]$ to an input $u[n]$ is given by

$$y[n] = \frac{u[n] + 2u[n-1] + u[n-2]}{4} \qquad \text{for } n \in \mathbb{Z}.$$

The output at time n is apparently a weighted average of the input $u[n]$ at times n, $n-1, n-2$. We verify the linearity of this system as follows:

$$au[n] + bv[n] \mapsto$$

$$\frac{au[n] + bv[n] + 2au[n-1] + 2bv[n-1] + au[n-2] + bv[n-2]}{4}$$

$$= \frac{a(u[n] + 2u[n-1] + u[n-2])}{4} + \frac{b(v[n] + 2v[n-1] + v[n-2])}{4}$$

$$= a(\mathrm{L}u)[n] + b(\mathrm{L}v)[n].$$

◄

The second property, the so-called time-invariance, has to do with the behaviour of a system with respect to time-delays, or, more generally, shifts in the variable t or n. When a system has the property that a time-shift in the input results in the same time-shift in the output, then the system is called time-invariant. A precise description is given in the following definition.

DEFINITION 1.2
Time-invariant system

A continuous-time system is called time-invariant *if for each input $u(t)$ and each $t_0 \in \mathbb{R}$ one has:*

$$\text{if} \qquad u(t) \mapsto y(t) \qquad \text{then} \qquad u(t - t_0) \mapsto y(t - t_0). \tag{1.5}$$

A discrete-time system is called time-invariant *if for each input $u[n]$ and each $n_0 \in \mathbb{Z}$ one has:*

$$\text{if} \qquad u[n] \mapsto y[n] \qquad \text{then} \qquad u[n - n_0] \mapsto y[n - n_0]. \tag{1.6}$$

EXAMPLE 1.3

Once again we consider the continuous-time system described by (1.1). This is a time-invariant system, which can be verified as follows:

$$u(t - t_0) \mapsto \frac{1}{L} \int_{-\infty}^{t} e^{-(t-\tau)R/L} u(\tau - t_0)\, d\tau$$

$$= \frac{1}{L} \int_{-\infty}^{t - t_0} e^{-(t - t_0 - \xi)R/L} u(\xi)\, d\xi = y(t - t_0).$$

In this calculation a new integration variable $\xi = \tau - t_0$ was introduced.   ◄

EXAMPLE 1.4

Again consider the discrete-time system given in example 1.2. This system is time-invariant as well. This immediately follows from condition (1.6):

$$u[n - n_0] \mapsto \frac{u[n - n_0] + 2u[n - 1 - n_0] + u[n - 2 - n_0]}{4} = y[n - n_0].$$

◄

Linear time-invariant system

A system which is both linear and time-invariant is called a *linear time-invariant system*. It is precisely these linear time-invariant systems for which the Fourier and Laplace transforms form a very attractive tool. These systems have the nice property that the response to a time-harmonic signal, whenever this response *exists*, is again a time-harmonic signal with the same frequency. However, the existence of the response to a time-harmonic signal is not ensured. This has to do with the eigenfrequencies and the stability of the system. We will discuss the role of stability following our next theorem. Treatment of eigenfrequencies will be postponed until chapter 5.

THEOREM 1.1

Let L be a linear time-invariant system and u a time-harmonic input with frequency $\omega \in \mathbb{R}$ for which the response exists. Then the output y is also a time-harmonic signal with the same frequency ω.

Proof
We will only prove the case of a continuous-time system. The proof for a discrete-time system can be given analogously. Let $u(t)$ be a time-harmonic input with frequency ω and let $y(t)$ be the corresponding output. Hence, $u(t) = ce^{i\omega t}$, where c is a complex constant and $\omega \in \mathbb{R}$. The system is time-invariant and so one has for each $\tau \in \mathbb{R}$:

$$ce^{i\omega(t-\tau)} \mapsto y(t - \tau).$$

On the other hand one has that

$$u(t - \tau) = ce^{i\omega(t-\tau)} = ce^{-i\omega\tau}e^{i\omega t} = e^{-i\omega\tau}u(t).$$

Because of the linearity of the system, the response to $u(t - \tau)$ is then also equal to $e^{-i\omega\tau}y(t)$. We conclude that for each $t \in \mathbb{R}$ and $\tau \in \mathbb{R}$ one has

$$y(t - \tau) = e^{-i\omega\tau}y(t).$$

Substitution of $t = 0$, and then replacing $-\tau$ by t, leads to $y(t) = y(0)e^{i\omega t}$, so $y(t)$ is again a time-harmonic signal with frequency ω. $\blacksquare$

From the preceding proof it follows that the response $y(t)$ to the input $u(t) = e^{i\omega t}$ equals $Ce^{i\omega t}$ for some complex constant C, which can still depend on ω, meaning that C is a function of ω. We call this function the *frequency response* of the system. Often one also uses the term *system function* or *transfer function*. For continuous-time systems the frequency response will be denoted by $H(\omega)$ and for discrete-time systems by $H(e^{i\omega})$. The reason for the different notation in the discrete case will not be explained until chapter 19.

The frequency response of a linear time-invariant system is thus defined by the following relations:

$$e^{i\omega t} \mapsto H(\omega)e^{i\omega t} \tag{1.7}$$

for a continuous-time system and

$$e^{i\omega n} \mapsto H(e^{i\omega})e^{i\omega n} \tag{1.8}$$

for a discrete-time system. The frequency response $H(\omega)$ is complex and so can be written in the form

$$H(\omega) = |H(\omega)|\, e^{i\Phi(\omega)}.$$

Here $|H(\omega)|$ and $\Phi(\omega)$ are, respectively, the modulus and the argument of $H(\omega)$. The function $|H(\omega)|$ is called the *amplitude response* and $\Phi(\omega)$ the *phase response*.

Once more we consider the system described by (1.1), which originated from the RL-network of figure 1.1. The response $y(t)$ to the input $u(t) = e^{i\omega t}$ equals

$$y(t) = \frac{1}{L}\int_{-\infty}^{t} e^{-(t-\tau)R/L}e^{i\omega\tau}\, d\tau = \frac{1}{L}\int_{0}^{\infty} e^{-\xi R/L}e^{i\omega(t-\xi)}\, d\xi$$

$$= \left(\frac{1}{L}\int_{0}^{\infty} e^{-\xi R/L}e^{-i\omega\xi}\, d\xi\right) e^{i\omega t}.$$

Frequency response

System function

Transfer function

Amplitude response

Phase response

EXAMPLE 1.5

This expression already shows that the response is again a time-harmonic signal with the same frequency ω. The frequency response in this example equals

$$H(\omega) = \frac{1}{L} \int_0^\infty e^{-\xi R/L} e^{-i\omega\xi} \, d\xi \, .$$

In chapter 6 we will learn how to calculate this kind of integral. This is because we are already dealing here with a Fourier transform. The result is:

$$H(\omega) = \frac{1}{R + i\omega L} \, .$$

◀

EXAMPLE 1.6

We consider the discrete-time system given in example 1.2, and calculate the response $y[n]$ to the input $u[n] = e^{i\omega n}$ as follows:

$$y[n] = \frac{e^{i\omega n} + 2e^{i\omega(n-1)} + e^{i\omega(n-2)}}{4} = \frac{e^{i\omega n}(1 + 2e^{-i\omega} + e^{-2i\omega})}{4} \, .$$

Again we see that the response is a time-harmonic signal. Apparently the frequency response equals

$$H(e^{i\omega}) = \frac{(1 + 2e^{-i\omega} + e^{-2i\omega})}{4} \, .$$

◀

1.3.3   Stable systems

Prior to theorem 1.1 we observed that the response to a time-harmonic signal doesn't always exist. For so-called stable systems, however, the response exists for all frequencies, and so the frequency response is defined for each ω. In order to describe what will be meant by a stable system, we first give the definition of a bounded signal.

A continuous-time signal $f(t)$, or a discrete-time signal $f[n]$ respectively, is called *bounded* if there exists a positive constant K such that

$$|f(t)| \leq K \qquad \text{for } t \in \mathbb{R} \text{ (continuous-time),}$$

$$|f[n]| \leq K \qquad \text{for } n \in \mathbb{N} \text{ (discrete-time).}$$

The definition of a stable system is now as follows.

DEFINITION 1.3
Stable system

A system L *is called* stable *if the response to each bounded signal is again bounded.*

A time-harmonic signal is an example of a bounded signal, since

$$\left| ce^{i\omega t} \right| = |c| \left| e^{i\omega t} \right| = |c| \qquad \text{for } t \in \mathbb{R}.$$

Hence, for a stable system the response to the input $ce^{i\omega t}$ exists and this response is bounded too.

1.3.4   Real systems

In our description of a system we assumed that the inputs and outputs are complex. In principle it is then possible that a real input leads to a complex output. A system is called *real* if the response to every real input is again real. Systems occurring in practice are mostly real.

If we apply a complex input u to a real linear system, so $u = u_1 + iu_2$ with u_1 and u_2 real signals, then, by the linearity property, the response y will equal $y_1 + iy_2$

where y_1 is the response to u_1 and y_2 the response to u_2. Since the system is real, the signals y_1 and y_2 are also real. For real linear systems one thus has the following property.

The response to the real part of an input u is equal to the real part of the output y and the response to the imaginary part of u is equal to the imaginary part of y.

One can use this property of real systems to calculate the response to a sinusoidal signal in a clever way in the case when the frequency response is known. For the continuous-time case this can be done as follows.

Let $u(t)$ be the given sinusoidal input $u(t) = A \cos(\omega t + \phi_0)$. Using Euler's formula we can consider the signal $u(t)$ as the real part of the time-harmonic signal $ce^{i\omega t}$ with $c = Ae^{i\phi_0}$. According to the definition of the frequency response, the response to this signal equals $cH(\omega)e^{i\omega t}$. The system being real, this implies that the response to the sinusoidal signal $u(t)$ is equal to the real part of $cH(\omega)e^{i\omega t}$. In order to calculate this real part, we write $H(\omega)$ in the form

$$H(\omega) = |H(\omega)| e^{i\Phi(\omega)},$$

where $|H(\omega)|$ is the modulus of $H(\omega)$ and $\Phi(\omega)$ the argument. For the response $y(t)$ to the input $u(t)$ we then find that

$$y(t) = \mathrm{Re}(Ae^{i\phi_0} |H(\omega)| e^{i\Phi(\omega)} e^{i\omega t}) = A|H(\omega)| \cos(\omega t + \phi_0 + \Phi(\omega)).$$

For real systems and real signals one can thus benefit from working with complex numbers. The response $y(t)$ is again a sinusoidal signal with amplitude $A|H(\omega)|$ and initial phase $\phi_0 + \Phi(\omega)$. The amplitude is multiplied by the factor $|H(\omega)|$ and one has a phase-shift $\Phi(\omega)$. On the basis of these properties it is clear why $|H(\omega)|$ is called the amplitude response and $\Phi(\omega)$ the phase response of the system.

1.3.5 Causal systems

A system for which the response to an input at any given time t_0 only depends on the input at times prior to t_0, hence, only on the 'past' of the input, is called a causal system. A precise formulation is as follows.

DEFINITION 1.4
Causal system

A continuous-time system L *is called* causal *if for each two inputs $u(t)$ and $v(t)$ and for each $t_0 \in \mathbb{R}$ one has:*

$$\text{if } u(t) = v(t) \text{ for } t < t_0, \text{ then } (\mathrm{L}u)(t) = (\mathrm{L}v)(t) \text{ for } t < t_0. \tag{1.9}$$

A discrete-time system L *is called* causal *if for each two inputs $u[n]$ and $v[n]$ and for each $n_0 \in \mathbb{Z}$ one has:*

$$\text{if } u[n] = v[n] \text{ for } n < n_0, \text{ then } (\mathrm{L}u)[n] = (\mathrm{L}v)[n] \text{ for } n < n_0. \tag{1.10}$$

Systems occurring in practice are mostly causal. The notion of causality can be simplified for linear time-invariant systems, since the following theorem holds for linear time-invariant systems.

THEOREM 1.2

A linear time-invariant system L *is causal if and only if the response to each causal input is again causal.*

Proof
Once again we confine ourselves to the case of continuous-time systems, since the proof for discrete-time systems is almost exactly the same and there are only some differences in notation.

Assume that the system L is causal and let $u(t)$ be a causal input. This means that $u(t) = 0$ for $t < 0$, so $u(t)$ equals the null-signal for $t < 0$. Since for linear systems the response to the null-signal is again the null-signal, it follows from (1.9) that the response $y(t)$ to $u(t)$ has to agree with the null-signal for $t < 0$, which means that $y(t)$ is causal.

Next assume that the response to each causal input is again a causal signal. Let $u(t)$ and $v(t)$ be two inputs for which $u(t) = v(t)$ for $t < t_0$. Now introduce $w(t) = u(t + t_0) - v(t + t_0)$. Then the signal $w(t)$, and so the response $(Lw)(t)$ as well, is causal. Since the system is linear and time-invariant, one has that $(Lw)(t) = (Lu)(t + t_0) - (Lv)(t + t_0)$. Hence, $(Lu)(t + t_0) = (Lv)(t + t_0)$ for $t < 0$, that is, $(Lu)(t) = (Lv)(t)$ for $t < t_0$. This finishes the proof. ∎

EXAMPLE 1.7   The linear time-invariant system described by (1.1) is causal. If we substitute a causal input $u(t)$ in (1.1), then for $t < 0$ the integrand equals 0 on the interval of integration $(-\infty, t]$, and so the integral also equals 0 for $t < 0$. ◀

EXAMPLE 1.8   The discrete-time system introduced in example 1.2 is causal. We have seen that the system is linear and time-invariant. Substitution of a causal signal $u[n]$ in the relation

$$y[n] = \frac{u[n] + 2u[n-1] + u[n-2]}{4}$$

leads for $n < 0$ to the value $y[n] = 0$. So the response is causal and hence the system is causal. ◀

1.3.6   *Systems described by differential equations*

For an important class of linear time-invariant continuous-time systems, the relation between the input $u(t)$ and the output $y(t)$ is described by a differential equation of the form

$$a_m \frac{d^m y}{dt^m} + a_{m-1} \frac{d^{m-1} y}{dt^{m-1}} + \cdots + a_1 \frac{dy}{dt} + a_0 y$$
$$= b_n \frac{d^n u}{dt^n} + b_{n-1} \frac{d^{n-1} u}{dt^{n-1}} + \cdots + b_1 \frac{du}{dt} + b_0 u.$$

For example, in electrical networks one can derive these differential equations from the so-called Kirchhoff laws. The differential equation above is called a linear differential equation with constant coefficients, and for these there exist general solution methods. In this chapter we will not pursue these matters any further. In chapters 5, 10 and 14, systems described by differential equations will be treated in more detail.

1.3.7   *Systems described by difference equations*

For the linear time-invariant discrete-time case, the role of differential equations is taken over by the so-called difference equations of the type

$$b_0 y[n] + b_1 y[n-1] + \cdots + b_M y[n - M]$$
$$= a_0 u[n] + a_1 u[n-1] + \cdots + a_N[n - N].$$

This equation for the input $u[n]$ and the output $y[n]$ is called a linear difference equation with constant coefficients. The systems described by difference equations are of major importance for the practical realization of systems. These will be discussed in detail in chapter 19.

EXERCISES

1.9 For a continuous-time system the response $y(t)$ to an input $u(t)$ is given by

$$y(t) = \int_{t-1}^{t} u(\tau)\,d\tau.$$

a Show that the system is real.
b Show that the system is stable.
c Show that the system is linear time-invariant.
d Calculate the response to the input $u(t) = \cos \omega t$.
e Calculate the response to the input $u(t) = \sin \omega t$.
f Calculate the amplitude response of the system.
g Calculate the frequency response of the system.

1.10 For a discrete-time system the response $y[n]$ to an input $u[n]$ is given by

$$y[n] = u[n-1] - 2u[n] + u[n+1].$$

a Show that the system is linear time-invariant.
b Is the system causal? Justify your answer.
c Is the system stable? Justify your answer.
d Calculate the frequency response of the system.

Cascade system

1.11 Two linear time-invariant continuous-time systems L_1 and L_2 are given with, respectively, frequency response $H_1(\omega)$ and $H_2(\omega)$, amplitude response $A_1(\omega)$ and $A_2(\omega)$ and phase response $\Phi_1(\omega)$ and $\Phi_2(\omega)$. The system L is a *cascade* connection of L_1 and L_2 as drawn below.

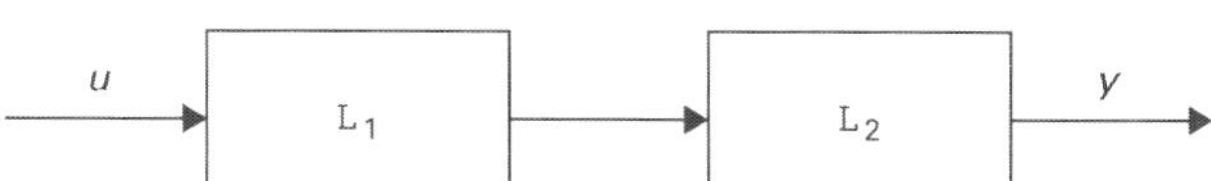

FIGURE 1.8
Cascade connection of L_1 and L_2.

a Determine the frequency response of L.
b Determine the amplitude response of L.
c Determine the phase response of L.

1.12 For a linear time-invariant discrete-time system the frequency response is given by
$H(e^{i\omega}) = (1 + i)e^{-2i\omega}$.
a Determine the amplitude response of the system.
b Determine the response to the input $u[n] = 1$ for all n.
c Determine the response to the input $u[n] = \cos \omega n$.
d Determine the response to the input $u[n] = \cos^2 2\omega n$.

SUMMARY

An important field for the applications of the Fourier and Laplace transforms is signal and systems theory. In this chapter we therefore introduced a number of important concepts relating to signals and systems.

Mathematically speaking, a system can be interpreted as a mapping which assigns in a unique way an output y to an input u. What matters here is the relation between input and output, not the physical realization of the system.

Mathematically, a signal is a function defined on $\mathbb{R}$ or $\mathbb{Z}$. The function values are allowed to be complex numbers.

In practice, various types of signal occur. Hence, the signals in this book were subdivided into continuous-time signals, which are defined on $\mathbb{R}$, and discrete-time signals, which are defined on $\mathbb{Z}$. An important class of signals is the periodic signals. Another subdivision is obtained by differentiating between energy- and power-signals. Signals occurring in practice are mostly real-valued. These are called real signals. An important real signal is the sinusoidal signal which, for a given frequency ω, initial phase ϕ_0 and amplitude A, can be written as $f(t) = A\cos(\omega t + \phi_0)$ in the continuous-time case and as $f[n] = A\cos(\omega n + \phi_0)$ in the discrete-time case. The sinusoidal signals are periodic in the continuous-time case. In general this is not true in the discrete-time case. A sinusoidal signal can be considered as the real part of a complex signal, the so-called time-harmonic signal $ce^{i\omega t}$ or $ce^{i\omega n}$, with frequency ω and complex constant c.

Time-harmonic signals play an important role, on the one hand in all of the Fourier transforms, and on the other hand in systems that are both linear and time-invariant. These are precisely the systems suitable for an analysis using Fourier and Laplace transforms, because these linear time-invariant systems have the property that time-harmonic input result in outputs which are again time-harmonic with the same frequency. For a linear time-invariant system, the relation between a time-harmonic input u and the response y can be expressed using the so-called frequency response $H(\omega)$ or $H(e^{i\omega})$ of the system:

$$e^{i\omega t} \mapsto H(\omega)e^{i\omega t} \qquad \text{(continuous-time system)},$$

$$e^{i\omega n} \mapsto H(e^{i\omega})e^{i\omega n} \qquad \text{(discrete-time system)}.$$

The modulus of the frequency response, $H(\omega)$ or $H(e^{i\omega})$ respectively, is called the amplitude response, while the argument of the frequency response is called the phase response of the system. Of practical importance are furthermore the real, the stable and the causal systems.

Real systems have the property that the response to a real input is again real. The response of a sinusoidal signal is then a sinusoidal signal as well, with the same frequency.

Stable systems have the property that bounded inputs result in outputs that are also bounded. For these systems the frequency response is well-defined for each ω.

The response of a causal system at a specific time t depends only on the input at earlier times, hence only on the 'past' of the input. For linear time-invariant systems causality means that the response to a causal input is causal too. Here a signal is called causal if it is switched on at time $t_0 \geq 0$.

SELFTEST

1.13   **a** Calculate the power of the signal $f(t) = A\cos\omega t + B\cos(\omega t + \phi_0)$.
      b Calculate the energy-content of the signal $f(t)$ given by

$$f(t) = \begin{cases} 0 & \text{for } t < 0, \\ \sin(\pi t) & \text{for } 0 \leq t < 1, \\ 0 & \text{for } t \geq 1. \end{cases}$$

1.14   Show that the power of the time-harmonic signal $f(t) = ce^{i\omega t}$ equals $|c|^2$.

1.15 **a** Calculate the power of the signal $f[n] = A\cos(\pi n/4) + B\sin(\pi n/2)$.
b Calculate the energy-content of the signal $f[n]$ given by

$$
f[n] = \begin{cases} 0 & \text{for } n < 0, \\ \left(\frac{1}{2}\right)^n & \text{for } n \geq 0. \end{cases}
$$

1.16 For a linear time-invariant continuous-time system the frequency response is given by

$$
H(\omega) = \frac{e^{i\omega}}{\omega^2 + 1} .
$$

a Calculate the amplitude and phase response of the system.
b The time-harmonic signal $u(t) = ie^{it}$ is applied to the system. Calculate the response $y(t)$ to $u(t)$.

1.17 For a real linear time-invariant discrete-time system the amplitude response $A(e^{i\omega})$ and phase response $\Phi(e^{i\omega})$ are given by $A(e^{i\omega}) = 1/(1 + \omega^2)$ and $\Phi(e^{i\omega}) = \omega$ respectively. To the system the sinusoidal signal $u[n] = \sin 2n$ is applied.
a Is the signal $u[n]$ periodic? Justify your answer.
b Show that the output is also a sinusoidal signal and determine the amplitude and initial phase of this signal.

1.18 For a continuous-time system the relation between the input $u(t)$ and the corresponding output $y(t)$ is given by

$$
y(t) = u(t - t_0) + \int_{t-1}^{t} u(\tau)\, d\tau.
$$

a For which values of t_0 is the system causal?
b Show that the system is stable.
c Is the system real? Justify your answer.
d Calculate the response to the sinusoidal signal $u(t) = \sin \pi t$.

1.19 For a discrete-time system the relation between the input $u[n]$ and the corresponding output $y[n]$ is given by

$$
y[n] = u[n - n_0] + \sum_{l=n-2}^{n} u[l] .
$$

a For which values of $n_0 \in \mathbb{Z}$ is the system causal?
b Show that the system is stable.
c Is the system real? Justify your answer.
d Calculate the response to the input $u[n] = \cos \pi n$.

Contents of Chapter 2

Mathematical prerequisites

INTRODUCTION

In this chapter we present an overview of the necessary basic knowledge that will be assumed as mathematical prerequisite in the chapters to follow. It is presupposed that the reader already has previous knowledge of the subject matter in this chapter. However, it is advisable to read this chapter thoroughly, and not only because one may discover, and fill in, possible gaps in mathematical knowledge. This is because in Fourier and Laplace transforms one uses the *complex numbers* quite extensively; in general the functions that occur are complex-valued, sequences and series are sequences and series of complex numbers or of complex-valued functions, and power series are in general complex power series. In introductory courses one usually restricts the treatment of these subjects to real numbers and real functions. This will not be the case in the present chapter. Complex numbers will play a prominent role.

In section 2.1 the principal properties of the complex numbers are discussed, as well as the significance of the complex numbers for the zeros of polynomials. In section 2.2 partial fraction expansions are treated, which is a technique to convert a rational function into a sum of simple fractions. Section 2.3 contains a short treatment of differential and integral calculus for complex-valued functions, that is, functions which are defined on the real numbers, but whose function values may indeed be complex numbers. One will find, however, that the differential and integral calculus for complex-valued functions hardly differs from the calculus of real-valued functions. In the same section we also introduce the class of piecewise continuous functions, and the class of piecewise smooth functions, which are of importance later on for the Fourier and Laplace transforms. In section 2.4 general convergence properties, and other fundamental properties, of sequences and series of complex numbers are considered. Again there will be some similarities with the theory of sequences and series of real numbers. Fourier series are series of complex-valued functions. Therefore some attention is paid to series of functions in section 2.4.3. Finally, we treat the complex power series in section 2.5. These possess almost identical properties to real power series.

LEARNING OBJECTIVES

After studying this chapter it is expected that you
- can perform calculations with complex numbers, in cartesian as well as in polar and exponential form
- know that by using complex numbers, a polynomial can be factorized entirely into linear factors and that you are able to perform this factorization in simple cases
- know what is meant by the nth roots of unity
- know the technique of (complex) partial fraction expansions
- can apply differential and integral calculus to complex-valued functions
- know what is meant by a piecewise continuous and a piecewise smooth function
- can apply the theory of sequences and series to sequences and series with complex terms
- are familiar with the concept of radius of convergence for complex power series and can calculate the radius of convergence in simple situations
- know the properties of the sum of a power series.

2.1   Complex numbers, polynomials and rational functions

2.1.1   *Elementary properties of complex numbers*

The complex numbers are necessary in order to determine the solutions of *all* quadratic equations. The equation $x^2 - 2x + 5 = 0$ has no solution in the real numbers. This is because by completing the square it follows that $x^2 - 2x + 5 = (x - 1)^2 + 4 > 0$ for all real x. If we now introduce the imaginary unit i, which by definition satisfies

$$i^2 = -1,$$

and subsequently the complex number $x = 1 + 2i$, then $(x - 1)^2 + 4 = (2i)^2 + 4 = -4 + 4 = 0$. Apparently the complex number $x = 1 + 2i$ is a solution of the given equation. Complex numbers are therefore defined as the numbers z that can be written as

$$z = x + iy \qquad \text{with } x, y \in \mathbb{R}. \tag{2.1}$$

Real part

Imaginary part

The collection of all these numbers is denoted by $\mathbb{C}$. The real number x is called the *real part* of z and denoted by $x = \operatorname{Re} z$. The real number y is called the *imaginary part* of z and denoted by $y = \operatorname{Im} z$. Two complex numbers are equal if the real parts and the imaginary parts are equal. For the complex number z one has that $z = 0$ if $\operatorname{Re} z = 0$ and $\operatorname{Im} z = 0$. For the addition and multiplication of two complex numbers $z = x + iy$ and $w = u + iv$ one has by definition that

$$z + w = (x + iy) + (u + iv) = (x + u) + i(y + v),$$

$$z \cdot w = (x + iy)(u + iv) = (xu - yv) + i(xv + yu).$$

For subtraction and division one subsequently finds:

$$z - w = (x - u) + i(y - v),$$

$$\frac{z}{w} = \frac{x + iy}{u + iv} = \frac{x + iy}{u + iv} \cdot \frac{u - iv}{u - iv} = \frac{xu + yv}{u^2 + v^2} + i\frac{yu - xv}{u^2 + v^2} \qquad \text{for } w \neq 0.$$

We see that both sum and product as well as difference and quotient of two complex numbers have been written in the form $z = \operatorname{Re} z + i\operatorname{Im} z$ again. The set $\mathbb{C}$ is an

Complex conjugate

extension of $\mathbb{R}$, since a real number x can be written as $x = x + 0 \cdot i$. The *complex conjugate* of a complex number z is defined as

$$\bar{z} = x - iy.$$

Note that $z\bar{z} = x^2 + y^2$ and $\bar{\bar{z}} = z$. One can easily check that the complex conjugate has the following properties:

$$\overline{z + w} = \bar{z} + \bar{w}, \qquad \overline{zw} = \bar{z}\bar{w}, \qquad \overline{\left(\frac{z}{w}\right)} = \frac{\bar{z}}{\bar{w}} \quad \text{if } w \neq 0.$$

Using the complex conjugate one can express the real and imaginary part of a complex number as follows:

$$\operatorname{Re} z = \frac{z + \bar{z}}{2} \qquad \text{and} \qquad \operatorname{Im} z = \frac{z - \bar{z}}{2i}.$$

Since we need two real numbers x and y to describe a complex number $z = x + iy$, one can assign to each complex number a *point in the plane* with rectangular coordinates x and y, as shown in figure 2.1. The coordinates x and y are called

Cartesian coordinates

Complex plane

Real axis

Imaginary axis

the *cartesian coordinates*. Figure 2.1 moreover shows the complex conjugate of z. The plane in figure 2.1 is called the *complex plane* and the axes the *real axis* and the *imaginary axis*. In this figure we see how the location of z can also be determined

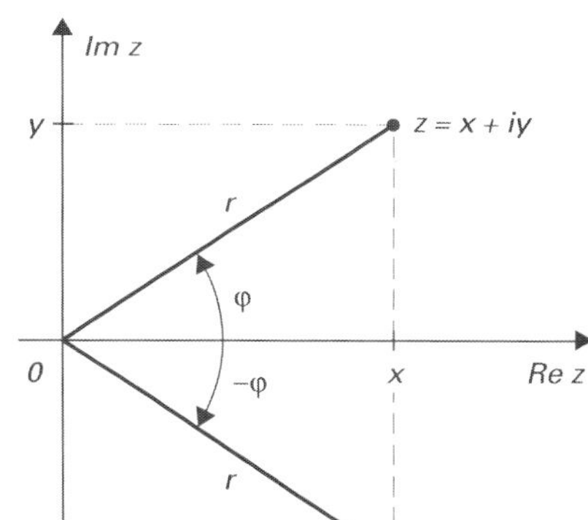

FIGURE 2.1
The complex number z and its complex conjugate $\bar{z}$.

Polar coordinates

by using *polar coordinates* r and ϕ. Here r is the distance from z to the origin and ϕ is the angle, expressed in radians, between the positive real axis and the vector from the origin to z. Since $x = r \cos \phi$ and $y = r \sin \phi$ one has that

$$z = r(\cos \phi + i \sin \phi). \tag{2.2}$$

Cartesian form

Polar form

Modulus

Unit circle

Expressing z as in (2.1) is called the *cartesian form* of z, while expressing it as in (2.2) is called the *polar form*. As we shall see in a moment, multiplication of complex numbers is much more convenient in polar form than in cartesian form. The number r, the distance from z to the origin, is called the *modulus* or *absolute value* of z and is denoted by $|z|$. The complex numbers with $|z| = 1$ all have distance 1 to the origin and so they form a circle with radius 1 and the origin as centre. This circle is called the *unit circle*. The modulus is a generalization of the absolute value for real numbers. For the modulus of $z = x + iy$ one has (see figure 2.1) $r^2 = |z|^2 = x^2 + y^2$, hence $r = |z| = \sqrt{x^2 + y^2}$. When z is real, $|z|$ coincides with the well-known absolute value of a real number: $|z| = |x| = \sqrt{x^2}$. If $z = x + iy$ and $w = u + iv$, then $|z - w|^2 = (x - u)^2 + (y - v)^2$. In the complex plane $|z - w|$ is thus equal to the distance between the complex numbers z and w. Using the definition of modulus it is easy to verify the following properties:

$$|z| = |\bar{z}|, \qquad z\bar{z} = |z|^2, \qquad \frac{1}{z} = \frac{\bar{z}}{|z|^2} \quad \text{if } z \neq 0. \tag{2.3}$$

An important property of the modulus is the so-called *triangle inequality*.

THEOREM 2.1
Triangle inequality

For complex numbers z and w one has

$$|z + w| \leq |z| + |w|.$$

Proof
If $z = x + iy$, then $|z|^2 = x^2 + y^2 \geq x^2 = |x|^2 = |\operatorname{Re} z|^2$. Hence, $|z| \geq |\operatorname{Re} z|$. In the same way one can show that $|z| \geq |\operatorname{Im} z|$. With these inequalities we can prove the triangle inequality as follows:

$$
\begin{aligned}
|z + w|^2 &= (z + w)(\bar{z} + \bar{w}) = z\bar{z} + (z\bar{w} + \bar{z}w) + w\bar{w} \\
&= z\bar{z} + 2\operatorname{Re}(z\bar{w}) + w\bar{w} \leq |z|^2 + 2|z\bar{w}| + |w|^2 \\
&= |z|^2 + 2|z||w| + |w|^2 = (|z| + |w|)^2.
\end{aligned}
$$

Hence, $|z + w| \leq |z| + |w|$.   ■

Argument

    The angle ϕ in the polar form of a complex number is called the *argument* of z and is denoted by $\phi = \arg z$. The argument of z is by no means uniquely determined since $\phi + 2k\pi$ (k an integer) is also an argument of z. If z is a positive number, then $\arg z = 0 + 2k\pi$. If z is a negative number, then $\arg z = \pi + 2k\pi$. From figure 2.1 one immediately infers that the property

$$\arg \bar{z} = -\arg z$$

holds up to an integer multiple of 2π. It will be agreed upon that all relations involving arguments have to be read with the following clause: they hold *up to an integer multiple of* 2π. In particular this also holds for the property in the following theorem.

THEOREM 2.2

For complex numbers z and w one has

$$\arg(zw) = \arg z + \arg w.$$

Proof
If $z = r(\cos \phi + i \sin \phi)$ and $w = s(\cos \psi + i \sin \psi)$, then we find by a straightforward multiplication, and using the formulas $\cos(\alpha + \beta) = \cos \alpha \cos \beta - \sin \alpha \sin \beta$ and $\sin(\alpha + \beta) = \sin \alpha \cos \beta + \cos \alpha \sin \beta$, that

$$
\begin{aligned}
zw &= rs(\cos \phi \cos \psi - \sin \phi \sin \psi + i(\cos \phi \sin \psi + \sin \phi \cos \psi)) \\
&= rs(\cos(\phi + \psi) + i \sin(\phi + \psi)).
\end{aligned}
$$

One thus has that $\arg(zw) = \phi + \psi = \arg z + \arg w$.   ■

    In the proof of theorem 2.2 we also see that $|zw| = |rs| = |r| \cdot |s| = |z| \cdot |w|$. Apparently a multiplication of complex numbers in polar form is simple: the arguments have to be added and the moduli multiplied. Similarly one can show that division is easy as well, when using the polar form. We summarize the properties of multiplication and division as follows:

$$\arg(zw) = \arg z + \arg w, \qquad |zw| = |z| \cdot |w|, \tag{2.4}$$

$$\arg\left(\frac{z}{w}\right) = \arg z - \arg w, \qquad \left|\frac{z}{w}\right| = \frac{|z|}{|w|} \quad \text{for } w \neq 0. \tag{2.5}$$

De Moivre's formula

From (2.4) it follows that for all integer n, so also for negative n, one has the so-called *De Moivre's formula*:

$$(\cos\phi + i\sin\phi)^n = \cos n\phi + i\sin n\phi \qquad \text{for } \phi \in \mathbb{R}.$$

Since multiplication of complex numbers means adding arguments, it is quite natural to use an exponential notation for the argument. For this we use Euler's formula.

DEFINITION 2.1
Euler's formula

For each $\phi \in \mathbb{R}$ the complex number $e^{i\phi}$ is defined as

$$e^{i\phi} = \cos\phi + i\sin\phi.$$

For $z = x + iy$ one defines e^z by

$$e^z = e^x e^{iy} = e^x(\cos y + i\sin y). \tag{2.6}$$

One can easily check that $\left|e^{i\phi}\right| = 1$ and $\arg(e^{i\phi}) = \phi$ for all $\phi \in \mathbb{R}$. Hence $e^{i\phi}$ lies on the unit circle for each $\phi \in \mathbb{R}$. Furthermore we have

$$\mathrm{Re}(e^z) = e^x \cos y, \quad \mathrm{Im}(e^z) = e^x \sin y,$$

$$\arg(e^z) = \mathrm{Im}\, z, \qquad \left|e^z\right| = e^x = e^{\mathrm{Re}\, z}.$$

Exponential form

Since each complex number z can be written as $z = r(\cos\phi + i\sin\phi)$ with $r = |z|$, one can also write z in the so-called *exponential form*

$$z = re^{i\phi} = |z|\, e^{i\phi}.$$

Analogous to the proof of theorem 2.2 it now follows for $\phi, \psi \in \mathbb{R}$ that

$$\begin{aligned} e^{i\phi} e^{i\psi} &= (\cos\phi + i\sin\phi)(\cos\psi + i\sin\psi) \\ &= \cos(\phi + \psi) + i\sin(\phi + \psi) = e^{i(\phi+\psi)}. \end{aligned} \tag{2.7}$$

The product of two complex numbers can now be written as

$$zw = |z|\, e^{i\phi}\, |w|\, e^{i\psi} = |z|\,|w|\, e^{i(\phi+\psi)}.$$

We close this subsection with a list of frequently used properties involving the exponential form:

$$\left|e^{i\phi}\right| = 1 \qquad \text{for } \phi \in \mathbb{R}, \tag{2.8}$$

$$\cos\phi = \mathrm{Re}(e^{i\phi}), \qquad \sin\phi = \mathrm{Im}(e^{i\phi}), \tag{2.9}$$

$$\overline{e^{i\phi}} = e^{-i\phi}, \tag{2.10}$$

$$\cos\phi = \frac{e^{i\phi} + e^{-i\phi}}{2} \qquad \text{and} \qquad \sin\phi = \frac{e^{i\phi} - e^{-i\phi}}{2i}, \tag{2.11}$$

$$e^{2\pi i k} = 1 \qquad \text{for } k \in \mathbb{Z}, \tag{2.12}$$

$$e^{i(\phi + 2k\pi)} = e^{i\phi} \qquad \text{for } k \in \mathbb{Z}, \tag{2.13}$$

$$\left(e^{i\phi}\right)^k = e^{ik\phi} \qquad \text{for } k \in \mathbb{Z}. \tag{2.14}$$

2.1.2   *Zeros of polynomials*

We know that quadratic equations do not always admit real solutions. A quadratic equation is an example of an equation of the more general type

$$a_n z^n + a_{n-1} z^{n-1} + \cdots + a_1 z + a_0 = 0 \tag{2.15}$$

in the unknown z, where we assume that the coefficients $a_0, a_1, \ldots, a_n$ may be complex with $a_n \neq 0$. In this subsection we will pay special attention to the solutions or the *roots* of this equation. The left-hand side will be denoted by $P(z)$, so $P(z) = a_n z^n + a_{n-1} z^{n-1} + \cdots + a_1 z + a_0$, and is called a *polynomial* of degree n. Hence, solving equation (2.15) means determining the *zeros* of a polynomial. Using algebra, one can show that if $z = a$ is a zero of $P(z)$ (where a may also be complex), then the polynomial $P(z)$ can be written as $P(z) = (z - a)Q_1(z)$ for some polynomial $Q_1(z)$. We then say that the linear factor $z - a$ *divides* $P(z)$. If $Q_1(a) = 0$ as well, then we can write $P(z)$ as $P(z) = (z - a)^2 Q_2(z)$. Of course, we can continue in this way if $Q_2(a) = 0$ as well. Ultimately, this leads to the following statement.

If $z = a$ is a zero of a polynomial $P(z)$, then there exists a positive integer ν such that

$$P(z) = (z - a)^\nu Q(z)$$

for some polynomial $Q(z)$ with $Q(a) \neq 0$. The number ν is called the *multiplicity* of the zero a.

Let $P(z)$ be the polynomial of degree four given by

$$P(z) = z^4 - 2z^3 + 5z^2 - 8z + 4.$$

Now $z = 1$ is a zero of $P(z)$. This implies that we can divide by the factor $z - 1$, which results in $P(z) = (z - 1)(z^3 - z^2 + 4z - 4)$. However, $z = 1$ is also a zero of $z^3 - z^2 + 4z - 4$. Again dividing by $z - 1$ results in $P(z) = (z - 1)^2(z^2 + 4)$. Since $z = 1$ is no longer a zero of $z^2 + 4$, $z = 1$ is a zero of $P(z)$ of multiplicity 2. We also note that $z = 2i$ and $z = -2i$ are zeros of $P(z)$. Factorizing $z^2 + 4$ gives

$$P(z) = (z - 1)^2(z - 2i)(z + 2i).$$

The polynomial has a zero $z = 1$ of multiplicity 2 and zeros $z = 2i$ and $z = -2i$ of multiplicity 1.   ◄

Zeros of multiplicity 1 are also called *simple zeros*. In the preceding example we saw that in the complex plane the polynomial $P(z)$ is factorized entirely into linear factors. This is a major advantage of the introduction of complex numbers. The fact is that *any* arbitrary polynomial can be factorized entirely into linear factors. This statement is based on the so-called fundamental theorem of algebra, which states that (2.15) always has a solution in $\mathbb{C}$. The treatment of the fundamental theorem of algebra falls outside the scope of this book. We thus have the following important property:

Let $P(z) = a_n z^n + a_{n-1} z^{n-1} + \cdots + a_1 z + a_0$ be a polynomial of degree n. Then $P(z)$ can be written as

$$P(z) = a_n (z - z_1)^{\nu_1} \cdots (z - z_k)^{\nu_k}. \tag{2.16}$$

Here $z_1, \ldots, z_k$ are the distinct zeros of $P(z)$ in $\mathbb{C}$ with their respective multiplicities.

From (2.16) we can immediately conclude that the degree of $P(z)$ equals the sum of the multiplicities. If we count the number of zeros of a polynomial, with each zero counted according to its multiplicity, then it follows that a polynomial of degree n has precisely n zeros.

When in particular the coefficients of $P(z)$ are real, then

$$\overline{P(z)} = a_n \bar{z}^n + a_{n-1}\bar{z}^{n-1} + \cdots + a_1\bar{z} + a_0 = P(\bar{z}).$$

Hence, if $P(a) = 0$, then $P(\bar{a}) = 0$ as well. Now if a is a non-real zero, that is if $a \neq \bar{a}$, then

$$P(z) = (z - a)(z - \bar{a})Q(z) = (z^2 - (a + \bar{a})z + a\bar{a})Q(z)$$
$$= \left(z^2 - (2\operatorname{Re} a)z + |a|^2\right)Q(z).$$

Apparently the polynomial $P(z)$ contains a quadratic factor with real coefficients $z^2 - (2\operatorname{Re} a)z + |a|^2$, which cannot be factorized any further into real linear factors. As a consequence we have:

A polynomial with real coefficients can always be factorized into factors which are linear or quadratic and having real coefficients. The zeros are real or they occur in pairs of complex conjugates.

EXAMPLE 2.2 The polynomial $z^4 + 4$ is a polynomial with real coefficients. The complex zeros are (see exercise 2.6a): $1 + i$, $1 - i$, $-(1 + i)$, $-(1 - i)$. So $z^4 + 4 = (z - 1 - i)(z - 1 + i)(z + 1 + i)(z + 1 - i)$. Since $(z - 1 - i)(z - 1 + i) = (z - 1)^2 + 1 = z^2 - 2z + 2$ and $(z + 1 + i)(z + 1 - i) = (z + 1)^2 + 1 = z^2 + 2z + 2$, one has that

$$z^4 + 4 = (z^2 - 2z + 2)(z^2 + 2z + 2).$$

This factorizes the polynomial $z^4 + 4$ into two factors with real coefficients, which can still be factorized into linear factors, but then these will no longer have real coefficients. ◀

In the theory of the discrete Fourier transform, an important role is played by the roots of the equation

$$z^n = 1.$$

*n*th *roots of unity* These roots are called the *n*th *roots of unity*. We know that the number of zeros of $z^n - 1$ is equal to n. First we determine the moduli of the zeros. From (2.4) follows that $|z^n| = |z|^n$ and since $z^n = 1$, we obtain that $|z| = 1$. From this we conclude that all solutions lie on the unit circle. The arguments of the solutions can be found as follows:

$$n \arg z = \arg(z^n) = \arg 1 = 2k\pi.$$

Dividing by n leads to $\arg z = 2k\pi/n$. For $k = 0, 1, \ldots, n - 1$ this gives n distinct solutions. The *n*th roots of unity are thus

$$z_k = e^{i2k\pi/n} = \cos\left(\frac{2k\pi}{n}\right) + i\sin\left(\frac{2k\pi}{n}\right) \quad \text{for } k = 0, 1, \ldots, n - 1.$$

These solutions are drawn in figure 2.2 for the case $n = 5$. Note that the roots display a symmetry with respect to the real axis. This is a consequence of the fact that the polynomial $z^n - 1$ has real coefficients and that thus the complex conjugate of a zero is a zero as well.

The method described for solving the equation $z^n = 1$ can be extended to equations of the type $z^n = a$, where a is an arbitrary complex number. We will illustrate this using the following example.

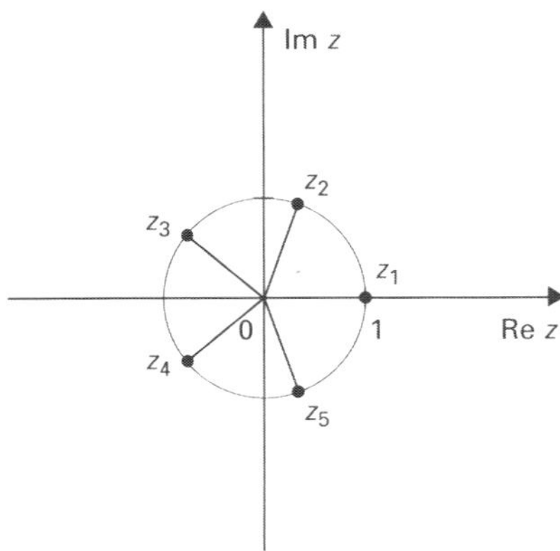

FIGURE 2.2
The solutions of $z^5 = 1$.

EXAMPLE 2.3

We determine the roots of the equation $z^3 = 8i$. First we note that $\left| z \right|^3 = \left| z^3 \right| = |8i| = 8$. From this it follows that $|z| = 2$, so the roots will lie on a circle with radius 2. For the argument one has: $3 \arg z = \arg(z^3) = \arg(8i) = \pi/2 + 2k\pi$. So $\arg z = \pi/6 + 2k\pi/3$ $(k = 0, 1, 2)$. Hence, the roots are

$$z_k = 2e^{i(\pi/6 + 2k\pi/3)} = 2\left(\cos\left(\tfrac{1}{6}\pi + \tfrac{2}{3}k\pi\right) + i\sin\left(\tfrac{1}{6}\pi + \tfrac{2}{3}k\pi\right)\right),$$

for $k = 0, 1, 2$. See figure 2.3. ◀

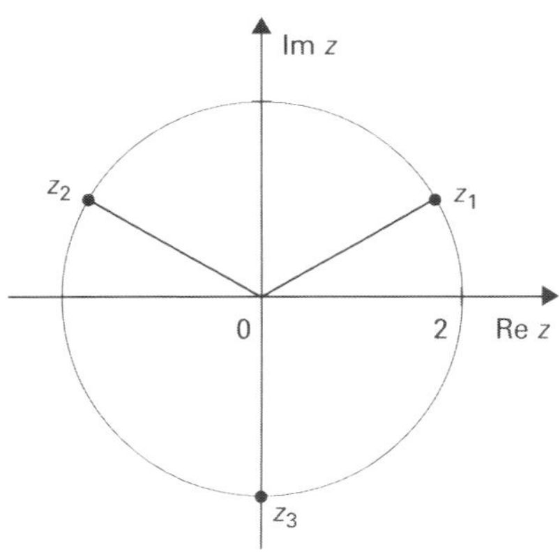

FIGURE 2.3
The solutions of $z^3 = 8i$.

By completing the square one can reduce a quadratic equation to an equation of the type $z^2 = a$, which can subsequently be solved using the method above. To that end we consider the following example.

EXAMPLE 2.4

Let the quadratic equation $z^2 + 2iz - 1 + i = 0$ be given. Completing the square leads to $(z + i)^2 + i = 0$. Put $w = z + i$. Then w satisfies the equation $w^2 = -i$. Hence, $|w^2| = 1$, implying that $|w| = 1$. For the argument one has: $2 \arg w =$

$\arg(w^2) = \arg(-i) = 3\pi/2 + 2k\pi$. From this it follows that $\arg w = 3\pi/4 + k\pi$ ($k = 0, 1$). We find the two roots $w_1 = \cos(3\pi/4) + i\sin(3\pi/4) = (-1+i)/\sqrt{2}$ and $w_2 = -w_1$ and subsequently the two roots z_1 and z_2 we were looking for:

$$z_1 = w_1 - i = -\frac{1}{2}\sqrt{2} + i(\frac{1}{2}\sqrt{2} - 1),$$

$$z_2 = w_2 - i = \frac{1}{2}\sqrt{2} - i(\frac{1}{2}\sqrt{2} + 1).$$

◀

EXERCISES

2.1   **a**   Determine argument and modulus of $z = -1 + i$, $z = 2i$, $z = -3$ and $z = -1 - i\sqrt{3}$.
        b   Write $z = 2 + 2i$, $z = -\sqrt{3} + i$ and $z = -3i$ in exponential form.

2.2   Prove that for all complex z and w one has:

$$| z \pm w | \geq || z | - | w ||.$$

2.3   Prove the properties (2.8) up to (2.14).

2.4   Let the complex numbers $z_1 = 4 - 4i$ and $z_2 = -2 + 2i\sqrt{3}$ be given. Give the exponential form of z_1/z_2, $z_1^2 z_2^3$ and z_1^2/z_2^3.

2.5   Draw in each of the following cases the set of complex numbers z in the complex plane satisfying:
        a $0 < | z | < 3$,
        b $| z | < 2$,
        c $4 < | z | < 5$,
        d $| z - (1 + 2i) | < 3/2$,
        e $\operatorname{Re} z < 3$,
        f $| \operatorname{Re} z | < 2$,
        g $\operatorname{Re} z > 2\frac{1}{2}$ and $| z - 2 | < 1$.

2.6   Solve the following equations:
        a $z^4 + 4 = 0$,
        b $z^6 + 1 = i\sqrt{3}$,
        c $5z^2 + 2z + 10 = 0$.

2.7   Give a factorization of the polynomial $z^5 - z^4 + z - 1$ into linear factors only, and also into factors with real coefficients only and which cannot be factorized any further into factors with real coefficients.

2.8   Determine the solutions z of the equation $e^z = 2i$.

2.2   Partial fraction expansions

The partial fraction expansion of rational functions is a technique which is used to determine a primitive of a rational function. In the chapters on the Laplace transform and the z-transform we shall see that the same technique can be used to determine the inverse transform of a rational function. The technique will be explained in the present section.

A rational function $F(z)$ is a function which can be written in the following form:

$$F(z) = \frac{a_n z^n + a_{n-1} z^{n-1} + \cdots + a_1 z + a_0}{b_m z^m + b_{m-1} z^{m-1} + \cdots + b_1 z + b_0}. \tag{2.17}$$

Pole

The numerator and denominator consist of polynomials in the complex variable z. Denote the numerator by $P(z)$ and the denominator by $Q(z)$. We assume that $b_m \neq 0$ and $a_n \neq 0$. The degree of the numerator is n and the degree of the denominator is m. The zeros of $Q(z)$ are called the *poles* of $F(z)$. In section 2.1.2 we noted that, as a consequence of the fundamental theorem of algebra, in the complex plane each polynomial can be factorized entirely into linear factors. The denominator $Q(z)$ can thus be written as

$$Q(z) = b_m (z - z_1)^{\nu_1} (z - z_2)^{\nu_2} \cdots (z - z_k)^{\nu_k},$$

where $z_1, z_2, \ldots, z_k$ are the distinct zeros of $Q(z)$. The point z_j is then called a *pole of order ν_j* of $F(z)$.

Order of pole

Before starting with a partial fraction expansion, one should first check whether or not the degree of the numerator is *smaller* than the degree of the denominator. If this is not the case, then we first divide by the denominator, that is, we determine polynomials $D(z)$ and $R(z)$ such that $P(z) = D(z)Q(z) + R(z)$, where the degree of $R(z)$ is smaller than the degree of the denominator $Q(z)$. As a consequence, $F(z)$ can be written as

$$F(z) = D(z) + \frac{R(z)}{Q(z)}.$$

The rational function $R(z)/Q(z)$ in the right-hand side now *does* have the property that the degree of the numerator is smaller than the degree of the denominator. We shall illustrate this using an example.

EXAMPLE 2.5

The rational function $F(z)$ is given by

$$F(z) = \frac{z^4 + 1}{z^2 - 1}.$$

The degree of the numerator is greater than the degree of the denominator. We therefore perform the following long division:

$$
\begin{array}{l}
z^2 - 1 \, / \, z^4 \qquad\quad + 1 \, \backslash \, z^2 + 1 \\
\qquad\quad\, z^4 \quad - z^2 \\
\qquad\quad \overline{} \\
\qquad\qquad\quad z^2 + 1 \\
\qquad\qquad\quad z^2 - 1 \\
\qquad\qquad\quad \overline{} \\
\qquad\qquad\qquad\; 2
\end{array}
$$

From this it follows that $z^4 + 1 = (z^2 + 1)(z^2 - 1) + 2$ and hence

$$F(z) = z^2 + 1 + \frac{2}{z^2 - 1}.$$

◀

Henceforth we will assume that we are dealing with a rational function $F(z)$ as given by (2.17) but with the additional condition that $n < m$. The purpose of a partial fraction expansion is then to write this rational function as a sum of fractions with the numerators being (complex) constants and the denominators $z - z_1, z - z_2, \ldots, z - z_k$. If the order ν_j of a pole is greater than 1, then denominators $(z - z_j)^2$, $(z - z_j)^3, \ldots, (z - z_j)^{\nu_j}$ also occur. If the coefficients of $Q(z)$ are real, then $Q(z)$ can be written as a product of linear and quadratic factors with real coefficients, and $F(z)$ can then be expanded into fractions with (powers of) linear and quadratic denominators with real coefficients. For the quadratic denominators, the numerators may be linear.

The first step in the technique of the partial fraction expansion consists of a factorization of the denominator $Q(z)$. This means determining the zeros of $Q(z)$ together with their multiplicities. Next, the actual partial fraction expansion takes place. The following two examples will show how partial fraction expansions are carried out in the case where $Q(z)$ is factorized entirely into linear factors.

EXAMPLE 2.6

Let the rational function be given by

$$F(z) = \frac{z}{(z-1)^2(z^2+1)}.$$

The denominator $Q(z)$ can be factorized into linear factors as follows: $Q(z) = (z-1)^2(z-i)(z+i)$. There is a zero $z = 1$ with multiplicity 2 and there are simple zeros for $z = \pm i$. A partial fraction expansion is aimed at writing $F(z)$ in the following form:

$$F(z) = \frac{A}{z-1} + \frac{B}{(z-1)^2} + \frac{C}{z-i} + \frac{D}{z+i}.$$

Here A, B, C and D are constants, still to be determined. Note that the zero of multiplicity 2 has *two* fractions linked to it, while the simple zeros have only *one*. The constants can be calculated as follows. Multiplying the expansion above by the denominator $Q(z)$ gives the identity

$$z = A(z-1)(z^2+1) + B(z^2+1) + C(z-1)^2(z+i) + D(z-1)^2(z-i). \tag{2.18}$$

Formally one should exclude the zeros of $Q(z)$ in this identity, since these are the poles of $F(z)$. However, the right-hand and left-hand sides contain polynomials and by a limit transition one can prove that for these values of z the identity remains valid. Substituting the zeros $z = 1$, $z = i$ and $z = -i$ of $Q(z)$ in (2.18) gives the following results:

$$1 = 2B, \qquad i = 4C, \qquad -i = 4D.$$

We still lack one equation. To find it, we use yet another property of polynomials. Namely, two polynomials in z are equal if and only if the coefficients of corresponding powers are equal. Comparing the coefficients of z^3 in both sides of equation (2.18) establishes that

$$0 = A + C + D.$$

From the preceding equations it follows easily that $A = 0$, $B = 1/2$, $C = i/4$, $D = -i/4$. Ultimately, the partial fraction expansion is as follows:

$$F(z) = \frac{1/2}{(z-1)^2} + \frac{i/4}{z-i} - \frac{i/4}{z+i} = \frac{1}{2(z-1)^2} + \frac{i}{4(z-i)} - \frac{i}{4(z+i)}.$$

◀

EXAMPLE 2.7

Let the function $F(z)$ be given by

$$F(z) = \frac{z}{z^2 - 6iz - 8}.$$

We factorize the denominator $Q(z)$ by first completing the square: $Q(z) = (z-3i)^2 + 1 = (z-3i-i)(z-3i+i) = (z-4i)(z-2i)$. There are simple zeros at $z = 4i$ and at $z = 2i$. The rational function can then be expanded as follows:

$$F(z) = \frac{A}{z-4i} + \frac{B}{z-2i}.$$

The constants A and B can be found by first multiplying by the denominator and then substituting its zeros. It then follows that $A = 2$ and $B = -1$. The partial fraction expansion is as follows:

$$F(z) = \frac{2}{z - 4i} - \frac{1}{z - 2i}.$$

$\blacktriangleleft$

For some applications it is more convenient to obtain, starting from a rational function with real coefficients in the numerator as well as in the denominator, a partial fraction expansion where only fractions having real coefficients occur as well. When all zeros of the denominator $Q(z)$ are real, this is no problem; the partial fraction expansion can then be performed as in the previous examples. If, however, $Q(z)$ has also non-real zeros, then linear factors will no longer suffice. In a real factorization of $Q(z)$, *quadratic factors* will then appear as well. In the following examples we will show how in these circumstances one can determine a real expansion.

EXAMPLE 2.8   Let the function $F(z)$ from example 2.6 be given:

$$F(z) = \frac{z}{(z - 1)^2(z^2 + 1)}.$$

The denominator contains a quadratic factor having no factorization into linear factors with real coefficients. With this factor we associate a fraction of the form

$$\frac{Az + B}{z^2 + 1}$$

with real coefficients A and B. The partial fraction expansion now looks like this:

$$\frac{z}{(z - 1)^2(z^2 + 1)} = \frac{A}{z - 1} + \frac{B}{(z - 1)^2} + \frac{Cz + D}{z^2 + 1}.$$

Multiplying by the denominator of $F(z)$ leads to the following identity:

$$z = A(z - 1)(z^2 + 1) + B(z^2 + 1) + (Cz + D)(z - 1)^2.$$

Substitution of $z = 1$ gives $B = 1/2$. Next we equate coefficients of corresponding powers of z. For the coefficient of z^3, z^2 and z it subsequently follows that

$$0 = A + C, \qquad 0 = -A + B - 2C + D, \qquad 1 = A + C - 2D.$$

The solution to this system of equations is $A = 0$, $B = 1/2$, $C = 0$, $D = -1/2$ and so the real partial fraction expansion looks like this:

$$\frac{z}{(z - 1)^2(z^2 + 1)} = \frac{1/2}{(z - 1)^2} - \frac{1/2}{z^2 + 1} = \frac{1}{2(z - 1)^2} - \frac{1}{2(z^2 + 1)}.$$

$\blacktriangleleft$

We finish with an example where a quadratic factor occurs twice in the denominator of $F(z)$.

EXAMPLE 2.9   Let the function $F(z)$ be given by

$$F(z) = \frac{z^2 + 3z + 3}{(z^2 + 2z + 4)^2}.$$

The quadratic factor in the denominator cannot be factorized into linear factors with real coefficients. Since the quadratic factor occurs twice, the partial fraction expansion has the following form:

$$\frac{z^2 + 3z + 3}{(z^2 + 2z + 4)^2} = \frac{Az + B}{z^2 + 2z + 4} + \frac{Cz + D}{(z^2 + 2z + 4)^2}.$$

Multiplying by the denominator of $F(z)$ gives

$$z^2 + 3z + 3 = (Az + B)(z^2 + 2z + 4) + (Cz + D).$$

Equating the coefficients of z^0, z^1, z^2 and z^3 gives, respectively,

$$3 = 4B + D, \qquad 3 = 4A + 2B + C, \qquad 1 = 2A + B, \qquad 0 = A.$$

The solution to this system is $A = 0$, $B = 1$, $C = 1$, $D = -1$. The partial fraction expansion is then as follows:

$$\frac{z^2 + 3z + 3}{(z^2 + 2z + 4)^2} = \frac{1}{z^2 + 2z + 4} + \frac{z - 1}{(z^2 + 2z + 4)^2}.$$

◀

EXERCISES

2.9 Determine the partial fraction expansion of $F(z)$ given by

$$F(z) = \frac{z}{(z - 1/2)(z - 2)}.$$

2.10 Determine the partial fraction expansion, into fractions with linear denominators, of the function $F(z)$ given by

$$F(z) = \frac{1}{z^2 + 4z + 8}.$$

2.11 Determine the partial fraction expansion of the function $F(z)$ given by

a $\quad F(z) = \dfrac{z^2}{(z + 1)^2(z + 3)}$,

b $\quad F(z) = \dfrac{z^2 + 1}{(z + 1)^3}.$

2.12 Determine the partial fraction expansion, into fractions with denominators having real coefficients, of the function $F(z)$ given by

$$F(z) = \frac{2z^2 - 5z + 11}{z^3 - 3z^2 + 7z - 5}.$$

2.3 Complex-valued functions

Complex-valued function

In general, the functions in this book are prescriptions associating a complex number with each real variable t from the domain. We call this a *complex-valued function*. An important example is the function $f(t) = e^{i\omega t}$, where ω is a real constant. This function is called a time-harmonic signal with frequency ω (see chapter 1). Note that according to definition 2.1 one has that $e^{i\omega t} = \cos \omega t + i \sin \omega t$.

A complex-valued function can be written as

$$f(t) = u(t) + iv(t),$$

where u and v are real-valued functions of the real argument t. Analogous to complex numbers, we use the concepts real and imaginary part of $f(t)$:

$$u(t) = \operatorname{Re} f(t), \qquad v(t) = \operatorname{Im} f(t).$$

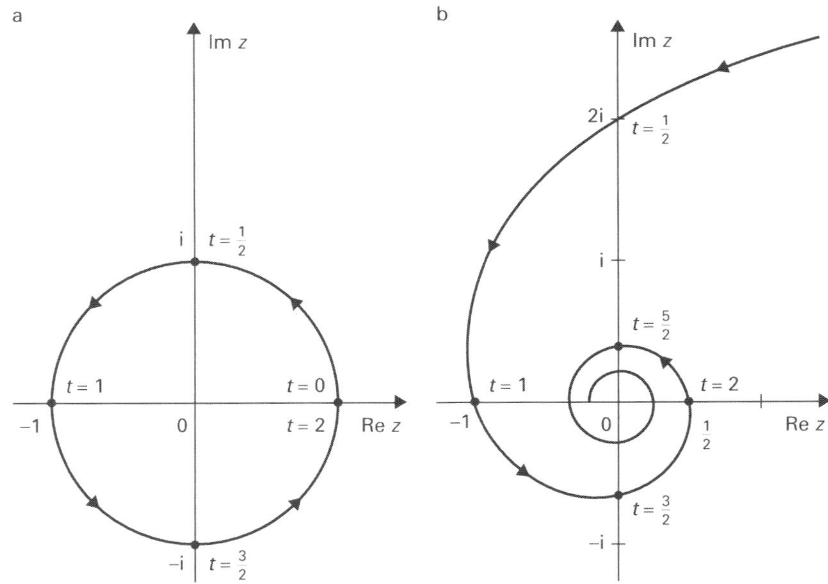

FIGURE 2.4
Range of the functions $e^{i\pi t}$ and $e^{i\pi t}/t$.

A complex-valued function can be represented as a graph by drawing its range in the complex plane. This range is a curve in the complex plane having parametric representation $(u(t), v(t))$. For the functions $e^{i\pi t}$ and $e^{i\pi t}/t$ (a part of) the range is shown in figure 2.4.

For complex-valued functions the notion of a limit can simply be defined by starting from limits of real-valued functions. We utilize the following definition.

DEFINITION 2.2
Limit of a complex-valued function

Let L be a complex number and $f(t)$ a complex-valued function with real part $u(t)$ and imaginary part $v(t)$. Then

$$\lim_{t \to a} f(t) = L \qquad \text{if and only if} \qquad \lim_{t \to a} u(t) = \operatorname{Re} L \text{ and } \lim_{t \to a} v(t) = \operatorname{Im} L.$$

A consequence of this definition is the following theorem.

THEOREM 2.3

Let L be a complex number and $f(t)$ a complex-valued function. Then $\lim_{t \to a} f(t) = L$ if and only if $\lim_{t \to a} |f(t) - L| = 0$.

Proof
We first prove that $\lim_{t \to a} f(t) = L$ implies that $\lim_{t \to a} |f(t) - L| = 0$. If $\lim_{t \to a} f(t) = L$, then by definition $\lim_{t \to a} u(t) = \operatorname{Re} L$ and $\lim_{t \to a} v(t) = \operatorname{Im} L$ and hence

$$\lim_{t \to a} |f(t) - L| = \lim_{t \to a} \sqrt{(u(t) - \operatorname{Re} L)^2 + (v(t) - \operatorname{Im} L)^2} = \sqrt{0 + 0} = 0.$$

Next we prove that $\lim_{t \to a} |f(t) - L| = 0$ implies that $\lim_{t \to a} f(t) = L$. One has that:

$$|u(t) - \operatorname{Re} L| = \sqrt{(u(t) - \operatorname{Re} L)^2} \leq \sqrt{(u(t) - \operatorname{Re} L)^2 + (v(t) - \operatorname{Im} L)^2}$$
$$= |f(t) - L|.$$

From this it follows that if $\lim_{t \to a} | f(t) - L | = 0$, then $\lim_{t \to a} u(t) = \operatorname{Re} L$. Similarly one proves that $\lim_{t \to a} v(t) = \operatorname{Im} L$. This completes the proof. ■

EXAMPLE 2.10

We show that for each real ω one has: $\lim_{t \to \infty} e^{i\omega t}/t = 0$. Since $|e^{i\omega t}/t| = 1/t$ and $\lim_{t \to \infty} 1/t = 0$ we have $\lim_{t \to \infty} |e^{i\omega t}/t| = 0$ and hence, according to theorem 2.3, $\lim_{t \to \infty} e^{i\omega t}/t = 0$. ◀

Continuity

Differentiability

Concepts like *continuity* and *differentiability* of a function are defined using limits. For instance, a function is continuous at $t = a$ if $\lim_{t \to a} f(t) = f(a)$ and differentiable at $t = a$ if $\lim_{t \to a}(f(t) - f(a))/(t - a)$ exists. One can show (this is not very hard, but it will be omitted here) that for a complex-valued function $f(t) = u(t) + iv(t)$, the continuity of $f(t)$ is equivalent to the continuity of both the real part $u(t)$ and the imaginary part $v(t)$, and also that the differentiability of $f(t)$ is equivalent to the differentiability of $u(t)$ and $v(t)$. Moreover, one has for the derivative at a point t that

$$f'(t) = u'(t) + iv'(t). \tag{2.19}$$

Consequently, for the differentiation of complex-valued functions the same rules apply as for real-valued functions. Complex numbers may be considered as constants here.

EXAMPLE 2.11

If $f(t) = e^{at}$ with $a \in \mathbb{C}$, then $f'(t) = ae^{at}$. We can show this as follows: put $a = x + iy$ and write $f(t)$ as $f(t) = e^{xt}e^{iyt} = e^{xt}\cos yt + ie^{xt}\sin yt$. The real and imaginary parts are differentiable everywhere with derivatives $e^{xt}(x\cos yt - y\sin yt)$ and $e^{xt}(x\sin yt + y\cos yt)$ respectively. So

$$\begin{aligned}
f'(t) &= xe^{xt}(\cos yt + i\sin yt) + iye^{xt}(\cos yt + i\sin yt) \\
&= (x + iy)e^{xt}e^{iyt} = ae^{at}.
\end{aligned}$$

◀

Chain rule

With the chain rule we can differentiate a composition $f(g(t))$ of two functions. The function $f(t)$, however, is defined on (a part of) $\mathbb{R}$. Hence, in the composition $g(t)$ should also be a real-valued function. The *chain rule* then has the usual form

$$\frac{d}{dt} f(g(t)) = f'(g(t))g'(t).$$

A consequence of this is:

$$\frac{d}{dt}[f(t)]^n = n[f(t)]^{n-1} f'(t) \qquad \text{for } n = 1, 2, \ldots.$$

Now that the concepts continuity and differentiability of complex-valued functions have been introduced, we will proceed with the introduction of two classes of functions that will play an important role in theorems on Fourier series, Fourier integrals and Laplace transforms.

The first class consists of the so-called piecewise continuous functions. To start with, we define in the usual manner the *left-hand limit* $f(t-)$ and *right-hand limit* $f(t+)$ of a function at the point t:

Left-hand limit

Right-hand limit

$$f(t-) = \lim_{h \downarrow 0} f(t - h) \qquad \text{and} \qquad f(t+) = \lim_{h \downarrow 0} f(t + h),$$

provided these limits exist.

DEFINITION 2.3
Piecewise continuous function

A function $f(t)$ is called piecewise continuous *on the interval $[a, b]$ if $f(t)$ is continuous at each point of (a, b), except possibly in a finite number of points $t_1, t_2, \ldots, t_n$. Moreover, $f(a+)$, $f(b-)$ and $f(t_i+)$, $f(t_i-)$ should exist for $i = 1, \ldots, n$.*

A function $f(t)$ is called piecewise continuous on $\mathbb{R}$ if $f(t)$ is piecewise continuous on each subinterval $[a, b]$ of $\mathbb{R}$.

One can show that a function $f(t)$ which is piecewise continuous on an interval $[a, b]$ is also bounded on $[a, b]$, that is to say: there exists a constant $M > 0$ such that for all t in $[a, b]$ one has $| f(t) | \leq M$. Functions that possess a real or imaginary part with a vertical asymptote in $[a, b]$ are thus not piecewise continuous on $[a, b]$. Another example is the function $f(t) = \sin(1/t)$ for $t \neq 0$ and $f(0) = 0$. This function is continuous everywhere except at $t = 0$. Since $f(0+)$ does not exist, this function is not piecewise continuous on $[0, 1]$ according to our definition.

Note that the function value $f(t)$ at a point t of discontinuity doesn't necessarily have to equal $f(t+)$ or $f(t-)$.

A second class of functions to be introduced is the so-called piecewise smooth functions. This property is linked to the derivative of the function. For a piecewise continuous function, we will mean by f' the derivative of f at all points where it exists.

DEFINITION 2.4
Piecewise smooth function

A piecewise continuous function $f(t)$ on the interval $[a, b]$ is called piecewise smooth if its derivative $f'(t)$ is piecewise continuous.
A function is called piecewise smooth on $\mathbb{R}$ if this function is piecewise smooth on each subinterval $[a, b]$ of $\mathbb{R}$.

In figure 2.5 a graph is drawn of a real-valued piecewise smooth function.

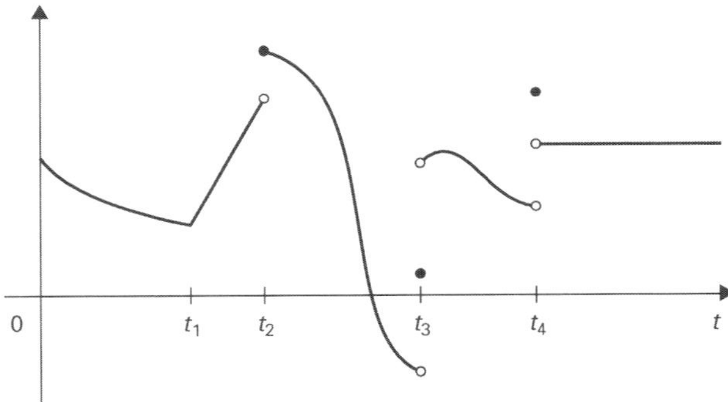

FIGURE 2.5
A piecewise smooth function.

There are now two possible ways of looking at the derivative in the neighbourhood of a point: on the one hand as the limits $f'(t+)$ and $f'(t-)$; on the other hand by defining a *left-hand derivative* $f'_{-}(t)$ and a *right-hand derivative* $f'_{+}(t)$ as follows:

Left-hand derivative

Right-hand derivative

$$f'_{-}(t) = \lim_{h \uparrow 0} \frac{f(t+h) - f(t-)}{h}, \qquad f'_{+}(t) = \lim_{h \downarrow 0} \frac{f(t+h) - f(t+)}{h}, \qquad (2.20)$$

provided these limits exist. Note that in this definition of left-hand and right-hand derivative one does not use the function value at t, since $f(t)$ need not exist at the point t. Often it is the case that $f'_{-}(t) = f'(t-)$ and $f'_{+}(t) = f'(t+)$. This holds in particular for piecewise smooth functions, notably at the points of discontinuity of f, as is proven in the next theorem.

THEOREM 2.4

Let $f(t)$ be a piecewise smooth function on the interval $[a, b]$. Then $f'_+(a) = f'(a+)$, $f'_-(b) = f'(b-)$ and for all $a < t < b$ one has, moreover, that $f'_-(t) = f'(t-)$ and $f'_+(t) = f'(t+)$.

Proof
We present the proof for right-hand limits. The proof for left-hand limits is analogous. So let $t \in [a, b)$ be arbitrary. Using the mean value theorem from calculus we will show that the existence of $f'(t+)$ implies that $f'_+(t)$ exists and that $f'(t+) = f'_+(t)$. Since f and f' are both piecewise continuous, there exists an $h > 0$ such that f and f' have no point of discontinuity on $(t, t + h]$. Possibly f has a discontinuity at t. If we now redefine f at t as $f(t+)$, then f is continuous on $[t, t + h]$. Moreover, f is differentiable on $(t, t + h)$. According to the mean value theorem there then exists a $\xi \in (t, t + h)$ such that

$$\frac{f(t + h) - f(t+)}{h} = f'(\xi).$$

Now let $h \downarrow 0$, then $\xi \downarrow t$. Since $f'(t+) = \lim_{\xi \downarrow t} f'(\xi)$ exists, it follows from (2.20) that $f'_+(t)$ exists and that $f'(t+) = f'_+(t)$. $\blacksquare$

When a function is not piecewise smooth, the left- and right-hand derivatives will not always be equal to the left- and right-hand limits of the derivative, as the following example shows.

EXAMPLE

Let the function $f(t)$ be given by

$$f(t) = \begin{cases} t^2 \sin(1/t) & \text{for } t \neq 0, \\ 0 & \text{for } t = 0. \end{cases}$$

The left- and right-hand derivatives of $f(t)$ at $t = 0$, calculated according to (2.20), exist and are equal to 0. However, if we first calculate the derivative $f'(t)$, then

$$f'(t) = \begin{cases} 2t \sin(1/t) - \cos(1/t) & \text{for } t \neq 0, \\ 0 & \text{for } t = 0. \end{cases}$$

It then turns out that $f'(0+)$ and $f'(0-)$ do not exist, since $\cos(1/t)$ has no left- or right-hand limit at $t = 0$. $\blacktriangleleft$

The theory of the Riemann integral and the improper Riemann integral for real-valued functions can easily be extended to complex-valued functions as well. For complex-valued functions the (improper) Riemann integral exists on an interval if and only if both the (improper) Riemann integral of the real part $u(t)$ and the imaginary part $v(t)$ exist on that interval. Moreover, one has

Definite integral

$$\int_a^b f(t)\, dt = \int_a^b u(t)\, dt + i \int_a^b v(t)\, dt. \tag{2.21}$$

Here $a = -\infty$ or $b = \infty$ is also allowed. We recall that the value of an integral does not change by altering the value of the function at the possible jump discontinuities. From (2.21) the following properties of definite integrals for complex-valued functions immediately follow:

$$\overline{\int_a^b f(t)\, dt} = \int_a^b \overline{f(t)}\, dt,$$

$$\text{Re} \int_a^b f(t)\, dt = \int_a^b \text{Re}\, f(t)\, dt \quad \text{and} \quad \text{Im} \int_a^b f(t)\, dt = \int_a^b \text{Im}\, f(t)\, dt.$$

The fundamental theorem of calculus for complex-valued functions is no different from the one for real-valued functions. Here we will formulate the fundamental theorem for piecewise continuous functions, for which we state without proof that they are Riemann integrable.

THEOREM 2.5

Let $f(t)$ be a piecewise continuous function defined on the interval $[a, b]$. Let $F(x) = \int_a^x f(t)\,dt$. Then F is continuous and piecewise smooth on $[a, b]$.

From the preceding theorem one can then derive that for continuous and piecewise smooth complex-valued functions, the rule for integration by parts can be applied in the same way as the rule for integration by parts of real-valued functions.

EXAMPLE 2.12

We calculate the integral $\int_0^\pi te^{2it}\,dt$ by applying the rule for integration by parts.

$$
\begin{aligned}
\int_0^\pi te^{2it}\,dt &= \frac{1}{2i}\int_0^\pi t(e^{2it})'\,dt = \left[\frac{1}{2i}te^{2it}\right]_0^\pi - \frac{1}{2i}\int_0^\pi e^{2it}(t)'\,dt \\
&= \frac{1}{2i}\pi e^{2\pi i} - \frac{1}{2i}\int_0^\pi e^{2it}\,dt = \frac{\pi}{2i} + \left[\frac{1}{4}e^{2it}\right]_0^\pi \\
&= \frac{\pi}{2i} + \frac{1}{4}(e^{2\pi i} - 1) = \frac{\pi}{2i}.
\end{aligned}
$$

◄

The following inequality is often applied to estimate integrals:

$$
\left|\int_a^b f(t)\,dt\right| \le \int_a^b |f(t)|\,dt \qquad \text{for } b \ge a. \tag{2.22}
$$

For real-valued functions this is a well-known inequality. Here we omit the proof for complex-valued functions. In fact one can consider this inequality as a generalization of the triangle inequality (theorem 2.1). A direct consequence of inequality (2.22) is the following inequality. If $|f(t)| \le M$ on the integration interval $[a, b]$, then

$$
\left|\int_a^b f(t)\,dt\right| \le \int_a^b |f(t)|\,dt \le \int_a^b M\,dt = M(b - a).
$$

EXAMPLE 2.13

For all complex s with $\mathrm{Re}\,s \ne 0$ and real $T > 0$ one has

$$
\left|\int_0^T e^{-st}\,dt\right| \le \int_0^T |e^{-st}|\,dt = \int_0^T e^{-(\mathrm{Re}\,s)t}\,dt = \frac{1 - e^{-(\mathrm{Re}\,s)T}}{\mathrm{Re}\,s}.
$$

◄

EXERCISES

2.13   Determine the derivative of the following functions:

     a $f(t) = \dfrac{1}{1 + it}$,

     b $f(t) = e^{-it^2}$.

2.14   Let a positive real number ω_0 be given. Put $T = 2\pi/\omega_0$. Calculate $\int_0^T t^2 e^{i\omega_0 t}\,dt$.

2.15   Show that

$$
\left|\int_0^1 \frac{1}{2 - e^{it}}\,dt\right| \le 1.
$$

2.4 Sequences and series

2.4.1 Basic properties

The concept of an infinite series plays a predominant role in chapters 3 to 5 and will also return regularly later on. We will assume that the reader is already acquainted with the theory of sequences and series, as far as the terms of the sequence or the series consist of real numbers. In this section the theory is extended to complex numbers. In general, the terms of a sequence

$$(a_n) \qquad \text{with } n = 0, 1, 2, \ldots$$

will then be complex numbers, as is the case for the terms of a series

$$\sum_{n=0}^{\infty} a_n = a_0 + a_1 + \cdots.$$

For limits of sequences of complex numbers we follow the same line as for limits of complex-valued functions. Assuming that the concept of convergence of a sequence of real numbers is known, we start with the following definition of convergence of a sequence of complex numbers.

DEFINITION 2.5
Convergence of sequences

A sequence (a_n) of complex numbers with $u_n = \operatorname{Re} a_n$ and $v_n = \operatorname{Im} a_n$ converges if both the sequence of real numbers (u_n) and the sequence of real numbers (v_n) converge. Moreover, the limit of the sequence (a_n) then equals

$$\lim_{n \to \infty} a_n = \lim_{n \to \infty} u_n + i \lim_{n \to \infty} v_n.$$

EXAMPLE 2.14

Let the sequence (a_n) be given by

$$a_n = n(e^{i/n} - 1) \qquad \text{with } n = 0, 1, 2, \ldots.$$

Since $e^{i/n} = \cos(1/n) + i \sin(1/n)$ one has $u_n = \operatorname{Re} a_n = n(\cos(1/n) - 1)$ and $v_n = \operatorname{Im} a_n = n \sin(1/n)$. Verify for yourself that $\lim_{n \to \infty} u_n = 0$ and $\lim_{n \to \infty} v_n = 1$. Hence, the sequence (a_n) converges and $\lim_{n \to \infty} a_n = i$. ◀

Our next theorem resembles theorem 2.3 and can also be proven in the same way.

THEOREM 2.6

A sequence (a_n) converges and has limit a if and only if $\lim_{n \to \infty} |a_n - a| = 0$.

EXAMPLE 2.15

For complex z one has

$$\lim_{n \to \infty} z^n = 0 \qquad \text{if } |z| < 1.$$

We know that for real r with $-1 < r < 1$ one has $\lim_{n \to \infty} r^n = 0$. So if $|z| < 1$, then $\lim_{n \to \infty} |z^n| = \lim_{n \to \infty} |z|^n = 0$. Using theorem 2.6 with $a = 0$ we conclude that $\lim_{n \to \infty} z^n = 0$. ◀

In the complex plane $|a_n - a|$ is the distance from a_n to a. Theorem 2.6 states that for convergent sequences this distance tends to zero as n tends to infinity.

Divergence of sequences
A sequence *diverges*, or is called *divergent*, if the sequence does not converge. All kinds of properties that are valid for convergent sequences of real numbers are valid for convergent sequences of complex numbers as well. We will formulate the following properties, which we immediately recognize from convergent sequences of real numbers:

Let (a_n) and (b_n) be two convergent sequences such that $\lim_{n \to \infty} a_n = a$ and $\lim_{n \to \infty} b_n = b$. Then

a $\lim_{n\to\infty}(\alpha a_n + \beta b_n) = \alpha a + \beta b$ for all $\alpha, \beta \in \mathbb{C}$,
b $\lim_{n\to\infty} a_n b_n = ab$,
c $\lim_{n\to\infty} a_n/b_n = a/b$ if $b \neq 0$.

Now that we know what a convergent sequence is, we can define convergence of a series in the usual way. For this we use the *partial sums s_n* of the sequence (a_n):

Partial sum

$$s_n = \sum_{k=0}^{n} a_k = a_0 + a_1 + \cdots + a_n.$$

DEFINITION 2.6
Convergence of a series

A series $\sum_{n=0}^{\infty} a_n$ is called convergent *if and only if the sequence of partial sums (s_n) converges.*

Sum of a series

When $s = \lim_{n\to\infty} s_n$, we call s the *sum* of the series. For a convergent series the sum is also denoted by $\sum_{n=0}^{\infty} a_n$, so $s = \sum_{n=0}^{\infty} a_n$.

EXAMPLE 2.16
Geometric series

Consider the geometric series $\sum_{n=0}^{\infty} z^n$ with ratio $z \in \mathbb{C}$. The partial sum s_n is equal to $s_n = 1 + z + z^2 + \cdots + z^n$. Note that this is a polynomial of degree n. When $z = 1$, then we see by direct substitution that $s_n = n + 1$. Hence, the geometric series diverges for $z = 1$, since $\lim_{n\to\infty} s_n = \infty$. Multiplying s_n by the factor $1 - z$ gives

$$(1 - z)s_n = 1 + z + z^2 + \cdots + z^n - z(1 + z + z^2 + \cdots + z^n) = 1 - z^{n+1}.$$

For $z \neq 1$ one thus has

$$s_n = \frac{1 - z^{n+1}}{1 - z}.$$

For $|z| < 1$ we have seen that $\lim_{n\to\infty} z^n = 0$ (see example 2.15); so then the series converges with sum equal to $1/(1 - z)$. We write

$$\sum_{n=0}^{\infty} z^n = \frac{1}{1 - z} \qquad \text{if } |z| < 1.$$

Since for $|z| \geq 1$ the sequence with terms z^n does not tend to zero, the series diverges for these values of z.   ◄

Just as for sequences, the convergence of a series can be verified on the basis of the real and imaginary parts of the terms. For sequences we used this as a definition. For series we formulate it as a theorem.

THEOREM 2.7

Let (a_n) be a sequence of numbers and $u_n = \mathrm{Re}\, a_n$ and $v_n = \mathrm{Im}\, a_n$. Then the series $\sum_{n=0}^{\infty} a_n$ converges if and only if both the series $\sum_{n=0}^{\infty} u_n$ and the series $\sum_{n=0}^{\infty} v_n$ converge.

Proof
Let (s_n), (r_n) and (t_n) be the partial sums of the series with terms a_n, u_n and v_n respectively. Note that $s_n = r_n + it_n$. According to the definition of convergence of a sequence, the sequence (s_n) converges if and only if both the sequence (r_n) and the sequence (t_n) converge. This is precisely the definition of convergence for the series with terms a_n, u_n and v_n.   ■

From the preceding theorem we also conclude that for a convergent series with terms $a_n = u_n + iv_n$ one has

$$\sum_{n=0}^{\infty} a_n = \sum_{n=0}^{\infty} u_n + i \sum_{n=0}^{\infty} v_n.$$

For convergent series with complex terms, the same properties hold as for convergent series with real terms. Here we formulate the linearity property, which is a direct consequence of definition 2.6 and the linearity property for series with real terms.

THEOREM 2.8

Let $\sum_{n=0}^{\infty} a_n$ and $\sum_{n=0}^{\infty} b_n$ be convergent series with sum s and t respectively. Then

$$\sum_{n=0}^{\infty} (\alpha a_n + \beta b_n) = \alpha s + \beta t \qquad \textit{for all } \alpha, \beta \in \mathbb{C}.$$

The next property formulates a necessary condition for a series to converge.

THEOREM 2.9

If the series $\sum_{n=0}^{\infty} a_n$ converges, then $\lim_{n \to \infty} a_n = 0$.

Proof
If the series $\sum_{n=0}^{\infty} a_n$ converges and has sum s, then

$$\lim_{n \to \infty} a_n = \lim_{n \to \infty} (s_n - s_{n-1}) = \lim_{n \to \infty} s_n - \lim_{n \to \infty} s_{n-1} = s - s = 0.$$

$\blacksquare$

As a consequence we have that $\lim_{n \to \infty} a_n \neq 0$ excludes the convergence of the series. The theorem only gives a necessary and not a sufficient condition for convergence. To show this, we consider the harmonic series.

EXAMPLE 2.17
Harmonic series

Series of the type $\sum_{n=1}^{\infty} 1/n^p$, with p a real constant, are called *harmonic series*. From the theory of series with real terms it is known that for $p > 1$ the harmonic series is convergent, while for $p \leq 1$ the harmonic series is divergent. Hence, for $0 < p \leq 1$ we obtain a divergent series with terms that do tend to zero. ◀

2.4.2 Absolute convergence and convergence tests

For practical applications one usually needs a strong form of convergence of a series. Convergence itself is not enough, and usually one requires in addition the convergence of the series of absolute values, or moduli, of the terms.

DEFINITION 2.7
Absolute convergence

A series $\sum_{n=0}^{\infty} a_n$ is called absolutely convergent *if the series $\sum_{n=0}^{\infty} |a_n|$ converges.*

The series of the absolute values is a series with non-negative real terms. Convergence of series with non-negative terms can be verified using the following test, which is known as the *comparison test.*

THEOREM 2.10
Comparison test

When (a_n) and (b_n) are sequences of real numbers with $0 \leq a_n \leq b_n$ for $n = 0, 1, 2, \ldots$, and the series $\sum_{n=0}^{\infty} b_n$ converges, then the series $\sum_{n=0}^{\infty} a_n$ converges as well.

Proof
Let (s_n) be the partial sums of the series with terms a_n. Then $s_{n+1} - s_n = a_{n+1} \geq 0$. So $s_{n+1} \geq s_n$, which means that the sequence (s_n) is a non-decreasing sequence. It is known, and this is based on fundamental properties of the real numbers, that such a sequence converges whenever it has a upper bound. This means that there should be a constant c such that $s_n \leq c$ for all $n = 0, 1, 2, \ldots$. For the sequence (s_n) this is easy to show, since it follows from $a_n \leq b_n$ that

$$\begin{aligned}
s_n &= a_0 + a_1 + \cdots + a_n \leq b_0 + b_1 + \cdots + b_n \\
&\leq b_0 + b_1 + \cdots + b_n + b_{n+1} + \cdots.
\end{aligned}$$

The sequence (s_n) apparently has as upper bound the sum of the series with terms b_n. This sum exists since it is given that this series converges. This proves the theorem. ∎

Absolutely convergent series with _real_ terms are convergent. One can show this as follows. Write $a_n = b_n - c_n$ with $b_n = |a_n|$ and $c_n = |a_n| - a_n$. It is given that the series with terms b_n converges. The terms c_n satisfy the inequality $0 \leq c_n \leq 2|a_n|$. According to the comparison test, the series with terms c_n then converges as well. Since $a_n = b_n - c_n$, the series with terms a_n thus converges. For series with _complex_ terms, the statement is a consequence of the next theorem.

THEOREM 2.11

Let (a_n) be a sequence of numbers and let $u_n = \mathrm{Re}\, a_n$ and $v_n = \mathrm{Im}\, a_n$. The series $\sum_{n=0}^{\infty} a_n$ converges absolutely if and only if both the series $\sum_{n=0}^{\infty} u_n$ and $\sum_{n=0}^{\infty} v_n$ converge absolutely.

Proof
For the terms u_n and v_n one has the inequalities $|u_n| \leq |a_n|$, $|v_n| \leq |a_n|$. If the series with the non-negative terms $|a_n|$ converges, then we know from the comparison test that the series with, respectively, terms u_n and v_n converge absolutely. Conversely, if the series with, respectively, terms u_n and v_n converge absolutely, then the series with the non-negative terms $|u_n| + |v_n|$ also converges. It then follows from the inequality $|a_n| = \sqrt{u_n^2 + v_n^2} \leq |u_n| + |v_n|$ and again the comparison test that the series with terms $|a_n|$ converges. This means that the series with terms a_n converges absolutely. ∎

THEOREM 2.12

An absolutely convergent series is convergent.

Proof
Above we sketched the proof of this statement for series with real terms. For series with, in general, complex terms, the statement follows from the preceding theorem. Specifically, when the series with complex terms is absolutely convergent, then the series consisting of the real and the imaginary parts also converge absolutely. These are series of real numbers and therefore convergent. Next we can apply theorem 2.7, resulting in the convergence of the series with the complex terms. ∎

Of course, introducing absolute convergence only makes sense when there are convergent series which are not absolutely convergent. An example of this is the series $\sum_{n=1}^{\infty} (-1)^n/n$. This series converges and has sum $-\ln 2$ (see (2.26) with $t = 1$), while the series of the absolute values is the divergent harmonic series with $p = 1$ (see example 2.17).

We now present some convergence tests already known for series with real terms, but which remain valid for series with complex terms.

THEOREM 2.13

_If $|a_n| \leq b_n$ for $n = 0, 1, \ldots$ and $\sum_{n=0}^{\infty} b_n$ converges, then $\sum_{n=0}^{\infty} a_n$ converges._

Proof
According to the comparison test (theorem 2.10), the series with terms $|a_n|$ converges. The series with terms a_n is thus absolutely convergent and hence convergent. ∎

Of course, in order to use the preceding theorem, one should first have available a convergent series with non-negative terms. Suitable candidates are the harmonic series with $p > 1$ and the geometric series with a positive ratio r satisfying $0 < r < 1$ (see example 2.16), and all linear combinations of these two as well.

EXAMPLE 2.18

Consider the series $\sum_{n=1}^{\infty} e^{in\omega}/(n^2 + n)$ where ω is a real constant. The series converges absolutely since

$$\left| \frac{e^{in\omega}}{n^2 + n} \right| = \frac{1}{n^2 + n} \leq \frac{1}{n^2}$$

and the harmonic series with terms $1/n^2$ converges. ◀

A geometric series has the property that the ratio a_{n+1}/a_n of two consecutive terms is a constant. If, more generally, a sequence has terms with the property that

$$\lim_{n \to \infty} \left| \frac{a_{n+1}}{a_n} \right| = L$$

for some L, then one may conclude, as for geometric series, that the series is absolutely convergent if $L < 1$ and divergent if $L > 1$. In the case $L = 1$, however, one cannot draw any conclusion. We summarize this, without proof, in the next theorem.

THEOREM 2.14
D'Alembert's ratio test

Let (a_n) be a sequence of terms unequal to zero with $\lim_{n \to \infty} \left| a_{n+1}/a_n \right| = L$ for some L. Then one has:
a *if $L < 1$, then the series with terms a_n converges absolutely;*
b *if $L > 1$, then the series with terms a_n diverges.*

EXAMPLE 2.19

Consider the series $\sum_{n=1}^{\infty} z^n/n^p$. Here z is a complex number and p an arbitrary real constant. Put $a_n = z^n/n^p$, then

$$\lim_{n \to \infty} \left| \frac{a_{n+1}}{a_n} \right| = \lim_{n \to \infty} \left| \frac{n^p z^{n+1}}{(n+1)^p z^n} \right| = \lim_{n \to \infty} \left(\frac{n}{n+1} \right)^p |z| = |z|.$$

Hence, the series is absolutely convergent for $|z| < 1$ and divergent for $|z| > 1$. If $|z| = 1$, then no conclusions can be drawn from the ratio test. For $p > 1$ and $|z| = 1$ we are dealing with a convergent harmonic series and so the given series converges absolutely. For $p \leq 1$ we are dealing with a divergent harmonic series and so the series of absolute values diverges. From this we may *not* conclude that the series itself diverges. Take for example $p = 1$ and $z = -1$, then one obtains the series with terms $(-1)^n/n$, which is convergent. ◀

2.4.3 Series of functions

In the theory of Fourier series in part 2, and of the z-transform in chapter 18, we will encounter series having terms a_n that still depend on a variable. The geometric series in example 2.16 can again serve as an example. Other examples are

$$\sum_{n=0}^{\infty} \frac{z^n}{n!} = 1 + z + \frac{z^2}{2!} + \frac{z^3}{3!} + \cdots + \frac{z^n}{n!} + \cdots \qquad \text{for } z \in \mathbb{C},$$

$$\sum_{n=1}^{\infty} \frac{\cos nt}{n^2} = \cos t + \frac{\cos 2t}{4} + \frac{\cos 3t}{9} + \cdots + \frac{\cos nt}{n^2} + \cdots \qquad \text{for } t \in \mathbb{R}.$$

The first series is an example of a power series. The partial sums of this series are polynomials in z, so functions defined on $\mathbb{C}$. In section 2.5 we will study these more closely. In this section we will confine ourselves to series of the second type, where the functions are defined on $\mathbb{R}$ or on a part of $\mathbb{R}$. We will thus consider series of the type

$$\sum_{n=0}^{\infty} f_n(t).$$

Convergence of such a series depends, of course, on the value of t and then the sum will in general depend on t as well. For the values of t for which the series converges, the sum will be denoted by the function $f(t)$. In this case we write

$$f(t) = \sum_{n=0}^{\infty} f_n(t)$$

Pointwise convergence

and call this *pointwise convergence*. For each value of t one has a different series for which, in principle, one should verify the convergence. It turns out, however, that in many cases it is possible to analyse the convergence for an entire interval.

EXAMPLE 2.20

Let $f_n(t) = t^n$. In example 2.16 it was already shown that $\sum_{n=0}^{\infty} t^n$ converges for $|t| < 1$, with sum $1/(1 - t)$. This means that the series $\sum_{n=0}^{\infty} f_n(t)$ converges on the interval $(-1, 1)$ and that $f(t) = 1/(1 - t)$. Outside this interval, the series diverges. ◀

One would like to derive properties of $f(t)$ directly from the properties of the functions $f_n(t)$, without knowing the function $f(t)$ explicitly as a function of t. One could wonder, for example, whether a series may be differentiated term-by-term, so whether $f'(t) = \sum f_n'(t)$ if $f(t) = \sum f_n(t)$. A simple example will show that this is not always permitted.

EXAMPLE 2.21

Let, for example, $f_n(t) = \sin(nt)/n^2$, then $f_n'(t) = \cos(nt)/n$. So for each $n > 0$ the derivative exists for all $t \in \mathbb{R}$. However, if we now look at $\sum_{n=1}^{\infty} f_n'(t)$ at $t = 0$, then this equals $\sum_{n=1}^{\infty} 1/n$, which is a divergent harmonic series, as we have seen in example 2.17. Although all functions $f_n(t)$ are differentiable, $f(t)$ is not. ◀

One should also be careful with integration. When, for instance, the functions $f_n(t)$ are integrable, then one would like to conclude from this that $f(t)$ is also integrable and that

$$\int f(t)\, dt = \int \sum f_n(t)\, dt = \sum \left(\int f_n(t)\, dt \right).$$

This is not always the case, as our next example will show.

EXAMPLE 2.22

Let $u_n(t) = nt e^{-nt^2}$ for $n = 0, 1, 2, \ldots$ and let $f_n(t) = u_n(t) - u_{n-1}(t)$ for $n = 1, 2, 3, \ldots$ and $f_0(t) = 0$. Then one has for the partial sums $s_n(t)$:

$$\begin{aligned}
s_n(t) &= f_1(t) + \cdots + f_n(t) \\
&= u_1(t) - u_0(t) + u_2(t) - u_1(t) + \cdots + u_n(t) - u_{n-1}(t) = u_n(t).
\end{aligned}$$

The sequence of partial sums converges and has limit

$$f(t) = \lim_{n \to \infty} s_n(t) = \lim_{n \to \infty} nt e^{-nt^2} = 0.$$

On the interval $(0, 1)$ one thus has, on the one hand,

$$\int_0^1 \sum_{n=0}^{\infty} f_n(t)\, dt = \int_0^1 f(t)\, dt = 0,$$

while on the other hand

$$\sum_{n=0}^{\infty} \int_0^1 f_n(t)\,dt = \sum_{n=1}^{\infty} \int_0^1 \left(u_n(t) - u_{n-1}(t)\right) dt$$

$$= \sum_{n=1}^{\infty} \left[-\tfrac{1}{2}e^{-nt^2} + \tfrac{1}{2}e^{-(n-1)t^2}\right]_0^1 = \tfrac{1}{2}(e-1)\sum_{n=1}^{\infty} e^{-n} = \tfrac{1}{2}.$$

These results are unequal, hence

$$\int_0^1 \sum_{n=0}^{\infty} f_n(t)\,dt \neq \sum_{n=0}^{\infty} \int_0^1 f_n(t)\,dt.$$

◀

In order to define conditions such that properties like interchanging the order of summation and integration are valid, one could for example introduce the notion of uniform convergence. This is outside the scope of this book. We will therefore always confine ourselves to pointwise convergence, and in the case when one of the properties mentioned above is used, we will always state explicitly whether it is allowed to do so.

EXERCISES

2.16 Use the comparison test to prove the convergence of:

a $\displaystyle\sum_{n=0}^{\infty} \frac{1}{n^3 + i}$,

b $\displaystyle\sum_{n=1}^{\infty} \frac{\sin n}{n^2}$,

c $\displaystyle\sum_{n=1}^{\infty} \frac{e^{-n(1+i)}}{n}$.

2.17 Determine which of the following series converge. Justify each of your answers.

a $\displaystyle\sum_{n=1}^{\infty} \frac{(-i)^n}{n!}$,

b $\displaystyle\sum_{n=1}^{\infty} \frac{2^n + 1}{3^n + n}$,

c $\displaystyle\sum_{n=1}^{\infty} \frac{n}{(1+i)^n}$.

2.18 Show that the following series of functions converges absolutely for all t:

$$\sum_{n=0}^{\infty} \frac{e^{2int}}{2n^4 + 1}.$$

2.5   Power series

As final subject of this chapter, we consider some properties of complex power series. Power series were already introduced in section 2.4.3. These series have a simple structure. Let us start with a definition.

DEFINITION 2.8
Power series

A power series in the complex variable z and with complex coefficients c_0, c_1, c_2, ..., is a series of the form

$$\sum_{n=0}^{\infty} c_n z^n = c_0 + c_1 z + c_2 z^2 + \cdots + c_n z^n + \cdots.$$

Apparently, a partial sum $s_n = c_0 + c_1 z + c_2 z^2 + \cdots + c_n z^n$ is a polynomial in z of degree at most n. The geometric series in example 2.16 is an example of a power series. Other examples arise from so-called Taylor-series expansions of a real-valued function $f(t)$ at the real variable $t = 0$. Such a Taylor-series expansion looks like this:

$$\sum_{n=0}^{\infty} \frac{f^{(n)}(0)}{n!} t^n.$$

Here $f^{(n)}(0)$ is the value at $t = 0$ of the nth derivative. In this case we are dealing with a real power series. For a large number of functions, the Taylor-series expansion at $t = 0$ is explicitly known and in the case of convergence the sum of the Taylor-series often represents the function itself. Well-known examples are:

$$e^t = \sum_{n=0}^{\infty} \frac{t^n}{n!} = 1 + t + \frac{t^2}{2!} + \frac{t^3}{3!} + \cdots \qquad \text{for all } t, \tag{2.23}$$

$$\sin t = \sum_{n=0}^{\infty} (-1)^n \frac{t^{2n+1}}{(2n+1)!} = t - \frac{t^3}{3!} + \frac{t^5}{5!} - \cdots \qquad \text{for all } t, \tag{2.24}$$

$$\cos t = \sum_{n=0}^{\infty} (-1)^n \frac{t^{2n}}{(2n)!} = 1 - \frac{t^2}{2!} + \frac{t^4}{4!} - \cdots \qquad \text{for all } t, \tag{2.25}$$

$$\ln(1+t) = \sum_{n=1}^{\infty} (-1)^{n+1} \frac{t^n}{n} = t - \frac{t^2}{2} + \frac{t^3}{3} - \cdots \qquad \text{for } -1 < t \le 1. \tag{2.26}$$

The series above are power series in the real variable t. If we replace the real variable t by a complex variable z, then complex power series arise, for which we first of all ask ourselves: for which complex z does the power series converge and, subsequently, what is the sum of that power series? If, for example, we replace the real variable t in (2.23) by the complex variable z, then one can wonder if the series converges for all z as well, and if its sum is then still equal to the exponential function e^z. The answer is affirmative, but the treatment of functions defined on $\mathbb{C}$ will be postponed until chapter 11. In this section we will only go into the question for which values of z a power series converges. We first present some examples.

EXAMPLE 2.23

Given is the power series

$$\sum_{n=0}^{\infty} \frac{z^n}{n!} = 1 + z + \frac{z^2}{2!} + \frac{z^3}{3!} + \cdots.$$

To investigate the values of z for which this series converges, we put $a_n = z^n/n!$ and apply the ratio test:

$$\left| \frac{a_{n+1}}{a_n} \right| = \frac{n!}{(n+1)!} |z| = \frac{|z|}{n+1}.$$

We see that $\lim_{n\to\infty} |a_{n+1}/a_n| = 0 < 1$ for all z. The series thus converges for all z. ◀

EXAMPLE 2.24

Given is the power series

$$\sum_{n=1}^{\infty} \frac{(-1)^{n+1}}{n} z^n = z - \frac{z^2}{2} + \frac{z^3}{3} - \frac{z^4}{4} + \cdots .$$

To investigate the values of z for which this series converges, we put $a_n = (-1)^{n+1} z^n / n$ and apply the ratio test:

$$\left| \frac{a_{n+1}}{a_n} \right| = \frac{n}{n+1} |z| .$$

We see that $\lim_{n\to\infty} |a_{n+1}/a_n| = |z|$ and so the series converges absolutely for $|z| < 1$ and diverges for $|z| > 1$. For $|z| = 1$ no conclusion can be drawn from the ratio test. For $z = 1$ we know from (2.26) (substitute $t = 1$) that the series converges and has sum $\ln 2$, while for $z = -1$ we know that the series diverges. For all other values of z on the unit circle one can show, with quite some effort, that the series converges. ◀

EXAMPLE 2.25

Given is the power series

$$\sum_{n=0}^{\infty} n! z^n = 1 + z + 2! z^2 + \cdots .$$

To investigate the values of z for which this series converges, we put $a_n = n! z^n$ and apply the ratio test:

$$\left| \frac{a_{n+1}}{a_n} \right| = \frac{(n+1)!}{n!} |z| = (n+1) |z| .$$

We see that $\lim_{n\to\infty} |a_{n+1}/a_n| = \infty$ for $z \neq 0$ and so the series diverges for all $z \neq 0$ and it converges only for $z = 0$. ◀

EXAMPLE 2.26

Given is the power series

$$\sum_{n=0}^{\infty} n^2 2^n z^n = 2z + 16z^2 + \cdots .$$

To investigate the values of z for which this series converges, we put $a_n = n^2 2^n z^n$ and apply the ratio test:

$$\left| \frac{a_{n+1}}{a_n} \right| = \left(\frac{n+1}{n} \right)^2 |2z| .$$

We see that $\lim_{n\to\infty} |a_{n+1}/a_n| = 2|z|$ and so the series converges absolutely for $|z| < \frac{1}{2}$ and it diverges for $|z| > \frac{1}{2}$. If $|z| = \frac{1}{2}$, then $|a_n| = n^2$ and this sequence does not tend to zero. Hence, on the circle with radius $\frac{1}{2}$ the series diverges. ◀

The previous examples suggest that for each power series there exists a number R such that the power series converges absolutely for $|z| < R$ and diverges for $|z| > R$. This is indeed the case. The proof will be omitted. Usually one can find this number R with the ratio test, as in the previous examples. Summarizing, we now have the following.

For a power series in z one of the following three statements is valid.

a   The power series converges only for $z = 0$.

b   There exists a number $R > 0$ such that the power series converges absolutely for all z with $|z| < R$ and diverges for $|z| > R$.

c   The power series converges for all z.

Radius of convergence

The number R is called the *radius of convergence*. In case **a** we put $R = 0$ and in case **c** we put $R = \infty$. The radii of convergence of the power series in examples 2.23 up to 2.26 are, respectively, $R = \infty$, $R = 1$, $R = 0$, $R = \frac{1}{2}$.

If R is the radius of convergence of a power series, then this power series has a sum $f(z)$ for $|z| < R$. In the complex plane the points with $|z| = R$ form a circle of radius R and with the origin as centre, and this is sometimes called the *circle of*

Circle of convergence

convergence. For example, the geometric series in example 2.16 has sum $1/(1 - z)$ for $|z| < 1$.

Continuity and differentiability of functions on $\mathbb{C}$ will be defined in chapter 11. As far as the technique of differentiation is concerned, there is no difference in differentiating with respect to a complex variable or a real variable. Hence, the derivative of $1/(1 - z)$ is equal to $1/(1 - z)^2$. One can prove that the sum of a power series is a differentiable function within the circle of convergence. Indeed, we have the following theorem.

THEOREM 2.15

The power series $\sum_{n=0}^{\infty} c_n z^n$ and $\sum_{n=1}^{\infty} n c_n z^{n-1}$ have the same radius of convergence R. Moreover, one has: if $f(z) = \sum_{n=0}^{\infty} c_n z^n$ for $|z| < R$, then $f'(z) = \sum_{n=1}^{\infty} n c_n z^{n-1}$ for $|z| < R$.

EXAMPLE 2.27

We know that $\sum_{n=0}^{\infty} z^n = 1/(1 - z)$ for $|z| < 1$. Applying theorem 2.15 gives

$$\sum_{n=1}^{\infty} n z^{n-1} = \frac{1}{(1 - z)^2} \qquad \text{for } |z| < 1.$$

We can apply this theorem repeatedly (say k times) to obtain the following result:

$$\sum_{n=k}^{\infty} n(n - 1) \cdots (n - k + 1) z^{n-k} = \frac{k!}{(1 - z)^{k+1}}.$$

Using binomial coefficients this can be written as

$$\sum_{n=k}^{\infty} \binom{n}{k} z^{n-k} = \sum_{n=0}^{\infty} \binom{n + k}{k} z^n = \frac{1}{(1 - z)^{k+1}} \qquad \text{for } |z| < 1.$$

◀

EXERCISES

2.19

Determine the radius of convergence of the series $\displaystyle\sum_{n=0}^{\infty} \frac{2^n}{n^2 + 1} z^{2n}$.

2.20

Determine the values of z for which the following series converges, and, moreover, determine the sum for these values.

$$\sum_{n=0}^{\infty} \frac{1}{1 - i} (z - i)^n.$$

2.21

Show that the sum $f(z)$ of $\sum_{n=0}^{\infty} z^{2n}/n!$ satisfies $f'(z) = 2z f(z)$.

2.22

For which values of z does the series $\sum_{n=1}^{\infty} 2^n z^{-n}$ converge absolutely?

SUMMARY

Complex numbers play a fundamental role in the treatment of Fourier and Laplace transforms. The functions that occur are mostly functions defined on (a part of) $\mathbb{R}$ or on (a part of) the complex plane $\mathbb{C}$, with function values being complex numbers. Important examples are the time-harmonic functions $e^{i\omega t}$ with frequency ω and defined on $\mathbb{R}$, and rational functions defined on $\mathbb{C}$.

Using complex numbers one can factorize polynomials entirely into linear factors. As a consequence, when allowing complex factors, the partial fraction expansion of a rational function (for which the degree of the numerator is smaller than the degree of the denominator) will consist of fractions with numerators being just constants and denominators being polynomials of degree one or powers thereof, depending on the multiplicity of the various zeros. If the rational function has real coefficients, then one can also expand it as a sum of fractions with real coefficients and having denominators which are (powers of) linear and/or quadratic polynomials. The numerators associated with the quadratic denominators may then be linear.

The differential and integral calculus for complex-valued functions and for real-valued functions are very similar. If a complex-valued function $f(t) = u(t) + iv(t)$ with $t \in \mathbb{R}$ has a certain property, like continuity or differentiability, then this means that both the real part $u(t)$ and the imaginary part $v(t)$ have this property. The derivative $f'(t)$ of a complex-valued function equals $u'(t) + iv'(t)$. As a result, the existing rules for differentiation and integration of real-valued functions are also valid for complex-valued functions. Classes of complex-valued functions that may appear in theorems on Fourier and Laplace transforms are the class of piecewise continuous functions and the class of piecewise smooth functions.

The theory of sequences and series of real numbers can easily be extended to a theory of sequences and series of complex numbers. All kinds of properties, such as convergence and absolute convergence of a series with terms $a_n = u_n + iv_n$, can immediately be deduced from the same properties for the series with real terms u_n and v_n. Convergence tests, such as the ratio test, are the same for series with complex terms and for series with real terms.

Just as real power series, complex power series have a radius of convergence R. A power series in the complex variable z converges absolutely for $|z| < R$, that is within a circle in the complex plane with radius R (the circle of convergence), and diverges for $|z| > R$. Within the circle of convergence, the sum of a power series can be differentiated an arbitrary number of times. The derivative can be determined by differentiating the power series term-by-term.

SELFTEST

2.23   Determine the (complex) zeros and their multiplicities for the following polynomials $P(z)$:

a $P(z) = z^3 - 1$,
b $P(z) = (z^2 + i)^2 + 1$,
c $P(z) = z^5 + 8z^3 + 16z$.

2.24   Determine the partial fraction expansion of

$$F(z) = \frac{z^2 + z - 2}{(z + 1)^3}.$$

2.25   Determine the partial fraction expansion, into fractions with real coefficients, of

$$F(z) = \frac{z^2 - 6z + 7}{(z^2 - 4z + 5)^2}.$$

2.26   Calculate the integral $\int_0^{2\pi} e^{it} \cos t \, dt$.

2.27   Find out if the following series converge:

a $\displaystyle\sum_{n=1}^{\infty} n \left(\frac{2-i}{3}\right)^n$,

b $\displaystyle\sum_{n=1}^{\infty} \frac{n + i^n}{n^2}$.

2.28   Given is the series of functions

$$\sum_{n=1}^{\infty} \frac{2^{-n}}{n} e^{int} \sin t.$$

Show that this series converges absolutely for all t.

2.29   Show that if the power series $\sum_{n=0}^{\infty} c_n z^n$ has radius of convergence R, then the power series $\sum_{n=0}^{\infty} c_n z^{2n}$ has radius of convergence $R^{1/2}$.

2.30   **a** Calculate the radius of convergence R of the power series

$$\sum_{n=0}^{\infty} \frac{(1+i)^{2n}}{n+1} z^n.$$

b Let $f(z)$ be the sum of this power series. Calculate $zf'(z) + f(z)$.

Part 2
Fourier series

The Fourier series that we will encounter in this part are a tool to analyse numerous problems in mathematics, in the natural sciences and in engineering. For this it is essential that periodic functions can be written as sums of infinitely many sine and cosine functions of different frequencies. Such sums are called Fourier series.

In chapter 3 we will examine how, for a given periodic function, a Fourier series can be obtained, and which properties it possesses. In chapter 4 the conditions will be established under which the Fourier series give an exact representation of the periodic functions. In the final chapter the theory of the Fourier series is used to analyse the behaviour of systems, as defined in chapter 1, and to solve differential equations. The description of the heat distribution in objects and of the vibrations of strings are among the oldest applications from which the theory of Fourier series has arisen. Together with the Fourier integrals for non-periodic functions from part 3, this theory as a whole is referred to as *Fourier analysis*.

Jean-Baptiste Joseph Fourier (1768 – 1830) was born in Auxerre, France, as the son of a tailor. He was educated by Benedictine monks at a school where, after finishing his studies, he became a mathematics teacher himself. In 1794 he went to Paris, where he became mathematics teacher at the Ecole Normale. He declined a professorial chair offered to him by the famous Ecole Polytechnique in order to join Napoleon on his expedition to Egypt. In 1789 he was appointed governor of part of Egypt. Ousted by the English, he left Egypt again in 1801 and became prefect of Grenoble. Here he started with heat experiments and their mathematical analysis.

Fourier's mathematical ideas were not entirely new, but were built on earlier work by Bernoulli and Euler. Fourier was, however, the first to boldly state that *any* function could be developed into a series of sine and cosine functions. At first, his contemporaries refused to accept this, and publication of his work was held up for several years by the members of the Paris *Académie des Sciences*. The problem was that his ideas were considered to be insufficiently precise. And indeed, Fourier could not prove that for an arbitrary function the series would always converge pointwise to the function values. Dirichlet was one of the first to find proper conditions under which a Fourier series would converge pointwise to a periodic function. For the further development of Fourier analysis, additional fundamentally new results were required, like set-theory and the Lebesgue integral, which was developed in the one and a half centuries following Fourier.

Historically, Fourier's work has contributed enormously to the development of mathematics. Fourier set down his work in a book on the theory of heat, *Théorie analytique de la chaleur*, published in 1822. The heat or diffusion equation occurring here, as well as the wave equation for the vibrating string, can be solved, under the most frequently occurring additional conditions, using Fourier series. The methods that were used turned out to be much more widely applicable. Thereafter, applying Fourier series would produce fruitful results in many different fields, even though the mathematical theory had not yet fully crystallized. By now, the Fourier theory has become a very versatile mathematical tool. From the times of Fourier up to the present day, research has been carried out in this field, both concrete and abstract, and new applications are being developed.

Contents of Chapter 3

Fourier series: definition and properties

Fourier series: definition and properties

INTRODUCTION

Many phenomena in the applications of the natural and engineering sciences are *periodic* in nature. Examples are the vibrations of strings, springs and other objects, rotating parts in machines, the movement of the planets around the sun, the tides of the sea, the movement of a pendulum in a clock, the voltages and currents in electrical networks, electromagnetic signals emitted by transmittters in satellites, light signals transmitted through glassfibers, etc. Seemingly, all these systems operate in complicated ways; the phenomena that can be observed often behave in an erratic way. In many cases, however, they do show some kind of repetition. In order to analyse these systems, one can make use of elementary periodic functions or signals from mathematics, the sine and cosine functions. For many systems, the response or behaviour can be completely calculated or measured, by exposing them to influences or inputs given by these elementary functions. When, moreover, these systems are *linear*, then one can also calculate the response to a linear combination of such influences, since this will result in the same linear combination of responses.

Hence, for the study of the aforementioned phenomena, two matters are of importance.

On the one hand one should look at *how systems behave* under influences that can be described by elementary mathematical functions. Such an analysis will in general require specific knowledge of the system being studied. This may involve knowledge about how forces, resistances, and inertias influence each other in mechanical systems, how fluids move under the influence of external forces, or how voltages, currents and magnetic fields are mutually interrelated in electrical applications. In this book we will not go into these analyses, but the results, mostly in terms of mathematical formulations, will often be chosen as a starting point for further considerations.

On the other hand it is of importance to examine if and how an arbitrary periodic function can be described as a *linear combination of elementary sine and cosine functions*. This is the central theme of the theory of Fourier series: determine the conditions under which periodic functions can be represented as linear combinations of sine and cosine functions. In this chapter we study such linear combinations (also with infinitely many functions). These combinations are called *Fourier series* and the coefficients that occur are the *Fourier coefficients*. We will also determine the Fourier series and the Fourier coefficients for a number of standard functions and treat a number of properties of Fourier series.

In the next chapter we will examine the conditions under which a Fourier series gives an exact representation of the original function.

LEARNING OBJECTIVES

After studying this chapter it is expected that you
- know what trigonometric polynomials and series are, and know how to determine their coefficients
- know the definitions of the real and complex Fourier coefficients and Fourier series
- can determine the real and complex Fourier series for a given periodic function
- can calculate and interpret the spectrum of a periodic function
- can determine the Fourier series for some standard functions
- know and can apply the most important properties of Fourier series
- can develop a function on a given interval into a Fourier cosine or a Fourier sine series.

3.1   Trigonometric polynomials and series

The central problem of the theory of Fourier series is, how arbitrary periodic functions or signals might be written as a series of sine and cosine functions. The sine and cosine functions are also called *sinusoidal functions*. (See section 1.2.2 for a description of periodic functions or signals and section 2.4.3 for a description of series of functions.) In this section we will first look at the functions that can be constructed if we start from the sine and cosine functions. Next we will examine how, given such a function, one can recover the sinusoidal functions from which it is build up. In the next section this will lead us to the definition of the Fourier coefficients and the Fourier series for arbitrary periodic functions.

The period of periodic functions will always be denoted by T. We would like to approximate arbitrary periodic functions with linear combinations of sine and cosine functions. These sine and cosine functions must then have period T as well. One can easily check that the functions $\sin(2\pi t/T)$, $\cos(2\pi t/T)$, $\sin(4\pi t/T)$, $\cos(4\pi t/T)$, $\sin(6\pi t/T)$, $\cos(6\pi t/T)$ and so on all have period T. The constant function also has period T. Jointly, these functions can be represented by $\sin(2\pi nt/T)$ and $\cos(2\pi nt/T)$, where $n \in \mathbb{N}$. Instead of $2\pi/T$ one often writes ω_0, which means that the functions can be denoted by $\sin n\omega_0 t$ and $\cos n\omega_0 t$, where $n \in \mathbb{N}$. All these functions are periodic with period T. In this context, the constant ω_0 is called the *fundamental frequency*: $\sin \omega_0 t$ and $\cos \omega_0 t$ will complete exactly one cycle on an interval of length T, while all functions $\sin n\omega_0 t$ and $\cos n\omega_0 t$ with $n > 1$ will complete several cycles. The frequencies of these functions are thus all integer multiples of ω_0. See figure 3.1, where the functions $\sin n\omega_0 t$ and $\cos n\omega_0 t$ are sketched for $n = 1$, 2 and 3. Linear combinations, also called superpositions, of the functions

Sinusoidal function

Fundamental frequency

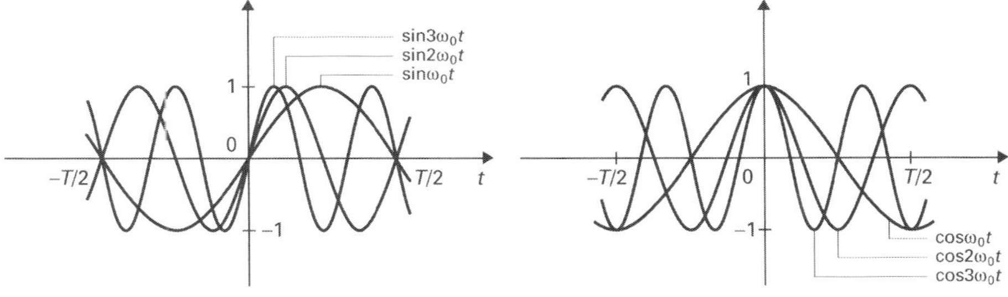

FIGURE 3.1

The sinusoidal functions $\sin n\omega_0 t$ and $\cos n\omega_0 t$ for $n = 1$, 2 and 3.

Trigonometric polynomial

$\sin n\omega_0 t$ and $\cos n\omega_0 t$ are again periodic with period T. If in such a combination we include a finite number of terms, then the expression is called a *trigonometric polynomial*. Besides the sinusoidal terms, a constant term may also occur here. Hence, a trigonometric polynomial $f(t)$ with period T can be written as

$$f(t) = A + a_1 \cos \omega_0 t + b_1 \sin \omega_0 t + a_2 \cos 2\omega_0 t + b_2 \sin 2\omega_0 t$$
$$+ \cdots + a_n \cos n\omega_0 t + b_n \sin n\omega_0 t \qquad \text{with } \omega_0 = \frac{2\pi}{T}.$$

In figure 3.2a some examples of trigonometric polynomials are shown with $\omega_0 = 1$ and so $T = 2\pi$. The polynomials shown are

$$f_1(t) = 2\sin t,$$
$$f_2(t) = 2(\sin t - \tfrac{1}{2}\sin 2t),$$
$$f_3(t) = 2(\sin t - \tfrac{1}{2}\sin 2t + \tfrac{1}{3}\sin 3t),$$
$$f_4(t) = 2(\sin t - \tfrac{1}{2}\sin 2t + \tfrac{1}{3}\sin 3t - \tfrac{1}{4}\sin 4t).$$

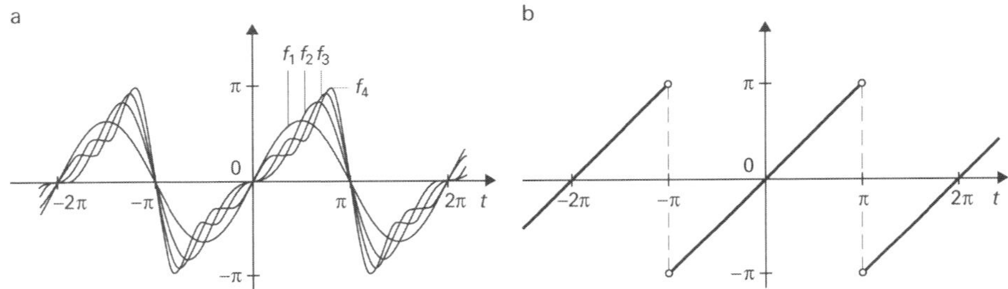

FIGURE 3.2
Some trigonometric polynomials (a) and the sawtooth function (b).

Periodic extension

In figure 3.2b the sawtooth function is drawn. It is defined as follows. On the interval $(-T/2, T/2) = (-\pi, \pi)$ one has $f(t) = t$, while elsewhere the function is extended periodically, which means that it is defined by $f(t + kT) = f(t)$ for all $k \in \mathbb{Z}$. The function $f(t)$ is then periodic with period T and is called the *periodic extension* of the function $f(t) = t$. The function values at the endpoints of the interval $(-T/2, T/2)$ are not of importance for the time being and are thus not taken into account for the moment. Comparing the figures 3.2a and 3.2b suggests that the sawtooth function, a periodic function not resembling a sinusoidal function at all, can in this case be approximated by a linear combination of sine functions only. The trigonometric polynomials f_1, f_2, f_3 and f_4 above, are partial sums of the infinite series $\sum_{n=1}^{\infty} (-1)^{n-1}(2/n)\sin nt$. It turns out that as more terms are being included in the partial sums, the approximations improve. When an infinite number of terms is included, one no longer speaks of trigonometric polynomials, but of *trigonometric series*. The most important aspect of such series is, of course, how well they can approximate an *arbitrary* periodic function. In the next chapter it will be shown that for a piecewise smooth periodic function it is indeed possible to find a trigonometric series whose sum converges at the points of continuity and is equal to the function.

Trigonometric series

At this point it suffices to observe that in this way a large class of periodic functions can be constructed, namely the trigonometric polynomials and series, all based upon the functions $\sin n\omega_0 t$ and $\cos n\omega_0 t$. All functions f which can be obtained as linear combinations or superpositions of the constant function and the sinusoidal functions with period T can be represented as follows:

$$f(t) = A + \sum_{n=1}^{\infty} (a_n \cos n\omega_0 t + b_n \sin n\omega_0 t) \qquad \text{with } \omega_0 = \frac{2\pi}{T}. \tag{3.1}$$

This, of course, only holds under the assumption that the right-hand side actually exists, that is, converges for all t.

Let us now assume that a function from the previously described class is given, but that the values of the coefficients are unknown. We thus assume that the right-hand side of (3.1) exists for all t. It is then relatively easy to recover these coefficients. In doing so, we will use the trigonometric identities

$$\sin \alpha \cos \beta = \tfrac{1}{2} \left(\sin(\alpha + \beta) + \sin(\alpha - \beta) \right),$$

$$\cos \alpha \cos \beta = \tfrac{1}{2} \left(\cos(\alpha + \beta) + \cos(\alpha - \beta) \right),$$

$$\sin \alpha \sin \beta = \tfrac{1}{2} \left(\cos(\alpha - \beta) - \cos(\alpha + \beta) \right).$$

Using these formulas one can derive the following results for $n, m \in \mathbb{N}$ with $n \neq 0$.

$$\int_{-T/2}^{T/2} \cos n\omega_0 t \, dt = \left[\frac{\sin n\omega_0 t}{n\omega_0} \right]_{-T/2}^{T/2} = 0,$$

$$\int_{-T/2}^{T/2} \sin n\omega_0 t \, dt = \left[-\frac{\cos n\omega_0 t}{n\omega_0} \right]_{-T/2}^{T/2} = 0,$$

$$\int_{-T/2}^{T/2} \cos^2 n\omega_0 t \, dt = \frac{1}{2} \int_{-T/2}^{T/2} (1 + \cos 2n\omega_0 t) dt$$

$$= \frac{1}{2} \left[t + \frac{\sin 2n\omega_0 t}{2n\omega_0} \right]_{-T/2}^{T/2} = \frac{T}{2},$$

$$\int_{-T/2}^{T/2} \sin^2 n\omega_0 t \, dt = \frac{1}{2} \int_{-T/2}^{T/2} (1 - \cos 2n\omega_0 t) dt$$

$$= \frac{1}{2} \left[t - \frac{\sin 2n\omega_0 t}{2n\omega_0} \right]_{-T/2}^{T/2} = \frac{T}{2},$$

$$\int_{-T/2}^{T/2} \sin n\omega_0 t \cos m\omega_0 t \, dt$$

$$= \frac{1}{2} \int_{-T/2}^{T/2} (\sin(n + m)\omega_0 t + \sin(n - m)\omega_0 t) \, dt = 0.$$

For $n, m \in \mathbb{N}$ with $n \neq m$ one has, moreover, that

$$\int_{-T/2}^{T/2} \cos n\omega_0 t \cos m\omega_0 t \, dt$$

$$= \frac{1}{2} \int_{-T/2}^{T/2} (\cos(n + m)\omega_0 t + \cos(n - m)\omega_0 t) \, dt = 0,$$

$$\int_{-T/2}^{T/2} \sin n\omega_0 t \sin m\omega_0 t \, dt$$

$$= \frac{1}{2} \int_{-T/2}^{T/2} (\cos(n - m)\omega_0 t - \cos(n + m)\omega_0 t) \, dt = 0.$$

On the basis of the last three equations it is said that the functions from the set $\{\sin n\omega_0 t$ and $\cos n\omega_0 t$ with $n \in \mathbb{N}\}$ are *orthogonal*: the integral of a product of two distinct functions over one period is equal to 0.

After this enumeration of results, we now return to (3.1) and try to determine the unknown coefficients A, a_n and b_n for a given $f(t)$. To this end we multiply the left-hand and right-hand side of (3.1) by $\cos m\omega_0 t$ and then integrate over the interval $(-T/2, T/2)$. It then follows that

$$\int_{-T/2}^{T/2} f(t) \cos m\omega_0 t \, dt$$

$$= \int_{-T/2}^{T/2} \left(A + \sum_{n=1}^{\infty} (a_n \cos n\omega_0 t + b_n \sin n\omega_0 t) \right) \cos m\omega_0 t \, dt$$

$$= A \int_{-T/2}^{T/2} \cos m\omega_0 t \, dt + \sum_{n=1}^{\infty} a_n \int_{-T/2}^{T/2} \cos n\omega_0 t \cos m\omega_0 t \, dt$$

$$+ \sum_{n=1}^{\infty} b_n \int_{-T/2}^{T/2} \sin n\omega_0 t \cos m\omega_0 t \, dt.$$

In this calculation we assume, for the sake of convenience, that the integral of the series may be calculated by integrating each term in the series separately. We note here that in general this has to be justified. If we now use the results stated above, then all the terms will equal 0 except for the term with $\cos n\omega_0 t \cos m\omega_0 t$, where n equals m. The integral in this term has value $T/2$, and so

$$\int_{-T/2}^{T/2} f(t) \cos m\omega_0 t \, dt = a_m \frac{T}{2},$$

or

$$a_m = \frac{2}{T} \int_{-T/2}^{T/2} f(t) \cos m\omega_0 t \, dt \qquad \text{for } m = 1, 2, \ldots. \tag{3.2}$$

This means that for a given $f(t)$, it is possible to determine a_m using (3.2). In an analogous way an expression can be found for b_m. Multiplying (3.1) by $\sin m\omega_0 t$ and again integrating over the interval $(-T/2, T/2)$, one obtains an expression for b_m (also see exercise 3.2).

A direct integration of (3.1) over $(-T/2, T/2)$ gives an expression for the constant A:

$$\int_{-T/2}^{T/2} f(t) \, dt = \int_{-T/2}^{T/2} \left(A + \sum_{n=1}^{\infty} (a_n \cos n\omega_0 t + b_n \sin n\omega_0 t) \right) dt$$

$$= \int_{-T/2}^{T/2} A \, dt = TA$$

and so

$$A = \frac{1}{T} \int_{-T/2}^{T/2} f(t) \, dt.$$

The right-hand side of this equality is, up to a factor 2, equal to the right-hand side of (3.2) for $m = 0$, because $\cos 0\omega_0 t = 1$. Hence, instead of A one usually takes $a_0/2$:

$$a_0 = 2A = \frac{2}{T} \int_{-T/2}^{T/2} f(t) \, dt.$$

All coefficients in (3.1) can thus be determined if $f(t)$ is a given trigonometric polynomial or series. The calculations are summarized in the following two expressions, from which the coefficients can be found for all functions in the class of trigonometric polynomials and series, in so far as these coefficients exist and interchanging the order of summation and integration, mentioned above, is allowed:

$$a_n = \frac{2}{T} \int_{-T/2}^{T/2} f(t) \cos n\omega_0 t \, dt \qquad \text{for } n = 0, 1, 2, \ldots, \tag{3.3}$$

$$b_n = \frac{2}{T} \int_{-T/2}^{T/2} f(t) \sin n\omega_0 t \, dt \qquad \text{for } n = 1, 2, \ldots. \tag{3.4}$$

In these equations, the interval of integration is $(-T/2, T/2)$. This interval is precisely of length one period. To determine the coefficients a_n and b_n, one can in general integrate over any other arbitrary interval of length T. Sometimes the interval $(0, T)$ is chosen (also see exercise 3.4).

EXERCISES

3.1   Verify that all functions $\sin n\omega_0 t$ and $\cos n\omega_0 t$ with $n \in \mathbb{N}$ and $\omega_0 = 2\pi/T$ have period T.

3.2   Prove that if $f(t)$ is a trigonometric polynomial with period T, then b_n can indeed be found using (3.4).

3.3   In (3.3) and (3.4) the a_n and b_n are defined for, respectively, $n = 0, 1, 2, \ldots$ and $n = 1, 2, \ldots$. Why isn't it very useful to include b_0 in these expressions?

3.4   Verify that we obtain the same values for a_n if we integrate over the interval $(0, T)$ in (3.3).

3.2   Definition of Fourier series

In the previous section we demonstrated how, starting from a collection of elementary periodic functions, one can construct new periodic functions by taking linear combinations. The coefficients in this combination could be recovered using formulas (3.3) and (3.4). These formulas can in principle be applied to *any* arbitrary periodic function with period T, provided that the integrals exist. This is an important step: the starting point is now an *arbitrary periodic function*. To it, we then apply formulas (3.3) and (3.4), which were originally only intended for trigonometric polynomials and series. The coefficients a_n and b_n thus defined are called the Fourier coefficients. The series in (3.1), which is determined by these coefficients, is called the Fourier series.

For functions that are piecewise smooth, the integrals in (3.3) and (3.4) exist. One can even show that such a function is equal to the Fourier series in (3.1) at the points of continuity. The proof of this is postponed until chapter 4. But at present, we will give the formal definitions of the Fourier coefficients and the Fourier series of a periodic function. In section 3.2.1 we define the Fourier series using the trigonometric functions $\sin n\omega_0 t$ and $\cos n\omega_0 t$, in accordance with (3.1). In many cases it is easier to work with a Fourier series with functions $e^{in\omega_0 t}$ (the time-harmonic signals, as in section 1.2.2). This complex Fourier series is introduced in section 3.2.2. Through Euler's formula, these two expressions for the Fourier series are immediately related to each other.

3.2.1 *Fourier series*

If, for an arbitrary periodic function f, the coefficients a_n and b_n, as defined by (3.3) and (3.4), can be calculated, then these coefficients are called the Fourier coefficients of the function f.

DEFINITION 3.1
Fourier coefficients

Let $f(t)$ be a periodic function with period T and fundamental frequency $\omega_0 = 2\pi/T$, then the Fourier coefficients a_n and b_n of $f(t)$, if they exist, are defined by

$$a_n = \frac{2}{T} \int_{-T/2}^{T/2} f(t) \cos n\omega_0 t \, dt \qquad for \ n = 0, 1, 2, \ldots, \tag{3.5}$$

$$b_n = \frac{2}{T} \int_{-T/2}^{T/2} f(t) \sin n\omega_0 t \, dt \qquad for \ n = 1, 2, \ldots. \tag{3.6}$$

In definition 3.1 the integration is over the interval $(-T/2, T/2)$. One can, however, integrate over any arbitrary interval of length T. The only thing that matters is that the length of the interval of integration is exactly one period (also see exercise 3.4).

In fact, in definition 3.1 a *mapping* or *transformation* is defined from functions to number sequences. This is also denoted as a transformation pair:

$$f(t) \leftrightarrow a_n, b_n.$$

One should pronounce this as: 'to the function $f(t)$ belong the Fourier coefficients a_n and b_n'. This mapping is the *Fourier transform* for periodic functions. The function $f(t)$ can be complex-valued. In that case, the coefficients a_n and b_n will also be complex. Using definition 3.1 one can now define the Fourier series associated with a function $f(t)$.

Fourier transform

DEFINITION 3.2
Fourier series

When a_n and b_n are the Fourier coefficients of the periodic function $f(t)$ with period T and fundamental frequency $\omega_0 = 2\pi/T$, then the Fourier series of $f(t)$ is defined by

$$\frac{a_0}{2} + \sum_{n=1}^{\infty} (a_n \cos n\omega_0 t + b_n \sin n\omega_0 t). \tag{3.7}$$

We do emphasize here that for arbitrary periodic functions the Fourier series will not necessarily converge for all t, and in case of convergence will not always equal $f(t)$. In chapter 4 it will be proven that for piecewise smooth functions the series does equal $f(t)$ at the points of continuity.

EXAMPLE 3.1

In section 3.1 it was suggested that the series $\sum_{n=1}^{\infty} (-1)^{n-1} (2/n) \sin nt$ approximates the sawtooth function $f(t)$, given by $f(t) = t$ for $t \in (-\pi, \pi)$ and having period 2π. We will now check that the Fourier coefficients of the sawtooth function are indeed equal to the coefficients in this series. In the present situation we have $T = 2\pi$, so $\omega_0 = 2\pi/T = 1$. The definition of Fourier coefficients can immediately be applied to the function $f(t)$. Using integration by parts it follows for $n \geq 1$ that

$$a_n = \frac{2}{T} \int_{-T/2}^{T/2} f(t) \cos n\omega_0 t \, dt = \frac{1}{\pi} \int_{-\pi}^{\pi} t \cos nt \, dt = \frac{1}{n\pi} \int_{-\pi}^{\pi} t (\sin nt)' \, dt$$

$$= \frac{1}{n\pi} [t \sin nt]_{-\pi}^{\pi} - \frac{1}{n\pi} \int_{-\pi}^{\pi} \sin nt \, dt = \frac{1}{n^2\pi} [\cos nt]_{-\pi}^{\pi} = 0.$$

For $n = 0$ we have

$$a_0 = \frac{2}{T} \int_{-T/2}^{T/2} f(t)\, dt = \frac{1}{\pi} \int_{-\pi}^{\pi} t\, dt = \frac{1}{\pi} \left[\frac{1}{2} t^2 \right]_{-\pi}^{\pi} = 0.$$

For the coefficients b_n we have that

$$
\begin{aligned}
b_n &= \frac{2}{T} \int_{-T/2}^{T/2} f(t) \sin n\omega_0 t\, dt = \frac{1}{\pi} \int_{-\pi}^{\pi} t \sin nt\, dt = -\frac{1}{n\pi} \int_{-\pi}^{\pi} t (\cos nt)'\, dt \\
&= -\frac{1}{n\pi} \left[t \cos nt \right]_{-\pi}^{\pi} + \frac{1}{n\pi} \int_{-\pi}^{\pi} \cos nt\, dt \\
&= -\frac{1}{n\pi} \left(\pi \cos n\pi - (-\pi) \cos(-n\pi) \right) + \frac{1}{n^2 \pi} \left[\sin nt \right]_{-\pi}^{\pi} \\
&= -\frac{2\pi}{n\pi} \cos n\pi = -\frac{2}{n} (-1)^n = (-1)^{n-1} \frac{2}{n}.
\end{aligned}
$$

Here we used that $\cos n\pi = (-1)^n$ for $n \in \mathbb{N}$. Hence, the Fourier coefficients a_n are all equal to zero, while the coefficients b_n are equal to $2(-1)^{n-1}/n$. The Fourier series of the sawtooth function is thus indeed equal to $\sum_{n=1}^{\infty} (-1)^{n-1}(2/n) \sin nt$. That the partial sums of the series are a good approximation of the sawtooth function can be seen in figure 3.3, where $\sum_{n=1}^{10} (-1)^{n-1}(2/n) \sin nt$ is sketched. ◀

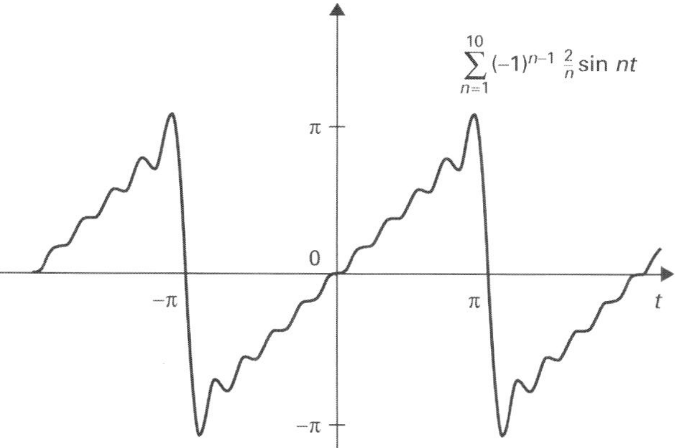

FIGURE 3.3
Partial sums of the Fourier series approximating the sawtooth function.

nth harmonic

When the periodic function f is real, and thus the coefficients a_n and b_n as well, then the nth term in the Fourier series, $a_n \cos n\omega_0 t + b_n \sin n\omega_0 t$, is called the nth *harmonic*. Instead of the sum of a cosine and a sine of equal frequency, one can also write it as a single cosine, in which case, however, a constant will occur in the argument. One then has for $a_n, b_n \in \mathbb{R}$:

$$a_n \cos n\omega_0 t + b_n \sin n\omega_0 t = \sqrt{a_n^2 + b_n^2}\, \cos(n\omega_0 t + \phi_n)$$

where

$$\tan \phi_n = -\frac{b_n}{a_n} \qquad \text{if } a_n \neq 0,$$

$$\phi_n = -\frac{\pi}{2} \qquad \text{if } a_n = 0.$$

The factor $\sqrt{a_n^2 + b_n^2}$ is the *amplitude* of the nth harmonic, ϕ_n the *initial phase*. Hence, the Fourier series can also be written as the sum of infinitely many harmonics, written exclusively in cosines. The amplitude of the nth harmonic tells us its weight in the Fourier series. From the initial phase one can deduce how far the nth harmonic is shifted relative to $\cos n\omega_0 t$.

EXAMPLE

Suppose that a function with period $T = 2\pi$ has Fourier coefficients $a_1 = 1$, $a_2 = 1/2$, $b_2 = 1/2$ and that all other coefficients are 0. Since $\omega_0 = 2\pi/T = 1$, the Fourier series is then

$$\cos t + \frac{1}{2}\cos 2t + \frac{1}{2}\sin 2t.$$

The first harmonic is $\cos t$, with amplitude 1 and initial phase 0. The amplitude of the second harmonic is $\sqrt{(1/2)^2 + (1/2)^2} = \sqrt{1/2} = \sqrt{2}/2$, while its initial phase follows from $\tan \phi_2 = -1$, so $\phi_2 = -\pi/4$. For the second harmonic we thus have

$$\frac{1}{2}\cos 2t + \frac{1}{2}\sin 2t = \frac{1}{2}\sqrt{2}\cos\left(2t - \frac{\pi}{4}\right).$$

◀

3.2.2 Complex Fourier series

In many cases it is easier to work with another representation of the Fourier series. One then doesn't use the functions $\sin n\omega_0 t$ and $\cos n\omega_0 t$, but instead the functions $e^{in\omega_0 t}$. Euler's formula gives the connection between these functions, making it possible to derive one formulation of the Fourier series from the other. According to (2.11) one has

$$\cos n\omega_0 t = \frac{e^{in\omega_0 t} + e^{-in\omega_0 t}}{2} \quad \text{and} \quad \sin n\omega_0 t = \frac{e^{in\omega_0 t} - e^{-in\omega_0 t}}{2i}.$$

If we substitute this into (3.7), it follows that

$$\frac{a_0}{2} + \sum_{n=1}^{\infty}(a_n \cos n\omega_0 t + b_n \sin n\omega_0 t)$$

$$= \frac{a_0}{2} + \sum_{n=1}^{\infty}\left(a_n \frac{e^{in\omega_0 t} + e^{-in\omega_0 t}}{2} - ib_n \frac{e^{in\omega_0 t} - e^{-in\omega_0 t}}{2}\right)$$

$$= \frac{a_0}{2} + \sum_{n=1}^{\infty}\left(\frac{1}{2}(a_n - ib_n)e^{in\omega_0 t} + \frac{1}{2}(a_n + ib_n)e^{-in\omega_0 t}\right)$$

$$= c_0 + \sum_{n=1}^{\infty}\left(c_n e^{in\omega_0 t} + c_{-n}e^{-in\omega_0 t}\right) = \sum_{n=-\infty}^{\infty} c_n e^{in\omega_0 t}.$$

Here the coefficients c_n are defined as follows:

$$c_0 = \frac{a_0}{2}, \qquad c_n = \frac{1}{2}(a_n - ib_n), \qquad c_{-n} = \frac{1}{2}(a_n + ib_n) \qquad \text{for } n \in \mathbb{N}. \qquad (3.8)$$

Instead of a Fourier series with coefficients a_n and b_n and the functions $\cos n\omega_0 t$ and $\sin n\omega_0 t$ with $n \in \mathbb{N}$, one can thus also construct, for a periodic function $f(t)$, a series with (complex) coefficients c_n and time-harmonic functions $e^{in\omega_0 t}$ with $n \in \mathbb{Z}$. The coefficients c_n are the complex Fourier coefficients. They can be calculated from the coefficients a_n and b_n using (3.8), but they can also be derived directly from the function $f(t)$. To this end, one should substitute for a_n and b_n

in (3.8) the definitions (3.5) and (3.6) (see also exercise 3.5). This leads to the following definition for the complex Fourier coefficients.

DEFINITION 3.3
Complex Fourier coefficients

Let $f(t)$ be a periodic function with period T and fundamental frequency $\omega_0 = 2\pi/T$. Then the complex Fourier coefficients c_n of $f(t)$, whenever they exist, are defined by

$$c_n = \frac{1}{T} \int_{-T/2}^{T/2} f(t) e^{-in\omega_0 t}\, dt \qquad \text{for } n \in \mathbb{Z}. \tag{3.9}$$

The complex Fourier coefficients have the prefix 'complex' since they've been determined using complex exponentials, namely, the time-harmonic signals. This prefix has thus nothing to do with the coefficients being themselves complex or not.

Like the Fourier coefficients from definition 3.1, the complex Fourier coefficients from definition 3.3 can also be calculated with an integral over an interval that differs from $(-T/2, T/2)$, as long as the interval has length T.

The mapping defined by (3.9) will also be denoted by the transformation pair

$$f(t) \leftrightarrow c_n.$$

Using the complex Fourier coefficients thus defined, we can now introduce the complex Fourier series associated with a periodic function $f(t)$.

DEFINITION 3.4
Complex Fourier series

When c_n are the complex Fourier coefficients of the periodic function $f(t)$ with period T and fundamental frequency $\omega_0 = 2\pi/T$, then the complex Fourier series of $f(t)$ is defined by

$$\sum_{n=-\infty}^{\infty} c_n e^{in\omega_0 t}. \tag{3.10}$$

Hence, for periodic functions for which the complex Fourier coefficients exist, a complex Fourier series exists as well. In chapter 4 it will be proven that for piecewise smooth functions the Fourier series converges to the function at the points of continuity.

In (3.8) the complex Fourier coefficients were derived from the real ones. Conversely one can derive the coefficients a_n and b_n from c_n using

$$a_n = c_n + c_{-n} \qquad \text{and} \qquad b_n = i(c_n - c_{-n}). \tag{3.11}$$

Therefore, when determining the Fourier series one has a choice between the real and the complex form. The coefficients can always be expressed in each other using (3.8) and (3.11). From (3.5) and (3.6) it follows that for *real* periodic functions the coefficients a_n and b_n assume real values. From (3.8) it can then immediately be deduced that c_n and c_{-n} are each other's complex conjugates:

$$c_{-n} = \overline{c_n} \qquad \text{when } f \text{ is real.} \tag{3.12}$$

Since for real functions c_n and c_{-n} are each other's complex conjugates, we obtain from (3.11) that

$$a_n = 2\operatorname{Re} c_n \quad \text{and} \quad b_n = -2\operatorname{Im} c_n \qquad \text{when } f \text{ is real.} \tag{3.13}$$

In the next example we calculate for the sawtooth function, which we already encountered in section 3.1 and example 3.1, the complex Fourier coefficients in a direct way. Moreover, we will verify that the coefficients a_n, b_n and c_n can indeed be obtained from each other.

EXAMPLE 3.2

On the interval $(-\pi, \pi)$ the sawtooth function with period 2π is given by $f(t) = t$. One has $T = 2\pi$ and $\omega_0 = 1$. The complex Fourier coefficients can be calculated directly using definition 3.3. For $n \neq 0$ it follows from integration by parts that

$$
\begin{aligned}
c_n &= \frac{1}{T} \int_{-T/2}^{T/2} f(t) e^{-in\omega_0 t}\, dt = \frac{1}{2\pi} \int_{-\pi}^{\pi} t e^{-int}\, dt \\
&= \frac{-1}{2in\pi} \left\{ \left[t e^{-int} \right]_{-\pi}^{\pi} - \int_{-\pi}^{\pi} e^{-int}\, dt \right\} \\
&= \frac{-1}{2in\pi} \left\{ \pi e^{-in\pi} + \pi e^{in\pi} + \frac{1}{in} (e^{-in\pi} - e^{in\pi}) \right\} \\
&= \frac{-1}{2in\pi} \left\{ 2\pi \cos n\pi - \frac{2}{n} \sin n\pi \right\} = \frac{-1}{in} \cos n\pi = \frac{i}{n} (-1)^n.
\end{aligned}
$$

For $n = 0$ one has

$$
c_0 = \frac{1}{T} \int_{-T/2}^{T/2} f(t)\, dt = \frac{1}{2\pi} \int_{-\pi}^{\pi} t\, dt = 0.
$$

The complex Fourier series of the sawtooth function $f(t) = t$ with period 2π is thus equal to $\sum_{n=-\infty, n\neq 0}^{\infty} (i/n)(-1)^n e^{int}$. That these complex Fourier coefficients coincide with the coefficients from example 3.1 can be verified with (3.13) (the sawtooth function is real):

$$
\begin{aligned}
a_n &= 2\operatorname{Re} c_n = 0 \qquad \text{for } n = 1, 2, 3, \ldots, \\
b_n &= -2\operatorname{Im} c_n = -2\frac{1}{n}(-1)^n \qquad \text{for } n = 1, 2, 3, \ldots, \\
a_0 &= 2c_0 = 0.
\end{aligned}
$$

Comparing these coefficients with the ones from example 3.1 will show that they are equal. ◄

EXERCISES

3.5

Use the definitions of a_n and b_n to verify that for a periodic function $f(t)$ with period T it follows from (3.8) that

$$
c_n = \frac{1}{T} \int_{-T/2}^{T/2} f(t) e^{-in\omega_0 t}\, dt \qquad \text{for } n \in \mathbb{Z}.
$$

3.6

Sketch the periodic function $f(t)$ with period 2π given by $f(t) = |t|$ for $t \in (-\pi, \pi)$ and determine its Fourier series.

3.7

Sketch the periodic function $g(t)$ with period 2 and determine its complex Fourier series when $g(t)$ is given for $-1 < t < 1$ by

$$
g(t) = \begin{cases} 0 & \text{for } -1 < t < 0, \\ e^{-t} & \text{for } 0 \leq t < 1. \end{cases}
$$

3.8

Sketch the periodic function $f(t)$ with period 4 and determine its Fourier series when $f(t)$ is given for $-2 < t < 2$ by

$$
f(t) = \begin{cases} 2 & \text{for } -2 < t < 0, \\ t & \text{for } 0 \leq t < 2. \end{cases}
$$

3.9

Determine the complex Fourier series of the periodic complex-valued function $g(t)$ given by $g(t) = t^2 + it$ for $t \in (-1, 1)$ and having period 2.

3.3   The spectrum of periodic functions

Time domain

The periodic functions, or periodic signals, that we have considered so far, both the real and the complex ones, were all defined for $t \in \mathbb{R}$. Here the variable t is often interpreted as a time-variable. We say that these functions are defined in the *time domain*. For these functions the Fourier coefficients c_n (or a_n and b_n) can be determined. Through the Fourier series, each of these coefficients is associated with a function of a specific frequency $n\omega_0$. The values of the Fourier coefficients tell us the weight of the function with frequency $n\omega_0$ in the Fourier series. For piecewise smooth functions we will establish in chapter 4 that the Fourier series is equal to the function. This means that these functions are completely determined by their Fourier coefficients. Since the coefficients c_n (or a_n and b_n) are associated with frequency $n\omega_0$, we then say that the function $f(t)$ is described by the Fourier coefficients in

Frequency domain

the *frequency domain*. As soon as the values c_n are known, the original function in the time domain is also fixed.

In daily life as well, we often interpret signals in terms of frequencies. Sound and light are quantities that are expressed in terms of frequencies, and we observe these as pitch and colour.

Spectrum
Discrete spectrum
Line spectrum
Amplitude spectrum
Phase spectrum

The sequence of Fourier coefficients c_n with $n \in \mathbb{Z}$, which thus describe a function in the frequency domain, is called the *spectrum* of the function. Since n assumes only integer values, the spectrum is called a *discrete* or a *line spectrum*. Often, not the spectrum itself is given, but instead the *amplitude spectrum* $|c_n|$ and the *phase spectrum* $\arg(c_n)$. Hence, the amplitude and phase spectrum are defined as soon as the complex Fourier coefficients exist, also in the case when the function $f(t)$ is complex-valued. This definition of amplitude and phase is thus more general than the one for the nth harmonic, which only existed in the case when $f(t)$ was real.

EXAMPLE 3.3

Figure 3.4 shows the amplitude and phase spectrum of the sawtooth function, for which we deduced in example 3.2 that the complex Fourier coefficients c_n are given by $c_n = (-1)^n (i/n)$ for $n \neq 0$ and $c_0 = 0$. The amplitude spectrum is thus given by $|c_n| = 1/|n|$ for $n \neq 0$ and $|c_0| = 0$, while the phase spectrum is given by $\arg(c_n) = (-1)^n (\pi/2)$ for $n > 0$ and by $\arg(c_n) = (-1)^{n-1} (\pi/2)$ for $n < 0$, and is undefined for $n = 0$.   ◄

a   b

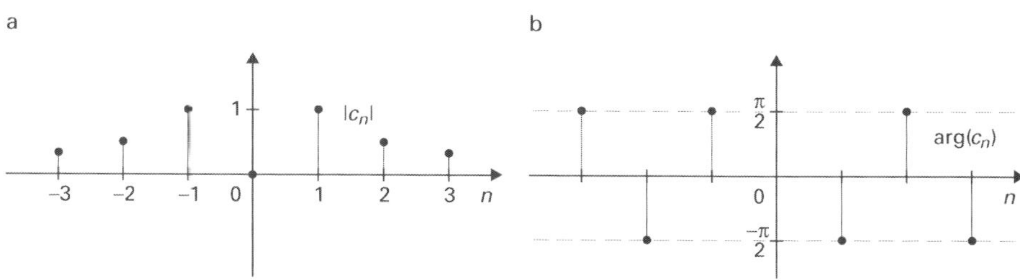

FIGURE 3.4
The amplitude spectrum (a) and phase spectrum (b) of the sawtooth function.

EXERCISES

3.10

Determine and sketch the amplitude and phase spectra of the functions from exercises 3.6 to 3.9.

3.4   Fourier series for some standard functions

In the preceding sections the Fourier coefficients of the sawtooth function have been determined. It will be convenient to know the Fourier series for some other standard functions as well. Together with the properties, to be treated in the next section, this will enable us to determine the Fourier series for quite a number of periodic functions relatively easily. In this section we determine the Fourier series for a number of functions. These standard functions and their Fourier coefficients are also included in table 1 at the back of the book. The first two functions that will
Even function   be treated are *even*, that is, $f(t) = f(-t)$. The third function will be *odd*, that is,
Odd function $f(t) = -f(-t)$.

3.4.1   *The periodic block function*

The periodic function $p_{a,T}(t)$ with period $T > 0$ and $0 \le a \le T$ and having value
Periodic block function   1 for $|t| \le a/2 \le T/2$ and value 0 for $a/2 < |t| \le T/2$ is called the *periodic block function*. Its graph is sketched in figure 3.5. The complex Fourier coefficients

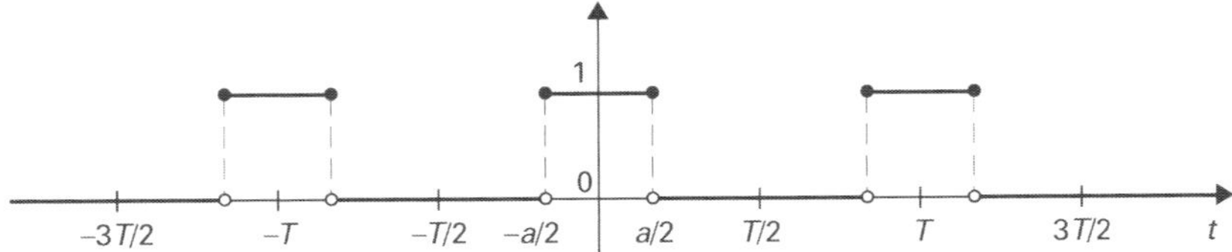

FIGURE 3.5
The periodic block function $p_{a,T}(t)$.

of the periodic block function can be calculated for $n \ne 0$ using (3.9):

$$c_n = \frac{1}{T} \int_{-T/2}^{T/2} p_{a,T}(t) e^{-in\omega_0 t}\, dt = \frac{1}{T} \int_{-a/2}^{a/2} e^{-in\omega_0 t}\, dt = \frac{1}{T} \left[\frac{e^{-in\omega_0 t}}{-in\omega_0} \right]_{-a/2}^{a/2}$$

$$= \frac{2}{Tn\omega_0} \left(\frac{e^{in\omega_0 a/2} - e^{-in\omega_0 a/2}}{2i} \right) = \frac{2}{T} \frac{\sin(n\omega_0 a/2)}{n\omega_0}.$$

For $n = 0$ it follows that

$$c_0 = \frac{1}{T} \int_{-T/2}^{T/2} p_{a,T}(t)\, dt = \frac{1}{T} \int_{-a/2}^{a/2} 1\, dt = \frac{a}{T}.$$

For a given value of a, the Fourier coefficients for $n \ne 0$ are thus equal to $2\sin(n\omega_0 a/2)/Tn\omega_0$, which are precisely the values of $2\sin(ax/2)/Tx$ evaluated at $x = n\omega_0$, where n runs through the integers. In figure 3.6 the function $f(x) = 2\sin(ax/2)/Tx$ is drawn; for $x = 0$ the function is defined by $\lim_{x \to 0} f(x) = a/T = c_0$. From this we can obtain the Fourier coefficients of the periodic block function by evaluating the function values at $n\omega_0$ for $n \in \mathbb{Z}$. One thus has for the periodic block function:

$$p_{a,T}(t) \leftrightarrow \frac{2}{T} \frac{\sin(n\omega_0 a/2)}{n\omega_0}. \tag{3.14}$$

Here one should take the value $\lim_{x \to 0} 2\sin(ax/2)/Tx$ for $n = 0$.

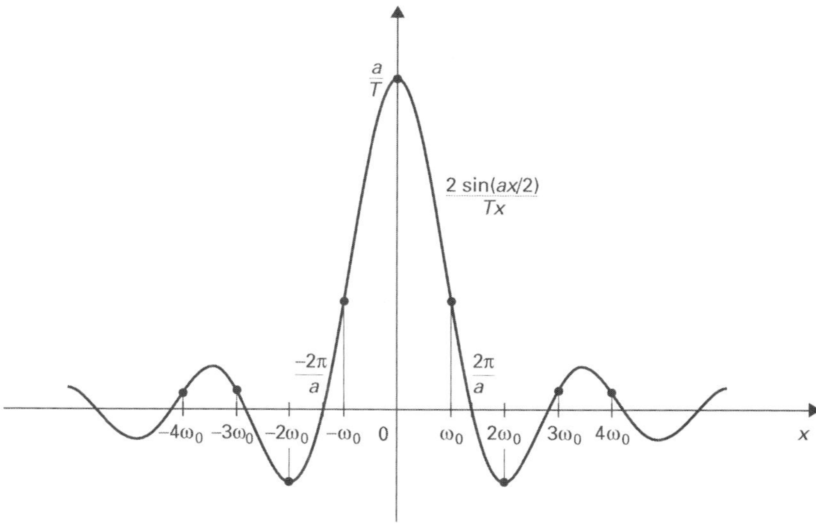

FIGURE 3.6
Evaluating the function values of $2\sin(ax/2)/Tx$ at $x = n\omega_0$ for $n \in \mathbb{Z}$ gives the Fourier coefficients of the periodic block function.

EXAMPLE 3.4

When $a = T/2$, the Fourier coefficients of the periodic block function are

$$c_n = \frac{1}{n\pi}\sin(n\pi/2) \qquad \text{for } n \neq 0, \qquad c_0 = \frac{1}{2}.$$

For even n (with the exception of 0) the Fourier coefficients are thus equal to 0 and for odd n they are equal to $(-1)^{(n-1)/2}/n\pi$. The amplitude spectrum $|c_n|$ of this periodic block function is drawn in figure 3.7. ◀

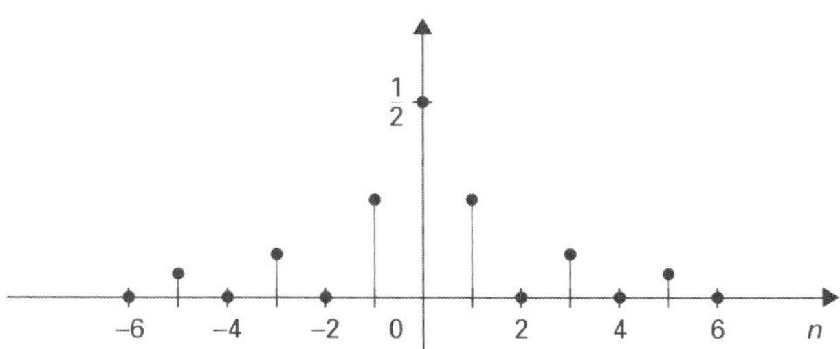

FIGURE 3.7
Amplitude spectrum of the periodic block function for $a = T/2$.

The partial sums $\sum_{n=-m}^{m} c_n e^{in\omega_0 t}$ of the Fourier series give a better approximation of the periodic block function as we increase the numbers of terms included. To illustrate this, the periodic block function for $a = T/2$ is drawn in figure 3.8, together with the partial sums of the Fourier series for $m = 0, 1, 3, 5$ and 7. To

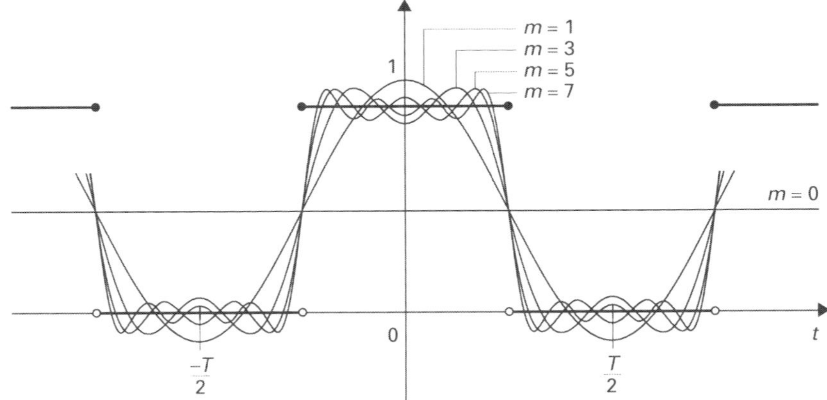

FIGURE 3.8
The periodic block function for $a = T/2$ and approximations using the partial sums
of the Fourier series.

do so, the partial sums have been rewritten as follows:

$$
\sum_{n=-m}^{m} c_n e^{in\omega_0 t} = c_0 + \sum_{n=1}^{m} (c_n e^{in\omega_0 t} + c_{-n} e^{-in\omega_0 t})
$$

$$
= c_0 + \sum_{n=1}^{m} 2c_n \frac{e^{in\omega_0 t} + e^{-in\omega_0 t}}{2} = c_0 + \sum_{n=1}^{m} 2c_n \cos n\omega_0 t
$$

$$
= \frac{1}{2} + \sum_{\substack{n=1 \\ n \text{ odd}}}^{m} \frac{2}{n\pi} (-1)^{(n-1)/2} \cos n\omega_0 t.
$$

Hence, the Fourier series of the periodic block function contains only cosine terms.

3.4.2　The periodic triangle function

The periodic triangle function $q_{a,T}(t)$ with period T is defined for $0 < a \leq T/2$.
On $(-T/2, T/2)$ it is defined by:

$$
q_{a,T}(t) = \begin{cases} 1 - \dfrac{|t|}{a} & \text{for } |t| \leq a, \\ 0 & \text{for } a < |t| \leq T/2. \end{cases}
$$

The graph of the periodic triangle function is sketched in figure 3.9. For $n \neq 0$
the complex Fourier coefficients of the periodic triangle function can be calculated
using (3.9):

$$
c_n = \frac{1}{T} \int_{-T/2}^{T/2} q_{a,T}(t) e^{-in\omega_0 t}\, dt
$$

$$
= \frac{1}{T} \int_{-a}^{0} (1 + \frac{t}{a}) e^{-in\omega_0 t}\, dt + \frac{1}{T} \int_{0}^{a} (1 - \frac{t}{a}) e^{-in\omega_0 t}\, dt
$$

$$
= \frac{1}{T} \int_{0}^{a} (1 - \frac{t}{a}) e^{-in\omega_0 t}\, dt + \frac{1}{T} \int_{0}^{a} (1 - \frac{t}{a}) e^{-in\omega_0 t}\, dt
$$

$$
= \frac{2}{T} \int_{0}^{a} (1 - \frac{t}{a}) \cos n\omega_0 t\, dt.
$$

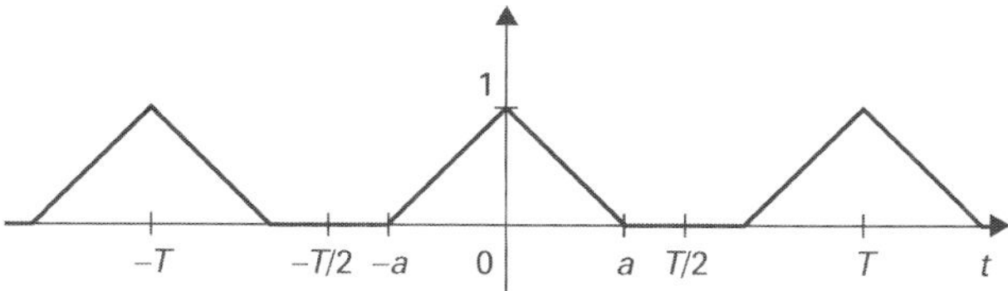

FIGURE 3.9
The periodic triangle function $q_{a,T}(t)$.

From integration by parts it then follows that

$$
c_n = \frac{2}{T}\frac{1}{n\omega_0}\left\{\left[(1-\frac{t}{a})\sin n\omega_0 t\right]_0^a + \frac{1}{a}\int_0^a \sin n\omega_0 t\, dt\right\}
$$

$$
= \frac{2}{Tn\omega_0}\frac{1}{a}\frac{-1}{n\omega_0}\left[\cos n\omega_0 t\right]_0^a = \frac{2(1-\cos n\omega_0 a)}{n^2\omega_0^2 aT}.
$$

Because $1 - \cos n\omega_0 a = 2\sin^2(n\omega_0 a/2)$, we thus have

$$
c_n = \frac{4\sin^2(n\omega_0 a/2)}{n^2\omega_0^2 aT}.
$$

Since $\int_{-a}^{a}(1 - |t|/a)\,dt = a$, it immediately follows that $c_0 = a/T$. For a given a, the Fourier coefficients for $n \neq 0$ equal $4\sin^2(n\omega_0 a/2)/n^2\omega_0^2 aT$, which are precisely the function values of $4\sin^2(ax/2)/ax^2 T$, evaluated at $x = n\omega_0$, where n runs through the integers. In figure 3.10 the function $f(x) = 4\sin^2(ax/2)/ax^2 T$ is drawn. For $x = 0$ the function is defined by $\lim_{x\to 0} f(x) = a/T = c_0$. By evaluating the function values at $n\omega_0$ for $n \in \mathbb{Z}$, one can thus derive the Fourier coefficients from this function. For the periodic triangle function we thus have

$$
q_{a,T}(t) \leftrightarrow \frac{4\sin^2(n\omega_0 a/2)}{n^2\omega_0^2 aT}, \tag{3.15}
$$

where for $n = 0$ one has to take the value $\lim_{x\to 0} 4\sin^2(ax/2)/ax^2 T$.

3.4.3   *The sawtooth function*

We have already discussed the sawtooth function several times in this chapter. For an arbitrary period T it is defined by $f(t) = 2t/T$ on the interval $(-T/2, T/2)$ and extended periodically elsewhere. The 'standard' sawtooth function thus varies between -1 and $+1$. Analogous to the previous examples one can derive that the complex Fourier coefficients are equal to $c_n = i(-1)^n/\pi n$ for $n \neq 0$ and $c_0 = 0$.

EXERCISES

3.11

In example 3.4 the Fourier coefficients of the periodic block function have been determined for $a = T/2$.
a Determine the Fourier coefficients for $a = T/4$ and sketch the amplitude spectrum.

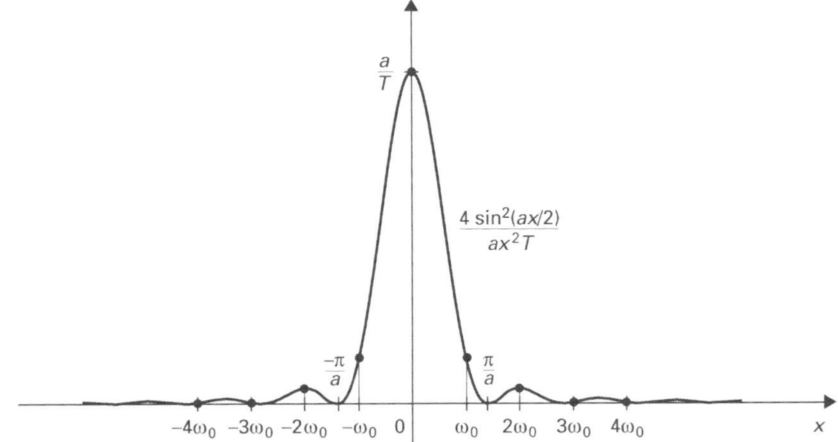

FIGURE 3.10
Evaluating the function values of $4\sin^2(ax/2)/ax^2T$ at $x = n\omega_0$ for $n \in \mathbb{Z}$ gives
the Fourier coefficients of the periodic triangle function.

b   Now determine the Fourier coefficients for $a = T$. What do these Fourier coefficients imply for the Fourier series?

3.12   Determine for $a = T/2$ the Fourier coefficients and the amplitude spectrum of the periodic triangle function from section 3.4.2.

3.13   Determine the Fourier coefficients of the sawtooth function given by $f(t) = 2t/T$ on the interval $(-T/2, T/2)$ and extended periodically elsewhere, and sketch the amplitude and phase spectrum.

3.5   Properties of Fourier series

In the previous section Fourier series were determined for a number of standard functions. In the same way one can, in principle, determine the Fourier series for many more periodic functions. This, however, is quite cumbersome. By using a number of properties of Fourier series one can determine in a relatively simple way the Fourier series of a large number of periodic functions. These properties have also been included in table 2 at the back of the book.

3.5.1   *Linearity*

Fourier coefficients of linear combinations of functions are equal to the same linear combination of the Fourier coefficients of the individual functions. This property is formulated in the following theorem.

THEOREM 3.1
*Linearity of the Fourier
transform*

When the complex Fourier coefficients of $f(t)$ and $g(t)$ are f_n and g_n respectively, then one has for $a, b \in \mathbb{C}$:

$$af(t) + bg(t) \leftrightarrow af_n + bg_n.$$

Proof
The proof of this theorem is a straightforward application of the linearity of integration. When c_n denotes the Fourier coefficients of $af(t) + bg(t)$, then

$$c_n = \frac{1}{T} \int_{-T/2}^{T/2} (af(t) + bg(t))e^{-in\omega_0 t}\, dt$$

$$= \frac{a}{T} \int_{-T/2}^{T/2} f(t)e^{-in\omega_0 t}\, dt + \frac{b}{T} \int_{-T/2}^{T/2} g(t)e^{-in\omega_0 t}\, dt = af_n + bg_n.$$

$\blacksquare$

EXAMPLE 3.5

With the linearity property one can easily determine the Fourier coefficients of linear combinations of functions whose individual Fourier coefficients are already known. Let f be the periodic function with period 6 as sketched in figure 3.11. The function

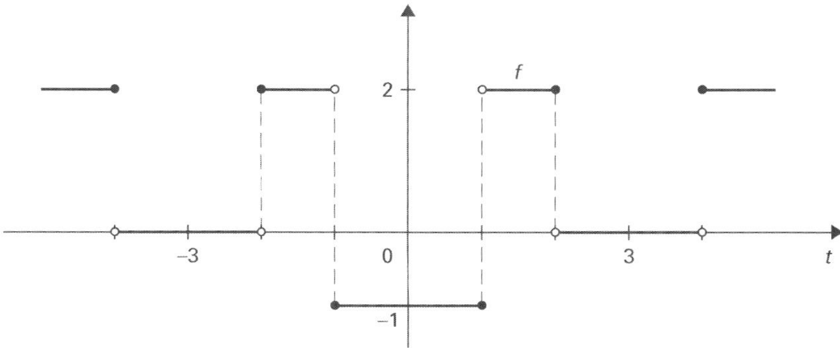

FIGURE 3.11
Periodic function as a combination of periodic block functions.

f is then equal to $2g - 3h$, where g and h are periodic block functions, as defined in section 3.4.1, with period 6 and, respectively, $a = 4$ and $a = 2$. Using (3.14) or table 1 and applying theorem 3.1, it then follows that the Fourier coefficients are given by

$$c_n = \frac{4}{6} \frac{\sin(n\omega_0 4/2)}{n\omega_0} - \frac{6}{6} \frac{\sin(n\omega_0 2/2)}{n\omega_0}$$

$$= \frac{2}{3} \frac{\sin(n2\pi/3)}{n2\pi/6} - \frac{\sin(n\pi/3)}{n2\pi/6} = \frac{2\sin(2n\pi/3) - 3\sin(n\pi/3)}{n\pi}.$$

$\blacktriangleleft$

3.5.2   Conjugation

The Fourier coefficients of the complex conjugate of f can be derived from the Fourier coefficients of the function itself. How this can be done is the subject of our next theorem.

THEOREM 3.2
Fourier coefficients of a conjugate

When the Fourier coefficients of $f(t)$ are equal to c_n, then

$$\overline{f(t)} \leftrightarrow \overline{c_{-n}}.$$

Proof
Since $\overline{e^{in\omega_0 t}} = e^{-in\omega_0 t}$, it follows by direct calculation of the Fourier coefficients of $\overline{f(t)}$ that

$$\frac{1}{T}\int_{-T/2}^{T/2} \overline{f(t)}e^{-in\omega_0 t}\,dt = \overline{\frac{1}{T}\int_{-T/2}^{T/2} f(t)e^{in\omega_0 t}\,dt}$$

$$= \overline{\frac{1}{T}\int_{-T/2}^{T/2} f(t)e^{-i(-n)\omega_0 t}\,dt} = \overline{c_{-n}}.$$

∎

This property has a special consequence when $f(t)$ is real. This is because we then have $f(t) = \overline{f(t)}$. The Fourier coefficients of $f(t)$ and $\overline{f(t)}$, whenever these exist at least, must then also be equal and hence $c_n = \overline{c_{-n}}$. This result has been derived before, see (3.12). Furthermore, one has for the moduli that $|c_n| = |\overline{c_{-n}}|$ and since the moduli of complex conjugates are the same (see (2.3)), this in turn equals $|c_{-n}|$. For a real function it thus follows that $|c_n| = |c_{-n}|$, which means that the amplitude spectrum is even. We also know that the arguments of complex conjugates are each other's opposite, and so $\arg(c_n) = \arg(\overline{c_{-n}}) = -\arg(c_{-n})$. Hence, the phase spectrum is odd.

EXAMPLE

The standard functions treated in section 3.4 are all real. One can easily check that the amplitude spectra are indeed even. For the sawtooth function one can check moreover that the phase spectrum is odd, while the phase spectra of the periodic block and triangle functions are zero, and so odd as well. ◀

3.5.3   Shift in time

The standard functions treated in section 3.4 were all neatly 'centred' around $t = 0$. From these one can, by a shift in time, obtain functions that are, of course, no longer centred around $t = 0$. When the shift equals t_0, then the new function will be given by $f(t - t_0)$. The Fourier coefficients of the shifted function can immediately be obtained from the Fourier coefficients of the original function.

THEOREM 3.3
Shift in time

When c_n are the Fourier coefficients of $f(t)$, then

$$f(t - t_0) \leftrightarrow e^{-in\omega_0 t_0} c_n.$$

Proof
The Fourier coefficients of $f(t - t_0)$ can be calculated using the definition. In this calculation we introduce the new variable $\tau = t - t_0$ and we integrate over $(-T/2, T/2)$ instead of $((-T/2)+t_0, (T/2)+t_0)$, since this gives the same result:

$$\frac{1}{T}\int_{-T/2}^{T/2} f(t - t_0)e^{-in\omega_0 t}\,dt$$

$$= e^{-in\omega_0 t_0}\frac{1}{T}\int_{(-T/2)+t_0}^{(T/2)+t_0} f(t - t_0)e^{-in\omega_0(t-t_0)}\,d(t - t_0)$$

$$= e^{-in\omega_0 t_0}\frac{1}{T}\int_{-T/2}^{T/2} f(\tau)e^{-in\omega_0 \tau}\,d\tau = e^{-in\omega_0 t_0}\cdot c_n.$$

∎

It follows immediately from theorem 3.3 that the amplitude spectra of $f(t)$ and $f(t - t_0)$ are the same: $|e^{-in\omega_0 t_0}c_n| = |c_n|$. Hence, the amplitude spectrum of a function does not change when the function is shifted in time. For the phase spectrum one has

$$\arg\left(e^{-in\omega_0 t_0}c_n\right) = \arg\left(e^{-in\omega_0 t_0}\right) + \arg(c_n) = n\arg\left(e^{-i\omega_0 t_0}\right) + \arg(c_n).$$

The phase spectrum thus changes linearly with n, apart from the fact that the argument can always be reduced to a value in the interval $[-\pi, \pi]$.

EXAMPLE

The periodic block function from example 3.4 is centred around $t = 0$. The Fourier coefficients of the periodic function with $f(t) = 1$ for $0 \le t < T/2$ and $f(t) = 0$ for $-T/2 < t < 0$, see figure 3.12, can easily be derived from this. For this periodic

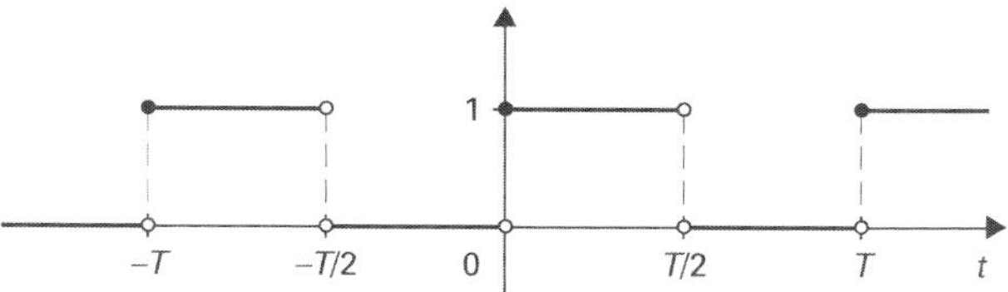

FIGURE 3.12
The shifted periodic block function.

block function one has $t_0 = T/4$ and $a = T/2$, so the Fourier coefficients are equal to

$$c_n = e^{-in\omega_0 t_0}\frac{1}{n\pi}\sin(n\pi/2) = e^{-in\pi/2}\frac{1}{n\pi}\frac{e^{in\pi/2} - e^{-in\pi/2}}{2i}$$

$$= \frac{1 - e^{-in\pi}}{2in\pi} = \frac{(-1)^n - 1}{2n\pi}i.$$

Hence, for even n and $n \ne 0$ one has $c_n = 0$ and for odd n one has $c_n = -i/n\pi$. Furthermore, $c_0 = 1/2$. The amplitude spectrum is thus equal to that of the periodic block function in example 3.4. ◀

3.5.4 *Time reversal*

The process of changing from the variable t to the variable $-t$ is called *time reversal*. In this case there is again a simple relationship between the Fourier coefficients of the functions.

THEOREM 3.4
Time reversal

When c_n are the Fourier coefficients of $f(t)$, then

$$f(-t) \leftrightarrow c_{-n}.$$

Proof
A direct application of the definition and changing from the variable $-t$ to τ gives the proof:

$$\frac{1}{T}\int_{-T/2}^{T/2} f(-t)e^{-in\omega_0 t}\,dt = \frac{1}{T}\int_{T/2}^{-T/2} f(\tau)e^{-in\omega_0(-\tau)}\,d(-\tau)$$

$$= \frac{1}{T}\int_{-T/2}^{T/2} f(\tau)e^{-i(-n)\omega_0 \tau}\,d\tau = c_{-n}.$$

■

A direct consequence of this theorem is that if $f(t)$ is even, so $f(t) = f(-t)$ for all t, then $c_n = c_{-n}$. In this case the amplitude spectrum as well as the phase spectrum are even. Furthermore, it follows from (3.8) that the coefficients b_n of the ordinary Fourier series are all 0. Thus, the Fourier series of an even function contains only cosine terms. This result is easily understood: the sines are odd, while the cosines are even. A series containing sine functions will never be even. When, moreover, $f(t)$ is real, then c_n and c_{-n} are each other's complex conjugate (see (3.12)) and in that case the coefficients will be real as well.

EXAMPLE

The periodic block function and the periodic triangle function from sections 3.4.1 and 3.4.2 are even and real. The spectra are also even and real.   ◄

When $f(t)$ is odd, so $f(t) = -f(-t)$, it follows that $c_n = -c_{-n}$ and so the spectrum is odd. Since $c_{-n} = \overline{c_n}$ for $f(t)$ real, the Fourier coefficients are purely imaginary. The spectrum of a real and odd function is thus odd and purely imaginary. Moreover, in the case of an odd function it follows from (3.8) that the coefficients a_n are 0 and that the Fourier series consists of sine functions only.

EXAMPLE

The periodic sawtooth function is a real and odd function. The complex Fourier coefficients are odd and purely imaginary, while the Fourier series contains only sine functions.   ◄

EXERCISES

3.14   The periodic function f with period 4 is given by $f(t) = 1 + |t|$ for $|t| \leq 1$ and $f(t) = 0$ for $1 < |t| < 2$. Sketch the graph of the function and determine its Fourier coefficients.

3.15   Determine the Fourier coefficients of the periodic function with period T defined by $f(t) = t$ on the interval $(0, T)$.

3.16   Let the complex-valued function $f(t) = u(t) + iv(t)$ be given, where $u(t)$ and $v(t)$ are real functions with Fourier coefficients u_n and v_n.
a   Determine the Fourier coefficients of $f(t)$ and of $\overline{f(t)}$.
b   Suppose that $f(t)$ is even, but not real. Will the Fourier coefficients of $f(t)$ be even and real then?

3.17   The amplitude spectrum of a function does not change when a shift in time is applied. For which shifts does the phase spectrum remains unchanged as well?

3.18   In section 3.5.4 we derived that for even functions the ordinary Fourier series contains only cosine terms. Show that this also follows directly from (3.5) and (3.6).

3.6   Fourier cosine and Fourier sine series

In section 3.5.4 we showed that the ordinary Fourier series of an even periodic function contains only cosine terms and that the Fourier series of an odd periodic function contains only sine terms. For the standard functions we have seen that the periodic block function and the periodic triangle function, which are even, do indeed contain cosine terms only and that the sawtooth function, which is odd, contains sine terms only. Sometimes it is desirable to obtain for an *arbitrary* function on the interval $(0, T)$ a Fourier series containing only sine terms or containing only cosine terms. Such series are called Fourier sine series and Fourier cosine series. In order

Fourier sine series
Fourier cosine series

to find a Fourier cosine series for a function defined on the interval $(0, T)$, we extend the function to an even function on the interval $(-T, T)$ by defining $f(-t) = f(t)$ for $-T < t < 0$ and subsequently extending the function periodically with period

$2T$. The function thus created is now an even function and its ordinary Fourier series will contain only cosine terms, while the function is equal to the original function on the interval $(0, T)$.

In a similar way one can construct a Fourier sine series for a function by extending the function defined on the interval $(0, T)$ to an odd function on the interval $(-T, T)$ and subsequently extending it periodically with period $2T$. Such an odd function will have an ordinary Fourier series containing only sine terms.

Forced series development

Determining a Fourier sine series or a Fourier cosine series in the way described above is sometimes called a *forced series development*.

EXAMPLE

Let the function $f(t)$ be given by $f(t) = t^2$ on the interval $(0, 1)$. We wish to obtain a Fourier sine series for this function. We then first extend it to an odd function on the interval $(-1, 1)$ and subsequently extend it periodically with period 2. The function and its odd and periodic extension are drawn in figure 3.13. The ordinary

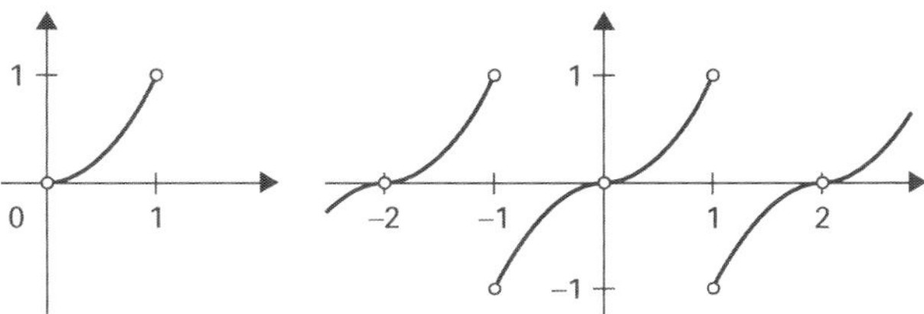

FIGURE 3.13
The function $f(t) = t^2$ on the interval $(0, 1)$ and its odd and periodic extension.

Fourier coefficients of the function thus created can be calculated using (3.5) and (3.6). Since the function is odd, all coefficients a_n will equal 0. For b_n we have

$$
b_n = \frac{2}{T} \int_{-T/2}^{T/2} f(t) \sin n\omega_0 t \, dt = \int_{-1}^{0} (-t^2) \sin n\pi t \, dt + \int_{0}^{1} t^2 \sin n\pi t \, dt
$$
$$
= 2 \int_{0}^{1} t^2 \sin n\pi t \, dt.
$$

Applying integration by parts twice, it follows that

$$
b_n = \frac{-2}{n\pi} \left\{ \left[t^2 \cos n\pi t \right]_0^1 - \frac{2}{n\pi} \left[t \sin n\pi t \right]_0^1 - \frac{2}{n^2\pi^2} \left[\cos n\pi t \right]_0^1 \right\}
$$
$$
= \frac{2}{n\pi} \left(\frac{2(\cos n\pi - 1)}{n^2\pi^2} - \cos n\pi \right) = \frac{2}{n\pi} \left(\frac{2((-1)^n - 1)}{n^2\pi^2} - (-1)^n \right).
$$

The Fourier sine series of $f(t) = t^2$ on the interval $(0, 1)$ is thus equal to

$$
\sum_{n=1}^{\infty} \frac{2}{n\pi} \left(\frac{2((-1)^n - 1)}{n^2\pi^2} - (-1)^n \right) \sin n\pi t.
$$

◄

EXAMPLE

In this final example we will show that one can even obtain a Fourier cosine series for the sine function on the interval $(0, \pi)$. To this end we first extend $\sin t$ to an even function on the interval $(-\pi, \pi)$ and then extend it periodically with period 2π; see figure 3.14. The ordinary Fourier coefficients of the function thus created can be calculated using (3.5) and (3.6). Since the function is even, all coefficients

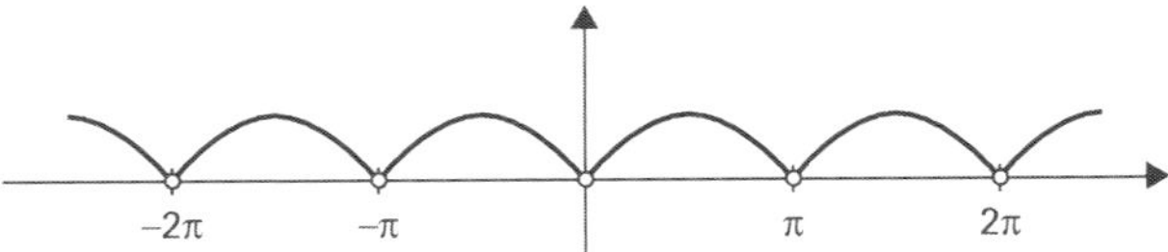

FIGURE 3.14
The even and periodic extension of the function $f(t) = \sin t$ on the interval $(0, \pi)$.

b_n will be 0. For a_n one has

$$
\begin{aligned}
a_n &= \frac{2}{T} \int_{-T/2}^{T/2} f(t) \cos n\omega_0 t \, dt \\
&= \frac{1}{\pi} \left(\int_{-\pi}^{0} (-\sin t) \cos nt \, dt + \int_{0}^{\pi} \sin t \cos nt \, dt \right) \\
&= \frac{2}{\pi} \int_{0}^{\pi} \sin t \cos nt \, dt.
\end{aligned}
$$

Applying the trigonometric formula $\sin t \cos nt = (\sin(1+n)t + \sin(1-n)t)/2$ then gives for a_n with $n \neq 1$:

$$
\begin{aligned}
a_n &= \frac{1}{\pi} \int_{0}^{\pi} (\sin(1+n)t + \sin(1-n)t) \, dt \\
&= \frac{1}{\pi} \left[\frac{-1}{1+n} \cos(1+n)t + \frac{-1}{1-n} \cos(1-n)t \right]_{0}^{\pi} \\
&= \frac{1}{\pi} \left(\frac{1-(-1)^{n-1}}{1+n} + \frac{1-(-1)^{n-1}}{1-n} \right) = \frac{2(1-(-1)^{n-1})}{\pi(1-n^2)}.
\end{aligned}
$$

It is easy to check that $a_0 = 4/\pi$ and $a_1 = 0$. The Fourier cosine series of the function $f(t) = \sin t$ on the interval $(0, \pi)$ is thus equal to

$$
\frac{2}{\pi} + \sum_{n=2}^{\infty} \frac{2(1-(-1)^{n-1})}{\pi(1-n^2)} \cos nt.
$$

◀

EXERCISES

3.19   Determine the Fourier sine series and the Fourier cosine series on $(0, 4)$ for the function $f(t)$ given for $0 < t < 4$ by

$$
f(t) = \begin{cases} t & \text{for } 0 < t \leq 2, \\ 2 & \text{for } 2 < t < 4. \end{cases}
$$

3.20   Determine a Fourier sine series of $\cos t$ on the interval $(0, \pi)$.

3.21   Determine a Fourier sine series and a Fourier cosine series of the function $f(t) = t(t-4)$ on the interval $(0, 4)$.

3.22   Determine a Fourier sine series of the function $f(t)$ defined on the interval $(0, T/2)$ by $f(t) = 1/2$ for $0 \leq t < T/2$.

SUMMARY

Trigonometric polynomials and series are, respectively, finite and infinite linear combinations of the functions $\cos n\omega_0 t$ and $\sin n\omega_0 t$ with $n \in \mathbb{N}$. They are all periodic with period $T = 2\pi/\omega_0$. When a trigonometric polynomial $f(t)$ is given, the coefficients in the linear combination can be calculated using the formulas

$$a_n = \frac{2}{T} \int_{-T/2}^{T/2} f(t) \cos n\omega_0 t \, dt \qquad \text{for } n = 0, 1, 2, \ldots,$$

$$b_n = \frac{2}{T} \int_{-T/2}^{T/2} f(t) \sin n\omega_0 t \, dt \qquad \text{for } n = 1, 2, \ldots.$$

These formulas can be applied to any arbitrary periodic function, provided that the integrals exist. The numbers a_n and b_n are called the Fourier coefficients of the function $f(t)$. Using these coefficients one can then form a Fourier series of the function $f(t)$:

$$\frac{a_0}{2} + \sum_{n=1}^{\infty} (a_n \cos n\omega_0 t + b_n \sin n\omega_0 t).$$

Instead of a Fourier series with functions $\cos n\omega_0 t$ and $\sin n\omega_0 t$ one can also obtain a complex Fourier series with the time-harmonic functions $e^{in\omega_0 t}$. The complex Fourier coefficients can then be calculated using

$$c_n = \frac{1}{T} \int_{-T/2}^{T/2} f(t) e^{-in\omega_0 t} \, dt \qquad \text{for } n \in \mathbb{Z},$$

while the complex Fourier series has the form

$$\sum_{n=-\infty}^{\infty} c_n e^{in\omega_0 t}.$$

These two Fourier series can immediately be converted into each other and, depending on the application, one may chose either one of these forms.

The sequence of Fourier coefficients (c_n) is called the spectrum of the function. This is usually split into the amplitude spectrum $|c_n|$ and the phase spectrum $\arg(c_n)$.

For some standard functions the Fourier coefficients have been determined. Moreover, a number of properties were derived making it possible to find the Fourier coefficients for far more functions and in a much simpler way than by a direct calculation.

Even functions have Fourier series containing cosine terms only. Series like this are called Fourier cosine series. Odd functions have Fourier sine series. When desired, one can extend a function, given on a certain interval, in an even or an odd way, so that they can be forced into a Fourier cosine or a Fourier sine series.

SELFTEST

3.23   The function $f(t)$ is periodic with period 10 and is drawn on the interval $(-5, 5)$ in figure 3.15. Determine the ordinary and complex Fourier coefficients of f.

3.24   Show that when for a real function f the complex Fourier coefficients are real, f has to be even, and when the complex Fourier coefficients are purely imaginary, f has to be odd.

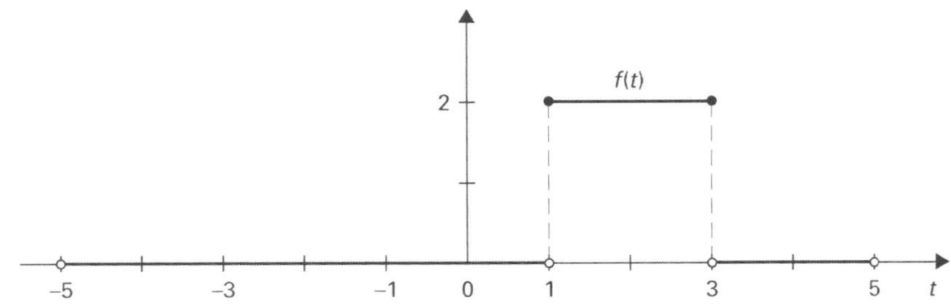

FIGURE 3.15
The periodic function $f(t)$ from exercise 3.23.

3.25   Determine the Fourier series of the periodic function $f(t)$ with period T, when for $-T/2 < t < T/2$ the function $f(t)$ is given by

$$f(t) = \begin{cases} 0 & \text{for } -T/2 < t < 0, \\ \sin \omega_0 t & \text{for } 0 < t < T/2. \end{cases}$$

3.26   Calculate and sketch the amplitude and phase spectrum of the periodic function $f(t)$, when $f(t)$ has period 2π and is given for $-\pi < t < \pi$ by

$$f(t) = \begin{cases} 0 & \text{for } -\pi < t < 0, \\ t & \text{for } 0 \leq t < \pi. \end{cases}$$

3.27   Consider the function $f(t)$ defined by:

$$f(t) = \begin{cases} \dfrac{2b}{a}t & \text{for } 0 < t < \dfrac{a}{2}, \\ \dfrac{2b}{a}(a - t) & \text{for } \dfrac{a}{2} \leq t < a. \end{cases}$$

a   Sketch the graph of $f(t)$, of its odd and of its even periodic extension.

b   Give a development of $f(t)$ on $(0, a)$ as a Fourier cosine series and also as a Fourier sine series.

Contents of Chapter 4

The fundamental theorem of Fourier series

The fundamental theorem of Fourier series

INTRODUCTION

Chapter 3 has been a first introduction to Fourier series. These series can be associated with periodic functions. We also noted in chapter 3 that if the function satisfies certain conditions, the Fourier series converges to the periodic function. What these specific conditions should be has not been analysed in chapter 3. The conditions that will be imposed in this book imply that the function should be piecewise smooth. In this chapter we will prove that a Fourier series of a piecewise smooth periodic function converges pointwise to the periodic function. We stress here that this condition is *sufficient*: when it holds, the series is pointwise convergent. This condition does not cover all cases of pointwise convergence and is thus *not necessary* for convergence.

In the first section of this chapter we derive a number of properties of Fourier coefficients that will be useful in the second section, where we prove the fundamental theorem. In the fundamental theorem we prove that for a piecewise smooth periodic function the Fourier series converges to the function. In the third section we then derive some further properties of Fourier series: product and convolution, Parseval's theorem (which has applications in the analysis of systems and signals), and integration and differentiation of Fourier series. We end this chapter with the treatment of Gibbs' phenomenon, which describes the convergence behaviour of the Fourier series at a jump discontinuity. This is then also an appropriate occasion to introduce the function $\mathrm{Si}(x)$, the sine integral. This function will re-appear in other chapters as well.

LEARNING OBJECTIVES
After studying this chapter it is expected that you
- can formulate Bessel's inequality and the Riemann–Lebesgue lemma
- can formulate and apply the fundamental theorem of Fourier series
- can calculate the sum of a number of special series using the fundamental theorem
- can determine the Fourier series of the product and the convolution of two periodic functions
- can formulate and apply Parseval's theorem
- can integrate and differentiate Fourier series
- know the definition of the sine integral and know its limit
- can explain Gibbs' phenomenon*.

4.1 Bessel's inequality and Riemann–Lebesgue lemma

In chapter 3 we always assumed in the definition of the Fourier coefficients that the integrals, necessary for the calculation of the coefficients, existed. As such, this is not a very important problem: if the integrals do not exist, then the Fourier coefficients do not exist and a further Fourier analysis is then impossible. In this chapter we confine ourselves to piecewise continuous periodic functions and for these it is easy to verify that the Fourier coefficients exist; see section 2.3.

As soon as the Fourier coefficients of a periodic function exist, it is, however, by no means obvious that the Fourier series converges to the function. In this chapter we will present conditions under which convergence is assured.

In the present section we will first treat some properties of the Fourier coefficients that will be needed later on. First we show that the sum of the squares of the Fourier coefficients of a piecewise continuous function is finite. This is called Bessel's inequality. Next we show that the Fourier coefficients c_n of a piecewise continuous function tend to 0 as $n \to \pm\infty$. This is called the Riemann–Lebesgue lemma and is needed in the next section to prove the fundamental theorem of Fourier series.

THEOREM 4.1
Bessel's inequality

When c_n are the Fourier coefficients of a piecewise continuous periodic function $f(t)$ with period T, then

$$\sum_{n=-\infty}^{\infty} |c_n|^2 \le \frac{1}{T} \int_{-T/2}^{T/2} |f(t)|^2 \, dt. \tag{4.1}$$

Proof
In the proof we use the partial sums of the Fourier series of $f(t)$ with Fourier coefficients c_k. We denote these by $s_n(t)$, so

$$s_n(t) = \sum_{k=-n}^{n} c_k e^{ik\omega_0 t}. \tag{4.2}$$

For a fixed value of n with $-n \le k \le n$ we now calculate

$$\frac{1}{T} \int_{-T/2}^{T/2} (f(t) - s_n(t)) \, e^{-ik\omega_0 t} \, dt$$

$$= \frac{1}{T} \int_{-T/2}^{T/2} f(t) e^{-ik\omega_0 t} \, dt - \frac{1}{T} \sum_{l=-n}^{n} c_l \int_{-T/2}^{T/2} e^{i(l-k)\omega_0 t} \, dt.$$

The first integral in the right-hand side is precisely the definition of the Fourier coefficient c_k. The integrals in the sum in the right-hand side are all equal to 0 for $l \ne k$. When $l = k$, the integrand is 1 and the integral T, so ultimately the sum equals c_k. We thus have

$$\frac{1}{T} \int_{-T/2}^{T/2} (f(t) - s_n(t)) \, e^{-ik\omega_0 t} \, dt = c_k - c_k = 0 \qquad \text{for } -n \le k \le n.$$

Using this result we calculate the following integral:

$$\int_{-T/2}^{T/2} (f(t) - s_n(t)) \, \overline{s_n(t)} \, dt = \sum_{k=-n}^{n} \overline{c_k} \int_{-T/2}^{T/2} (f(t) - s_n(t)) \, e^{-ik\omega_0 t} \, dt = 0.$$

If we now multiply $f(t) - s_n(t)$ in the integral in the left-hand side not by $\overline{s_n(t)}$, but by $\overline{f(t) - s_n(t)}$, it follows that

$$\int_{-T/2}^{T/2} (f(t) - s_n(t)) \, \overline{(f(t) - s_n(t))} \, dt = \int_{-T/2}^{T/2} (f(t) - s_n(t)) \, \overline{f(t)} \, dt$$

$$= \int_{-T/2}^{T/2} f(t) \overline{f(t)} \, dt - \int_{-T/2}^{T/2} s_n(t) \overline{f(t)} \, dt$$

$$= \int_{-T/2}^{T/2} |f(t)|^2 \, dt - \sum_{k=-n}^{n} c_k \int_{-T/2}^{T/2} e^{ik\omega_0 t} \overline{f(t)} \, dt$$

$$= \int_{-T/2}^{T/2} |f(t)|^2 \, dt - \sum_{k=-n}^{n} c_k \overline{\int_{-T/2}^{T/2} e^{-ik\omega_0 t} f(t) \, dt}$$

$$= \int_{-T/2}^{T/2} |f(t)|^2 \, dt - \sum_{k=-n}^{n} c_k T \, \overline{c_k} = \int_{-T/2}^{T/2} |f(t)|^2 \, dt - T \sum_{k=-n}^{n} |c_k|^2.$$

The first term in this series of equalities is greater than or equal to 0, since $(f(t) - s_n(t)) \overline{(f(t) - s_n(t))} = |f(t) - s_n(t)|^2 \geq 0$. The last term must then also be greater than or equal to 0, which means that

$$T \sum_{k=-n}^{n} |c_k|^2 \leq \int_{-T/2}^{T/2} |f(t)|^2 \, dt.$$

This inequality holds for any $n \in \mathbb{N}$, while the right-hand side is independent of n. One thus has

$$\sum_{n=-\infty}^{\infty} |c_n|^2 = \lim_{n \to \infty} \sum_{k=-n}^{n} |c_k|^2 \leq \frac{1}{T} \int_{-T/2}^{T/2} |f(t)|^2 \, dt.$$

■

If $f(t)$ is a piecewise continuous function, then $|f(t)|^2$ is one as well, and so the right-hand side of inequality (4.1) is finite. In particular it then follows that the series $\sum_{n=-\infty}^{\infty} |c_n|^2$ converges. Hence, we must have $c_n \to 0$ as $n \to \pm\infty$. This result is known as the *Riemann–Lebesgue lemma*.

THEOREM 4.2
Riemann–Lebesgue lemma

If $f(t)$ is a piecewise continuous periodic function with Fourier coefficients c_n, then

$$\lim_{n \to \infty} c_n = \lim_{n \to -\infty} c_n = 0. \tag{4.3}$$

Theorem 4.2 can be interpreted as follows. In order to calculate the coefficients c_n, the function $f(t)$ is multiplied by $e^{-in\omega_0 t}$ and integrated over one period. For increasing n, the frequency of the corresponding sine and cosine functions keeps increasing. Now consider two consecutive intervals such that, for example, the sine function is negative in the first interval and positive in the second. For increasing n, and hence for ever smaller intervals, the value of f in the first and in the second interval will differ less and less. Multiplied by first a negative and then a positive sine function, the contributions to the integral will cancel each other better and better for increasing n. In this way the coefficients will eventually converge to 0.

EXAMPLE

The periodic block function, introduced in section 3.4.1, is piecewise continuous. The Fourier coefficients, which have also been calculated there, are equal to $2\sin(n\omega_0 a/2)/Tn\omega_0$. The numerator ranges for increasing n between -1 and 1, while the denominator tends to infinity as $n \to \infty$. For the Fourier coefficients we thus have

$$\lim_{n \to \infty} \frac{2}{T} \frac{\sin(n\omega_0 a/2)}{n\omega_0} = 0.$$

The Fourier coefficients of the periodic block function thus tend to 0 as $n \to \infty$. Similarly one can check that the same is true for the Fourier coefficients of the periodic triangle function and the sawtooth function from sections 3.4.2 and 3.4.3.

◀

EXERCISES

4.1   **a** Check for the periodic block function and the periodic triangle function from section 3.4 whether or not they are piecewise smooth and whether or not the Fourier coefficients c_n exist.

b Show for the functions from part a that $\sum_{n=-\infty}^{\infty} |c_n|^2$ is convergent.

4.2   The ordinary Fourier coefficients of a piecewise continuous periodic function are a_n and b_n (see definition 3.1).

a Prove that $\lim_{n \to \infty} a_n = \lim_{n \to \infty} b_n = 0$. To do so, start from theorem 4.2.

b Now assume that $\lim_{n \to \infty} a_n = \lim_{n \to \infty} b_n = 0$. Use this to prove that $\lim_{n \to \pm\infty} c_n = 0$.

4.3   Check that the Fourier coefficients c_n of the periodic sawtooth function tend to 0 as $n \to \infty$, in accordance with the Riemann–Lebesgue lemma.

4.2   The fundamental theorem

In chapter 3 we have noted more than once that at the points of continuity, the Fourier series of a piecewise smooth periodic function is equal to that function. This statement, formulated in the fundamental theorem of Fourier series, will be proven in this section. This will involve pointwise convergence. Before we go into this, we first introduce the so-called Dirichlet kernel. This is a function that will be needed in the proof of the fundamental theorem. We will also deduce some properties of the Dirichlet kernel.

DEFINITION 4.1
Dirichlet kernel

The Dirichlet kernel $D_n(x)$ *is defined by*

$$D_n(x) = \sum_{k=-n}^{n} e^{-ik\omega_0 x} = e^{in\omega_0 x} + e^{i(n-1)\omega_0 x} + e^{i(n-2)\omega_0 x} + \cdots + e^{-in\omega_0 x}.$$

The Dirichlet kernel is a periodic function, which can be considered as a geometric series with $2n + 1$ terms, starting with $e^{in\omega_0 x}$ and having ratio $e^{-i\omega_0 x}$. In example 2.16 the sum of a geometric series was determined. From this it follows that for $e^{-i\omega_0 x} \neq 1$ one has that

$$D_n(x) = \frac{e^{in\omega_0 x}(1 - e^{-i\omega_0(2n+1)x})}{1 - e^{-i\omega_0 x}}. \tag{4.4}$$

Performing the multiplication in the numerator and multiplying numerator and denominator by $e^{i\omega_0 x/2}$, it follows upon using (2.11) that

$$D_n(x) = \frac{e^{i(n+1/2)\omega_0 x} - e^{-i(n+1/2)\omega_0 x}}{e^{i\omega_0 x/2} - e^{-i\omega_0 x/2}} = \frac{\sin((n+1/2)\omega_0 x)}{\sin(\omega_0 x/2)}. \tag{4.5}$$

The above is valid for $e^{-i\omega_0 x} \neq 1$, that is to say, for $\omega_0 x \neq k \cdot 2\pi$, or $x \neq k \cdot T$ with $k \in \mathbb{Z}$. If, however, we do have $x = k \cdot T$, then $e^{-ik\omega_0 x} = e^{-ik^2 2\pi} = 1$. From the definition of the Dirichlet kernel it then immediately follows that $D_n(k \cdot T) = 2n+1$ for $k \in \mathbb{Z}$. Furthermore, it is easy to see that the Dirichlet kernel is an even function. In figure 4.1 the graph of the Dirichlet kernel is sketched for $n = 6$. When n increases, the number of oscillations per period increases. The peaks at the points $x = kT$ continue to exist and increase in value.

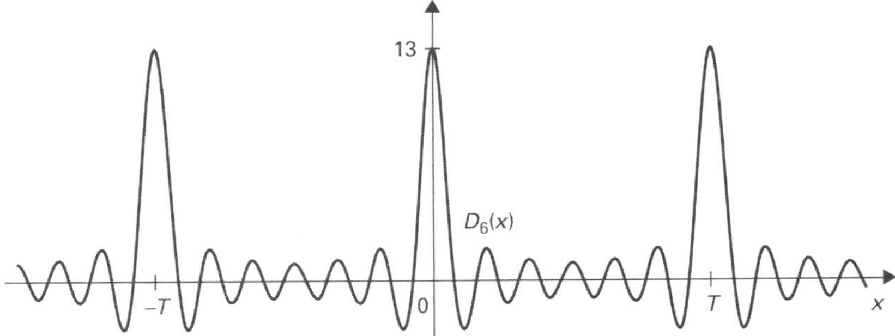

FIGURE 4.1
The Dirichlet kernel for $n = 6$.

The integral of $D_n(x)$ over one period is independent of n. In fact, since $\int_{-T/2}^{T/2} e^{-ik\omega_0 x}\, dx = 0$ for $k \neq 0$, it follows that

$$\int_{-T/2}^{T/2} D_n(x)\, dx = \sum_{k=-n}^{n} \int_{-T/2}^{T/2} e^{-ik\omega_0 x}\, dx = \int_{-T/2}^{T/2} 1\, dx = T. \tag{4.6}$$

Since $D_n(x)$ is an even function, it moreover follows that

$$\int_0^{T/2} D_n(x)\, dx = \frac{T}{2}. \tag{4.7}$$

We have now established enough properties of the Dirichlet kernel to enable us to formulate and prove the fundamental theorem. According to the fundamental theorem, the Fourier series converges to the function at each point of continuity of a piecewise smooth periodic function. At a point where the function is *discontinuous*, the Fourier series converges to the *average* of the left- and right-hand limits at that point. Hence, both at the points of continuity and at the points of discontinuity the series converges to $(f(t+) + f(t-))/2$. The fundamental theorem now reads as follows.

THEOREM 4.3
Fundamental theorem of
Fourier series

Let $f(t)$ be a piecewise smooth periodic function on $\mathbb{R}$ with Fourier coefficients c_n. Then one has for any $t \in \mathbb{R}$:

$$\sum_{n=-\infty}^{\infty} c_n e^{in\omega_0 t} = \frac{1}{2}\left(f(t+) + f(t-)\right).$$

Proof
For the proof of this theorem we start from the partial sums $s_n(t)$ of the Fourier series as defined in (4.2). Replacing c_k by the integral defining the Fourier coefficients we obtain

$$s_n(t) = \sum_{k=-n}^{n} c_k e^{ik\omega_0 t} = \sum_{k=-n}^{n} \left(\frac{1}{T}\int_{-T/2}^{T/2} f(u) e^{-ik\omega_0 u}\, du\right) e^{ik\omega_0 t}$$

$$= \frac{1}{T}\int_{-T/2}^{T/2} f(u) \sum_{k=-n}^{n} e^{-ik\omega_0(u-t)}\, du.$$

Interchanging the order of integration and summation is allowed here, since the sum contains a finite number of terms. Substitution of $u - t = x$ then gives

$$s_n(t) = \frac{1}{T} \int_{-T/2-t}^{T/2-t} f(x+t) \sum_{k=-n}^{n} e^{-ik\omega_0 x} \, dx$$

$$= \frac{1}{T} \int_{-T/2}^{T/2} f(x+t) D_n(x) \, dx. \tag{4.8}$$

We are allowed to change the integration interval in the last step, since the integrand is periodic with period T. In the last step we also introduced the Dirichlet kernel. If we take the limit $n \to \infty$ in (4.8), then $s_n(t)$ will become the Fourier series of $f(t)$. Hence, if we can show that the final term in (4.8) converges to $(f(t+) + f(t-))/2$, then we've completed the proof. If we assume for the moment that f is continuous at t, then it is plausible that this final term will indeed be equal to $f(t)$. For, according to (4.6) one has $\int_{-T/2}^{T/2} D_n(x) \, dx = T$. Moreover, $D_n(0)$ keeps increasing for increasing n, while for $x \neq 0$ the oscillations of $D_n(x)$ become more frequent. Since eventually $f(x+t)$ will vary less rapidly, the consecutive oscillations of $f(x+t)D_n(x)$ will cancel each other more and more, and hence only the value of $f(x+t)$ for $x = 0$, so $f(t)$, will remain. To prove this, we will now first of all split the integration interval $(-T/2, T/2)$ in two parts and replace the variable x by $-x$ on $(-T/2, 0)$. Since $D_n(x)$ is even, it then follows that

$$s_n(t) = \frac{1}{T} \int_{-T/2}^{0} f(t+x) D_n(x) \, dx + \frac{1}{T} \int_{0}^{T/2} f(t+x) D_n(x) \, dx$$

$$= \frac{1}{T} \int_{0}^{T/2} f(t-x) D_n(-x) \, dx + \frac{1}{T} \int_{0}^{T/2} f(t+x) D_n(x) \, dx$$

$$= \frac{1}{T} \int_{0}^{T/2} (f(t+x) + f(t-x)) D_n(x) \, dx.$$

We now subtract $f(t+) + f(t-)$ from the factor $f(t+x) + f(t-x)$ in the integrand. In order not to change $s_n(t)$, we have to add this term as well, which will be done in a separate integral. We then get:

$$s_n(t) = \frac{1}{T} \int_{0}^{T/2} (f(t+x) - f(t+) + f(t-x) - f(t-)) D_n(x) \, dx$$

$$+ \frac{1}{T} \int_{0}^{T/2} (f(t+) + f(t-)) D_n(x) \, dx.$$

According to (4.7), the second term equals $(f(t+) + f(t-))/2$. The first term will be called $I_n(t)$. If we can show that this term tends to 0 as $n \to \infty$, then we have finished the proof. To this end we use (4.5) to write $I_n(t)$ as

$$I_n(t)$$

$$= \frac{1}{T} \int_{0}^{T/2} \frac{f(t+x) - f(t+) + f(t-x) - f(t-)}{x} \cdot \frac{x \sin((n+\frac{1}{2})\omega_0 x)}{\sin(\omega_0 x/2)} \, dx$$

$$= \frac{1}{T} \int_{0}^{T/2} Q(x) \sin((n+\tfrac{1}{2})\omega_0 x) \, dx. \tag{4.9}$$

Here $Q(x)$ is given by

$$Q(x) = \frac{f(t+x) - f(t+) + f(t-x) - f(t-)}{x} \cdot \frac{x}{\sin(\omega_0 x/2)}.$$

For $x = 0$ the integrand of $I_n(t)$ is not defined, since then the denominator of $Q(x)$ equals 0. However, since $f(t)$ is a piecewise smooth function, the limits for $x \to 0$ of $(f(t+x) - f(t+))/x$ and $(f(t-x) - f(t-))/x$, which occur in $Q(x)$, do exist according to theorem 2.4, and they equal $f'(t+)$ and $f'(t-)$ respectively. Since the limit for $x \to 0$ of $x/\sin(\omega_0 x/2)$ exists as well, it follows that $Q(x)$ is a piecewise continuous function. Furthermore, we note that $Q(x)$ is odd. And since $\sin((n + \frac{1}{2})\omega_0 x)$ is also odd, the integrand in (4.9) is an even function. We can therefore extend the integration interval to $[-T/2, T/2]$. Using the trigonometric identity $\sin(\alpha + \beta) = \sin\alpha \cos\beta + \cos\alpha \sin\beta$ we then obtain:

$$
\begin{aligned}
I_n(t) &= \frac{1}{2T} \int_{-T/2}^{T/2} Q(x) \sin((n + \tfrac{1}{2})\omega_0 x) \, dx \\
&= \frac{1}{2T} \int_{-T/2}^{T/2} Q(x) \Big(\sin(n\omega_0 x) \cos(\omega_0 x/2) + \cos(n\omega_0 x) \sin(\omega_0 x/2) \Big) dx \\
&= \frac{1}{4} \cdot \frac{2}{T} \int_{-T/2}^{T/2} Q(x) \cos(\omega_0 x/2) \sin(n\omega_0 x) \, dx \\
&\quad + \frac{1}{4} \cdot \frac{2}{T} \int_{-T/2}^{T/2} Q(x) \sin(\omega_0 x/2) \cos(n\omega_0 x) \, dx.
\end{aligned}
$$

The two integrals in the final expression are precisely the formulas for the ordinary Fourier coefficients of the function $Q(x) \cos(\omega_0 x/2)$ and the function $Q(x) \sin(\omega_0 x/2)$; see definition 3.1. Since $Q(x)$ is piecewise continuous, so are the functions $Q(x) \cos(\omega_0 x/2)$ and $Q(x) \sin(\omega_0 x/2)$, and hence one can apply the Riemann–Lebesgue lemma from theorem 4.2. We then see that indeed $I_n(t)$ tends to 0 as $n \to \infty$. This completes the proof. ∎

Having established the fundamental theorem, it is now a proven fact that Fourier series of piecewise smooth functions converge. At the points of continuity of the function, the Fourier series converges to the function value and at the points of discontinuity to the average of the left- and right-hand limits $(f(t+) + f(t-))/2$. For example, the Fourier series of the periodic block function from section 3.4.1 will converge to the function with graph given by figure 4.2. At the points of discontinuity the function value is 1/2.

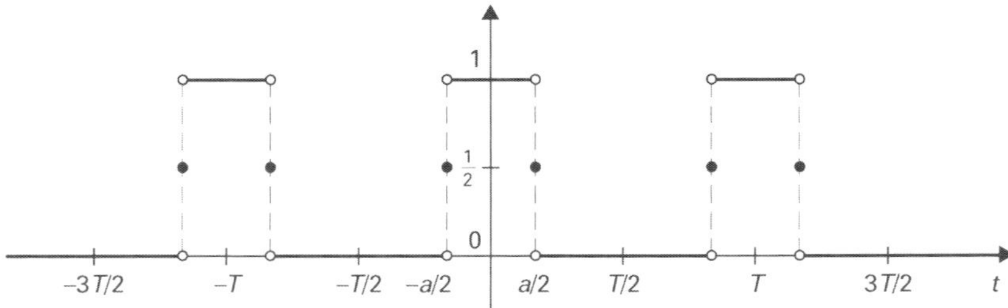

FIGURE 4.2
Limit of the Fourier series of the periodic block function.

From the fundamental theorem it immediately follows that if two periodic functions have the same Fourier series, then these functions must be equal at all points of continuity. Moreover, it follows from the definition of the Fourier coefficients that the values of the functions at the points of discontinuity have no influence on the coefficients. This is precisely the reason why we didn't pay any attention to the values of functions at the points of discontinuity in chapter 3. In the end, the Fourier series will always converge to the average of the left- and right-hand limit. Apart from the points of discontinuity, functions with equal Fourier series are the same. We formulate this result in the following *uniqueness theorem.*

THEOREM 4.4
Uniqueness theorem

Let $f(t)$ and $g(t)$ be piecewise smooth periodic functions with Fourier coefficients f_n and g_n. If $f_n = g_n$ for all $n \in \mathbb{Z}$, then $f(t) = g(t)$ at all points where f and g are continuous.

From the fundamental theorem it follows that the Fourier series of a function converges pointwise to the function at the points of continuity. Some general remarks can be made about the rate of convergence. If we compare the Fourier coefficients of the periodic block function and the periodic triangle function from sections 3.4.1 and 3.4.2, then we observe that the coefficients of the discontinuous block function decrease proportional to $1/n$, while the coefficients of the continuous triangle function decrease proportional to $1/n^2$. The feature that Fourier coefficients of continuous functions decrease more rapidly compared to discontinuous functions is true in general. If the derivative is continuous as well, then the Fourier coefficients decrease even more rapidly. As higher derivatives are continuous as well, the Fourier coefficients decrease ever more rapidly. Hence, *the smoother the function, the smaller the contribution of high frequency components in the Fourier series.* We summarize this in the following statements.

a If the function $f(t)$ is piecewise continuous, then the Fourier coefficients tend to zero (this is the Riemann–Lebesgue lemma).
b If $f(t)$ is continuous and $f'(t)$ piecewise continuous, then the Fourier coefficients decrease as $1/n$, so $\lim_{n \to \pm\infty} nc_n = 0$.
c If $f(t)$ and its derivatives up to the $(k-1)$th order are continuous and $f^{(k)}(t)$ is piecewise continuous, then the Fourier coefficients decrease as $1/n^k$, so $\lim_{n \to \pm\infty} n^k c_n = 0$.

These statements will not be proven here (but see the remarks following theorem 4.10). From these statements it follows that in comparison to smooth functions, for less smooth functions one needs more terms from the series in order to achieve the same degree of accuracy. The statements also hold in the opposite direction: the faster the Fourier coefficients decrease, the smoother the function. An example is the following result, which will be used in chapter 7 and is stated without proof here.

THEOREM 4.5

Let a sequence of numbers c_n be given for which $\sum_{n=-\infty}^{\infty} |c_n| < \infty$. Then the series $\sum_{n=-\infty}^{\infty} c_n e^{in\omega_0 t}$ converges to the continuous function $f(t)$ having Fourier coefficients c_n.

For functions with discontinuities it is the case that – no matter how many terms one includes – in a small neighbourhood left and right of a discontinuity, *any* approximation will 'overshoot' the function value on one side and 'undershoot' the function value on the other side. It is even the case that the values of these overshoots are a *fixed percentage* of the difference $|f(t+) - f(t-)|$. These overshoots do get closer and closer to the point of discontinuity, as more terms are being included. This curious phenomenon is called Gibbs' phenomenon and will be discussed in section 4.4.2.

An important side result of the fundamental theorem is the fact that *sums* can be calculated for many of the series for which, up till now, we could only establish the *convergence* using the tests from chapter 2. Below we present some examples of such calculations.

EXAMPLE 4.1

Consider the periodic triangle function $q_{a,T}$ for $a = T/2$, whose Fourier coefficients have been calculated in section 3.4.2: $c_n = 2\sin^2(n\pi/2)/n^2\pi^2$ for $n \neq 0$ and $c_0 = 1/2$. For even n and $n \neq 0$ the Fourier coefficients are thus all zero. For odd n they equal $2/n^2\pi^2$. Since the periodic triangle function is a continuous piecewise smooth function, we may apply the fundamental theorem of Fourier series and thus one has for all t:

$$q_{T/2,T}(t) = \sum_{n=-\infty}^{\infty} c_n e^{in\omega_0 t} = \frac{1}{2} + \sum_{n=-\infty}^{\infty} \frac{2}{(2n-1)^2\pi^2} e^{i(2n-1)\omega_0 t}.$$

In particular this equality holds for $t = 0$. The triangle function then has value 1, while $e^{i(2n-1)\omega_0 t}$ equals 1 as well. Hence,

$$1 = \frac{1}{2} + \sum_{n=-\infty}^{\infty} \frac{2}{(2n-1)^2\pi^2} = \frac{1}{2} + \sum_{n=1}^{\infty} \frac{4}{(2n-1)^2\pi^2}.$$

Now take 1/2 to the other side of the equality-sign and multiply left and right by $\pi^2/4$, then

$$\frac{\pi^2}{8} = \sum_{n=1}^{\infty} \frac{1}{(2n-1)^2} = 1 + \frac{1}{9} + \frac{1}{25} + \frac{1}{49} + \cdots. \tag{4.10}$$

◀

EXAMPLE 4.2

The sawtooth function $2t/T$, as defined in section 3.4.3, is a piecewise smooth function having Fourier coefficients $i(-1)^n/\pi n$ for $n \neq 0$ and $c_0 = 0$. The fundamental theorem is applicable and for $-T/2 < t < T/2$ one has

$$\frac{2t}{T} = \sum_{n=-\infty}^{\infty} c_n e^{in\omega_0 t} = \sum_{n=1}^{\infty} \left(c_n e^{in\omega_0 t} + c_{-n} e^{-in\omega_0 t} \right)$$

$$= \sum_{n=1}^{\infty} \left((-1)^n \frac{i}{\pi n} e^{in\omega_0 t} + (-1)^{-n} \frac{i}{-\pi n} e^{-in\omega_0 t} \right)$$

$$= \sum_{n=1}^{\infty} (-1)^n \frac{i}{\pi n} 2i \left(\frac{e^{in\omega_0 t} - e^{-in\omega_0 t}}{2i} \right) = \sum_{n=1}^{\infty} (-1)^n \frac{-2}{\pi n} \sin n\omega_0 t.$$

This equality holds in particular for $t = T/4$. For this value of t one has that $\sin n\omega_0 t = \sin(n\pi/2)$. This equals 0 for even n, so

$$\frac{1}{2} = \sum_{n=1}^{\infty} (-1)^{2n-1} \frac{-2}{\pi(2n-1)} \sin \frac{(2n-1)\pi}{2} = \sum_{n=1}^{\infty} \frac{2}{\pi(2n-1)} (-1)^{n-1}.$$

If we multiply this by $\pi/2$, then we obtain

$$\frac{\pi}{4} = \sum_{n=1}^{\infty} \frac{(-1)^{n-1}}{2n-1} = 1 - \frac{1}{3} + \frac{1}{5} - \frac{1}{7} + \cdots. \tag{4.11}$$

◀

EXERCISES

4.4

Calculate the sum of the Fourier series of the sawtooth function at a point of discontinuity and use this to verify the fundamental theorem at that point.

4.5

Let $f(t)$ be an odd piecewise smooth periodic function with period T. Show that

$$\sum_{n=1}^{\infty} b_n \sin n\omega_0 t = \tfrac{1}{2}(f(t+) + f(t-)).$$

4.6   **a**   Determine the Fourier series of the periodic function $f(t)$ with period 2π defined for $-\pi \leq t < \pi$ by

$$f(t) = \begin{cases} 0 & \text{for } -\pi \leq t < 0, \\ 1 & \text{for } 0 \leq t < \pi. \end{cases}$$

b   Verify that $f(t)$ satisfies the conditions of the fundamental theorem. Then take $t = \pi/2$ and use this to derive (4.11) again.

4.7   **a**   Determine the Fourier series of the periodic function $f(t)$ with period 2π defined for $-\pi \leq t < \pi$ by

$$f(t) = \begin{cases} 0 & \text{for } -\pi \leq t < 0, \\ t & \text{for } 0 \leq t < \pi. \end{cases}$$

b   To which value does the Fourier series converge at $t = \pi$ according to the fundamental theorem? Use this to derive (4.10) again.

4.8   Let $f(t)$ be the periodic function with period 1 defined for $-1/2 \leq t < 1/2$ by $f(t) = t^2$.

a   Verify that f is continuous and that f' is piecewise continuous. Determine the Fourier coefficients and check that these decrease as $1/n^2$.

b   Apply the fundamental theorem at $t = 0$ in order to derive that

$$\sum_{n=1}^{\infty} \frac{(-1)^{n+1}}{n^2} = 1 - \frac{1}{4} + \frac{1}{9} - \frac{1}{16} + \cdots = \frac{\pi^2}{12}.$$

c   Show that

$$\sum_{n=1}^{\infty} \frac{1}{n^2} = 1 + \frac{1}{4} + \frac{1}{9} + \frac{1}{16} + \cdots = \frac{\pi^2}{6}.$$

4.9   Let $f(t)$ be the periodic function with period 2 defined for $-1 < t \leq 1$ by

$$f(t) = \begin{cases} t^2 + t & \text{for } -1 < t \leq 0, \\ -t^2 + t & \text{for } 0 < t \leq 1. \end{cases}$$

a   Verify that f and f' are continuous, while f'' is discontinuous.

b   Determine the Fourier coefficients and check that these decrease as $1/n^3$.

c   Show that

$$\sum_{n=0}^{\infty} \frac{(-1)^n}{(2n+1)^3} = 1 - \frac{1}{27} + \frac{1}{125} - \cdots = \frac{\pi^3}{32}.$$

4.3   Further properties of Fourier series

Now that we have proven, in the fundamental theorem of Fourier series, that for piecewise smooth functions the Fourier series converges to the function, one can derive some additional properties. These are properties of Fourier series with respect to products and convolutions of functions, the Parseval identity, and the integration and differentiation of Fourier series.

4.3.1   Product and convolution

In this section we consider the relations that emerge when functions are multiplied, or when Fourier coefficients are multiplied. First we consider how one can determine the Fourier coefficients of a function which is the product of two functions.

THEOREM 4.6
Fourier series of a product of functions

Let $f(t)$ and $g(t)$ be piecewise smooth periodic functions with Fourier coefficients f_n and g_n. When $h(t) = f(t)g(t)$, then $h(t)$ has a convergent Fourier series with Fourier coefficients h_n given by

$$h_n = \sum_{k=-\infty}^{\infty} f_k \cdot g_{n-k}. \tag{4.12}$$

Proof
Since $f(t)$ and $g(t)$ are piecewise smooth periodic functions, so is $h(t)$ and thus it has a convergent Fourier series. According to definition 3.3 one has for the coefficients h_n:

$$h_n = \frac{1}{T} \int_{-T/2}^{T/2} f(t)g(t)e^{-in\omega_0 t}\, dt.$$

Since $f(t)$ is a piecewise smooth periodic function, one can, according to the fundamental theorem, replace $f(t)$ by its Fourier series at all points of continuity. Since we are integrating, the value at the points of discontinuity are of no importance. Hence

$$h_n = \frac{1}{T} \int_{-T/2}^{T/2} \sum_{k=-\infty}^{\infty} f_k e^{ik\omega_0 t} g(t)e^{-in\omega_0 t}\, dt.$$

Under the conditions of the theorem, one may interchange the order of integration and summation. Using the definition of the $(n - k)$th Fourier coefficient of $g(t)$ it then follows that

$$h_n = \sum_{k=-\infty}^{\infty} f_k \frac{1}{T} \int_{-T/2}^{T/2} g(t)e^{-i(n-k)\omega_0 t}\, dt = \sum_{k=-\infty}^{\infty} f_k \cdot g_{n-k}. \qquad \blacksquare$$

EXAMPLE 4.3

Consider the periodic functions f, g and h with period 2 defined on the interval $(-1, 1)$ by $f(t) = g(t) = t$ and $h(t) = t^2$. In exercise 3.9 it was deduced that the Fourier coefficients f_n, g_n and h_n of these functions are given by

$$f_n = g_n = \begin{cases} \dfrac{i}{n\pi}(-1)^n & \text{for } n \neq 0, \\[2mm] 0 & \text{for } n = 0, \end{cases}$$

$$h_n = \begin{cases} \dfrac{2(-1)^n}{n^2\pi^2} & \text{for } n \neq 0, \\[2mm] \dfrac{1}{3} & \text{for } n = 0. \end{cases}$$

Since the functions are piecewise smooth and $h(t) = f(t)g(t)$, one can also obtain the coefficients of $h(t)$ from f_n and g_n using theorem 4.6. We first calculate h_0.

$$h_0 = \sum_{k=-\infty}^{\infty} f_k \cdot g_{-k} = 2\sum_{k=1}^{\infty} \frac{i}{k\pi}(-1)^k \cdot \frac{i}{-k\pi}(-1)^{-k} = \frac{2}{\pi^2} \sum_{k=1}^{\infty} \frac{1}{k^2}$$

$$= \frac{2}{\pi^2}\frac{\pi^2}{6} = \frac{1}{3}.$$

Here we used the result $\sum_{k=1}^{\infty} 1/k^2 = \pi^2/6$ from exercise 4.8c. We now calculate h_n.

$$
h_n = \sum_{k=-\infty}^{\infty} f_k \cdot g_{n-k} = \sum_{\substack{k=-\infty \\ k\neq 0, k\neq n}}^{\infty} \frac{i}{k\pi}(-1)^k \frac{i}{(n-k)\pi}(-1)^{n-k}
$$

$$
= \frac{-(-1)^n}{\pi^2} \sum_{\substack{k=-\infty \\ k\neq 0, k\neq n}}^{\infty} \frac{1}{(n-k)k}.
$$

This last sum is somewhat difficult to calculate. Using a partial fraction expansion it follows that

$$
\sum_{\substack{k=-\infty \\ k\neq 0, k\neq n}}^{\infty} \frac{1}{(n-k)k} = \frac{1}{n} \sum_{\substack{k=-\infty \\ k\neq 0, k\neq n}}^{\infty} \left(\frac{1}{k} + \frac{1}{n-k} \right)
$$

$$
= \frac{1}{n} \left(-\frac{1}{n} + \sum_{\substack{k=-\infty \\ k\neq 0}}^{\infty} \frac{1}{k} \right) + \frac{1}{n} \left(-\frac{1}{n} + \sum_{\substack{k=-\infty \\ k\neq n}}^{\infty} \frac{1}{n-k} \right).
$$

Since the sums in the right-hand side always contain terms having opposite signs, these terms will all cancel each other and so these sums are 0, implying that

$$
\sum_{\substack{k=-\infty \\ k\neq 0, k\neq n}}^{\infty} \frac{1}{(n-k)k} = -\frac{2}{n^2}.
$$

For h_n it now follows that

$$
h_n = \frac{-(-1)^n}{\pi^2} \cdot \frac{-2}{n^2} = \frac{2(-1)^n}{n^2\pi^2}.
$$

Hence, the expressions for h_0 and h_n, calculated using theorem 4.6, coincide with the direct calculation of the Fourier coefficients in exercise 3.9. ◄

We have seen that the Fourier coefficients of the product of two functions can be calculated using the Fourier coefficients of the individual functions. To do so, one has to form a sum of products of the Fourier coefficients. Another nice relationship arises if we examine what kind of functions emerge if we multiply the Fourier coefficients of two functions. The resulting function is called the convolution product. This convolution product will play an important role in systems analysis.

DEFINITION 4.2
Convolution of periodic functions

The convolution product *of two piecewise smooth periodic functions f and g, both with period T, is denoted by $f * g$ and is defined by*

$$
(f * g)(t) = \frac{1}{T} \int_{-T/2}^{T/2} f(t-\tau)g(\tau)\, d\tau.
$$

When f and g are periodic with period T, then so is the convolution product. Specifically, for all $k \in \mathbb{Z}$ one has

$$
(f * g)(t + kT) = \frac{1}{T} \int_{-T/2}^{T/2} f(t + kT - \tau)g(\tau)\, d\tau
$$

$$
= \frac{1}{T} \int_{-T/2}^{T/2} f(t - \tau)g(\tau)\, d\tau = (f * g)(t).
$$

If we now multiply the Fourier coefficients of two functions, then the resulting numbers are the Fourier coefficients of the convolution product of the two functions.

THEOREM 4.7
Fourier coefficients of a convolution product

Let $f(t)$ and $g(t)$ be piecewise smooth periodic functions with Fourier coefficients f_n and g_n. Then $(f * g)(t)$ has a convergent Fourier series with Fourier coefficients $(f * g)_n$ satisfying

$$(f * g)_n = f_n g_n.$$

Proof
Since f and g are piecewise smooth periodic functions, it follows that the convolution product $f * g$ is also a piecewise smooth function. We state this without proof. Hence, $f * g$ has a convergent Fourier series. For the Fourier coefficients in this series one has

$$(f * g)_n = \frac{1}{T} \int_{-T/2}^{T/2} (f * g)(t) e^{-in\omega_0 t} \, dt$$

$$= \frac{1}{T^2} \int_{-T/2}^{T/2} \left(\int_{-T/2}^{T/2} f(t - \tau) g(\tau) \, d\tau \right) e^{-in\omega_0 t} \, dt.$$

Under the conditions of the theorem one may interchange the order of integration and so

$$(f * g)_n = \frac{1}{T} \int_{-T/2}^{T/2} \left(\frac{1}{T} \int_{-T/2}^{T/2} f(t - \tau) e^{-in\omega_0 (t-\tau)} \, dt \right) g(\tau) e^{-in\omega_0 \tau} \, d\tau.$$

The expression in parentheses in the right-hand side is the nth Fourier coefficient of f, and by applying once again the definition of the Fourier coefficients for g, the proof is completed:

$$(f * g)_n = f_n \frac{1}{T} \int_{-T/2}^{T/2} g(\tau) e^{-in\omega_0 \tau} \, d\tau = f_n g_n.$$

$\blacksquare$

EXAMPLE 4.4

In this example we will see how the convolution of two periodic block functions gives rise to a periodic triangle function, whereas the Fourier coefficients of the triangle functions can be obtained by multiplying the Fourier coefficients of the two block functions. In doing so, we verify theorem 4.7.

Consider the periodic block function f with period 2 and $a = 1$ from example 3.4. The function equals 1 for $|t| \leq \frac{1}{2}$ and 0 for $\frac{1}{2} < |t| \leq 1$. Calculating the convolution of f with f according to definition 4.2 gives

$$(f * f)(t) = \frac{1}{2} \int_{-1}^{1} f(t - \tau) f(\tau) \, d\tau = \frac{1}{2} \int_{-1/2}^{1/2} f(t - \tau) \, d\tau.$$

One has that $f(t - \tau) = 1$ for $-\frac{1}{2} \leq t - \tau \leq \frac{1}{2}$, that is, for $t - \frac{1}{2} \leq \tau \leq t + \frac{1}{2}$. This means that the integral equals

$$(f * f)(t) = \frac{1}{2} \int_{t-1/2}^{1/2} 1 \, d\tau = \frac{1 - t}{2} \qquad \text{for } 0 \leq t < 1,$$

$$(f * f)(t) = \frac{1}{2} \int_{-1/2}^{t+1/2} 1 \, d\tau = \frac{1 + t}{2} \qquad \text{for } -1 \leq t < 0.$$

Hence, up to a factor $\frac{1}{2}$, the convolution is precisely the periodic triangle function f with period 2 and $a = 2$ from section 3.4.2. Furthermore, according to example 3.4 one has for the Fourier coefficients of the block function that $c_n = \sin(n\pi/2)/n\pi$

for $n \neq 0$ and $c_0 = \frac{1}{2}$. The squares of these coefficients are $\sin^2(n\pi/2)/n^2\pi^2$ for $n \neq 0$ and $\frac{1}{4}$ for $n = 0$. Multiplied by 2 these are exactly the Fourier coefficients of the periodic triangle function from section 3.4.2 for $a = 1$, $\omega_0 = \pi$ and $T = 2$. ◂

4.3.2 Parseval's identity

In this subsection we will show that for a large class of functions, Bessel's inequality from section 4.1 is in fact an equality. Somewhat more general is the following Parseval identity, which has important applications in the analysis of signals and systems.

THEOREM 4.8
Parseval's identity

Let $f(t)$ and $g(t)$ be piecewise smooth periodic functions with Fourier coefficients f_n and g_n. Then

$$\frac{1}{T} \int_{-T/2}^{T/2} f(t)\overline{g(t)}\, dt = \sum_{n=-\infty}^{\infty} f_n \overline{g_n}. \tag{4.13}$$

Proof
According to theorem 4.6, the Fourier coefficients of the product h of two functions f and g are given by $h_k = \sum_{n=-\infty}^{\infty} f_n g_{k-n}$. In particular this holds for the Fourier coefficient h_0, for which, moreover, one has by definition that $h_0 = (1/T) \int_{-T/2}^{T/2} f(t)g(t)\, dt$. Combining these facts, it follows that

$$\frac{1}{T} \int_{-T/2}^{T/2} f(t)g(t)\, dt = h_0 = \sum_{n=-\infty}^{\infty} f_n g_{-n}.$$

Instead of the function g we now take the conjugate function $\overline{g}$. According to theorem 3.2 the Fourier coefficients of $\overline{g}$ are $\overline{g_{-n}}$, proving (4.13). ∎

It is now just a small step to prove that the Bessel inequality is an *equality* for piecewise smooth functions. In order to do so, we take $g(t)$ in theorem 4.8 equal to the function $f(t)$. The Fourier coefficients of $f(t)$ will be denoted by c_n again. We then obtain

$$\frac{1}{T} \int_{-T/2}^{T/2} |f(t)|^2\, dt = \frac{1}{T} \int_{-T/2}^{T/2} f(t)\overline{f(t)}\, dt = \sum_{n=-\infty}^{\infty} c_n \overline{c_n}$$

$$= \sum_{n=-\infty}^{\infty} |c_n|^2. \tag{4.14}$$

In section 1.2.3 the power P of a periodic time-continuous signal was defined as $(1/T) \int_{-T/2}^{T/2} |f(t)|^2\, dt$. If $f(t)$ is a piecewise smooth periodic function, then according to (4.14) the power can also be calculated using the Fourier coefficients:

Power of piecewise smooth periodic function

$$P = \frac{1}{T} \int_{-T/2}^{T/2} |f(t)|^2\, dt = \sum_{n=-\infty}^{\infty} |c_n|^2.$$

4.3.3 Integration

Using Parseval's identity from the previous subsection, one can derive a result concerning the relationship between the Fourier coefficients of the integral of a periodic function and the Fourier coefficients of the periodic function itself. We thus want to find an expression for the Fourier coefficients of $\int_{-T/2}^{t} f(\tau)\,d\tau$. If this integral is to be a periodic function in t again, one should have $\int_{-T/2}^{T/2} f(\tau)\,d\tau = 0$, that is, $c_0 = 0$.

THEOREM 4.9
Integration of Fourier series

Let $f(t)$ be a piecewise smooth periodic function with period T and Fourier coefficients c_n for which $c_0 = 0$. Then the function $h(t)$ defined by $h(t) = \int_{-T/2}^{t} f(\tau)\,d\tau$ is also periodic with period T and one has for all $t \in \mathbb{R}$:

$$h(t) = \sum_{n=-\infty}^{\infty} h_n e^{in\omega_0 t}, \tag{4.15}$$

with

$$h_n = \frac{c_n}{in\omega_0} \qquad \text{for } n \neq 0,$$

$$h_0 = -\sum_{\substack{n=-\infty \\ n\neq 0}}^{\infty} \frac{(-1)^n c_n}{in\omega_0}.$$

Proof
Since $f(t)$ is a piecewise smooth periodic function with $c_0 = 0$, the function $h(t) = \int_{-T/2}^{t} f(\tau)\,d\tau$ is also a piecewise smooth periodic function and thus equal to its Fourier series. In order to determine the Fourier coefficients of $h(t)$, we introduce a piecewise smooth function $g(\tau)$ on the interval $(-T/2, T/2)$ satisfying

$$g(\tau) = \begin{cases} 1 & \text{for } -T/2 < \tau \leq t, \\ 0 & \text{for } t < \tau \leq T/2. \end{cases}$$

With the function $g(\tau)$ thus defined it follows that

$$h(t) = \int_{-T/2}^{t} f(\tau)\,d\tau = \int_{-T/2}^{T/2} f(\tau)\overline{g(\tau)}\,d\tau.$$

Parseval's identity may now be applied to the functions $f(\tau)$ and $g(\tau)$. If, moreover, we substitute for g_n the definition of the nth Fourier coefficient of $g(\tau)$, then we obtain

$$h(t) = T \sum_{n=-\infty}^{\infty} c_n \overline{g_n} = \sum_{n=-\infty}^{\infty} c_n \overline{\int_{-T/2}^{T/2} g(\tau)e^{-in\omega_0\tau}\,d\tau}$$

$$= \sum_{n=-\infty}^{\infty} c_n \int_{-T/2}^{t} e^{in\omega_0\tau}\,d\tau.$$

Since $c_0 = 0$ it follows that

$$h(t) = \sum_{\substack{n=-\infty \\ n\neq 0}}^{\infty} c_n \int_{-T/2}^{t} e^{in\omega_0\tau}\,d\tau$$

$$= \sum_{\substack{n=-\infty \\ n\neq 0}}^{\infty} \frac{c_n}{in\omega_0} e^{in\omega_0 t} - \sum_{\substack{n=-\infty \\ n\neq 0}}^{\infty} \frac{c_n(-1)^n}{in\omega_0}.$$

This final expression is the Fourier series of the function $h(t)$ having coefficients $h_n = c_n/in\omega_0$ for $n \neq 0$ and coefficient h_0 equal to the second series. ∎

From theorem 4.9 we can conclude that if the Fourier series is known for a piecewise smooth periodic function and if $c_0 = 0$, then instead of integrating the function one may integrate its Fourier series term-by-term. The resulting series converges to the integral of the function.

EXAMPLE 4.5

Consider the periodic function f with period 2 for which $f(t) = 1$ for $-1 \leq t < 0$ and $f(t) = -1$ for $0 \leq t < 1$. One can easily check (see exercise 3.22, if necessary) that the Fourier coefficients of this function are given by $c_0 = 0$ and $c_n = ((-1)^n - 1)/in\pi$. Furthermore, $\int_{-T/2}^{t} f(\tau)\,d\tau$ is a periodic triangle function with period 2 and $a = 1$. According to section 3.4.2 its Fourier coefficients are equal to $(1 - (-1)^n)/n^2\pi^2$, which indeed equals $c_n/in\pi$. The zeroth Fourier coefficient of this periodic triangle function equals 1/2. From (4.10) it immediately follows that indeed

$$-\sum_{\substack{n=-\infty \\ n \neq 0}}^{\infty} \frac{c_n(-1)^n}{in\pi} = \sum_{\substack{n=-\infty \\ n \neq 0}}^{\infty} \frac{1 - (-1)^n}{n^2\pi^2} = \frac{4}{\pi^2} \sum_{k=1}^{\infty} \frac{1}{(2k-1)^2} = \frac{1}{2}.$$

◀

4.3.4 Differentiation

In section 4.3.3 we have seen that the Fourier coefficients of the integral of a piece-wise smooth periodic function with $c_0 = 0$ can be derived quite easily from the Fourier coefficients of the function itself. The function could in fact be integrated term-by-term and the resulting series converged to the integral of the function. In this section we investigate under which conditions the term-by-term *derivative* of the Fourier series of a function converges to the derivative of the function itself. Term-by-term differentiation leads less often to a convergent series. It is not hard to understand the reason for this. In section 4.3.3 we have seen that integrating a Fourier series corresponds to a *division* of the nth term by a factor proportional to n, improving the rate of convergence of the series. However, differentiating a Fourier series corresponds to a *multiplication* of the nth term by a factor proportional to n, and this will diminish the rate of convergence.

EXAMPLE 4.6

Consider the sawtooth function $f(t) = 2t/T$ for $-T/2 < t < T/2$ from section 3.4.3. We have seen that the Fourier coefficients are equal to $i(-1)^n/n\pi$ and so the Fourier series is equal to $\sum_{n=-\infty}^{\infty} i(-1)^n e^{in\omega_0 t}/n\pi$. The sawtooth function has discontinuities at $t = \pm T/2, \pm 3T/2, \ldots$. If we differentiate the Fourier series of the sawtooth function term-by-term, then we find

$$\sum_{n=-\infty}^{\infty} i(-1)^n in\omega_0 e^{in\omega_0 t}/n\pi = \sum_{n=-\infty}^{\infty} (-1)^{n+1}\omega_0 e^{in\omega_0 t}/\pi.$$

This series does not converge, since the terms in the series do not tend to 0 for $n \to \infty$.

◀

It turns out that continuity of the periodic function is an important condition for the term-by-term differentiability. We formulate this in the following theorem.

THEOREM 4.10
Differentiation of Fourier series

Let $f(t)$ be a piecewise smooth periodic continuous function with Fourier coefficients c_n and for which $f'(t)$ is piecewise smooth as well. Then

$$\frac{1}{2}(f'(t+) + f'(t-)) = \sum_{n=-\infty}^{\infty} in\omega_0 c_n e^{in\omega_0 t}. \tag{4.16}$$

Proof
Since $f'(t)$ is piecewise smooth, $f'(t)$ has a convergent Fourier series. Let c_n' be the Fourier coefficients of $f'(t)$, then

$$c_n' = \frac{1}{T} \int_{-T/2}^{T/2} f'(t)e^{-in\omega_0 t}\, dt.$$

Since f is continuous, one can apply integration by parts. We then find:

$$c_n' = \frac{1}{T}\left[f(t)e^{-in\omega_0 t} \right]_{-T/2}^{T/2} + in\omega_0 \frac{1}{T} \int_{-T/2}^{T/2} f(t)e^{-in\omega_0 t}\, dt.$$

The first term in the right-hand side is 0 for all n, since $f(t)$ is periodic, so $f(-T/2) = f(T/2)$, and $e^{-in\omega_0 T/2} = e^{in\omega_0 T/2}$. The second term is, by definition, up to a factor $in\omega_0$, equal to c_n and hence

$$c_n' = in\omega_0 c_n.$$

If we now apply the fundamental theorem to the function f' it follows that

$$\frac{1}{2}(f'(t+) + f'(t-)) = \sum_{n=-\infty}^{\infty} c_n' e^{in\omega_0 t} = \sum_{n=-\infty}^{\infty} in\omega_0 c_n e^{in\omega_0 t}.$$

$\blacksquare$

In order to derive the equality $c_n' = in\omega_0 c_n$ in the proof of theorem 4.10, we did not use the assumption that $f'(t)$ was piecewise smooth. Hence, this equality also holds when $f'(t)$ is piecewise continuous. If we now apply the Riemann–Lebesgue lemma to the function $f'(t)$, it follows that $\lim_{n\to\pm\infty} c_n' = \lim_{n\to\pm\infty} in\omega_0 c_n = 0$. This proves statement b about the rate of convergence from section 4.2. By applying this repeatedly, statement c follows (see also exercise 4.20).

EXERCISES

4.10   Equation (4.14) has been stated for complex Fourier coefficients. Give the equation if one uses the ordinary Fourier coefficients.

4.11   An application of Parseval's identity can be found in electronics. Suppose that in an electric circuit the periodic voltage $v(t)$ gives rise to a periodic current $i(t)$, both with period T and both piecewise smooth. Let v_n and i_n be the Fourier coefficients of $v(t)$ and $i(t)$ respectively. Show that for the average power P over one period (see section 1.2.3) one has

$$P = \sum_{n=-\infty}^{\infty} v_n \overline{i_n} = \sum_{n=-\infty}^{\infty} v_n i_{-n}.$$

4.12   Consider the periodic block function $f(t)$ with period π and for some $a \in \mathbb{R}$ with $0 < a < \pi$, and the periodic block function $g(t)$ with period π and some $b \in \mathbb{R}$ with $0 < b < \pi$ (see section 3.4.1). Assume that $a \leq b$.
a   Use Parseval's identity to show that

$$\sum_{n=1}^{\infty} \frac{\sin na \sin nb}{n^2} = \frac{a(\pi - b)}{2}.$$

b   Choose $a = b = \pi/2$ and derive (4.10) again.

4.13 Consider the periodic triangle function from section 3.4.2 for $a = T/2$.

a Use (4.14) to obtain that

$$\sum_{n=0}^{\infty} \frac{1}{(2n+1)^4} = 1 + \frac{1}{3^4} + \frac{1}{5^4} + \frac{1}{7^4} + \cdots = \frac{\pi^4}{96}.$$

b Let $S = \sum_{n=1}^{\infty} 1/n^4$. Split this into a sum over the even and a sum over the odd positive integers and then show that $S = \pi^4/96 + S/16$. Conclude that

$$\sum_{n=1}^{\infty} \frac{1}{n^4} = \frac{\pi^4}{90}.$$

4.14 Let $f(t)$ be the periodic function with period 2 defined for $-1 < t \leq 1$ by

$$f(t) = \begin{cases} t^2 + t & \text{for } -1 < t \leq 0, \\ -t^2 + t & \text{for } 0 < t \leq 1. \end{cases}$$

a Use the results from exercise 4.9 to show that

$$\sum_{k=0}^{\infty} \frac{1}{(2k+1)^6} = \frac{\pi^6}{960}.$$

b Use part a and the method of exercise 4.13b to show that

$$\sum_{k=1}^{\infty} \frac{1}{k^6} = \frac{\pi^6}{945}.$$

4.15 Let $f(t)$ be a piecewise smooth periodic function with period T and Fourier series $\sum_{n=-\infty}^{\infty} c_n e^{in\omega_0 t}$, where $c_0 = 0$. Show that for $-T/2 \leq a \leq b$ one has

$$\int_a^b f(t)\, dt = \sum_{\substack{n=-\infty \\ n \neq 0}}^{\infty} \frac{c_n}{in\omega_0} (e^{in\omega_0 b} - e^{in\omega_0 a}).$$

4.16 Let $f(t)$ be a piecewise smooth periodic function with period T and with Fourier coefficients a_n and b_n, where $a_0 = 0$. Show that for $-T/2 \leq a \leq b$ one has

$$\int_a^b f(t)\, dt = \sum_{n=1}^{\infty} \frac{1}{n\omega_0} (a_n(\sin n\omega_0 b - \sin n\omega_0 a) - b_n(\cos n\omega_0 b - \cos n\omega_0 a)).$$

4.17 Consider the periodic function $f(t)$ with period 2π defined for $-\pi < t \leq \pi$ by

$$f(t) = \begin{cases} -1 & \text{for } -\pi < t \leq 0, \\ 1 & \text{for } 0 < t \leq \pi. \end{cases}$$

a Determine the ordinary Fourier coefficients of f and give the Fourier series of f.

b Integrate the series from part a over $[-\pi, t]$ and determine the resulting constant using (4.10).

c Show that the function represented by the series from part b is the periodic function with period 2π given by $g(t) = |t| - \pi$ for $-\pi < t \leq \pi$.

d Use exercise 3.6 to determine in a direct way the ordinary Fourier series of the function g from part c and use this to verify the result from part b.

4.18   Consider the periodic function $f(t)$ with period 2 from exercise 4.14. Use the integration theorem to show that

$$g(t) = \frac{4}{\pi^2} \sum_{n=-\infty}^{\infty} \frac{1}{(2n+1)^4} + \frac{4}{\pi^2} \sum_{n=-\infty}^{\infty} \frac{1}{(2n+1)^4} e^{\pi i n t},$$

where $g(t)$ is the periodic function with period 2 given for $-1 < t \leq 1$ by

$$g(t) = \tfrac{1}{3}|t|^3 - \tfrac{1}{2}t^2 + \tfrac{1}{6}.$$

Finally calculate the Fourier coefficient c_0 of $g(t)$ and check your answer using exercise 4.13a.

4.19   Show that theorem 4.10 reads as follows, when formulated in terms of the ordinary Fourier coefficients. Let $f(t)$ be a piecewise smooth periodic continuous function with ordinary Fourier coefficients a_n and b_n and for which $f'(t)$ is piecewise smooth. Then

$$\tfrac{1}{2}(f'(t+) + f'(t-)) = \sum_{n=1}^{\infty} (n\omega_0 b_n \cos n\omega_0 t - n\omega_0 a_n \sin n\omega_0 t).$$

4.20   Let $f(t)$ be a periodic continuous function with piecewise smooth continuous derivative $f'(t)$. Show that $\lim_{n \to \pm\infty} n^2 c_n = 0$.

4.21   In the example prior to theorem 4.10 we saw that the Fourier series of the periodic sawtooth function $f(t)$ with period T, given by $f(t) = 2t/T$ for $-T/2 < t \leq T/2$, could not be differentiated term-by-term. However, now consider the even function $g(t)$ with period T given by $f(t) = 2t/T$ for $0 \leq t \leq T/2$.
a   Determine the Fourier cosine series of the function $g(t)$.
b   Differentiate the series from part a term-by-term. Verify that the result is the Fourier sine series of the function $g'(t)$ and that this series converges to $g'(t)$ for all $t \neq nT/2$ $(n \in \mathbb{Z})$. How can this result be reconciled with theorem 4.10?

4.22   **a**   Determine the Fourier series of the periodic function with period 2π defined for $-\pi < t \leq \pi$ by

$$f(t) = \begin{cases} 0 & \text{for } -\pi < t \leq 0, \\ \sin t & \text{for } 0 < t \leq \pi. \end{cases}$$

b   Verify that we can differentiate f from part a by differentiating its Fourier series, except at $t = n\pi$ $(n \in \mathbb{Z})$. Describe the function that is represented by the differentiated series.

4.23   Formulate the convolution theorem (theorem 4.7) for the ordinary Fourier coefficients.

4.24   Let f be the periodic block function with period 2 and $a = 1$, so $f(t) = 1$ for $|t| \leq \tfrac{1}{2}$ and $f(t) = 0$ for $\tfrac{1}{2} < |t| \leq 1$. Let g be the periodic triangle function with period 2 and $a = \tfrac{1}{2}$, so $g(t) = 1 - 2|t|$ for $|t| \leq \tfrac{1}{2}$ and $g(t) = 0$ for $\tfrac{1}{2} < |t| \leq 1$.
a   Show that $f_1 * f_2$ is even when both f_1 and f_2 are even periodic functions with period T.

b Show that $f * g$ is the even periodic function with period 2 which for $0 \leq t \leq 1$ is given by

$$(f * g)(t) = \begin{cases} -\frac{1}{2}t^2 + \frac{1}{4} & \text{for } 0 \leq t \leq \frac{1}{2}, \\ \frac{1}{2}(t-1)^2 & \text{for } \frac{1}{2} < t \leq 1. \end{cases}$$

c Determine the Fourier series of $(f * g)(t)$ and verify that it converges to $(f * g)(t)$ for all $t \in \mathbb{R}$.
d Verify the constant in the Fourier series of $(f * g)(t)$ by calculating the zeroth Fourier coefficient in a direct way.

4.4 The sine integral and Gibbs' phenomenon

The fundamental theorem of Fourier series was formulated for piecewise smooth functions. According to this theorem, the series converges pointwise to the function. Possible points of discontinuity were excluded here. At these points, the series converges to the average value of the left- and right-hand limits of the function. Towards the end of section 4.2 we already noted that in a small neighbourhood of a discontinuity, the series will approximate the function much slower. This had already been observed by Wilbraham in 1848, but his results fell into oblivion. In 1898 the physicist Michaelson published an article in the magazine *Nature*, in which he doubted the fact that 'a real discontinuity (of a function f) can replace a sum of continuous curves' (i.e., the terms in the partial sums $s_n(t)$). This is because Michaelson had constructed a machine which calculated the nth partial sum of the Fourier series of a function up to $n = 80$. In a small neighbourhood of a discontinuity, the partial sums $s_n(t)$ did not behave as he had expected: the sums continued to deviate and the largest deviation, the so-called *overshoot* of $s_n(t)$ relative to $f(t)$, did not decrease with increasing n. In figure 4.3 this is illustrated by the graphs of the partial sums approximating the periodic block function for different values of n. We see that the overshoots get narrower with increasing n, but the magnitude remains the same. In

Overshoot

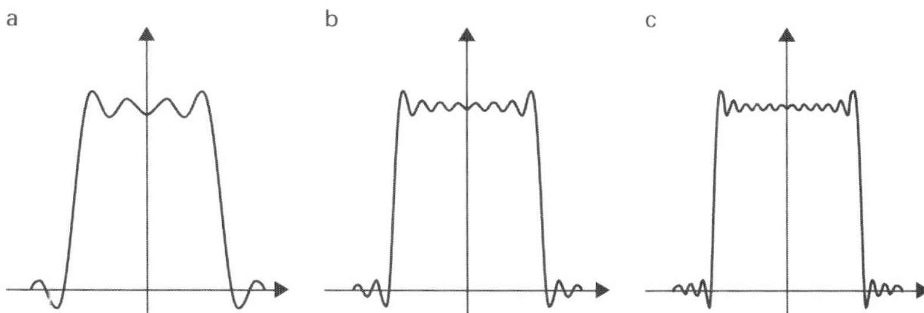

FIGURE 4.3
Partial sums of the periodic block function.

a letter to *Nature* from 1899, Gibbs explained this phenomenon and showed that $s_n(t)$ will always have an overshoot of about 9% of the magnitude of the jump at the discontinuity. We will investigate this so-called *Gibbs' phenomenon* more closely for the periodic block function. Before we do so, we first introduce the sine integral, a function that will be needed to determine Gibbs' phenomenon quantitatively. The sine integral will be encountered in later chapters as well.

4.4.1   The sine integral

The sine integral is a function which is used in several places. It is defined as follows.

DEFINITION 4.3
Sine integral

The sine integral *is the function* $\mathrm{Si}(x)$ *defined by*

$$\mathrm{Si}(x) = \int_0^x \frac{\sin t}{t}\, dt. \tag{4.17}$$

Since $|\sin t/t| \le 1$ for all $t \ne 0$, the integrand is bounded and the integral well-defined. Furthermore, we have $\mathrm{Si}(0) = 0$. The sine integral cannot be determined analytically, but there are tables containing function values. In particular one has

$$\mathrm{Si}(\pi) = \int_0^\pi \frac{\sin t}{t}\, dt = 1.852\ldots.$$

The definition of $\mathrm{Si}(x)$ can also be used for negative values of x, from which it follows that $\mathrm{Si}(x)$ is an odd function. Starting from the graph of $\sin t/t$, the graph of $\mathrm{Si}(x)$ can be sketched; see figure 4.4. Figure 4.4 seems to suggest that $\mathrm{Si}(x)$

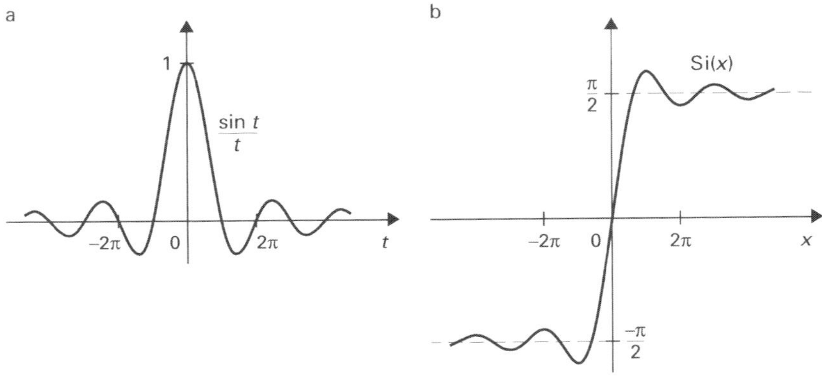

FIGURE 4.4
Graphs of the functions $\sin t/t$ (a) and $\mathrm{Si}(x)$ (b).

converges to $\pi/2$ for $x \to \infty$. Although $\mathrm{Si}(x)$ cannot be calculated analytically, one is able to determine its limit.

THEOREM 4.11

For the sine integral one has

$$\lim_{x \to \infty} \mathrm{Si}(x) = \int_0^\infty \frac{\sin t}{t}\, dt = \frac{\pi}{2}.$$

Proof
In order to prove this, we introduce for $p > 0$ the function $I(p)$ defined by

$$I(p) = \int_0^\infty e^{-pt}\sin t\, dt.$$

Using integration by parts twice we can derive that for $p > 0$ one has

$$I(p) = \int_0^\infty e^{-pt}\sin t\, dt = \left[\frac{-e^{-pt}}{p}\sin t\right]_0^\infty + \int_0^\infty \frac{e^{-pt}}{p}\cos t\, dt$$

$$= \left[\frac{-e^{-pt}}{p^2}\cos t\right]_0^\infty - \int_0^\infty \frac{e^{-pt}}{p^2}\sin t\, dt = \frac{1}{p^2} - \frac{1}{p^2}I(p).$$

The first and last term are equal. If we solve for $I(p)$, we obtain

$$I(p) = \frac{1}{1 + p^2}.$$

Next we integrate $I(p)$ over the interval $(0, \infty)$ to obtain

$$\int_0^\infty I(p)\, dp = \int_0^\infty \frac{1}{1 + p^2}\, dp = [\arctan p]_0^\infty = \frac{\pi}{2}. \tag{4.18}$$

On the other hand we find by interchanging the order of integration that

$$\int_0^\infty I(p)\, dp = \int_0^\infty \left(\int_0^\infty e^{-pt} \sin t\, dp \right) dt = \int_0^\infty \left[\frac{e^{-pt}}{-t} \right]_{p=0}^{p=\infty} \sin t\, dt$$

$$= \int_0^\infty \frac{\sin t}{t}\, dt. \tag{4.19}$$

We state without proof that interchanging the order of integration is allowed. With the equality of the right-hand sides of (4.18) and (4.19) the theorem is proved. ∎

4.4.2 Gibbs' phenomenon*

In this section we treat Gibbs' phenomenon. It is a rather technical treatment, which does not result in any specific new insight into Fourier series. This section may therefore be omitted without any consequences for the study of the remainder of the book.

We treat Gibbs' phenomenon using the periodic block function. Since it will result in much simpler formulas, we will not start from the periodic block function as defined in section 3.4.1, but instead from the periodic function $f(t)$ defined on the interval $(-T/2, T/2)$ by

$$f(t) = \begin{cases} \frac{1}{2} & \text{for } 0 < t < \frac{1}{2}T, \\[2mm] -\frac{1}{2} & \text{for } -\frac{1}{2}T < t < 0. \end{cases}$$

This odd function has a Fourier sine series whose coefficients have been calculated in exercise 3.22. The partial sums $s_n(t)$ of the Fourier series are

$$s_n(t) = \sum_{k=1}^n \frac{2}{(2k - 1)\pi} \sin((2k - 1)\omega_0 t). \tag{4.20}$$

The graph of the partial sum for $n = 12$ is drawn in figure 4.5. In it, Gibbs' phenomenon is clearly visible again: immediately next to a discontinuity of the function $f(t)$, the partial sum overshoots the values of $f(t)$. We will now calculate the overshoot, that is, the magnitude of the maximum difference between the function and the partial sums immediately next to the discontinuity. By determining the derivative of the partial sum, we can find out where the maximum difference occurs, and subsequently calculate its value. Differentiating $s_n(t)$ gives

$$s_n'(t) = \sum_{k=1}^n \frac{2}{\pi} \omega_0 \cos((2k - 1)\omega_0 t) = \sum_{k=1}^n \frac{4}{T} \cos((2k - 1)\omega_0 t). \tag{4.21}$$

In order to determine the zeros of the derivative, we rewrite the last sum. For this we use the trigonometric identity $\sin\alpha - \sin\beta = 2\sin((\alpha - \beta)/2)\cos((\alpha + \beta)/2)$.

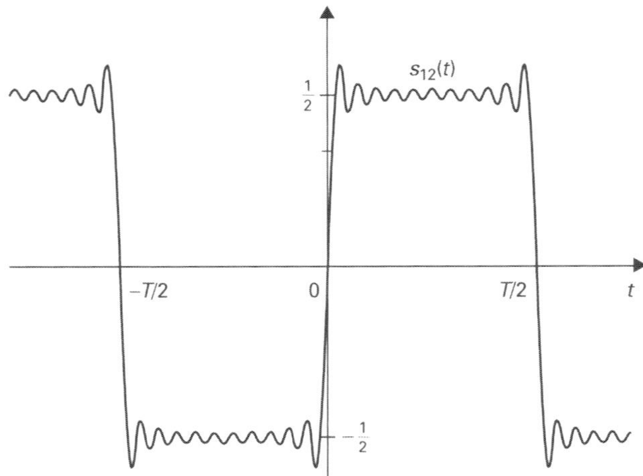

FIGURE 4.5
Overshoot of the partial sums approximating the block function.

With $\alpha = 2k\omega_0 t$ and $\beta = (2k-2)\omega_0 t$ it follows that

$$\cos((2k-1)\omega_0 t) = \frac{\sin 2k\omega_0 t - \sin((2k-2)\omega_0 t)}{2\sin\omega_0 t}.$$

By substituting this into expression (4.21) for the derivative $s_n'(t)$ it follows that

$$s_n'(t) = \frac{2}{T}\frac{\sin 2n\omega_0 t}{\sin\omega_0 t},$$

since consecutive terms cancel each other. The derivative is 0 if $2n\omega_0 t = k\pi$ for $k \in \mathbb{Z}$ and $k \neq 0$. We thus have extrema for $s_n(t)$ at $t = k\pi/2n\omega_0$. For $k = 1$ we find the first extremum immediately to the right of $t = 0$, that is, for $t = \pi/2n\omega_0$. The value at the extremum can be found by substituting $t = \pi/2n\omega_0$ in $s_n(t)$. This gives

$$s_n\left(\frac{\pi}{2n\omega_0}\right) = \sum_{k=1}^{n}\frac{2}{(2k-1)\pi}\sin\left((2k-1)\omega_0\frac{\pi}{2n\omega_0}\right)$$

$$= \sum_{k=1}^{n}\frac{2}{(2k-1)\pi}\sin((2k-1)\pi/2n).$$

The last sum can be rewritten as an expression which is a Riemann sum with stepsize π/n for the function $\sin x/x$:

$$s_n\left(\frac{\pi}{2n\omega_0}\right) = \frac{1}{\pi}\sum_{k=1}^{n}\frac{\pi}{n}\cdot\frac{\sin((2k-1)\pi/2n)}{(2k-1)\pi/2n}.$$

Taking the limit $n \to \infty$, the sums in the right-hand side converge to the integral $\int_0^{\pi}(\sin x/x)dx$. Note that for large values of n the contribution of the nth term gets smaller and smaller and will even tend to zero (since the series converges to the integral). The value of this integral was given in section 4.4.1 and hence

$$\lim_{n\to\infty} s_n\left(\frac{\pi}{2n\omega_0}\right) = \lim_{n\to\infty} \frac{1}{\pi}\sum_{k=1}^{n} \frac{\pi}{n}\frac{\sin((2k-1)\pi/2n)}{(2k-1)\pi/2n}$$

$$= \frac{1}{\pi}\int_0^{\pi}\frac{\sin x}{x}\,dx = \frac{1}{\pi}\,1.852\ldots = 0.589\ldots.$$

This establishes the value at the first maximum next to the jump. Since the jump has magnitude 1, the overshoot of the function value 0.5 is approximately 9 % of the jump. Since the additional contribution for large values of n gets increasingly smaller, this overshoot will remain almost constant with increasing n. Furthermore we see that the value of t where the extremum is attained is getting closer and closer to the point of discontinuity.

In this section we studied Gibbs' phenomenon using the periodic block function. However, the phenomenon occurs in a similar way for other piecewise smooth functions having points of discontinuity. There is always an overshoot of the partial sums immediately to the left and to the right of the points of discontinuity, with a value approximately equal to 9 % of the magnitude of the jump. As more terms are being included in the partial sums, the extrema are getting closer and closer to the point of discontinuity.

EXERCISES

4.25 Use the definition to verify that the sine integral $\mathrm{Si}(x)$ is an odd function.

4.26 The sine integral can be considered as an approximation of the function $f(x)$ given by $f(x) = \pi/2$ for $x > 0$ and $f(x) = -\pi/2$ for $x < 0$.
a What is the smallest value of $x > 0$ for which $\mathrm{Si}(x)$ has a maximum?
b What is the value of $\mathrm{Si}(x)$ at the first maximum and what percentage of the jump at $x = 0$ of the function $f(x)$ does the overshoot amount to?

4.27* Consider the periodic function with period T given by $f(t) = 2t/T - 1$ for $0 < t \leq T$. Let $s_n(t)$ be the nth partial sum of the Fourier series of f.
a Show that f arises from the sawtooth function from section 3.4.3 by a shift over $T/2$. Next determine $s_n(t)$.
b Show that

$$s_n'(t) = -\frac{2}{T}(D_n(t) - 1),$$

where D_n is the Dirichlet kernel from definition 4.1. Subsequently determine the value of t for which $s_n(t)$ has its first extreme value immediately to the left of the discontinuity at $t = 0$.
c Calculate the magnitude of the extremum from part b and show that the overshoot is again approximately equal to 9 % of the magnitude of the jump at $t = 0$.

SUMMARY

When a periodic function is piecewise continuous, its Fourier coefficients, as defined in chapter 3, exist. According to Bessel's inequality, the sum of the squares of the moduli of the Fourier coefficients of a periodic piecewise continuous function is finite. From this, the lemma of Riemann–Lebesgue follows, which states that the Fourier coefficients tend to 0 for $n \to \pm\infty$. The fundamental theorem of Fourier series has been formulated for piecewise smooth periodic functions. For such functions the Fourier series converges to the function at the points of continuity, and to the average of the left- and right-hand limit at the points of discontinuity. If f is a piecewise smooth periodic function with period T, then one has, according to the fundamental theorem,

$$\sum_{n=-\infty}^{\infty} c_n e^{in\omega_0 t} = \tfrac{1}{2}\left(f(t+) + f(t-)\right),$$

where

$$c_n = \frac{1}{T}\int_{-T/2}^{T/2} f(t)e^{-in\omega_0 t}\,dt \qquad \text{and} \qquad \omega_0 = \frac{2\pi}{T}.$$

As functions are smoother, meaning that higher derivatives do not contain discontinuities, the convergence of the Fourier series to the function is faster. Using the fundamental theorem, further properties of Fourier series were derived, such as series for products and convolutions, term-by-term differentiation and integration of Fourier series, and Parseval's identity

$$\frac{1}{T}\int_{-T/2}^{T/2} f(t)\overline{g(t)}\,dt = \sum_{n=-\infty}^{\infty} f_n\overline{g_n},$$

where f_n and g_n are the Fourier coefficients of f and g.

While the Fourier series of a piecewise smooth function with discontinuities does converge at each point of continuity to the function value, the series always has an overshoot of about 9 % of the magnitude of the jump immediately next to a point of discontinuity. This phenomenon is called Gibbs' phenomenon. As more terms are included in the partial sums, the overshoot shifts closer and closer to the point of discontinuity. For the analysis of this phenomenon we defined the sine integral $\mathrm{Si}(x)$, having as properties $\mathrm{Si}(\pi) = 1.852\ldots$ and $\lim_{x\to\infty}\mathrm{Si}(x) = \pi/2$.

SELFTEST

4.28   Let $f(t)$ be the odd periodic function with period 2π defined for $0 \le t < \pi$ by

$$f(t) = \begin{cases} \dfrac{2}{\pi}t & \text{for } 0 \le t < \dfrac{\pi}{2}, \\[2mm] 1 & \text{for } \dfrac{\pi}{2} \le t < \pi. \end{cases}$$

a   For which values of $t \in \mathbb{R}$ does the Fourier series of $f(t)$ converge? What is the sum for those values of t for which there is convergence?
b   Determine the Fourier series of f and verify the fundamental theorem of Fourier series for $t = 0$ and $t = \pi$.
c   Can one differentiate the function f by differentiating the Fourier series term-by-term? If not, explain. If so, give the function that is represented by the differentiated series.
d   Can one integrate the function f over $[-\pi, t]$ by integrating the Fourier series term-by-term? If not, explain. If so, give the function that is represented by the integrated series.

4.29   Let $a \in \mathbb{R}$ with $0 < a \le \pi/2$. Use the periodic block function $p_{a,\pi}$ and the periodic triangle function $q_{a,\pi}$ to show that

$$\sum_{n=1}^{\infty} \frac{\sin^3 na}{n^3} = \frac{a^2}{8}(3\pi - 4a).$$

Finally evaluate

$$\sum_{n=0}^{\infty} \frac{(-1)^n}{(2n+1)^3}.$$

4.30 Let $f(t)$ be the periodic function $f(t) = |\sin t|$ with period 2π.
a Find the Fourier series of f and verify that it converges to $f(t)$ for all $t \in \mathbb{R}$.
b Show that

$$\sum_{n=1}^{\infty} \frac{1}{4n^2 - 1} = \frac{1}{2} \quad \text{and} \quad \sum_{n=1}^{\infty} \frac{(-1)^n}{4n^2 - 1} = \frac{1}{2} - \frac{\pi}{4}.$$

c Show that

$$\sum_{n=1}^{\infty} \frac{1}{(4n^2 - 1)^2} = \frac{\pi^2}{16} - \frac{1}{2}.$$

4.31 Let f be the periodic sawtooth function with period 2 given by $f(t) = t$ for $-1 < t \leq 1$.
a Show that $f_1 * f_2$ is an even function if both f_1 and f_2 are odd periodic functions with period T.
b Show that $f * f$ is the even periodic function with period 2 which for $0 \leq t \leq 1$ is given by $(f * f)(t) = -\frac{1}{2}t^2 + t - \frac{1}{3}$. (Hint: how can one express f for $1 < t \leq 2$? Now split the integral in two parts.)
c Prove that for all $t \in \mathbb{R}$ we have

$$(f * f)(t) = -\frac{2}{\pi^2} \sum_{n=1}^{\infty} \frac{1}{n^2} \cos n\pi t.$$

d Show that for $t = 0$, the result from part c is equivalent to Parseval's identity (4.14) for the function f.
e Verify that term-by-term differentiation of $(f * f)(t)$ is allowed for $0 < |t| \leq 1$ and describe for all $t \in \mathbb{R}$ the function that is represented by the differentiated series.
f Determine in a direct way the zeroth Fourier coefficient of $(f * f)(t)$ and verify the answer using the result from part c. Next, verify that term-by-term integration over $[-1, t]$ is allowed and describe for all $t \in \mathbb{R}$ the function that is represented by the integrated series.

Contents of Chapter 5

Applications of Fourier series

Applications of Fourier series

INTRODUCTION

Applications of Fourier series can be found in numerous places in the natural sciences as well as in mathematics itself. In this chapter we confine ourselves to two kinds of applications, to be treated in sections 5.1 and 5.2. Section 5.1 explains how Fourier series can be used to determine the response of a linear time-invariant system to a periodic input. In section 5.2 we discuss the applications of Fourier series in solving partial differential equations, which often occur when physical processes, such as heat conduction or a vibrating string, are described mathematically.

The frequency response, introduced in chapter 1 using the response to the periodic time-harmonic signal $e^{i\omega t}$ with frequency ω, plays a central role in the calculation of the response of a linear time-invariant system to an arbitrary periodic signal. Specifically, a Fourier series shows how a periodic signal can be written as a superposition of time-harmonic signals with frequencies being an integer multiple of the fundamental frequency. By applying the so-called superposition rule for linear time-invariant systems, one can then easily find the Fourier series of the output. This is because the sequence of Fourier coefficients, or the line spectrum, of the *output* arises from the line spectrum of the *input* by a multiplication by the frequency response at the integer multiples of the fundamental frequency.

For stable systems which can be described by ordinary differential equations, which is almost any linear time-invariant system occurring in practice, we will see that the frequency response can easily be derived from the differential equation. The characteristic polynomial of the differential equation of a stable system has no zeros on the imaginary axis, and hence there are no periodic eigenfunctions. As a consequence, the response to a periodic signal is uniquely determined by the differential equation. If there are zeros $i\omega$ of the characteristic polynomial on the imaginary axis, then a periodic input may lead to resonance. For the corresponding frequencies ω, the frequency response is meaningless.

In the second and final section we treat applications of Fourier series in solving partial differential equations by separation of variables. This method is explained systematically in the case when the functions have one time-variable and one position-variable. We limit ourselves to simple examples of initial and boundary value problems, with the partial differential equation being either the one-dimensional diffusion or heat equation, or the one-dimensional wave equation.

LEARNING OBJECTIVES

After studying this chapter it is expected that you
- can express the line spectrum of the response of an LTC-system to a periodic input in terms of the frequency response and the line spectrum of the input
- know what eigenfunctions and eigenfrequencies are for a system described by a differential equation, and know the relevance of the zeros of the corresponding characteristic polynomial
- know when the periodic response of an LTC-system to a periodic input is uniquely determined by the differential equation, and know what causes resonance
- can determine the frequency response for stable LTC-systems described by a differential equation
- can use separation of variables and Fourier series to determine in a systematic way a formal solution of the one-dimensional heat equation and the one-dimensional wave equation, under certain conditions.

5.1   Linear time-invariant systems with periodic input

LTC-system

In this section we will occupy ourselves with the determination of the response of a linear time-invariant continuous-time system (LTC-system for short) to a periodic input. Calculating the response as a function of time or, put differently, determining the response in the *time domain*, is often quite difficult. If, however, we have the line spectrum of the input at our disposal, so if we know the sequence of Fourier coefficients, then it will turn out that by using the frequency response of the LTC-system it is easy to determine the line spectrum of the output. Apparently it is easy to calculate the response in the *frequency domain*.

The frequency response of an LTC-system was introduced in chapter 1 by the property

$$e^{i\omega t} \mapsto H(\omega)e^{i\omega t}.$$

That is to say, the response to the time-harmonic signal $e^{i\omega t}$ of frequency ω is equal to $H(\omega)e^{i\omega t}$. We assume that for periodic inputs $u(t)$ one has that

$$u(t) = \sum_{n=-\infty}^{\infty} u_n e^{in\omega_0 t}, \tag{5.1}$$

where $\omega_0 = 2\pi/T$ and u_n is the sequence of Fourier coefficients, or line spectrum, of $u(t)$. When $u(t)$ is a piecewise smooth function, then we know from the fundamental theorem of Fourier series that (5.1) holds everywhere if we assume that $u(t) = (u(t+) + u(t-))/2$ at the points of discontinuity of $u(t)$. In the present chapter this will always be tacitly assumed. One now has the following theorem.

THEOREM 5.1

Let $y(t)$ be the response of a stable LTC-system to a piecewise smooth and periodic input $u(t)$ with period T, fundamental frequency ω_0 and line spectrum u_n. Let $H(\omega)$ be the frequency response of the system. Then $y(t)$ is again periodic with period T and the line spectrum y_n of $y(t)$ is given by

$$y_n = H(n\omega_0)u_n \qquad \text{for } n = 0, \pm 1, \pm 2, \dots. \tag{5.2}$$

Proof

Let the line spectrum u_n of the input $u(t)$ be given. Then (5.1) holds, which represents $u(t)$ as a superposition of the time-harmonic signals $e^{in\omega_0 t}$ with frequencies $n\omega_0$. If only a finite number of Fourier coefficients u_n are unequal to zero, then $u(t)$ is a finite linear combination of time-harmonic signals. Because of the linearity of the system, in calculating the response it is sufficient first to determine the responses to the time-harmonic signals and then to take the same linear combination of the responses.

We now assume that for LTC-systems this property may be extended to infinite linear combinations of time-harmonic signals, so to series such as Fourier series from (5.1). We then say that for LTC-systems the *superposition rule* holds (also see chapter 10). This means that first one can determine the response to a time-harmonic signal with frequency $n\omega_0$ using the frequency response. Because of the stability of the LTC-system we know that $H(\omega)$ is defined for all values of ω (see section 1.3.3). On the basis of the superposition rule one thus has that

$$y(t) = \sum_{n=-\infty}^{\infty} u_n H(n\omega_0)e^{in\omega_0 t}.$$

We see that $y(t)$ is periodic with period T and, moreover, that the line spectrum y_n of the response satisfies (5.2). ∎

For a stable LTC-system the frequency response is given by

$$H(\omega) = \frac{1}{-\omega^2 + 3i\omega + 2}.$$

Consider the periodic input $u(t)$ with period 2π given on the interval $(-\pi, \pi)$ by $u(t) = t$. The line spectrum y_n of the response $y(t)$ satisfies (5.2). One has $H(n\omega_0) = 1/(2 + 3ni\omega_0 - n^2\omega_0^2)$. The line spectrum u_n of $u(t)$ can be obtained by a direct calculation of the Fourier coefficients. The result is: $u_n = (-1)^n i/n$ for $n \neq 0$, $u_0 = 0$. Hence,

$$y_n = \begin{cases} \dfrac{(-1)^n i}{n(2 + 3in\omega_0 - n^2\omega_0^2)} & \text{for } n \neq 0, \\[2ex] 0 & \text{for } n = 0. \end{cases}$$

◀

Systems that can be realized in practice are often described by differential equations. Well-known examples are electrical networks and mechanical systems. We will now examine how for such systems one can determine the frequency response.

5.1.1 Systems described by differential equations

In chapter 1 systems described by differential equations were briefly introduced. For such systems the relation between an input $u(t)$ and the corresponding output $y(t)$ is described by an ordinary differential equation with constant coefficients of the form

$$a_m \frac{d^m y}{dt^m} + a_{m-1} \frac{d^{m-1} y}{dt^{m-1}} + \cdots + a_1 \frac{dy}{dt} + a_0 y$$

$$= b_n \frac{d^n u}{dt^n} + b_{n-1} \frac{d^{n-1} u}{dt^{n-1}} + \cdots + b_1 \frac{du}{dt} + b_0 u \tag{5.3}$$

with $n \leq m$. Here $a_0, a_1, \ldots, a_m$ and $b_0, b_1, \ldots, b_n$ are constants with $a_m \neq 0$ and $b_n \neq 0$. The number m is called the *order of the differential equation*. An electric network with one source, a voltage generator or a current generator, and furthermore consisting of resistors, capacitors and inductors, can be considered as a system. The voltage of the voltage generator or the current from the current generator is then an input, with the output being, for example, the voltage across a certain element in the network or the current through a specific branch. One can derive the corresponding differential equation from Kirchhoff's laws and the voltage–current relations for the separate elements in the network. For a resistor of resistance R this is Ohm's law: $v(t) = Ri(t)$, where $i(t)$ is the current through the resistor and $v(t)$ the voltage across the resistor. For a capacitor with capacitance C and an inductor with self-inductance L these relations are, respectively,

$$v(t) = \frac{1}{C} \int_{-\infty}^{t} i(\tau)\,d\tau \qquad \text{and} \qquad v(t) = L\frac{d}{dt}i(t).$$

The theory of networks is not a subject of this book and thus we shall not occupy ourselves with the derivation of the differential equations. Hence, for the networks in the examples we will always state the differential equation explicitly. Readers with knowledge of network theory can derive these for themselves; others should consider the differential equation as being given and describing an LTC-system.

The same assumptions will be made with respect to mechanical systems. Here the differential equations follow from Newton's laws and the force–displacement relations for the mechanical components such as masses, springs and dampers. If we denote the force by F and the displacement by x, then one has for a spring with spring constant k that $F(t) = kx(t)$, for a damper with damping constant c that $F(t) = cdx/dt$, and for a mass m that $F(t) = md^2x/dt^2$. In these formulas the direction of the force and the displacement have not been taken into account.

The formulas for the mechanical systems are very similar to the formulas for the electrical networks when we replace a voltage $v(t)$ by a force $F(t)$ and a charge $Q(t) = \int_{-\infty}^{t} i(\tau)\,d\tau$ by a displacement. The latter means that a current $i(t)$ is replaced by a velocity dx/dt. The formula for the spring is then comparable to the formula for the capacitor, the formula for the damper with the one for the resistor and the formula for the mass with the inductor. This is listed in the following table.

electric network	mechanical system
$v(t) = Q(t)/C$ (capacitor)	$F(t) = kx(t)$ (spring)
$v(t) = RdQ/dt$ (resistor)	$F(t) = cdx/dt$ (damper)
$v(t) = Ld^2Q/dt^2$ (inductor)	$F(t) = md^2x/dt^2$ (mass)

In chapter 1 we already noted that LTC-systems which are equal in a mathematical sense can physically be realized in different ways. Hence, mathematically a mechanical network can be the same as an electric network.

For an LTC-system described by a differential equation, the frequency response $H(\omega)$ can easily be obtained. To this end we introduce the polynomials

$$A(s) = a_m s^m + a_{m-1}s^{m-1} + \cdots + a_1 s + a_0,$$
$$B(s) = b_n s^n + b_{n-1}s^{n-1} + \cdots + b_1 s + b_0.$$

Characteristic polynomial

The polynomial $A(s)$ is called the *characteristic polynomial* of differential equation (5.3). The following theorem shows how one can obtain $H(\omega)$ for those values of ω for which $A(i\omega) \neq 0$.

THEOREM 5.2

Let an LTC-system be described by differential equation (5.3) and have characteristic polynomial $A(s)$. Then one has for all ω with $A(i\omega) \neq 0$:

$$H(\omega) = \frac{B(i\omega)}{A(i\omega)}. \tag{5.4}$$

Proof
In order to find the frequency response, we substitute the input $u(t) = e^{i\omega t}$ into (5.3). The response $y(t)$ is then of the form $H(\omega)e^{i\omega t}$. Since the derivative of $e^{i\omega t}$ is $i\omega e^{i\omega t}$, substitution into (5.3) leads to $A(i\omega)H(\omega)e^{i\omega t} = B(i\omega)e^{i\omega t}$. From this it follows that $H(\omega) = B(i\omega)/A(i\omega)$, which proves the theorem. ∎

It is natural to examine the problems that may arise if $A(i\omega) = 0$ for a certain value of ω, in other words, if the characteristic polynomial has zeros on the imaginary axis. In order to do so, we will study in some more detail the solutions of differential equations with constant coefficients.

When the input $u(t)$ is known, the right-hand side of (5.3) is known and from the theory of ordinary differential equations we then know that several other solutions $y(t)$ exist. In fact, when one solution $y(t)$ of (5.3) is found, then *all* solutions can be obtained by adding to $y(t)$ an arbitrary solution $x(t)$ of the differential equation

$$a_m \frac{d^m x}{dt^m} + a_{m-1} \frac{d^{m-1} x}{dt^{m-1}} + \cdots + a_1 \frac{dx}{dt} + a_0 = 0. \tag{5.5}$$

Homogeneous solution

Eigenfunction

This differential equation is called the homogeneous differential equation corresponding to equation (5.3) and a solution of equation (5.5) is called a *homogeneous solution* or *eigenfunction*. These are thus the solutions with $u(t)$ being the null-signal. Of course, the null-function $x(t) = 0$ for all t satisfies the homogeneous differential equation. This homogeneous solution will be called the *trivial* homogeneous solution or trivial eigenfunction.

Particular solution

We will say that the general solution $y(t)$ of (5.3) can be written as a *particular solution* added to the general homogeneous solution:

general solution = particular solution + general homogeneous solution.

The general homogeneous solution can easily be determined using the *characteristic equation*. This equation arises by substituting $x(t) = e^{st}$ into (5.5), where s is a complex constant. From example 2.11 it follows that this is the result: $(a_m s^m + a_{m-1} s^{m-1} + \cdots + a_1 s + a_0)e^{st} = 0$. Since $e^{st} \neq 0$ one has

Characteristic equation

$$A(s) = a_m s^m + a_{m-1} s^{m-1} + \cdots + a_1 s + a_0 = 0. \tag{5.6}$$

To each zero s of the characteristic polynomial corresponds the homogeneous solution e^{st}. More generally, one can show that to a zero s with multiplicity k there also correspond k distinct homogeneous solutions, namely

$$e^{st}, \ te^{st}, \ \ldots, \ t^{k-1}e^{st}.$$

Now the sum of the multiplicities of the distinct zeros of the characteristic polynomial is equal to the degree m of the polynomial, which is the order of the differential equation. So in this way there is a total of m distinct homogeneous solutions that correspond to the zeros of the characteristic polynomial. We call these solutions

Fundamental homogeneous solution

the *fundamental homogeneous solutions*, since it follows from the theory of ordinary differential equations with constant coefficients that the general homogeneous solution can be written as a linear combination of the homogeneous solutions corresponding to the zeros of the characteristic polynomial.

To a zero $i\omega$ on the imaginary axis corresponds the time-harmonic fundamental solution $e^{i\omega t}$ of (5.5) with period $T = 2\pi/|\omega|$ if $\omega \neq 0$ and an arbitrary period

Eigenfrequency

if $\omega = 0$. In this case the value ω is called an *eigenfrequency*. Note that a time-harmonic fundamental solution with period T also has period $2T, 3T$, etc. So when there are zeros of the characteristic polynomial on the imaginary axis, then there exist non-trivial periodic eigenfunctions. One can show that the converse is also true, that is: if a non-trivial periodic eigenfunction with period T exists, then the characteristic polynomial has a zero lying on the imaginary axis and corresponding to this a time-harmonic fundamental solution having period T as well.

Hence, when $u(t)$ is a periodic input with period T and when, moreover, there exist non-trivial eigenfunctions having this same period, then a periodic solution of equation (5.3) with period T will certainly not be unique. Possibly, periodic solutions of (5.3) will not even exist. We will illustrate this in the next example.

EXAMPLE 5.2

Given is the differential equation

$$y'' + 4y = u.$$

The characteristic equation is $s^2 + 4 = 0$ and it has zeros $2i$ and $-2i$ on the imaginary axis. To these correspond the fundamental solutions e^{2it} and e^{-2it}. Hence, the general homogeneous solution is $x(t) = \alpha e^{2it} + \beta e^{-2it}$ for arbitrary complex α and β. So all eigenfunctions are periodic with period π. Now when $u(t) = 4\cos 2t$ is a periodic input with period π, then there is no periodic solution with period π. This is because one can show by substitution that $y(t) = t \sin 2t$ is a particular solution of the given differential equation for this u. Note that this solution is not bounded. The general solution is thus

$$y(t) = t \sin 2t + \alpha e^{2it} + \beta e^{-2it}.$$

Since all homogeneous solutions are bounded, while the particular solution is not, none of the solutions of the differential equation will be bounded, let alone periodic. The periodic input $u(t)$ gives rise to unbounded solutions here. This phenomenon is

Resonance

called *resonance*.   ◀

The preceding discussion has given us some insight into the problems that may arise when the characteristic polynomial has zeros on the imaginary axis. When eigenfunctions with period T occur, periodic inputs with period T can cause resonance. In the case when there are no eigenfunctions with period T, the theory of ordinary differential equations with constant coefficients states that for each periodic $u(t)$ in (5.3), having period T and an nth derivative which is piecewise continuous, there exists a periodic solution $y(t)$ with period T as well. This solution is then uniquely determined, since there are no periodic homogeneous solutions with period T. The solution $y(t)$ can then be determined using the frequency response. We will illustrate this in our next example.

EXAMPLE 5.3

Consider the differential equation

$$y'' + 3y' + 2y = \cos t.$$

Note that the right-hand side is periodic with period 2π. The corresponding homogeneous differential equation is

$$x'' + 3x' + 2x = 0.$$

The characteristic equation is $s^2 + 3s + 2 = 0$ and has zeros $s = -1$ and $s = -2$. There are thus no zeros on the imaginary axis and hence there are no periodic eigenfunctions, let alone periodic eigenfunctions with period 2π. As a consequence, there is exactly one periodic solution $y(t)$ with period 2π. This can be determined as follows. We consider the differential equation as an LTC-system with input $u(t) = \cos t$. Applying (5.4) gives the frequency response of the system:

$$H(\omega) = \frac{1}{2 + 3i\omega - \omega^2}.$$

Since $u(t) = (e^{it} + e^{-it})/2$, it is a linear combination of time-harmonic signals whose response can be found using the frequency response calculated above: $e^{it} \mapsto H(1)e^{it}$ and $e^{-it} \mapsto H(-1)e^{-it}$. Hence,

$$\cos t \mapsto \frac{H(1)e^{it} + H(-1)e^{-it}}{2} = \frac{e^{it}}{2(1+3i)} + \frac{e^{-it}}{2(1-3i)}$$
$$= \frac{1}{10}(\cos t + 3\sin t).$$

◀

In general we will consider *stable* systems. When a stable system is described by a differential equation of type (5.3), then one can prove (see chapter 10) that the real parts of the zeros of the characteristic polynomial are always *negative*. In other words, the zeros lie in the left-half plane of the complex plane and hence, none of the zeros lie on the imaginary axis. Then $A(i\omega) \neq 0$ for all ω and on the basis of theorem 5.2 $H(\omega)$ then exists for all ω. Using theorem 5.1 one can thus determine the line spectrum of the response for any periodic input.

Systems like electrical networks and mechanical systems are described by differential equations of the form (5.3) with the coefficients $a_0, a_1, \ldots, a_m$ and $b_0, b_1, \ldots, b_n$ being real numbers. The response to a real input is then also real. In chapter 1, section 1.3.4, we have called these systems *real systems*. A sinusoidal signal $A \cos(\omega t + \phi_0)$ with amplitude A and initial phase ϕ_0, for which the frequency ω is equal to an eigenfrequency of the system is then an eigenfunction. If the sinusoidal signal is not an eigenfunction, then $A(i\omega) \neq 0$ and so $H(\omega)$ exists. In section 1.3.4 we have seen that the response of a real system to the sinusoidal signal then equals

$$A \,|\, H(\omega) \,|\cos(\omega t + \phi_0 + \Phi(\omega)).$$

The amplitude is distorted with the factor $|\, H(\omega) \,|$, which is the amplitude response of the system, and the phase is shifted over $\Phi(\omega) = \arg H(\omega)$, which is the phase response of the system. Now an arbitrary piecewise smooth, real and periodic input can be written as a superposition of these sinusoidal signals:

$$u(t) = \sum_{n=0}^{\infty} A_n \cos(n\omega_0 t + \phi_n).$$

On the basis of the superposition rule, the response $y(t)$ of a stable LTC-system is then equal to

$$y(t) = \sum_{n=0}^{\infty} A_n \,|\, H(n\omega_0) \,|\cos(n\omega_0 t + \phi_n + \Phi(n\omega_0)).$$

Depending on the various frequencies contained in the input, amplitude distortions and phase shifts will occur.

If $H(n\omega_0) = 0$ for certain values of n, then we will say that the frequency $n\omega_0$ is blocked by, or will not pass, the system. Designing electrical networks with frequency responses that meet specific demands, or, put differently, designing *filters*, is an important part of network theory. Examples are low-pass filters, blocking almost all frequencies above a certain limit, high-pass filters, blocking almost all frequencies below a certain limit, or more general, band-pass filters, through which only frequencies in a specific band will pass.

Filter

We close this section with some examples.

EXAMPLE 5.4

In figure 5.1 an electrical network is drawn, which is considered as an LTC-system with input the voltage across the voltage generator and output the voltage across the

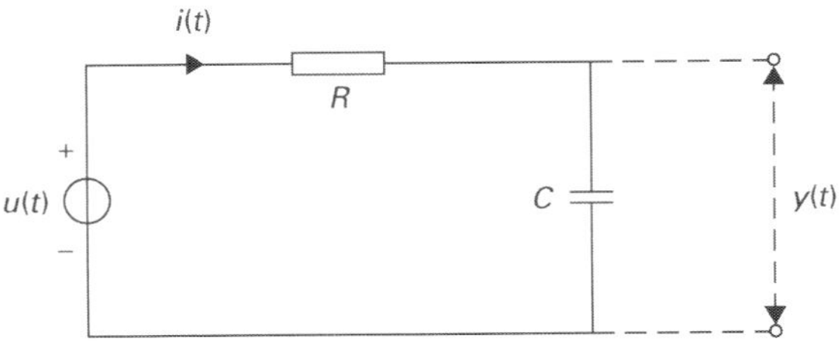

FIGURE 5.1
An RC-network.

capacitor. The corresponding differential equation reads as follows:

$$RCy'(t) + y(t) = u(t).$$

The characteristic equation $RCs + 1 = 0$ has one negative zero. The system is stable and has frequency response

$$H(\omega) = \frac{1}{i\omega RC + 1}.$$

Now let the sinusoidal input with frequency ω and initial phase ϕ_0 be given by $u(t) = a\cos(\omega t + \phi_0)$. In order to determine the response, we need the modulus and argument of the frequency response. Using the notation $RC = \tau$ one obtains that

$$|H(\omega)| = \left| \frac{1}{i\omega\tau + 1} \right| = \frac{1}{\sqrt{1 + \omega^2\tau^2}},$$

$$\arg(H(\omega)) = -\arg(i\omega\tau + 1) = -\arctan(\omega\tau).$$

The response of the system to $u(t)$ is thus

$$y(t) = \frac{a}{\sqrt{1 + \omega^2\tau^2}}\cos(\omega t - \arctan(\omega\tau) + \phi_0).$$

As we can see, the amplitude becomes small for large ω. One could consider this network as a low-pass filter.   ◄

EXAMPLE 5.5   In figure 5.2 a simple mechanical mass–spring system is drawn. An external force,

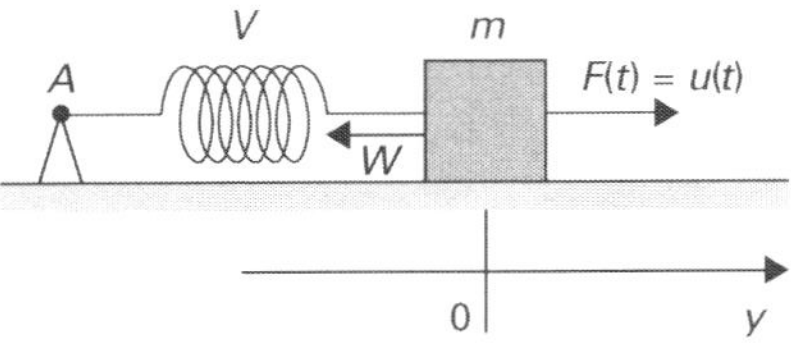

FIGURE 5.2
A simple mechanical mass–spring system.

the input $u(t)$, acts on a mass m, which can move in horizontal direction to the left and to the right over a horizontal plane. A spring V connects m with a fixed point A. Furthermore, a frictional force W acts on m. The displacement $y(t)$ is considered as output. The force that the spring exerts upon the mass m equals $-cy(t)$ (c is the spring constant). The frictional force is equal to $-\alpha y'(t)$. The corresponding differential equation is then

$$u(t) = my''(t) + \alpha y'(t) + cy(t),$$

having as characteristic equation $ms^2 + \alpha s + c = 0$. For $\alpha > 0$, the roots are real and negative or they have a negative real part $-\alpha/2m$. We are then dealing with a stable system. If, however, $\alpha = 0$, then the system has an eigenfrequency $\omega_r = \sqrt{c/m}$ and a periodic eigenfunction with frequency ω_r, and so the system is no longer stable. The response to the periodic input $\cos(\omega_r t)$ is not periodic. Resonance will then occur. ◀

As a final example we treat an application of Parseval's identity for periodic functions.

EXAMPLE 5.6

The electric network from figure 5.3 is considered as an LTC-system with input the voltage $u(t)$ across the voltage generator and output the voltage $y(t)$ across the resistor. In this network the quantities C, L, R satisfy the relation $R = \sqrt{L/C}$. The

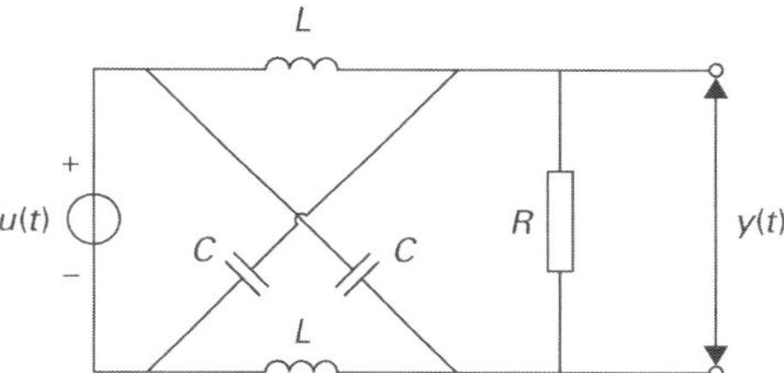

FIGURE 5.3
Electric network from example 5.6.

relation between the input $u(t)$ and the corresponding output $y(t)$ is given by

$$y'' - (2/RC)y' + (1/LC)y = u'' - (1/LC)u.$$

The frequency response follows immediately from (5.4). If we put $\alpha = 1/RC$, then it follows from $R = \sqrt{L/C}$ that $\alpha^2 = 1/LC$ and so

$$H(\omega) = \frac{(i\omega)^2 - \alpha^2}{(i\omega)^2 - 2i\alpha\omega + \alpha^2} = \frac{i\omega + \alpha}{i\omega - \alpha}.$$

If $u(t)$ is a periodic input with period T and line spectrum u_n, then, owing to $|a + ib| = |a - ib|$, the amplitude spectrum $|y_n|$ of the output is equal to

$$|y_n| = |H(n\omega_0)u_n| = \left| \frac{in\omega_0 + \alpha}{in\omega_0 - \alpha} u_n \right| = \left| \frac{in\omega_0 + \alpha}{in\omega_0 - \alpha} \right| |u_n| = |u_n|.$$

Apparently, the amplitude spectrum is not altered by the system. This then has consequences for the power of the output. Applying Parseval's identity for periodic functions, we can calculate the power P of $y(t)$ as follows:

$$P = \frac{1}{T} \int_0^T |y(t)|^2 \, dt = \sum_{n=-\infty}^{\infty} |y_n|^2 = \sum_{n=-\infty}^{\infty} |u_n|^2 = \frac{1}{T} \int_0^T |u(t)|^2 \, dt.$$

We see that the power of the output equals the power of the input. Systems having

All-pass system

this property are called *all-pass systems* (see also chapter 10). ◀

EXERCISES

5.1   A stable LTC-system is described by a differential equation of the form (5.3). Let $x(t)$ be an eigenfunction of the system. Show that $\lim_{t \to \infty} x(t) = 0$.

5.2   For an LTC-system the relation between an input $u(t)$ and the corresponding response $y(t)$ is described by $y' + y = u$. Let $u(t)$ be the periodic input with period 2π, which is given on the interval $(-\pi, \pi)$ by

$$u(t) = \begin{cases} 1 & \text{for } |t| < \pi/2, \\ 0 & \text{for } |t| > \pi/2. \end{cases}$$

Calculate the line spectrum of the output.

5.3   For an LTC-system the frequency response $H(\omega)$ is given by

$$H(\omega) = \begin{cases} 1 & \text{for } |\omega| \leq \pi, \\ 0 & \text{for } |\omega| > \pi. \end{cases}$$

a   Can the system be described by an ordinary differential equation of the form (5.3)? Justify your answer.
b   We apply the periodic signal of exercise 5.2 to the system. Calculate the power of the response.

5.4   To the network of example 5.6 we apply the signal $u(t) = |\sin t|$ as input. Calculate the integral $(1/\pi) \int_0^\pi y(t)\, dt$ of the corresponding output $y(t)$. This is the average value of $y(t)$ over one period.

5.5   For an LTC-system the relation between an input $u(t)$ and the output $y(t)$ is described by the differential equation

$$y'' + 2y' + 4y = u'' + u.$$

a   Which frequencies do not pass through the system?
b   Calculate the response to the input $u(t) = \sin t + \cos^2 2t$.

5.6   Given is the following differential equation:

$$y'' + \omega_0^2 y = u \qquad \text{with } |\omega_0| \neq 0, 1, 2, \ldots.$$

Here $u(t)$ is the periodic function with period 2π given by

$$u(t) = \begin{cases} t + \pi & \text{for } -\pi < t < 0, \\ -t + \pi & \text{for } 0 < t < \pi. \end{cases}$$

Does the differential equation have a unique periodic solution $y(t)$ with period 2π? If so, determine its line spectrum.

5.2   Partial differential equations

In this section we will see how Fourier series can be applied in solving partial differential equations. For this, we will introduce a method which will be explained systematically by using a number of examples wherein functions $u(x, t)$ occur, depending on a time-variable t and *one* position-variable x. However, this method can also be applied to problems with two, and often also three or four, position-variables. Here we will confine ourselves to the simple examples of the one-dimensional heat equation and the one-dimensional wave equation.

5.2.1　　The heat equation

In Fourier's time, around 1800, heat conduction was already a widely studied phenomenon, from a practical as well as from a scientific point of view. In the industry the phenomenon was important in the use of metals for machines, while in science heat conduction was an issue in determining the temperature of the earth's interior, in particular its variations in the course of time. The first problem that Fourier (1761 - 1830) addressed in his book *Théorie analytique de la chaleur* from 1822 was the determination of the temperature T in a solid as function of the position variables x, y, z and the time t. From physical principles he showed that the temperature $T(x, y, z, t)$ should satisfy the partial differential equation

$$\frac{\partial T}{\partial t} = k\left(\frac{\partial^2 T}{\partial x^2} + \frac{\partial^2 T}{\partial y^2} + \frac{\partial^2 T}{\partial z^2}\right).$$

Heat equation

Diffusion equation

Here k is a constant, whose value depends on the material of the solid. This equation is called the *heat equation*. The same equation also plays a role in the diffusion of gases and liquids. In that case the function $T(x, y, z, t)$ does not represent temperature, but the local concentration of the diffusing substances in a medium where the diffusion takes place. The constant is in that case the diffusion coefficient and the equation is then called the *diffusion equation*. We now look at the equation in a simplified situation.

EXAMPLE 5.7

Consider a thin rod of length L with a cylinder shaped cross-section and flat ends. The ends are kept at a temperature of $0°\,\mathrm{C}$ by cooling elements, while the side-surface (the mantle of the cylinder) is insulated, so that no heat flows through it. At time $t = 0$ the temperature distribution in the longitudinal direction of the rod (the x-direction; see figure 5.4) is given by a function $f(x)$. So for fixed value of x, the temperature in a cross-section of the rod is the same everywhere. This justifies a description of the problem using only x as a position variable. The variables y and z can be omitted, and so we can consider the temperature as function of x and t only: $T = T(x, t)$. The preceding equation then changes into

$$\frac{\partial T}{\partial t} = k\frac{\partial^2 T}{\partial x^2} \qquad \text{for } 0 < x < L \text{ and } t > 0.$$

We call this partial differential equation the one-dimensional heat equation.

In the remainder of this chapter we will denote the function that should satisfy a partial differential equation by $u(x, t)$. For the partial derivatives we introduce the following frequently used notation:

$$\frac{\partial u}{\partial x} = u_x, \qquad \frac{\partial^2 u}{\partial x^2} = u_{xx}, \qquad \frac{\partial u}{\partial t} = u_t, \qquad \frac{\partial^2 u}{\partial t^2} = u_{tt}. \tag{5.7}$$

With this notation the one-dimensional heat equation looks like this:

$$u_t = ku_{xx} \qquad \text{for } 0 < x < L \text{ and } t > 0. \tag{5.8}$$

Boundary condition

Since the temperature at both ends is kept at $0°\,\mathrm{C}$ for all time, starting from $t = 0$, one has the following two *boundary conditions*:

$$u(0, t) = 0, \qquad u(L, t) = 0 \qquad \text{for } t \geq 0. \tag{5.9}$$

Initial condition

Finally we formulate the situation for $t = 0$ as an *initial condition*:

$$u(x, 0) = f(x) \qquad \text{for } 0 \leq x \leq L. \tag{5.10}$$

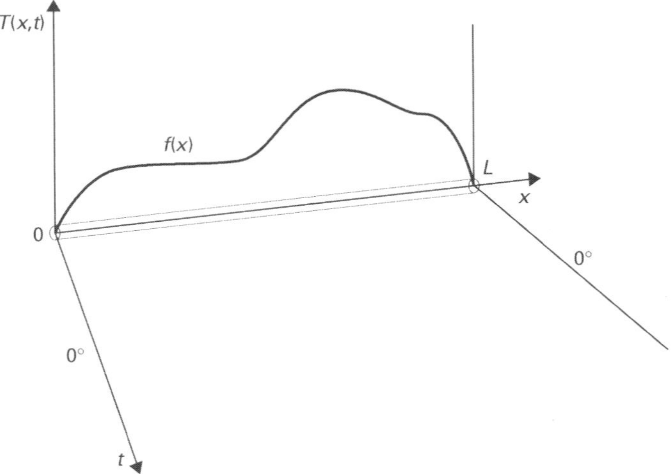

FIGURE 5.4
Thin rod with temperature distribution $f(x)$ at $t = 0$.

Here $f(x)$ is a piecewise smooth function. This situation is shown in figure 5.4. The partial differential equation (5.8) is an example of a linear homogeneous equation. That is to say, when two functions $u(x, t)$ and $v(x, t)$ satisfy this equation, then so does any linear combination of these two functions. In particular the null-function satisfies the equation. The boundary conditions (5.9) have the same property (verify this). This is why these conditions are also called *linear homogeneous conditions*. Constructing a solution of equation (5.8) satisfying conditions (5.9) and (5.10) will consist of three steps. In the first two steps we will be using separation of variables to construct a collection of functions which satisfy (5.8) as well as the linear homogeneous conditions (5.9). To this end we must solve a so-called eigenvalue problem, which will take place in the second step. Next we construct in the third step, by means of an infinite linear combination of the functions from this collection, or, put differently, by *superposition*, a solution which also satisfies the inhomogeneous condition (5.10). In this final step the Fourier series enter, and so we will have to deal with all kinds of convergence problems. If we ignore these problems during the construction, then it is said that we have obtained a *formal solution*, for which, in fact, one still has to show that it actually is a solution, or even a unique solution.

Separation of variables
Using separation of variables we will construct a collection of functions satisfying the partial differential equation (5.8) and the linear homogeneous conditions (5.9) and having the form

$$u(x, t) = X(x)T(t), \tag{5.11}$$

where $X(x)$ is a function of x only and $T(t)$ is a function of t only. If we substitute (5.11) into (5.8), then we obtain the relation

$$X(x)T'(t) = kX''(x)T(t).$$

After a division by $kX(x)T(t)$ the variables x and t are separated:

$$\frac{T'(t)}{kT(t)} = \frac{X''(x)}{X(x)}.$$

Here the left-hand side is a function of t and independent of x, while the right-hand side is a function of x and independent of t. Therefore, an equality can only occur if both sides are independent of x and t, and hence are equal to a constant. Call this constant c, then

$$\frac{T'(t)}{kT(t)} = \frac{X''(x)}{X(x)} = c.$$

The constant c is sometimes called the constant of separation. We thus have

$$X''(x) - cX(x) = 0, \tag{5.12}$$

$$T'(t) - ckT(t) = 0. \tag{5.13}$$

Substitution of (5.11) into the linear homogeneous boundary conditions (5.9) gives $u(0, t) = X(0)T(t) = 0$ and $u(L, t) = X(L)T(t) = 0$ for $t \geq 0$. We are not interested in the trivial solution $T(t) = 0$ and so $X(0) = 0$ and $X(L) = 0$. Together with (5.12) this leads to the problem

$$X'' - cX = 0 \qquad \text{for } 0 < x < L, \, X(0) = 0, \, X(L) = 0, \tag{5.14}$$

where $X(x)$ and c are to be determined. First we will solve problem (5.14) and subsequently (5.13). Problem (5.14) obviously has the trivial solution $X(x) = 0$, which is of no interest. We are therefore interested in those values of c for which there exists a non-trivial solution $X(x)$. These values are called *eigenvalues* and the corresponding non-trivial solutions *eigenfunctions*. Determining the eigenvalues and their corresponding eigenfunctions is the second step in our solution method.

Calculating eigenvalues and eigenfunctions
When we try to solve problem (5.14) we have to distinguish two cases, namely $c \neq 0$ and $c = 0$.
a For $c = 0$ equation (5.14) becomes $X'' = 0$, which has general solution $X(x) = \alpha x + \beta$. From the boundary conditions it follows that $X(0) = \beta = 0$ and $X(L) = \alpha L = 0$. Hence, $\beta = \alpha = 0$. We then obtain the trivial solution and this means that $c = 0$ is not an eigenvalue.
b For $c \neq 0$ the characteristic equation $s^2 - c = 0$ corresponding to (5.14) has two distinct roots s_1 and s_2 with $s_2 = -s_1$. Note that these roots may be complex. The general solution is then

$$X(x) = \alpha e^{s_1 x} + \beta e^{-s_1 x}.$$

The first boundary condition $X(0) = 0$ gives $\alpha + \beta = 0$, so $\beta = -\alpha$. Next we obtain from the second boundary condition $X(L) = 0$ the equation

$$\alpha(e^{s_1 L} - e^{-s_1 L}) = 0.$$

For $\alpha = 0$ we get the trivial solution again. So we must have $e^{s_1 L} - e^{-s_1 L} = 0$, implying that $e^{2s_1 L} = 1$. From this it follows that $s_1 = in\pi/L$, where n is an integer and $n \neq 0$. This gives us eigenvalues $c = s_1^2 = -(n\pi/L)^2$. The corresponding eigenfunction $X(x)$ is $X(x) = 2i\alpha \sin(n\pi x/L)$. However, since α is arbitrary, we can say that eigenvalue $c = -(n\pi/L)^2$ corresponds to the eigenfunction

$$X_n(x) = \sin(n\pi x/L),$$

where we may now assume that n is a positive integer.

For $c = -(n\pi/L)^2$ the first-order differential equation (5.13) has the fundamental solution

$$T_n(t) = e^{-(n\pi/L)^2 kt}.$$

Eigenvalue
Eigenfunction

Step 2

Fundamental solution

We have thus found the following collection of *fundamental solutions* satisfying equation (5.8) and conditions (5.9):

$$u_n(x,t) = T_n(t)X_n(x) = e^{-n^2\pi^2 kt/L^2}\sin(n\pi x/L) \quad \text{for } n = 1,2,\ldots. \tag{5.15}$$

Step 3

Superposition of fundamental solutions

Since (5.8) is a linear homogeneous differential equation and (5.9) are linear homogeneous conditions, it is possible to take linear combinations of fundamental solutions, or, as it is often put by scientists, to form new solutions of (5.8) and (5.9) by *superpositions*. Each finite linear combination $a_1 u_1 + a_2 u_2 + \cdots + a_n u_n$ of fundamental solutions also satisfies (5.8) and (5.9). However, in general we cannot expect that a suitable finite linear combination will give us a solution which also satisfies the remaining inhomogeneous condition. We therefore try an *infinite* linear combination of fundamental solutions, still called a superposition of fundamental solutions. It has the form

$$u(x,t) = \sum_{n=1}^{\infty} A_n e^{-n^2\pi^2 kt/L^2}\sin(n\pi x/L). \tag{5.16}$$

If this superposition is to satisfy the inhomogeneous condition (5.10), then $u(x,t)$ should be equal to the function $f(x)$ for $t = 0$ and hence

$$u(x,0) = \sum_{n=1}^{\infty} A_n \sin(n\pi x/L) = f(x) \qquad \text{for } 0 \le x \le L.$$

The coefficients A_n can thus be found by determining the Fourier sine series of $f(x)$. The result is

$$A_n = \frac{2}{L}\int_0^L f(x)\sin(n\pi x/L)\,dx.$$

Substitution of these coefficients in (5.16) finally gives us a formal solution of the heat conduction problem. Since we required $f(x)$ to be piecewise smooth, and we also assumed that $f(x)$ equals the average value of the left- and right-hand limit at jumps, $f(x)$ is equal to the sum of its Fourier sine series on the interval $[0, L]$. Hence, $u(x,0) = f(x)$. It it also easy to see that for $x = 0$ and $x = L$ the sum of the series in (5.16) equals 0, since all terms are 0 then. The homogeneous conditions are thus also satisfied. To show, however, that the series (5.16) also converges for other values of x and $t > 0$, and that its sum $u(x,t)$ satisfies differential equation (5.8), requires a detailed analysis of the convergence of the series in (5.16). We will content ourselves here with stating that in the case when $f(x)$ is piecewise smooth, one can prove that $u(x,t)$ found in this way is indeed a solution of the heat conduction problem that we have posed, and even that it is the *unique* solution. ◀

In the preceding example the temperature at both ends of the rod was kept at $0°$ C. This resulted in linear homogeneous conditions for the heat conduction problem. We will now look at what happens with the temperature of a rod whose ends are insulated.

EXAMPLE 5.8

Insulation of the ends can be expressed mathematically as $u_x(0,t) = 0$ and $u_x(L,t) = 0$. The heat conduction problem for this rod is thus as follows:

$$\begin{aligned} u_t &= ku_{xx} & &\text{for } 0 < x < L, t > 0, \\ u_x(0,t) &= 0, \quad u_x(L,t) = 0 & &\text{for } t \ge 0, \\ u(x,0) &= f(x) & &\text{for } 0 \le x \le L. \end{aligned}$$

Going through the three steps from example 5.7 again, we find a difference in step 1: instead of problem (5.14) we obtain the problem

$$X'' - cX = 0 \qquad \text{for } 0 \le x \le L,$$
$$X'(0) = 0, \quad X'(L) = 0.$$

For $c = 0$ we find that $X(x)$ is a constant. So now $c = 0$ is an eigenvalue as well, with eigenfunction a constant. For $c \ne 0$ we again find the eigenvalues $c = -(n\pi/L)^2$ with corresponding eigenfunctions $X_n(x) = \cos(n\pi x/L)$ for $n = 1, 2, \ldots$. Hence, the eigenfunctions are

$$X_n(x) = \cos(n\pi x/L) \qquad \text{for } n = 0, 1, 2, 3, \ldots.$$

Note that n now starts from $n = 0$. The remainder of the construction of the collection of fundamental solutions is entirely analogous to the previous example. The result is the collection of fundamental solutions

$$u_n(x, t) = e^{-n^2\pi^2 kt/L^2} \cos(n\pi x/L) \qquad \text{for } n = 0, 1, 2, \ldots.$$

Superposition of the fundamental solutions in step 3 gives

$$u(x, t) = \tfrac{1}{2}A_0 + \sum_{n=1}^{\infty} A_n e^{-n^2\pi^2 kt/L^2} \cos(n\pi x/L).$$

Since $u(x, 0) = f(x)$ for $0 \le x \le L$, the coefficients A_n are equal to

$$A_n = \frac{2}{L} \int_0^L f(x) \cos(n\pi x/L)\,dx \qquad \text{for } n = 0, 1, 2, 3, \ldots.$$

These Fourier coefficients arise by determining the Fourier cosine series of $f(x)$. Substitution of these coefficients in the series for $u(x, t)$ then gives a formal solution again. Since $f(x)$ is piecewise smooth, one can prove once more that this is a unique solution. ◀

If we look at the solution for $t \to \infty$, then all terms in the series for $u(x, t)$ tend to 0, except for the term $A_0/2$. One can indeed prove that $\lim_{t \to \infty} u(x, t) = A_0/2$. It is then said that in the *stationary phase* the temperature no longer depends on t and is everywhere equal to the average temperature

$$\frac{A_0}{2} = \frac{1}{L} \int_0^L f(x)\,dx$$

at $t = 0$ on the interval $[0, L]$. Figure 5.5 illustrates this result.

5.2.2 *The wave equation*

We now consider the example of a vibrating string of length L and having fixed ends. Just as in the heat conduction problem, we will not discuss the physical arguments needed to derive the wave equation.

EXAMPLE 5.9

The equation describing the vertical displacement $u(x, t)$ of a vibrating string is

$$u_{tt} = a^2 u_{xx} \qquad \text{for } 0 < x < L, t > 0. \tag{5.17}$$

Wave equation

Here a is a constant which is related to the tension in the string. This equation is called the *wave equation*. Since the ends of the string are fixed, one has the following boundary conditions:

$$u(0, t) = 0, \qquad u(L, t) = 0 \qquad \text{for } t \ge 0. \tag{5.18}$$

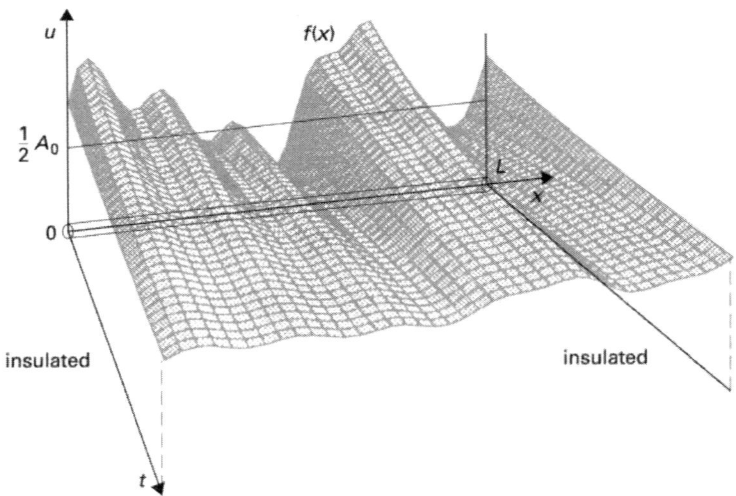

FIGURE 5.5
Temperature distribution in a rod with insulated ends.

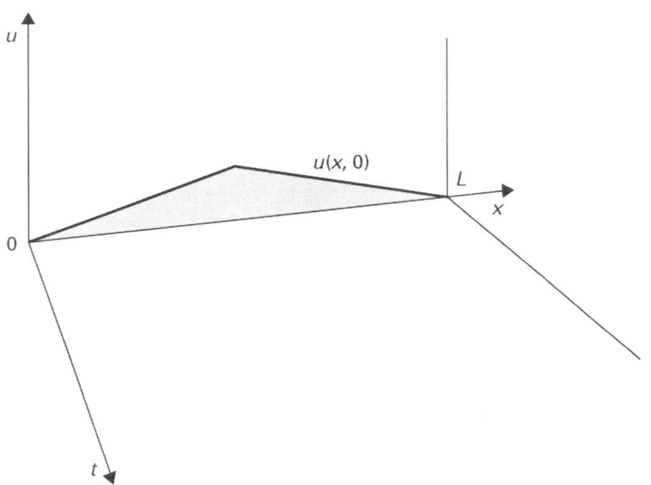

FIGURE 5.6
Displacement of a vibrating string.

In figure 5.6 we see the displacement of the string at time $t = 0$. We thus have the initial condition

$$u(x, 0) = f(x) \qquad \text{for } 0 \le x \le L. \tag{5.19}$$

Moreover, it is given that at $t = 0$ the string has no initial velocity. So as a second initial condition we have

$$u_t(x, 0) = 0 \qquad \text{for } 0 \le x \le L. \tag{5.20}$$

Analogous to the heat conduction problem in a rod we will derive a formal solution in three steps.

Step 1

Separation of variables
We first determine fundamental solutions of the form $u(x, t) = X(x)T(t)$ that satisfy the partial differential equation and the linear homogeneous conditions. Substitution of $u(x, t) = X(x)T(t)$ into (5.17) gives $XT'' = a^2 X''T$. Dividing by $a^2 XT$ leads to

$$\frac{T''}{a^2 T} = \frac{X''}{X} = c,$$

where c is the constant of separation. From this we obtain the two ordinary differential equations

$$X'' - cX = 0, \qquad T'' - ca^2 T = 0. \tag{5.21}$$

Conditions (5.18) and (5.20) are linear homogeneous conditions. Substitution of $u(x, t) = X(x)T(t)$ in condition (5.18), and using that $u(x, t)$ should not be the trivial solution, leads to the conditions $X(0) = 0$ and $X(L) = 0$. Subsequently substituting $u(x, t) = X(x)T(t)$ in condition (5.20) gives the relation $T'(0)X(x) = 0$ for $0 \leq x \leq L$. Hence $T'(0) = 0$.

Step 2

Calculating eigenvalues and eigenfunctions
For the function $X(x)$ we derived in step 1 the following differential equation with boundary conditions:

$$X'' - cX = 0 \qquad \text{for } 0 < x < L,$$
$$X(0) = 0, \qquad X(L) = 0.$$

So here we again find the eigenvalues $c = -(n\pi/L)^2$ for $n = 1, 2, \ldots$ and the corresponding eigenfunctions

$$X_n(x) = \sin(n\pi x/L) \qquad \text{for } n = 1, 2, 3, \ldots.$$

In order to find the fundamental solutions we still have to determine $T(t)$. From the eigenvalues we have found, we obtain from (5.21) the differential equation

$$T'' + \frac{n^2 \pi^2 a^2}{L^2} T = 0.$$

Its general solution is

$$T(t) = \alpha \cos(n\pi at/L) + \beta \sin(n\pi at/L),$$

which has as derivative

$$T'(t) = \frac{n\pi\alpha}{L} \left(-\alpha \sin(n\pi at/L) + \beta \cos(n\pi at/L) \right).$$

Substitution of the condition $T'(0) = 0$ gives $\beta = 0$ and so we obtain for the eigenvalue $-(n\pi/L)^2$ the following fundamental solution:

$$T_n(t) = \cos(n\pi at/L) \qquad \text{for } n = 1, 2, 3, \ldots.$$

We have thus found the following fundamental solutions:

$$u_n(x, t) = T_n(t)X_n(x) = \cos(n\pi at/L) \sin(n\pi x/L) \qquad \text{for } n = 1, 2, 3, \ldots.$$

Step 3

Superposition
Superposition of the fundamental solutions gives the series

$$u(x, t) = \sum_{n=1}^{\infty} A_n \cos(n\pi a t/L) \sin(n\pi x/L). \tag{5.22}$$

By substituting the remaining initial condition (5.19), a Fourier series arises:

$$u(x, 0) = f(x) = \sum_{n=1}^{\infty} A_n \sin(n\pi x/L) \qquad \text{for } 0 \le x \le L.$$

For the Fourier coefficients A_n one thus has

$$A_n = \frac{2}{L} \int_0^L f(x) \sin(n\pi x/L)\, dx.$$

The series (5.22) together with these coefficients give the formal solution of our problem. One can indeed show that the formal solution thus obtained is the *unique* solution, provided that $f(x)$ is piecewise smooth.   ◄

In many cases the assumption of an initial velocity $u_t(x, 0) = 0$ is artificial. This assumption made it possible for us to find a simple solution for $T_n(t)$. For a string which is struck from its resting position, one takes as initial conditions $u(x, 0) = 0$ and $u_t(x, 0) = g(x)$. When both the initial displacement and the initial velocity are unequal to 0, then we are dealing with a problem with *two* inhomogeneous conditions. As a consequence, the functions $T_n(t)$ will contain sine as well as cosine terms. We must then determine the coefficients in two distinct Fourier series. For detailed results we refer to *Fourier series, transforms and boundary value problems* by J. Hanna and J.H. Rowland, pages 228 – 233. In the same book, pages 219 – 227, one can also find the derivation of the wave equation and the heat equation, as well as the verification that the formal solutions constructed above are indeed solutions, which moreover are unique.

EXERCISES

5.7   A thin rod of length L with insulated sides has its ends kept at $0°$ C. The initial temperature is

$$u(x, 0) = \begin{cases} x & \text{for } 0 \le x \le L/2, \\ L - x & \text{for } L/2 \le x \le L. \end{cases}$$

Show that the temperature $u(x, t)$ is given by the series

$$u(x, t) = \frac{4L}{\pi^2} \sum_{n=0}^{\infty} \frac{(-1)^n}{(2n+1)^2} e^{-(2n+1)^2 \pi^2 k t/L^2} \sin((2n+1)\pi x/L).$$

5.8   Both ends and the sides of a thin rod of length L are insulated. The initial temperature of the rod is $u(x, 0) = 3\cos(8\pi x/L)$. Write down the heat equation for this situation and determine the initial and boundary conditions. Next determine the temperature $u(x, t)$.

5.9   For a thin rod of length L the end at $x = L$ is kept at $0°$ C, while the end at $x = 0$ is insulated (as well as the sides). The initial temperature of the rod is $u(x, 0) = 7\cos(5\pi x/2L)$. Write down the heat equation for this situation and determine the initial and boundary conditions. Next determine the temperature $u(x, t)$.

5.10   Determine the solution of the following initial and boundary value problem:

$$u_t = u_{xx} \qquad \text{for } 0 < x < 2, t > 0,$$
$$u_x(0, t) = 0, \quad u(2, t) = 0 \quad \text{for } t \geq 0,$$
$$u(x, 0) = \begin{cases} 1 & \text{for } 0 < x < 1, \\ 2 - x & \text{for } 1 \leq x < 2. \end{cases}$$

5.11   A thin rod of length L has initial temperature $u(x, 0) = f(x)$. The end at $x = 0$ is kept at $0°\,$C and the end at $x = L$ is insulated (as well as the sides). Write down the heat equation for this situation and determine the initial and boundary conditions. Next determine the temperature $u(x, t)$.

5.12   Determine the displacement $u(x, t)$ of a string of length L, with fixed ends and initial displacement $u(x, 0) = 0.05 \sin(4\pi x/L)$. At time $t = 0$ the string has no initial velocity.

5.13   A string is attached at the points $x = 0$ and $x = L$ and has as initial displacement

$$f(x) = \begin{cases} 0.02x & \text{for } 0 < x < L/2, \\ 0.02(L - x) & \text{for } L/2 \leq x < L. \end{cases}$$

At time $t = 0$ the string has no initial velocity. Write down the corresponding initial and boundary value problem and determine the solution. One could call this the problem of the 'plucked string': the initial *position* is unequal to 0 and the string is pulled at the point $x = L/2$, while the initial *velocity* is equal to 0.

5.14   A string is attached at the points $x = 0$ and $x = 2$ and has as initial displacement $u(x, 0) = 0$. The initial velocity is

$$u_t(x, 0) = g(x) = \begin{cases} 0.05x & \text{for } 0 < x < 1, \\ 0.05(2 - x) & \text{for } 1 < x < 2. \end{cases}$$

Write down the corresponding initial and boundary value problem and determine the solution. This problem could be called the problem of the 'struck string': the initial *position* is equal to 0, while the initial *velocity* is unequal to 0, and the string is struck at the midpoint.

5.15   Determine the solution of the following initial and boundary value problem, where k is a constant:

$$u_t = a^2 u_{xx} \qquad \text{for } 0 < x < \pi, t > 0,$$
$$u_x(0, t) = 0, \quad u_x(\pi, t) = 0 \quad \text{for } t > 0,$$
$$u_t(x, 0) = 0, \quad u(x, 0) = kx \quad \text{for } 0 < x < \pi.$$

SUMMARY

In this chapter Fourier series were first applied to determine the response of an LTC-system to a periodic input. Here the frequency response, introduced in chapter 1, played a central role. It determines the response to a time-harmonic input. Since the input can be represented as a superposition of time-harmonic signals, using Fourier series, one can easily determine the line spectrum of the output by applying the superposition rule. This line spectrum is obtained by multiplying the line spectrum of the input with the values of the frequency response at the integer multiples of the fundamental frequency of the input:

$$y_n = H(n\omega_0)u_n.$$

Here y_n is the line spectrum of the output $y(t)$, u_n the line spectrum of the input $u(t)$, $H(\omega)$ the frequency response, and ω_0 the fundamental frequency of the input. For real inputs and real systems the properties of the time-harmonic signals are taken over by the sinusoidal signals.

Systems occurring in practice are often described by differential equations of the form

$$a_m \frac{d^m y}{dt^m} + a_{m-1} \frac{d^{m-1} y}{dt^{m-1}} + \cdots + a_0 y = b_n \frac{d^n u}{dt^n} + b_{n-1} \frac{d^{n-1} u}{dt^{n-1}} + \cdots + b_0 u.$$

In order to determine a periodic solution $y(t)$ for a given periodic signal $u(t)$, it is important to know whether or not there are any periodic eigenfunctions. These are periodic solutions of the homogeneous differential equation arising from the differential equation above by taking the right-hand side equal to 0. Periodic eigenfunctions correspond to zeros $s = i\omega$ of the characteristic polynomial $A(s) = a_m s^m + a_{m-1} s^{m-1} + \cdots + a_0$, and these lie on the imaginary axis. When the period of a periodic input coincides with the period of a periodic eigenfunction, then resonance may occur, that is, for a given $u(t)$ the differential equation does not have periodic solutions, but instead unbounded solutions.

When the differential equation describes a stable system, then all zeros of $A(s)$ lie in the left-half plane and the frequency response is then for all ω equal to

$$H(\omega) = \frac{B(i\omega)}{A(i\omega)},$$

with $B(s) = b_n s^n + b_{n-1} s^{n-1} + \cdots + b_0$. For real systems this means that there are no sinusoidal eigenfunctions, that is, no sinusoidal signals with an eigenfrequency.

Secondly, Fourier series were applied in solving the one-dimensional heat equation

$$u_t = k u_{xx},$$

and the one-dimensional wave equation

$$u_{tt} = a^2 u_{xx}.$$

Using the method of separation of variables, and solving an eigenvalue problem, one can obtain a collection of fundamental solutions satisfying the partial differential equation under consideration, as well as the corresponding linear homogeneous conditions, but not yet the remaining inhomogeneous condition(s). By superposition of the fundamental solutions one can usually construct a formal solution which also satisfies the inhomogeneous condition(s). In most cases the formal solution is the solution of the problem being posed. In the superposition of the fundamental solutions lies the application of Fourier series. The fundamental solutions describe, in relation to one or several variables, sinusoidal functions with frequencies which are an integer multiple of a fundamental frequency. This fundamental frequency already emerges when one calculates the eigenvalues. The superposition is then a Fourier series whose coefficients can be determined by using the remaining inhomogeneous condition(s).

SELFTEST

5.16   For the frequency response of an LTC-system one has

$$H(\omega) = (1 - e^{-2i\omega})^2.$$

a Is the response to a real periodic input real again? Justify your answer.
b Calculate the response to the input $u(t) = \sin(\omega_0 t)$.
c What is the response to a periodic input with period 1?

5.17 For an LTC-system the relation between an input $u(t)$ and the corresponding output $y(t)$ is described by the differential equation $y'' + 4y' + 4y = u$. Let $u(t)$ be the periodic input with period 2π, given on the interval $(-\pi, \pi)$ by

$$u(t) = \begin{cases} \pi t + t^2 & \text{for } -\pi < t < 0, \\ \pi t - t^2 & \text{for } 0 < t < \pi. \end{cases}$$

Determine the first harmonic of the output $y(t)$.

5.18 For an LTC-system the relation between an input $u(t)$ and the output $y(t)$ is described by the differential equation $y''' + y'' + 4y' + 4y = u' + u$.
a Does the differential equation determine the periodic response to a periodic input uniquely? Justify your answer.
b Let $u(t) = \cos 3t$ and $y(t)$ the corresponding output. Calculate the power of the output.

5.19 A thin rod of length L has constant positive initial temperature $u(x, 0) = u_0$ for $0 < x < L$. The ends are kept at $0°$ C. The so-called heat-flux through a cross-section of the rod at position x_0 $(0 < x_0 < L)$ and at time $t > 0$ is by definition equal to $-Ku_x(x_0, t)$. Show that the heat-flux at the midpoint of the rod $(x_0 = L/2)$ equals 0.

5.20 Consider a thin rod for which one has the following equations:

$$\begin{aligned} u_t &= ku_{xx} && \text{for } 0 < x < L, t > 0, \\ u(0, t) &= 0, \quad u(L, t) = 0 && \text{for } t \geq 0, \end{aligned}$$

$$u(x, 0) = \begin{cases} a & \text{for } 0 \leq x \leq L/2, \\ 0 & \text{for } L/2 < x \leq L, \end{cases}$$

where a is a constant.
a Determine the solution $u(x, t)$.
b Two identical iron rods, each 20 cm in length, have their ends put against each other. Both of the remaining ends, at $x = 0$ and at $x = 40$ cm, are kept at $0°$ C. The left rod has a temperature of $100°$ C and the right rod a temperature of $0°$ C. Calculate for $k = 0.15 \, \text{cm}^2\text{s}^{-1}$ the temperature at the boundary layer of the two rods, 10 minutes after the rods made contact, and show that this value is approximately $36°$ C.
c Calculate approximately how many hours it will take to reach a temperature of $36°$ C at the boundary layer, when the rods are not made of iron, but concrete $(k = 0.005 \, \text{cm}^2\text{s}^{-1})$.

5.21 Given is the following initial and boundary value problem:

$$\begin{aligned} u_{tt} &= a^2 u_{xx} && \text{for } 0 < x < L, t > 0, \\ u(0, t) &= 0, \quad u(L, t) = 0 && \text{for } t > 0, \\ u(x, 0) &= \sin(\pi x/L) && \text{for } 0 < x < L, \\ u_t(x, 0) &= 7\sin(3\pi x/L) && \text{for } 0 < x < L. \end{aligned}$$

Show that the first two steps of the method described in section 5.2 lead to the collection of fundamental solutions

$$u_n(x, t) = (A_n \sin(n\pi at/L) + B_n \cos(n\pi at/L)) \sin(n\pi x/L),$$

and subsequently determine the formal solution which is adjusted to the given initial displacement and initial velocity.

Part 3

Fourier integrals and distributions

INTRODUCTION TO PART 3

In part 2 we have developed the Fourier analysis for periodic functions. To a periodic function $f(t)$ we assigned for each $n \in \mathbb{Z}$ a Fourier coefficient $c_n \in \mathbb{C}$. Using these Fourier coefficients we then defined the Fourier series, and under certain conditions on the function f this Fourier series converged to the function f. Schematically this can be represented as follows:

periodic function $f(t)$ with period T and frequency $\omega_0 = 2\pi / T$

$\downarrow$

Fourier coefficients $c_n = \dfrac{1}{T} \displaystyle\int_{-T/2}^{T/2} f(t) e^{-in\omega_0 t}\, dt$ for $n \in \mathbb{Z}$

$\downarrow$

Fourier series $f(t) = \displaystyle\sum_{n=-\infty}^{\infty} c_n e^{in\omega_0 t}$.

Often, however, we have to deal with non-periodic phenomena. In part 3 we now set up a similar kind of Fourier analysis for these non-periodic functions. To a non-periodic function we will assign for each $\omega \in \mathbb{R}$ a number $F(\omega) \in \mathbb{C}$. Instead of a sequence of numbers c_n, we thus obtain a *function $F(\omega)$* defined on $\mathbb{R}$. The function $F(\omega)$ is called the Fourier transform of the non-periodic function $f(t)$. Next, the Fourier series is replaced by the so-called Fourier integral: instead of a sum over $n \in \mathbb{Z}$ we take the integral over $\omega \in \mathbb{R}$. As for the Fourier series, this Fourier integral will represent the original function f again, under certain conditions on f. The scheme for periodic function will in part 3 be replaced by the following scheme for non-periodic functions:

non-periodic function $f(t)$

$\downarrow$

Fourier transform $F(\omega) = \displaystyle\int_{-\infty}^{\infty} f(t) e^{-i\omega t}\, dt$ for $\omega \in \mathbb{R}$

$\downarrow$

Fourier integral $f(t) = \dfrac{1}{2\pi} \displaystyle\int_{-\infty}^{\infty} F(\omega) e^{i\omega t}\, d\omega$.

We will start chapter 6 by showing that the transition from the Fourier series to the Fourier integral can be made quite plausible by taking the limit $T \to \infty$ of the period T in the theory of the Fourier series. Although this derivation is not mathematically correct, it does result in the right formulas and in particular it will show us precisely how the Fourier transform $F(\omega)$ should be defined. Following the formal definition of $F(\omega)$, we first calculate a number of standard examples of Fourier transforms. Next, we treat some fundamental properties of Fourier transforms.

In chapter 7 the fundamental theorem of the Fourier integral is proven: a function $f(t)$ can be recovered from its Fourier transform through the Fourier integral (compare this with the fundamental theorem of Fourier series from chapter 4). We finish the theory of the Fourier integral by deriving some important additional properties from the fundamental theorem, such as Parseval's identities for the Fourier integral.

A fundamental problem in the Fourier analysis of non-periodic functions is the fact that for very elementary functions, such as the constant function 1, the Fourier transform $F(\omega)$ does not exist (we will show this in chapter 6). In physics it turned out that useful results could be obtained by a symbolic manipulation with the Fourier transform of such functions. Eventually this led to new mathematical objects, called 'distributions'. Distributions form an extension of the concept of a function, just as the complex numbers form an extension of the real numbers. And just as the complex numbers, distributions have become an indispensable tool in the applications of Fourier analysis in, for example, systems theory and (partial) differential equations. In chapter 8 distributions are introduced and some basic properties of distributions are treated. The Fourier transform of distributions is examined in chapter 9.

Just as in part 2, the Fourier analysis of non-periodic functions and distributions is applied to the theory of linear systems and (partial) differential equations in the final chapter 10.

Contents of Chapter 6

Fourier integrals: definition and properties

Fourier integrals: definition and properties

INTRODUCTION

We start this chapter with an intuitive derivation of the main result for Fourier integrals from the fundamental theorem of Fourier series. A mathematical rigorous treatment of the results obtained is postponed until chapter 7. In the present chapter the Fourier integral will thus play a minor role. First we will concentrate ourselves on the Fourier transform of a non-periodic function, which will be introduced in section 6.2, motivated by our intuitive derivation. After discussing the existence of the Fourier transform, a number of frequently used and often recurring examples are treated in section 6.3. In section 6.4 we prove some basic properties of Fourier transforms. Subsequently, the concept of a 'rapidly decreasing function' is discussed in section 6.5; in fact this is a preparation for the distribution theory of chapters 8 and 9. The chapter closes with the treatment of convolution and the convolution theorem for non-periodic functions.

LEARNING OBJECTIVES
After studying this chapter it is expected that you
- know the definition of the Fourier transform
- can calculate elementary Fourier transforms
- know and can apply the properties of the Fourier transform
- know the concept of rapidly decreasing function
- know the definition of convolution and know and can apply the convolution theorem.

6.1 An intuitive derivation

In the introduction we already mentioned that in order to make the basic formulas of the Fourier analysis of non-periodic function plausible, we use the theory of Fourier series. We do emphasize that the derivation in this section is mathematically not correct. It does show which results are to be expected later on (in chapter 7). It moreover motivates the definition of the Fourier transform of a non-periodic function.

So let us start with a non-periodic function $f : \mathbb{R} \to \mathbb{C}$ which is piecewise smooth (see chapter 2 for 'piecewise smooth'). For an arbitrary $T > 0$ we now consider the function $f_T(t)$ on the interval $(-T/2, T/2)$ obtained from f by taking f equal to 0 outside this interval. The function values at the points $-T/2$ and $T/2$ are of no importance to us and are left undetermined. Next we extend $f_T(t)$ periodically to $\mathbb{R}$. See figure 6.1. In this way we obtain a function with period T to which we can apply the theory of Fourier series. Since this periodic function coincides with the original function f on $(-T/2, T/2)$, one thus has, according to

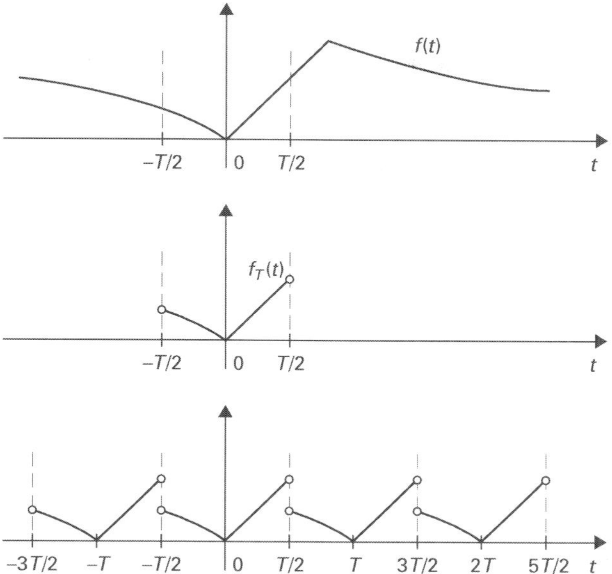

FIGURE 6.1
How a periodic function arises from a non-periodic function.

the fundamental theorem of Fourier series (see theorem 4.3),

$$f(t) = \sum_{n=-\infty}^{\infty} c_n e^{in\triangle\omega t} \qquad \text{for } t \in (-T/2, T/2) \text{ and with } \triangle\omega = \frac{2\pi}{T},$$

where c_n is the nth Fourier coefficient of f. Instead of ω_0, used in part 2, we write $\triangle\omega$ here. From the definition of c_n it then follows that

$$f(t) = \sum_{n=-\infty}^{\infty} \left(\frac{1}{T} \int_{-T/2}^{T/2} f(\tau)e^{-in\triangle\omega\tau}d\tau \right) e^{in\triangle\omega t}$$

and hence

$$f(t) = \sum_{n=-\infty}^{\infty} \frac{1}{T} \int_{-T/2}^{T/2} f(\tau)e^{in\triangle\omega(t-\tau)}d\tau, \tag{6.1}$$

where always $t \in (-T/2, T/2)$. Of this last expression we would like to determine the limit as $T \to \infty$, since in that case the identity will hold for all $t \in \mathbb{R}$, giving us a result for the original non-periodic function f. Let us write

$$G(\omega) = \int_{-\infty}^{\infty} f(\tau)e^{i\omega(t-\tau)}d\tau, \tag{6.2}$$

then

$$\frac{1}{2\pi} \sum_{n=-\infty}^{\infty} G(n\triangle\omega) \cdot \triangle\omega = \frac{\triangle\omega}{2\pi} \sum_{n=-\infty}^{\infty} \int_{-\infty}^{\infty} f(\tau)e^{in\triangle\omega(t-\tau)}d\tau \tag{6.3}$$

seems to be a good approximation for the right-hand side of (6.1). If we now let $T \to \infty$, so $\triangle\omega \to 0$, then the left-hand side of (6.3) looks like a Riemann sum,

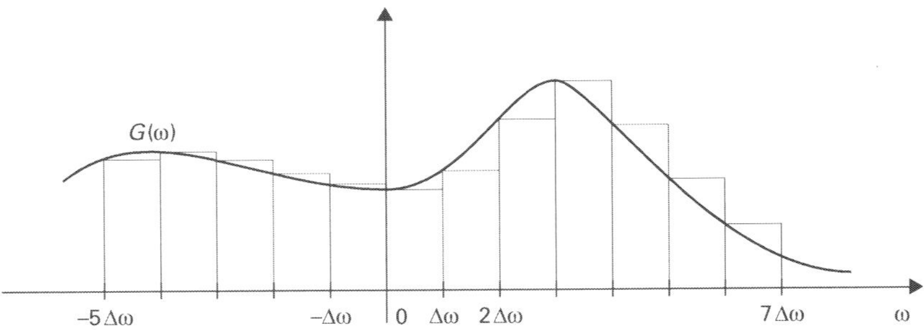

FIGURE 6.2
An approximation of $\frac{1}{2\pi}\int_{-\infty}^{\infty} G(\omega)\,d\omega$ by $\frac{1}{2\pi}\sum_{n=-\infty}^{\infty} G(n\Delta\omega)\cdot\Delta\omega$.

which will be a good approximation for

$$\frac{1}{2\pi}\int_{-\infty}^{\infty} G(\omega)\,d\omega. \tag{6.4}$$

This is illustrated in figure 6.2. Formulas (6.2) and (6.4) combined suggest that for $T \to \infty$ the identity (6.1) will transform into

$$f(t) = \frac{1}{2\pi}\int_{-\infty}^{\infty}\left(\int_{-\infty}^{\infty} f(\tau)e^{-i\omega\tau}\,d\tau\right)e^{i\omega t}\,d\omega. \tag{6.5}$$

Formula (6.5) can be interpreted as follows: when a function $f(\tau)$ is multiplied by a factor $e^{-i\omega\tau}$ and is then integrated over $\mathbb{R}$, and when subsequently the resulting function of ω is multiplied by $e^{i\omega t}$ and then again integrated over $\mathbb{R}$, then, up to a factor 2π, the original function f will result.

This important result will return in chapter 7 as the so-called fundamental theorem of the Fourier integral and it will also be proven there. This is because the intuitive derivation given here is incorrect in two ways. First, (6.3) is only an approximation of the right-hand side of (6.1) since $G(n\Delta\omega)$ is an integral over $\mathbb{R}$ instead of over $(-T/2, T/2)$, as was the case in the right-hand side of (6.1). Furthermore, the right-hand side of (6.3) may certainly *not* be considered as an approximating Riemann sum of the integral in (6.4). This is because if we consider the integral in (6.4) as an improper Riemann integral, then we recall that by definition this equals

$$\lim_{a\to-\infty,\,b\to\infty}\frac{1}{2\pi}\int_{a}^{b} G(\omega)\,d\omega.$$

Hence, the integral over $\mathbb{R}$ is not at all defined through Riemann sums, but using the *limit* above. There is thus no real justification why (6.3) should transform into (6.4) for $T \to \infty$. In chapter 7 we will see, however, that the important result (6.5) is indeed correct for a large class of functions.

6.2   The Fourier transform

Motivated by the results from the previous section we now define, for a function $f : \mathbb{R} \to \mathbb{C}$, a new function $F(\omega)$ (for $\omega \in \mathbb{R}$) by the inner integral in (6.5).

DEFINITION 6.1
Fourier transform or spectrum

For a given function $f : \mathbb{R} \to \mathbb{C}$ the function $F(\omega)$ (for $\omega \in \mathbb{R}$) is defined by

$$F(\omega) = \int_{-\infty}^{\infty} f(t)e^{-i\omega t}\, dt, \tag{6.6}$$

provided the integral exists as an improper Riemann integral. The function $F(\omega)$ is called the Fourier transform *or* spectrum *of $f(t)$.*

Sometimes $F(\omega)$ is called the *spectral density* of $f(t)$. The mapping assigning the new function $F(\omega)$ to $f(t)$ is called the *Fourier transform*. We will sometimes write $(\mathcal{F}f)(\omega)$ instead of $F(\omega)$, while the notation $(\mathcal{F}f(t))(\omega)$ will also be useful, though not very elegant. In the mathematical literature $F(\omega)$ is usually denoted by $\hat{f}(\omega)$. Often, $f(t)$ will represent a function depending on time, while $F(\omega)$ usually depends on frequency. Hence, it is said that $f(t)$ is defined in the *time domain* and that $F(\omega)$ is defined in the *frequency domain*. Since $e^{-i\omega t}$ is a complex-valued function, $F(\omega)$ will in general be a complex-valued function as well, so $F : \mathbb{R} \to \mathbb{C}$. Often $F(\omega)$ is then split up into a real part and an imaginary part. One also regularly studies $|F(\omega)|$, the so-called *amplitude spectrum* (sometimes called spectral amplitude density) of f, and $\arg F(\omega)$, the so-called *phase spectrum* of f. Finally, in signal theory one calls $|F(\omega)|^2$ the *energy spectrum* or *spectral energy density* of $f(t)$.

The definition of $F(\omega)$ closely resembles the one given for the Fourier coefficients c_n. Here, however, we take $\omega \in \mathbb{R}$ instead of $n \in \mathbb{Z}$ and in addition we integrate the function $f(t)$ over $\mathbb{R}$ instead of over a bounded interval (of length one period). The fact that we integrate over $\mathbb{R}$ causes lots of problems. For the function $f(t) = 1$ on $(-T/2, T/2)$ for example, one can determine the Fourier coefficients, while for the function $f(t) = 1$ on $\mathbb{R}$ the Fourier transform $F(\omega)$ does not exist. The function introduced in our next example, re-appearing quite regularly, also has no Fourier transform.

Fourier transform

Time domain
Frequency domain

Amplitude spectrum
Phase spectrum
Energy spectrum
Spectral energy density

EXAMPLE 6.1
Unit step function
Heaviside function

Let $\epsilon(t)$ be the function defined by

$$\epsilon(t) = \begin{cases} 1 & \text{for } t \geq 0, \\ 0 & \text{otherwise.} \end{cases}$$

See figure 6.3. This function is called the *unit step function* or *Heaviside function*. The Fourier transform of $\epsilon(t)$ does not exist, because

$$\int_{-\infty}^{\infty} \epsilon(t)e^{-i\omega t}\, dt = \int_{0}^{\infty} e^{-i\omega t}\, dt = \lim_{A \to \infty} \int_{0}^{A} e^{-i\omega t}\, dt.$$

Since $(e^{-i\omega t})' = -i\omega e^{-i\omega t}$ (see example 2.11), it follows that

$$\int_{-\infty}^{\infty} \epsilon(t)e^{-i\omega t}\, dt = \frac{1}{-i\omega} \lim_{A \to \infty} \left[e^{-i\omega t} \right]_{0}^{A} = \frac{i}{\omega} \left(\lim_{A \to \infty} e^{-i\omega A} - 1 \right).$$

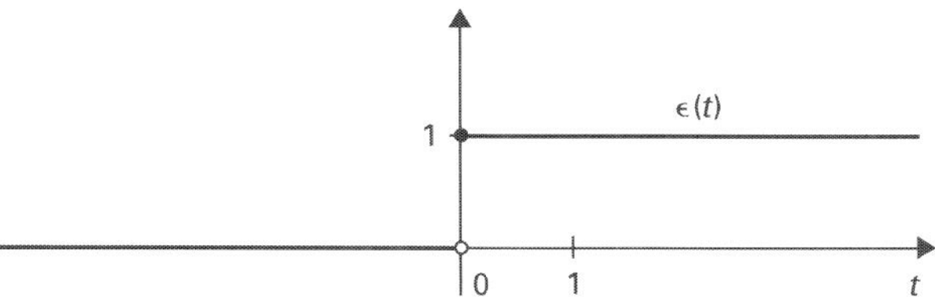

FIGURE 6.3
The unit step function $\epsilon(t)$.

However, the limit $\lim_{A\to\infty} e^{-i\omega A}$ does not exist, since $\lim_{A\to\infty} \sin A\omega$ (and also $\lim_{A\to\infty} \cos A\omega$) does not exist. ◀

Next we introduce an important class of functions for which $F(\omega)$ will certainly exist.

DEFINITION 6.2
Absolutely integrable

A function $f : \mathbb{R} \to \mathbb{C}$ *is called* absolutely integrable *(on* $\mathbb{R}$) *if* $\int_{-\infty}^{\infty} |f(t)|\,dt$ *exists as an improper Riemann integral.*

When $f(t)$ is absolutely integrable and $F(\omega)$ is as in (6.6), then

$$| F(\omega) | = \left| \int_{-\infty}^{\infty} f(t)e^{-i\omega t}\,dt \right| \leq \int_{-\infty}^{\infty} \left| f(t)e^{-i\omega t} \right|\,dt$$

$$= \int_{-\infty}^{\infty} | f(t) |\left| e^{-i\omega t} \right|\,dt,$$

and since $\left| e^{-i\omega t} \right| = 1$ it then follows from definition 6.2 that

$$| F(\omega) | \leq \int_{-\infty}^{\infty} | f(t) |\,dt < \infty. \tag{6.7}$$

This shows that $F(\omega)$ exists when $f(t)$ is absolutely integrable. Not all functions that we will need are absolutely integrable. This explains the somewhat weaker formulation of definition 6.1.

On the basis of the intuitive result (6.5) we expect that for each $t \in \mathbb{R}$ the function value $f(t)$ can be recovered from the spectrum $F(\omega)$ using the formula

$$f(t) = \frac{1}{2\pi} \int_{-\infty}^{\infty} F(\omega)e^{i\omega t}\,d\omega. \tag{6.8}$$

The right-hand side of (6.8) is called the Fourier integral of f. Formula (6.8) does however pose some problems. Even when $f(t)$ is absolutely integrable, $F(\omega)$ need not be absolutely integrable. In section 6.3.1 we will see an example. On the other hand we can still make sense of (6.8), even if $F(\omega)$ has no improper Riemann integral! And when the integral in (6.8) does exist (in some sense or other), then its value is not necessarily equal to $f(t)$. We will return to all of these problems in chapter 7. In the meantime we will agree to call the right-hand side of (6.8) the

Fourier integral

Fourier integral of $f(t)$, provided that it exists. In chapter 7 it will be shown that, just as for Fourier series, the Fourier integral does indeed exist for a large class of functions and that (6.8) is valid.

If we now look at the Fourier series of a periodic function again, then we see that only integer multiples $n\triangle\omega$ of the frequency $\triangle\omega = 2\pi/T$ occur (we write $\triangle\omega$ here, instead of ω_0 as used in part 2). In this case the spectrum is a function on $\mathbb{Z}$, which is a so-called discrete set. Therefore, it is said that a periodic function has a

Discrete or line spectrum

discrete or line spectrum (see also section 3.3). However, in the Fourier integral all frequencies ω occur, since we integrate over $\mathbb{R}$. Hence, a non-periodic function leads

Continuous spectrum

to a *continuous spectrum*. Note, though, that this does not mean that the function $F(\omega)$ is a continuous function, but only that $F(\omega)$ depends on a continuous variable ω.

The transition from a discrete to a continuous spectrum can be illustrated quite nicely with the same sort of process as in section 6.1. Let $f : \mathbb{R} \to \mathbb{C}$ be piecewise smooth and zero outside an interval $(-T/2, T/2)$ (using the terminology of section 6.1 we thus have $f_T(t) = f(t)$). Then the corresponding $F(\omega)$ certainly exists, since we only integrate over the bounded interval $(-T/2, T/2)$. If we now extend f periodically, precisely as we did in section 6.1, then we can determine the Fourier coefficients c_n ($n \in \mathbb{Z}$) of this periodic extension. Since the periodic extension coincides with f on $(-T/2, T/2)$, and f itself is zero outside $(-T/2, T/2)$, it follows that

$$c_n = \frac{1}{T} \int_{-T/2}^{T/2} f(t) e^{-in\omega_0 t} \, dt = \frac{1}{T} \int_{-\infty}^{\infty} f(t) e^{-in\omega_0 t} \, dt = \frac{1}{T} F(n\omega_0), \qquad (6.9)$$

where $\omega_0 = 2\pi/T$. We thus have

$$\frac{c_n}{\omega_0} = \frac{1}{2\pi} F(n\omega_0).$$

Hence, if we 'normalize' the Fourier coefficients with the factor $1/\omega_0$, we obtain for ever increasing values of T, so for ever decreasing values of ω_0, an approximation of the function $F(\omega)/2\pi$ that keeps improving. This is because we know the function values $F(n\omega_0)/2\pi$ at points that get closer and closer to each other. In figure 6.4 the functions $|F(\omega)|/2\pi$ and $|F(n\omega_0)|/2\pi = c_n/\omega_0$ are represented for decreasing values of ω_0. In the limit $T \to \infty$, or $\omega_0 \to 0$, we thus indeed expect that the discrete spectrum will change into the continuous spectrum $F(\omega)/2\pi$.

EXERCISE

6.1 Show that the function $f(t) = 1$ (for all $t \in \mathbb{R}$) has no Fourier transform.

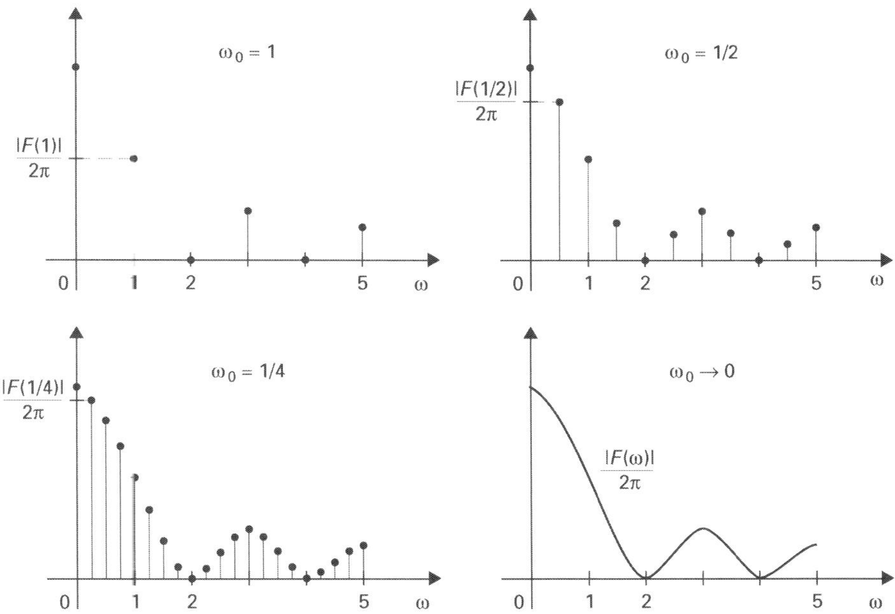

FIGURE 6.4
The normalized line spectrum approximates the continuous spectrum.

6.3   Some standard Fourier transforms

In this section the Fourier transforms of a number of frequently used functions are determined. The most important results of this section are included in table 3 at the back of the book.

6.3.1   *The block function*

Block function

Rectangular pulse function

As a first example we consider for a fixed $a > 0$ the *block function* or *rectangular pulse function* $p_a(t)$ of height 1 and duration a defined by

$$p_a(t) = \begin{cases} 1 & \text{for } |t| \leq a/2, \\ 0 & \text{otherwise.} \end{cases} \tag{6.10}$$

See figure 6.5. Note that $p_a(t)$ is certainly absolutely integrable. For $\omega \neq 0$ one has

$$(\mathcal{F}p_a)(\omega) = \int_{-\infty}^{\infty} p_a(t)e^{-i\omega t}\,dt = \int_{-a/2}^{a/2} e^{-i\omega t}\,dt = \left[\frac{-e^{-i\omega t}}{i\omega}\right]_{-a/2}^{a/2}$$

$$= \frac{e^{ia\omega/2} - e^{-ia\omega/2}}{i\omega} = \frac{2\sin(a\omega/2)}{\omega},$$

while for $\omega = 0$ one has

$$(\mathcal{F}p_a)(0) = \int_{-\infty}^{\infty} p_a(t)\,dt = \int_{-a/2}^{a/2} dt = a.$$

From the well-known limit $\lim_{x\to 0} \sin x/x = 1$ we obtain $\lim_{\omega\to 0}(\mathcal{F}p_a)(\omega) = \lim_{\omega\to 0} 2\sin(a\omega/2)/\omega = a$. Although $p_a(t)$ is itself not continuous, we see that

$$(\mathcal{F}p_a)(\omega) = \frac{2\sin(a\omega/2)}{\omega} \tag{6.11}$$

is continuous on $\mathbb{R}$. Also, $\mathcal{F}p_a$ is a real-valued function here. And finally we have that $\lim_{\omega\to\pm\infty}(\mathcal{F}p_a)(\omega) = 0$. All these remarks are precursors of general results to be treated in the chapters to come. In figure 6.5 we have sketched p_a and $\mathcal{F}p_a$. Finally we state without proof that the function $g(x) = \sin x/x$ (and so $F(\omega)$ as well) is *not* absolutely integrable (see, for example, *Fourier analysis* by

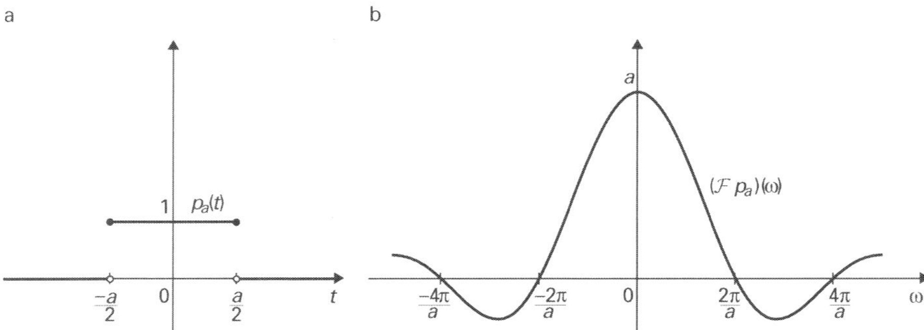

FIGURE 6.5
The block function $p_a(t)$ (a) and its spectrum (b).

T.W. Körner, Example 46.4). Hence, this is an example of an absolutely integrable function with a spectrum which is not absolutely integrable.

6.3.2 The triangle function

Triangle function

For a fixed $a > 0$ the *triangle function* $q_a(t)$ of height 1 and duration $2a$ is defined by

$$q_a(t) = \begin{cases} 1 - \dfrac{|t|}{a} & \text{for } |t| \le a, \\ 0 & \text{otherwise.} \end{cases} \tag{6.12}$$

See figure 6.6a. Note that $q_a(t)$ is absolutely integrable. We now have

$$(\mathcal{F}q_a)(\omega) = \int_{-\infty}^{\infty} q_a(t)e^{-i\omega t}\,dt = \int_0^{\infty} q_a(t)e^{-i\omega t}\,dt + \int_{-\infty}^{0} q_a(t)e^{-i\omega t}\,dt,$$

and since $q_a(t) = q_a(-t)$, we substitute $u = -t$ in the second integral and subsequently replace u by t again, which results in

$$\begin{aligned} (\mathcal{F}q_a)(\omega) &= \int_0^{\infty} q_a(t)e^{-i\omega t}\,dt + \int_0^{\infty} q_a(u)e^{i\omega u}\,du \\ &= \int_0^{\infty} q_a(t)(e^{-i\omega t} + e^{i\omega t})\,dt = 2\int_0^{\infty} q_a(t)\cos \omega t\,dt. \end{aligned}$$

From the definition of q_a in (6.12) it then follows that

$$(\mathcal{F}q_a)(\omega) = 2\int_0^{a} \left(1 - \frac{t}{a}\right)\cos \omega t\,dt.$$

For $\omega = 0$ we have $\cos \omega t = 1$ and so

$$(\mathcal{F}q_a)(0) = 2\left[t - \frac{1}{2a}t^2\right]_0^{a} = a.$$

For $\omega \ne 0$ it follows from integration by parts that

$$\begin{aligned} (\mathcal{F}q_a)(\omega) &= \frac{2}{\omega}\int_0^{a}\left(1 - \frac{t}{a}\right)(\sin \omega t)'\,dt \\ &= \frac{2}{\omega}\left[\left(1 - \frac{t}{a}\right)\sin \omega t\right]_0^{a} + \frac{2}{a\omega}\int_0^{a}\sin \omega t\,dt, \end{aligned}$$

since $(1 - t/a)' = -1/a$. The first term in this sum is zero and so

$$(\mathcal{F}q_a)(\omega) = \frac{2}{a\omega}\int_0^{a}\sin \omega t\,dt = -\frac{2}{a\omega^2}[\cos \omega t]_0^{a} = \frac{2}{a\omega^2}(1 - \cos a\omega).$$

But $1 - \cos a\omega = 2\sin^2(a\omega/2)$, so

$$(\mathcal{F}q_a)(\omega) = \frac{4\sin^2(a\omega/2)}{a\omega^2}. \tag{6.13}$$

As in the previous example, we have $\lim_{\omega \to 0}(\mathcal{F}q_a)(\omega) = a$, which means that $\mathcal{F}q_a$ is a continuous function on $\mathbb{R}$. Again $\mathcal{F}q_a$ is a real-valued function here and $\lim_{\omega \to \pm\infty}(\mathcal{F}q_a)(\omega) = 0$. In figure 6.6 q_a and $\mathcal{F}q_a$ have been drawn.

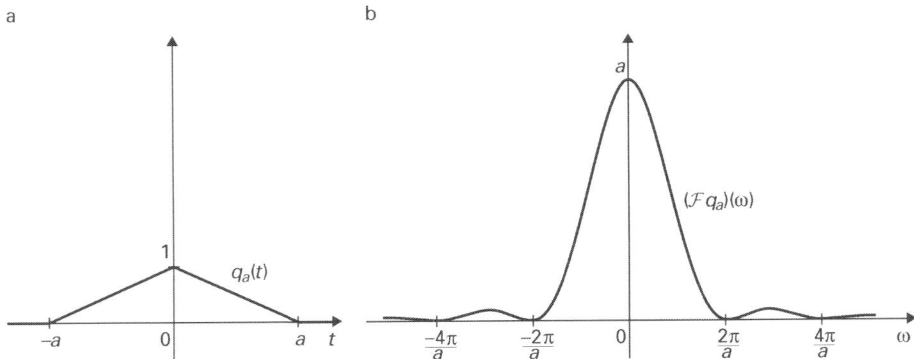

FIGURE 6.6
The triangle function $q_a(t)$ (a) and its spectrum (b).

6.3.3 The function $e^{-a|t|}$

We will now determine the spectrum $F(\omega)$ of the function $f(t) = e^{-a|t|}$, where $a > 0$ is fixed. Since for fixed x, $y \in \mathbb{R}$ one has $(e^{(x+iy)t})' = (x + iy)e^{(x+iy)t}$ (see example 2.11), it follows that

$$F(\omega) = \int_0^\infty e^{-at} e^{-i\omega t}\, dt + \int_{-\infty}^0 e^{at} e^{-i\omega t}\, dt$$

$$= -\left[\frac{e^{-(a+i\omega)t}}{a + i\omega}\right]_0^\infty + \left[\frac{e^{(a-i\omega)t}}{a - i\omega}\right]_{-\infty}^0.$$

Now $\lim_{R\to\infty} e^{-(a+i\omega)R} = \lim_{R\to\infty} e^{-aR} e^{-i\omega R} = 0$, since $|e^{-i\omega R}| = 1$ and $\lim_{R\to\infty} e^{-aR} = 0$ for $a > 0$. Similarly one has $\lim_{R\to-\infty} e^{(a-i\omega)R} = 0$. Hence

$$F(\omega) = \frac{1}{a + i\omega} + \frac{1}{a - i\omega} = \frac{2a}{a^2 + \omega^2}. \tag{6.14}$$

Again $F(\omega)$ is a continuous real-valued function with $\lim_{\omega\to\pm\infty} F(\omega) = 0$. Figure 6.7 shows $f(t)$ and $F(\omega)$.

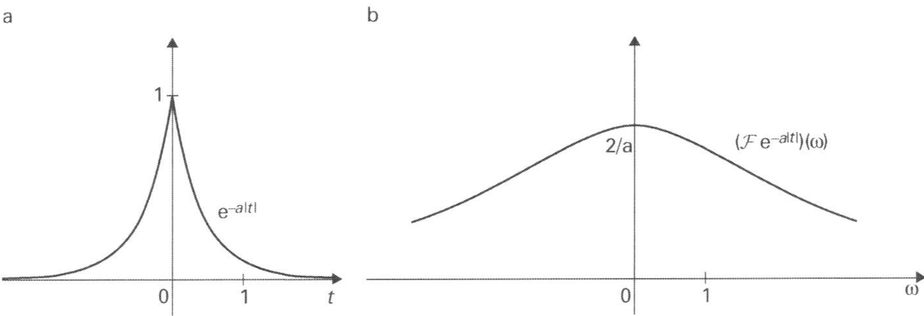

FIGURE 6.7
The function $e^{-a|t|}$ (a) and its spectrum (b).

Closely related to this example is the function $g(t)$ defined by

$$g(t) = \begin{cases} e^{-at} & \text{for } t \geq 0, \\ 0 & \text{otherwise,} \end{cases}$$

where $a \in \mathbb{C}$ with $\text{Re}\, a > 0$. A convenient way of expressing $g(t)$ uses the unit step function from example 6.1, since $g(t) = \epsilon(t)e^{-at}$. As before one quickly shows for $g(t)$ that

$$G(\omega) = \frac{1}{a + i\omega} = \frac{a - i\omega}{a^2 + \omega^2}.$$

The function $G(\omega)$ is now complex-valued. For $a > 0$ the function $G(\omega)$ has real part $a/(a^2+\omega^2) = F(\omega)/2$ and imaginary part $\omega/(a^2+\omega^2)$. Note that $\omega/(a^2+\omega^2)$ is *not* improper Riemann integrable.

6.3.4   The Gauss function

Gauss function

To conclude, we determine the spectrum $F(\omega)$ of the function $f(t) = e^{-at^2}$ for fixed $a > 0$. The function $f(t)$ is called the *Gauss function*. In order to determine $F(\omega)$ directly from the definition, one would need a part of complex function theory which falls outside the scope of this book. There is, however, a clever trick to find $F(\omega)$ in an indirect manner. To do so, we will assume the following fact:

$$\int_{-\infty}^{\infty} e^{-x^2}\, dx = \sqrt{\pi} \tag{6.15}$$

(for a proof see for example *Fourier analysis* by T.W. Körner, Lemma 48.10). As a matter of fact, (6.15) also shows immediately that the function $f(t)$ is absolutely integrable. Since $f(t) = f(-t)$ it follows as in section 6.3.2 that

$$F(\omega) = \int_{-\infty}^{\infty} e^{-at^2} e^{-i\omega t}\, dt = 2\int_{0}^{\infty} e^{-at^2} \cos \omega t\, dt.$$

We will now determine the derivative of $F(\omega)$. In doing so, we assume that the differentiation may be carried out *within* the integral. It would lead us too far to even *formulate* the theorem that would justify this step. The result of the differentiation with respect to ω within the integral is as follows:

$$F'(\omega) = -2\int_{0}^{\infty} t e^{-at^2} \sin \omega t\, dt.$$

Integrating by parts we obtain

$$F'(\omega) = \frac{1}{a}\int_{0}^{\infty} \left(e^{-at^2} \right)' \sin \omega t\, dt$$
$$= \frac{1}{a}\left[e^{-at^2} \sin \omega t \right]_{0}^{\infty} - \frac{\omega}{a}\int_{0}^{\infty} e^{-at^2} \cos \omega t\, dt.$$

The first term in this difference is equal to 0, while the second term equals $-\omega F(\omega)/2a$. Hence we obtain for $F(\omega)$ the (differential) equation

$$F'(\omega) = (-\omega/2a)F(\omega).$$

If we now divide left-hand and right-hand sides by $F(\omega)$, then

$$\frac{F'(\omega)}{F(\omega)} = -\frac{\omega}{2a}, \qquad \text{that is,} \qquad \frac{d}{d\omega} \ln |F(\omega)| = -\frac{\omega}{2a}.$$

But also $\frac{d}{d\omega}(-\omega^2/4a) = -\omega/2a$ and thus apparently $\ln|F(\omega)| = -\omega^2/4a + C$ for an arbitrary constant C. It then follows that

$$|F(\omega)| = e^C e^{-\omega^2/4a}.$$

This in fact states that $F(\omega) = De^{-\omega^2/4a}$, where D is an arbitrary constant (note that e^C is always positive). If we substitute $\omega = 0$, then we see that $D = F(0)$. Now change to the variable $x = t\sqrt{a}$ in (6.15), then it follows that

$$D = F(0) = \int_{-\infty}^{\infty} e^{-at^2}\, dt = \frac{1}{\sqrt{a}} \int_{-\infty}^{\infty} e^{-(t\sqrt{a})^2} \sqrt{a}\, dt = \sqrt{\frac{\pi}{a}}.$$

The spectrum of the Gauss function is thus given by

$$F(\omega) = \sqrt{\frac{\pi}{a}}\, e^{-\omega^2/4a}. \tag{6.16}$$

This is again a continuous real-valued function with $\lim_{\omega \to \pm\infty} F(\omega) = 0$. It is quite remarkable that apparently the spectrum of e^{-at^2} is of the same form as the original function. For $a = 1/2$ one has in particular that $(\mathcal{F}e^{-t^2/2})(\omega) = \sqrt{2\pi}\, e^{-\omega^2/2}$, so up to the factor $\sqrt{2\pi}$ exactly the same function. In figure 6.8 the Gauss function and its spectrum are drawn.

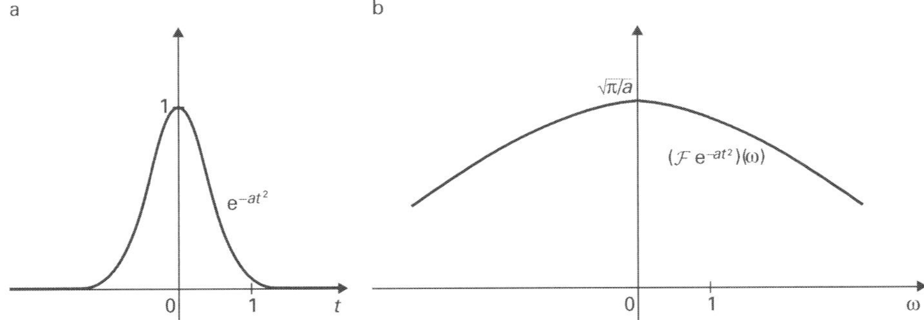

FIGURE 6.8
The Gauss function (a) and its spectrum (b).

For the moment this concludes our list of examples. The most important results have been included in table 3. In the next section some properties of the Fourier transform are established, enabling us, among other things, to calculate more Fourier transforms.

EXERCISES

6.2   Consider for fixed $a \in \mathbb{C}$ with $\operatorname{Re} a > 0$ the function $g(t)$ defined by $g(t) = \epsilon(t)e^{-at}$ (also see section 6.3.3).
a   Show that for the spectrum $G(\omega)$ one has: $G(\omega) = (a - i\omega)/(a^2 + \omega^2)$.
b   Take $a > 0$. Show that the Fourier integral for the imaginary part of $G(\omega)$ (and hence also for $G(\omega)$ itself) does not exist as an improper integral.
c   Verify that for the limit $a \to 0$ the function $g(t)$ transforms into $\epsilon(t)$, while $G(\omega)$ for $\omega \neq 0$ transforms into $-i/\omega$. This seems to suggest that $\epsilon(t)$ has the function $-i/\omega$ as its spectrum. This, however, contradicts the result from example 6.1. We will return to this in chapters 8 and 9 on distribution theory.

6.3   Determine the Fourier transform of the function

$$f(t) = \begin{cases} \cos t & \text{for } -\pi/2 \leq t \leq \pi/2, \\ 0 & \text{otherwise.} \end{cases}$$

6.4   Determine the spectrum $G(\omega)$ and verify that $G(\omega)$ is continuous when $g(t)$ is given by

$$g(t) = \begin{cases} |t| & \text{for } -1 \leq t \leq 1, \\ 0 & \text{otherwise.} \end{cases}$$

6.5   **a**   Let $a > 0$ be fixed. Determine the spectrum $F(\omega)$ of the function

$$f(t) = \begin{cases} 1 & \text{for } 0 \leq t \leq a/2, \\ -1 & \text{for } -a/2 \leq t < 0, \\ 0 & \text{otherwise.} \end{cases}$$

        b   Show that $F(\omega)$ is continuous at $\omega = 0$.

6.4   Properties of the Fourier transform

A large number of properties that we have established for Fourier series will return in this section. Often the proofs are slightly more difficult, since in Fourier analysis on $\mathbb{R}$ we always have to deal with *improper* (Riemann) integrals. The theory of improper integrals has quite a few difficult problems and we will not always formulate the precise theorems that will be needed. In section 6.3.4, for example, an improper integral and differentiation were 'interchanged', without formulating a theorem or giving conditions justifying this. A more frequently occurring problem is reversing the order of integration in a repeated integral. For example, even when

$$\int_b^\infty \left(\int_a^\infty f(x, y)\, dx \right) dy \qquad \text{and} \qquad \int_a^\infty \left(\int_b^\infty f(x, y)\, dy \right) dx$$

both exist, then still they do not necessarily have the same value (see exercise 6.6). Theorems on interchanging the order of integration will not be presented here. The interested reader can find such theorems in, for example, *Fourier analysis* by T.W. Körner, Chapters 47 & 48. When we interchange the order of integration in the proof of a theorem, we will always give sufficient conditions in the theorem such that the interchanging is allowed.

After these preliminary remarks we now start examining the properties of the Fourier transform.

6.4.1   *Linearity*

Linear combinations of functions are carried over by the Fourier transform into the same linear combination of the Fourier transforms of these functions. We formulate this linearity property in a precise manner in the following theorem.

THEOREM 6.1
Linearity of the Fourier transform

Let $f(t)$ and $g(t)$ be two functions with Fourier transforms $F(\omega)$ and $G(\omega)$ respectively. Then $aF(\omega) + bG(\omega)$ is the Fourier transform of $af(t) + bg(t)$.

Proof
This theorem follows immediately from the linearity of integration:

$$\int (af_1(t) + bf_2(t))\, dt = a \int f_1(t)\, dt + b \int f_2(t)\, dt$$

for arbitrary functions f_1 and f_2 and $a, b \in \mathbb{C}$. Now take $f_1(t) = f(t)e^{-i\omega t}$ and $f_2(t) = g(t)e^{-i\omega t}$. ∎

Because of this property, the Fourier transform is called a *linear transformation*.

6.4.2   Conjugation

For the complex conjugate of a function one has the following theorem.

THEOREM 6.2
Spectrum of the complex conjugate

Let $f(t)$ be a function with spectrum $F(\omega)$. Then the spectrum of the function $\overline{f(t)}$ is given by $\overline{F(-\omega)}$.

Proof
This result follows immediately from the properties of definite integrals of complex-valued functions (see section 2.3):

$$(\mathcal{F}\overline{f(t)})(\omega) = \int_{-\infty}^{\infty} \overline{f(t)}e^{-i\omega t}\, dt = \overline{\int_{-\infty}^{\infty} f(t)e^{i\omega t}\, dt} = \overline{F(-\omega)}.$$

∎

When in particular the function $f(t)$ is real-valued, so $\overline{f(t)} = f(t)$, then theorem 6.2 implies that $\overline{F(-\omega)} = F(\omega)$, or $F(-\omega) = \overline{F(\omega)}$.

6.4.3   Shift in the time domain

For a given function $f(t)$ and a fixed $a \in \mathbb{R}$, the function $f(t - a)$ is called the function *shifted over a*. There is simple relationship between the spectra of these two functions.

THEOREM 6.3
Shift in the time domain

Let $f(t)$ be a function with spectrum $F(\omega)$. Then one has for any fixed $a \in \mathbb{R}$ that $(\mathcal{F}f(t - a))(\omega) = e^{-i\omega a} F(\omega)$.

Proof
By changing to the new variable $\tau = t - a$ one obtains

$$(\mathcal{F}f(t - a))(\omega) = \int_{-\infty}^{\infty} f(t - a)e^{-i\omega t}\, dt = \int_{-\infty}^{\infty} f(\tau)e^{-i\omega(\tau + a)}\, d\tau$$

$$= e^{-i\omega a} F(\omega).$$

∎

So when a function is shifted over a in the time domain, its spectrum is multiplied by the factor $e^{-i\omega a}$. Note that this only changes the phase spectrum and not the amplitude spectrum. The factor $e^{-i\omega a}$ is called a *phase factor*.

Phase factor

6.4.4   Shift in the frequency domain

For a shift in the frequency domain there is a result similar to that of section 6.4.3.

THEOREM 6.4
Shift in the frequency domain

Let $f(t)$ be a function with spectrum $F(\omega)$. Then one has for $a \in \mathbb{R}$ that $(\mathcal{F}e^{iat} f(t))(\omega) = F(\omega - a)$.

Proof

$$F(\omega - a) = \int_{-\infty}^{\infty} f(t)e^{-i(\omega - a)t}\, dt = \int_{-\infty}^{\infty} e^{iat} f(t)e^{-i\omega t}\, dt = (\mathcal{F}e^{iat} f(t))(\omega).$$

∎

Modulation theorem

As an application of theorem 6.4 we mention the so-called *modulation theorem*: when $F(\omega)$ is the spectrum of $f(t)$, then

$$(\mathcal{F} f(t) \cos at)(\omega) = \frac{F(\omega - a)}{2} + \frac{F(\omega + a)}{2}. \tag{6.17}$$

Since $\cos at = e^{iat}/2 + e^{-iat}/2$, (6.17) follows from theorem 6.4 combined with the linearity from theorem 6.1: $(\mathcal{F} f(t) \cos at)(\omega) = ((\mathcal{F} f(t)e^{iat})(\omega) + (\mathcal{F} f(t)e^{-iat})(\omega))/2 = F(\omega - a)/2 + F(\omega + a)/2$. If $f(t)$ is a real-valued signal, then $f(t) \cos at$ describes a so-called amplitude modulated signal (this relates to AM on a radio: amplitude modulation).

EXAMPLE

For $f(t) = p_a(t)$ it follows from the results of section 6.3.1 that

$$(\mathcal{F} p_a(t) \cos bt)(\omega) = \frac{\sin(a(\omega - b)/2)}{\omega - b} + \frac{\sin(a(\omega + b)/2)}{\omega + b}. $$

◄

6.4.5 Scaling

We now describe the effect of scaling, or dilation, in the time domain.

THEOREM 6.5
Scaling in the time domain

Let $f(t)$ be a function with spectrum $F(\omega)$. Then one has for $c \in \mathbb{R}$ with $c \neq 0$ that $(\mathcal{F} f(ct))(\omega) = |c|^{-1} F(c^{-1}\omega)$.

Proof
We assume that $c > 0$ (give the proof yourself for the case $c < 0$). By substituting $\tau = ct$ it follows that

$$(\mathcal{F} f(ct))(\omega) = \int_{-\infty}^{\infty} f(ct)e^{-i\omega t}\, dt = c^{-1} \int_{-\infty}^{\infty} f(\tau)e^{-i\omega\tau/c}\, d\tau$$
$$= c^{-1} F(c^{-1}\omega). $$

■

EXAMPLE

Consider the block function $p_a(t)$ defined by (6.10), whose spectrum is given by (6.11). According to the scaling property one then has for $c > 0$

$$(\mathcal{F} p_a(ct))(\omega) = \frac{1}{c} \frac{2\sin(a\omega/2c)}{\omega/c} = \frac{2\sin(a\omega/2c)}{\omega}. $$

This also follows at once from the fact that $p_a(ct)$ equals $p_{a/c}(t)$. ◄

Time reversal

A special case of scaling is the so-called *time reversal*, that is, replacing t by $-t$ in the function $f(t)$. Applying theorem 6.5 with $c = -1$ we obtain that

$$(\mathcal{F} f(-t))(\omega) = F(-\omega). \tag{6.18}$$

6.4.6 Even and odd functions

A function $f : \mathbb{R} \to \mathbb{C}$ is called *even* when $f(-t) = f(t)$ for all $t \in \mathbb{R}$ and *odd* when $f(-t) = -f(t)$ for all $t \in \mathbb{R}$ (also see section 3.4). The function $\cos t$ and the functions p_a and q_a from sections 6.3.1 and 6.3.2 are examples of even functions, while the function $\sin t$ is an odd function. If we replace the function $q_a(t)$ in the first part of section 6.3.2 by an arbitrary *even* function $f(t)$, then we obtain

$$F(\omega) = 2 \int_0^{\infty} f(t) \cos \omega t\, dt. \tag{6.19}$$

Fourier cosine transform

One now calls the *integral* in the right-hand side (so without the factor 2) the *Fourier cosine transform* of the even function $f(t)$; we denote it by $F_c(\omega)$, so $F(\omega) = 2F_c(\omega)$. Since $\cos(-\omega t) = \cos \omega t$, the function $F(\omega)$ is in this case an even function as well. This also follows from (6.18) since $F(-\omega) = (\mathcal{F}f(-t))(\omega) = (\mathcal{F}f(t))(\omega) = F(\omega)$. If, in addition, we know that $f(t)$ is real-valued, then it follows from (6.19) that $F(\omega)$ is also real-valued. We already saw this in sections 6.3.1 and 6.3.2 for the functions p_a and q_a. Hence, for an even and real-valued function $f(t)$ we obtain that $F(\omega)$ is also even and real-valued.

When a function $f(t)$ is only defined for $t > 0$, then one can calculate the Fourier cosine transform of this function using the integral in (6.19). This is then in fact the ordinary Fourier transform (up to a factor 2) of the function that arises by extending f to an even function on $\mathbb{R}$; for $t < 0$ one thus defines f by $f(t) = f(-t)$. The value at $t = 0$ is usually taken to be 0, but this is hardly relevant.

There are similar results for *odd* functions. If $g : \mathbb{R} \to \mathbb{C}$ is odd and if we use the fact that $\sin \omega t = (e^{i\omega t} - e^{-i\omega t})/2i$, then we obtain, just as in the case of even functions,

$$G(\omega) = -2i \int_0^\infty g(t) \sin \omega t \, dt. \tag{6.20}$$

Fourier sine transform

One then calls the *integral* in the right-hand side (so without the factor $-2i$) the *Fourier sine transform* of the odd function $g(t)$; we denote it by $G_s(\omega)$, so $G(\omega) = -2iG_s(\omega)$. Now $G(\omega)$ is an odd function. Again this also follows from (6.18). If, in addition, we know that $g(t)$ is real-valued, then it follows from (6.20) that $G(\omega)$ can only assume purely imaginary values.

If a function $g(t)$ is only defined for $t > 0$, then one can calculate its Fourier sine transform using the integral in (6.20). This is then the ordinary Fourier transform (up to a factor $-2i$) of the function that arises by extending g to an odd function on $\mathbb{R}$; for $t < 0$ one defines g by $g(t) = -g(-t)$ and $g(0) = 0$.

EXAMPLE

Consider for $t > 0$ the function $f(t) = e^{-at}$, where $a > 0$ is fixed. Extend this function to an even function on $\mathbb{R}$. Then the Fourier transform of f equals the function $2a/(a^2 + \omega^2)$. This is because the even extension of f is the function $e^{-a|t|}$ and the Fourier transform of this function has been determined in section 6.3.3. Because of the factor 2, the Fourier cosine transform is given by $a/(a^2 + \omega^2)$. In exercise 6.14 you will be asked to determine the Fourier transform of the odd extension of f (for the case $a = 1$). ◀

6.4.7   Selfduality

The selfduality derived in this section is a preparation for the distribution theory in chapters 8 and 9. First we observe the following. Until now we have always regarded $f(t)$ as a function of the time t and $F(\omega)$ as a function of the frequency ω. In fact f and F are just two functions from $\mathbb{R}$ to $\mathbb{C}$ for which the name of the variable is irrelevant.

THEOREM 6.6
Selfduality

Let $f(t)$ and $g(t)$ be piecewise smooth and absolutely integrable functions with spectra $F(\omega)$ and $G(\omega)$ respectively. Then

$$\int_{-\infty}^\infty f(x)G(x) \, dx = \int_{-\infty}^\infty F(x)g(x) \, dx.$$

Proof
From the definition of spectrum it follows that

$$\int_{-\infty}^{\infty} f(x)G(x)\,dx = \int_{-\infty}^{\infty} f(x)\left(\int_{-\infty}^{\infty} g(y)e^{-ixy}\,dy\right)dx$$

$$= \int_{-\infty}^{\infty}\int_{-\infty}^{\infty} f(x)g(y)e^{-ixy}\,dy\,dx.$$

We mention without proof that under the conditions of theorem 6.6 one may interchange the order of integration. We then indeed obtain

$$\int_{-\infty}^{\infty} f(x)G(x)\,dx = \int_{-\infty}^{\infty} g(y)\left(\int_{-\infty}^{\infty} f(x)e^{-ixy}\,dx\right)dy = \int_{-\infty}^{\infty} g(y)F(y)\,dy.$$

$\blacksquare$

6.4.8   Differentiation in the time domain

Important for the application of Fourier analysis to, for example, differential equations is the relation between differentiation and Fourier transform. In this section we will see how the spectrum of a derivative f' can be obtained from the spectrum of f. In particular it will be assumed that $f(t)$ is continuously differentiable, so $f'(t)$ exists on $\mathbb{R}$ and is continuous.

THEOREM 6.7
Differentiation in time domain

Let $f(t)$ be a continuously differentiable function with spectrum $F(\omega)$ and assume that $\lim_{t\to\pm\infty} f(t) = 0$. Then the spectrum of $f'(t)$ exists and $(\mathcal{F}f')(\omega) = i\omega F(\omega)$.

Proof
Since f' is continuous it follows from integration by parts that

$$\lim_{A\to-\infty, B\to\infty}\int_A^B f'(t)e^{-i\omega t}\,dt$$

$$= \lim_{A\to-\infty, B\to\infty}\left[f(t)e^{-i\omega t}\right]_A^B + \lim_{A\to-\infty, B\to\infty} i\omega\int_A^B f(t)e^{-i\omega t}\,dt$$

$$= \lim_{B\to\infty} f(B)e^{-i\omega B} - \lim_{A\to-\infty} f(A)e^{-i\omega A} + i\omega F(\omega),$$

where in the last step we used that $F(\omega)$ exists. Since $\lim_{t\to\pm\infty} f(t) = 0$, it follows that $\lim_{B\to\infty} f(B)e^{-i\omega B} = 0$ and $\lim_{A\to-\infty} f(A)e^{-i\omega A} = 0$. Hence, $(\mathcal{F}f')(\omega)$ exists and we also see immediately that $(\mathcal{F}f')(\omega) = i\omega F(\omega)$. $\blacksquare$

Of course, theorem 6.7 can be applied repeatedly, provided that the conditions are satisfied in each case. If, for example, f is twice continuously differentiable (so f'' is now a continuous function) and both $\lim_{t\to\pm\infty} f(t) = 0$ and $\lim_{t\to\pm\infty} f'(t) = 0$, then theorem 6.7 can be applied twice and we obtain that $(\mathcal{F}f'')(\omega) = (i\omega)^2 F(\omega) = -\omega^2 F(\omega)$. In general one has: if f is m times continuously differentiable and $\lim_{t\to\pm\infty} f^{(k)}(t) = 0$ for each $k = 0, 1, 2, \ldots, m-1$ (where $f^{(k)}$ denotes the kth derivative of f and $f^{(0)} = f$), then

$$(\mathcal{F}f^{(m)})(\omega) = (i\omega)^m F(\omega). \tag{6.21}$$

We will use the Gauss function to illustrate theorem 6.7; it is the only continuously differentiable function among the examples in section 6.3.

EXAMPLE 6.2

The derivative of the function $f(t) = e^{-at^2}$ is given by the continuous function $f'(t) = -2ate^{-at^2}$. Moreover, $\lim_{t\to\pm\infty} e^{-at^2} = 0$. Theorem 6.7 can thus be applied and from (6.16) we obtain

$$(\mathcal{F}(-2at e^{-at^2}))(\omega) = i\omega\sqrt{\frac{\pi}{a}}\, e^{-\omega^2/4a}.$$

$\blacktriangleleft$

6.4.9 *Differentiation in the frequency domain*

In section 6.4.8 we have seen that differentiation in the time domain corresponds to multiplication in the frequency domain. The converse turns out to be true as well.

THEOREM 6.8
Differentiation in frequency domain

Let $f(t)$ be an absolutely integrable function with spectrum $F(\omega)$. If the function $tf(t)$ is absolutely integrable, then the spectrum $F(\omega)$ is differentiable and $F'(\omega) = -(\mathcal{F}itf(t))(\omega)$.

Proof
In order to show that $F(\omega)$ is differentiable, we determine

$$\lim_{h\to 0}\frac{F(\omega + h) - F(\omega)}{h} = \lim_{h\to 0}\int_{-\infty}^{\infty} f(t)\frac{e^{-i(\omega+h)t} - e^{-i\omega t}}{h}\, dt$$

$$= \lim_{h\to 0}\int_{-\infty}^{\infty} f(t)e^{-i\omega t}\frac{e^{-iht} - 1}{h}\, dt.$$

Again we mention without proof that the limit and the integral may be interchanged. Furthermore, one has

$$\lim_{h\to 0}\frac{e^{-iht} - 1}{h} = \lim_{h\to 0}\left(\frac{\cos ht - 1}{h} - i\frac{\sin ht}{h}\right) = -it$$

(apply for example De l'Hôpital's rule) and so

$$\lim_{h\to 0}\frac{F(\omega + h) - F(\omega)}{h} = \int_{-\infty}^{\infty}(-itf(t))e^{-i\omega t}\, dt.$$

According to our assumption, the function $tf(t)$ (and so $-itf(t)$ as well) is absolutely integrable, which implies that the limit indeed exists. This shows that $F(\omega)$ is differentiable and we also immediately obtain that $F'(\omega) = -(\mathcal{F}itf(t))(\omega)$. $\blacksquare$

Again, this rule can be applied repeatedly, assuming that in each case the conditions are met. When, for example, $f(t)$, $tf(t)$ and $t^2 f(t)$ are absolutely integrable, then $F(\omega)$ is twice differentiable and $F''(\omega) = (\mathcal{F}(-it)^2 f(t))(\omega)$. In general one has: if $t^k f(t)$ is absolutely integrable for $k = 0, 1, 2, \ldots, m$, then $F(\omega)$ is m times differentiable and

$$F^{(m)}(\omega) = (\mathcal{F}(-it)^m f(t))(\omega). \tag{6.22}$$

EXAMPLE 6.3

The function $tp_a(t)$ satisfies the conditions of theorem 6.8 and hence

$$(\mathcal{F}tp_a(t))(\omega) = -\frac{1}{i}F'(\omega) = i\left(\frac{2\sin(a\omega/2)}{\omega}\right)'$$

$$= i\frac{a\cos(a\omega/2)}{\omega} - i\frac{2\sin(a\omega/2)}{\omega^2}.$$

$\blacktriangleleft$

6.4.10   Integration

Finally we will use the differentiation rule in the time domain to derive a rule for integration in the time domain.

THEOREM 6.9
Integration in time domain

Let $f(t)$ be a continuous and absolutely integrable function with spectrum $F(\omega)$. Assume that $\lim_{t \to \infty} \int_{-\infty}^{t} f(\tau)\,d\tau = 0$. Then one has for $\omega \neq 0$ that

$$\left(\mathcal{F} \int_{-\infty}^{t} f(\tau)\,d\tau \right)(\omega) = \frac{F(\omega)}{i\omega}.$$

Proof
Put $g(t) = \int_{-\infty}^{t} f(\tau)\,d\tau$. Since f is continuous, it follows that g is a continuously differentiable function. Moreover, $g'(t) = f(t)$ (fundamental theorem of calculus). Since, according to our assumptions, $\lim_{t \to \pm\infty} g(t) = 0$, theorem 6.7 can now be applied to the function g. One then obtains that $(\mathcal{F}f)(\omega) = (\mathcal{F}g')(\omega) = i\omega(\mathcal{F}g)(\omega)$ and so the result follows by dividing by $i\omega$ for $\omega \neq 0$. ∎

Note that $\lim_{t \to \infty} g(t) = \int_{-\infty}^{\infty} f(\tau)\,d\tau = 0$. But $\int_{-\infty}^{\infty} f(\tau)\,d\tau$ is precisely $F(0)$, and so the conditions of theorem 6.9 apparently imply that $F(0) = 0$.

6.4.11   Continuity

We want to mention one final result which is somewhat separate from the rest of section 6.4. It is in agreement with a fact that can easily be observed: all spectra from section 6.3 are continuous functions on $\mathbb{R}$.

THEOREM 6.10
Continuity of spectra

Let $f(t)$ be an absolutely integrable function. Then the spectrum $F(\omega)$ is a continuous functions on $\mathbb{R}$.

Since theorem 6.10 will not be used in the future, we will not give a proof (see for example *Fourier analysis* by T.W. Körner, Lemma 46.3). For specific functions, like the functions from section 6.3, the theorem can usually be verified quite easily.

The function $\sin t / t$ is *not* absolutely integrable (this was noted in section 6.3.1). In exercise 7.5 we will show that the spectrum of this function is the *discontinuous* function $\pi p_2(\omega)$.

EXERCISES

6.6   In this exercise we will show that interchanging the order of integration is not always allowed. To do so, we consider the function $f(x, y)$ on $\{(x, y) \in \mathbb{R}^2 \mid x > 0$ and $y > 0\}$ given by $f(x, y) = (x - y)/(x + y)^3$. Show that $\int_{1}^{\infty} (\int_{1}^{\infty} f(x, y)\,dx)\,dy$ and $\int_{1}^{\infty} (\int_{1}^{\infty} f(x, y)\,dy)\,dx$ both exist, but are unequal.

6.7   Use the linearity property to determine the spectrum of the function $f(t) = 3e^{-2|t|} + 2iq_a(t)$, where $q_a(t)$ is the triangle function from (6.12).

6.8   Use the modulation theorem to determine the spectrum of the function $f(t) = e^{-7|t|} \cos \pi t$.

6.9   **a**   Let $F(\omega)$ be the spectrum of $f(t)$. What then is the spectrum of $f(t) \sin at$ $(a \in \mathbb{R})$?

      b   Determine the spectrum of

$$f(t) = \begin{cases} \sin t & \text{for } -\pi \leq t \leq \pi, \\ 0 & \text{otherwise.} \end{cases}$$

6.10   Let $b \in \mathbb{R}$ and $a \in \mathbb{C}$ with $\operatorname{Re} a > 0$ be fixed. Determine the Fourier transform of the function $f(t) = \epsilon(t)e^{-at}\cos bt$ and $g(t) = \epsilon(t)e^{-at}\sin bt$.

6.11   Prove the scaling property from theorem 6.5 for $c < 0$.

6.12   Verify that for an odd function $f(t)$ one has $F(\omega) = -2i \int_0^\infty f(t)\sin \omega t\, dt$.

6.13   **a**   Let $f(t)$ be a real-valued function and assume that the spectrum $F(\omega)$ is an even function. Show that $F(\omega)$ has to be real-valued.
b   Let $f(t)$ be a real-valued function. Show that $|F(\omega)|$ is even.

6.14   For $t > 0$ we define the function $f(t)$ by $f(t) = e^{-t}$. This function is extended to an odd function on $\mathbb{R}$, so $f(t) = -e^t$ for $t < 0$ and $f(0) = 0$. Determine the spectrum of f.

6.15   Consider for $a > 0$ fixed the function

$$f(t) = \begin{cases} 1 & \text{for } 0 < t \le a, \\ 0 & \text{for } t > a. \end{cases}$$

a   Determine the Fourier cosine transform of f.
b   Determine the Fourier sine transform of f.

6.16   Determine in a direct way, that is, using the definition of $F(\omega)$, the spectrum of the function $t p_a(t)$, where $p_a(t)$ is given by (6.10). Use this to check the result from example 6.3 in section 6.4.9.

6.17   Consider the Gauss function $f(t) = e^{-at^2}$.
a   Use the differentiation rule in the frequency domain to determine the spectrum of $t f(t)$.
b   Note that $t f(t) = -f'(t)/2a$. Show that the result in part a agrees with the result from example 6.2, which used the differentiation rule in the time domain.

6.18   Give at least three functions f whose Fourier transform $F(\omega)$ is equal to $k f(\omega)$, where k is a constant.

6.19   Determine the spectrum of the function $\epsilon(t)t e^{-at}$ (for $a \in \mathbb{C}$ with $\operatorname{Re} a > 0$).

6.5   Rapidly decreasing functions

In this section the results on differentiation in the time domain from section 6.4.8 are applied in preparation for chapters 8 and 9 on distributions. To this end we will introduce a collection of functions V which is invariant under the Fourier transform, that is to say: if $f(t) \in V$, then $F(\omega) \in V$. We recall that, for example, for the absolutely integrable functions this is not necessarily the case: if $f(t)$ is absolutely integrable, then $F(\omega)$ is not necessarily absolutely integrable. If we now look at the differentiation rules from sections 6.4.8 and 6.4.9, then we see that differentiation in one domain corresponds to multiplication in the other domain. We thus reach the conclusion that we should introduce the collection of so-called *rapidly decreasing functions*. It will turn out that this collection is indeed invariant under the Fourier transform. Let us write $f(t) \in C^\infty(\mathbb{R})$ to indicate that f can be differentiated arbitrarily many times, that is, $f^{(k)}(t)$ exists for each $k \in \mathbb{N}$. It is also said that f is *infinitely differentiable*.

DEFINITION 6.3
Rapidly decreasing function

A function $f : \mathbb{R} \to \mathbb{C}$ in $C^\infty(\mathbb{R})$ is called rapidly decreasing *if for each m and $n \in \mathbb{N}$ the function $t^n f^{(m)}(t)$ is bounded on $\mathbb{R}$, that is to say, there exists a constant $M > 0$ such that $|t^n f^{(m)}(t)| < M$ for all $t \in \mathbb{R}$.*

In this definition the constant M will of course depend on the value of m and n. The term 'rapidly decreasing' has a simple explanation: for large values of $n \in \mathbb{N}$ the functions $|t|^{-n}$ decrease rapidly for $t \to \pm\infty$ and from definition 6.3 it follows for $m = 0$ that $|f(t)| < M|t|^{-n}$ for each $n \in \mathbb{N}$. Hence, the function f (and even all of its derivatives) has to decrease quite rapidly for $t \to \pm\infty$.

The collection of all rapidly decreasing functions will be denoted by $\mathcal{S}(\mathbb{R})$ or simply by $\mathcal{S}$. For $f \in \mathcal{S}$ it follows immediately from the definition that $cf \in \mathcal{S}$ as well for an arbitrary constant $c \in \mathbb{C}$. And since

$$\left| t^n (f + g)^{(m)}(t) \right| = \left| t^n f^{(m)}(t) + t^n g^{(m)}(t) \right| \leq \left| t^n f^{(m)}(t) \right| + \left| t^n g^{(m)}(t) \right|,$$

it also follows that $f + g \in \mathcal{S}$ whenever $f \in \mathcal{S}$ and $g \in \mathcal{S}$.

Now this is all quite nice, but are there actually any functions at all that belong to $\mathcal{S}$, besides the function $f(t) = 0$ (for all $t \in \mathbb{R}$)?

THEOREM 6.11

The Gauss function $f(t) = e^{-at^2}$ $(a > 0)$ belongs to $\mathcal{S}$.

Proof
First note that f is infinitely differentiable because $f(t) = (h \circ g)(t)$, where $g(t) = -at^2$ and $h(s) = e^s$ are infinitely differentiable functions. From the chain rule it then follows that $f \in C^\infty(\mathbb{R})$. By repeatedly applying the product rule and the chain rule again, it follows that $(e^{-at^2})^{(m)}$ is a finite sum of terms of the form $ct^k e^{-at^2}$ ($k \in \mathbb{N}$ and c a constant). Hence, $t^n (e^{-at^2})^{(m)}$ is also a finite sum of terms of the form $ct^k e^{-at^2}$ for each m and $n \in \mathbb{N}$, and we must show that this is bounded on $\mathbb{R}$. It now suffices to show that $t^k e^{-at^2}$ is bounded on $\mathbb{R}$ for arbitrary $k \in \mathbb{N}$, since then any finite sum of such terms is bounded on $\mathbb{R}$. The boundedness of $t^k e^{-at^2}$ on $\mathbb{R}$ follows immediately from the fact that $t^k e^{-at^2}$ is a continuous function with $\lim_{t \to \pm\infty} t^k e^{-at^2} = 0$. $\blacksquare$

By multiplication and differentiation one can obtain new functions in $\mathcal{S}$: for $f(t) \in \mathcal{S}$ one has $t^n f(t) \in \mathcal{S}$ for each $n \in \mathbb{N}$ and even $(t^n f(t))^{(m)} \in \mathcal{S}$ for each m and $n \in \mathbb{N}$. For the proof of the latter statement one has to apply the product rule repeatedly again, resulting in a sum of terms of the form $t^k f^{(l)}(t)$, which all belong to $\mathcal{S}$ again because $f \in \mathcal{S}$. The same argument shows that the product $f \cdot g$ also belongs to $\mathcal{S}$ when $f, g \in \mathcal{S}$.

THEOREM 6.12
$\mathcal{S}$ is invariant under Fourier transform

If $f(t) \in \mathcal{S}$, then $F(\omega) \in \mathcal{S}$.

Proof
First we have to show that $F(\omega) \in C^\infty(R)$. But since $f \in \mathcal{S}$, it follows that $t^p f(t) \in \mathcal{S}$ for each $p \in \mathbb{N}$ and so, according to the remark above, $t^p f(t) \in \mathcal{S}$ is absolutely integrable for each $p \in \mathbb{N}$. The differentiation rule in the frequency domain can now be applied arbitrarily often. Hence, $F(\omega) \in C^\infty(R)$ and (6.22) holds. We still have to show that $\omega^n F^{(m)}(\omega)$ is bounded on $\mathbb{R}$ for every $m, n \in \mathbb{N}$. According to (6.22) one has

$$\omega^n F^{(m)}(\omega) = \omega^n (\mathcal{F}(-it)^m f(t))(\omega). \tag{6.23}$$

We now want to apply the differentiation rule in the time domain repeatedly. In order to do so, we first note that $t^m f(t) \in \mathcal{S}$, which implies that this function is at least n times continuously differentiable. Also $(t^m f(t))^{(k)} \in \mathcal{S}$ for each $k = 0, 1, \ldots, n - 1$ and so we certainly have $\lim_{t \to \pm\infty} (t^m f(t))^{(k)} = 0$. Hence, the differentiation rule in the time domain can indeed be applied repeatedly, and from (6.21) it then follows that

$$(i\omega)^n (\mathcal{F}g(t))(\omega) = (\mathcal{F}(g(t))^{(n)})(\omega), \tag{6.24}$$

where $g(t) = (-it)^m f(t)$. Combining equations (6.23) and (6.24) shows that $(i\omega)^n F^{(m)}(\omega)$ is the Fourier transform of $(g(t))^{(n)} = ((-it)^m f(t))^{(n)}$, which is absolutely integrable. The boundedness of $(i\omega)^n F^{(m)}(\omega)$, and so of $\omega^n F^{(m)}(\omega)$, then follows from the simple relationship (6.7), applied to the function $(g(t))^{(n)}$. This proves that $F(\omega) \in \mathcal{S}$. ∎

Let us take the Gauss function $f(t) = e^{-at^2} \in \mathcal{S}$ $(a > 0)$ as an example. From (6.16) it indeed follows that $F(\omega) \in \mathcal{S}$. This is because $F(\omega)$ has the same form as $f(t)$.

EXERCISES

6.20   Indicate why $e^{-a|t|}$ $(a > 0)$ and $(1 + t^2)^{-1}$ do not belong to $\mathcal{S}$.

6.21   Consider the Gauss function $f(t) = e^{-at^2}$ $(a > 0)$.
a   Verify that for arbitrary $k \in \mathbb{N}$ one has $\lim_{t \to \pm\infty} t^k e^{-at^2} = 0$.
b   Determine the first three derivatives of f and verify that these are a finite sum of terms of the form $ct^l f(t)$ $(l \in \mathbb{N}$ and c a constant). Conclude that $\lim_{t \to \pm\infty} f^{(k)}(t) = 0$ for $k = 1, 2, 3$.

6.22   Let f and g belong to $\mathcal{S}$. Show that $f \cdot g \in \mathcal{S}$.

6.6   Convolution

Convolution of periodic functions has been treated in chapter 4. In this section we study the concept of convolution for non-periodic functions. Convolution arises, for example, in the following situation. Let $f(t)$ be a function with spectrum $F(\omega)$. Then $F(\omega)$ is a function in the frequency domain. Often, such a function is multiplied by another function in the frequency domain. One should think for instance of a (sound) signal with a spectrum $F(\omega)$ containing undesirable high or low frequencies, which we then send through a filter in order to remove these frequencies. We may remove, for example, all frequencies above a fixed frequency ω_0 by multiplying $F(\omega)$ by the block function $p_{2\omega_0}(\omega)$. In general one will thus obtain a product function $F(\omega)G(\omega)$ in the frequency domain. Now the question is, how this alters our original signal $f(t)$. In other words: which function has as its spectrum the function $F(\omega)G(\omega)$? Is this simply the product function $f(t)g(t)$, where $g(t)$ is a function with spectrum $G(\omega)$? A very simple example shows that this cannot be the case. The product of the block function $p_a(t)$ (now in the *time* domain) with itself, that is, the function $(p_a(t))^2$, is just $p_a(t)$ again. However, the product of $(\mathcal{F}p_a)(\omega)$ (see (6.11)) with itself is $4\sin^2(a\omega/2)/\omega^2$, and this does not equal $(\mathcal{F}p_a)(\omega)$. We do recognize, however, the Fourier transform of the triangle function $q_a(t)$, up to a factor a (see (6.13)). Which operation can turn two block functions into a triangle function? The solution of this problem, in the general case as well, is given by *convolution*.

DEFINITION 6.4
Convolution

The convolution *product (or convolution for short) of two functions f and g, denoted by $f * g$, is defined by*

$$(f * g)(t) = \int_{-\infty}^{\infty} f(\tau)g(t - \tau)\, d\tau \qquad \text{for } t \in \mathbb{R},$$

provided the integral exists.

Before we discuss the existence of the convolution, we will first of all verify that $(p_a * p_a)(t)$ indeed results in the triangle function $q_a(t)$, up to the factor a.

EXAMPLE

One has

$$(p_a * p_a)(t) = \int_{-\infty}^{\infty} p_a(\tau) p_a(t - \tau)\, d\tau,$$

and this integral equals 0 if $\tau < -a/2$, $\tau > a/2$, $t - \tau < -a/2$ or $t - \tau > a/2$, so if $\tau < -a/2$, $\tau < t - a/2$, $\tau > a/2$ or $\tau > t + a/2$.

For $t > a$ or $t < -a$ these conditions are fulfilled for each $\tau \in \mathbb{R}$ and so $(p_a * p_a)(t) = 0$ in this case. In the boundary cases $t = a$ or $t = -a$ the conditions are fulfilled for each $\tau \in \mathbb{R}$, except for one point (when, for example, $t = a$ then we have the condition $\tau > a/2$ as well as $\tau < a/2$ and only $\tau = a/2$ does not satisfy this). Since one point will never contribute to an integral, we again have $(p_a * p_a)(t) = 0$.

Now if $0 \le t < a$, then

$$(p_a * p_a)(t) = \int_{t-a/2}^{a/2} d\tau = a - t = a\left(1 - \frac{t}{a}\right),$$

and if $-a < t \le 0$, then

$$(p_a * p_a)(t) = \int_{-a/2}^{t+a/2} d\tau = a + t = a\left(1 + \frac{t}{a}\right).$$

We thus have indeed $(p_a * p_a)(t) = a q_a(t)$. In figure 6.9 we show once more, for various 'typical' values of t, the intervals such that the functions $p_a(\tau)$ and $p_a(t-\tau)$ are non-zero. One only has a contribution to the integral $(p_a * p_a)(t)$ when the uppermost interval has an overlap with the other intervals sketched below it. ◀

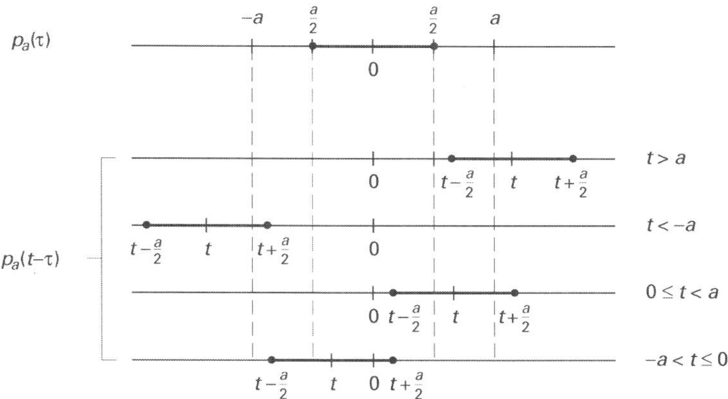

FIGURE 6.9

Integration intervals in the calculation of $(p_a * p_a)(t)$ for typical values of t.

If we now return to the general situation, then the convolution of two functions f and g will certainly exist if one of these functions, say f, is absolutely integrable, while the other function is bounded on $\mathbb{R}$. For in this case one has

$$|(f * g)(t)| = \left| \int_{-\infty}^{\infty} f(\tau) g(t - \tau)\, d\tau \right| \le \int_{-\infty}^{\infty} |f(\tau) g(t - \tau)|\, d\tau$$

$$\le M \int_{-\infty}^{\infty} |f(\tau)|\, d\tau,$$

where M is chosen such that $|g(t)| \leq M$ on $\mathbb{R}$. Since f is absolutely integrable, it then follows that $|(f * g)(t)| < \infty$, in other words, $f * g$ exists for each $t \in \mathbb{R}$. Moreover, we see that $f * g$ is bounded on $\mathbb{R}$. Note that the preceding conditions are certainly satisfied when f and g belong to S. Another condition which guarantees the existence of $f * g$ is that f and g should be absolutely integrable and that, in addition, $\int_{-\infty}^{\infty} |f(t)|^2 \, dt < \infty$ and $\int_{-\infty}^{\infty} |g(t)|^2 \, dt < \infty$. We will not prove this. Functions satisfying this condition form an important class of functions; we will return to this in chapter 7.

A simple property of convolution is commutativity, so $f * g = g * f$. This is because changing to the variable $u = t - \tau$ gives

$$(f * g)(t) = \int_{-\infty}^{\infty} f(\tau)g(t - \tau) \, d\tau = \int_{-\infty}^{\infty} f(t - u)g(u) \, du = (g * f)(t). \quad (6.25)$$

We will now prove the important result that the spectrum of $(f * g)(t)$ is the product $F(\omega)G(\omega)$. This result is called the *convolution theorem*.

THEOREM 6.13
Convolution theorem

Let $f(t)$ and $g(t)$ be piecewise continuous functions which, in addition, are absolutely integrable and bounded. Let $F(\omega)$ and $G(\omega)$ be the spectra of f and g. Then $f * g$ is absolutely integrable and

$$\mathcal{F}(f * g)(\omega) = F(\omega)G(\omega). \quad (6.26)$$

Proof
We have just seen that $f * g$ exists and is bounded on $\mathbb{R}$, since g is bounded and f is absolutely integrable. Next we note that

$$\int_{-\infty}^{\infty} \left(\int_{-\infty}^{\infty} |f(\tau)g(t - \tau)| \, dt \right) d\tau = \int_{-\infty}^{\infty} |f(\tau)| \left(\int_{-\infty}^{\infty} |g(t - \tau)| \, dt \right) d\tau$$

$$= \int_{-\infty}^{\infty} |f(\tau)| \, d\tau \int_{-\infty}^{\infty} |g(u)| \, du < \infty,$$

since f and g are absolutely integrable. Under the conditions of the theorem one may interchange the order of integration. We then obtain

$$\int_{-\infty}^{\infty} |(f * g)(t)| \, dt \leq \int_{-\infty}^{\infty} \left(\int_{-\infty}^{\infty} |f(\tau)g(t - \tau)| \, d\tau \right) dt < \infty,$$

in other words, $f * g$ is absolutely integrable. In particular, $\mathcal{F}(f * g)(\omega)$ exists and from the definitions we see that

$$\mathcal{F}(f * g)(\omega) = \int_{-\infty}^{\infty} (f * g)(t)e^{-i\omega t} \, dt$$

$$= \int_{-\infty}^{\infty} \left(\int_{-\infty}^{\infty} f(\tau)e^{-i\omega\tau}g(t - \tau)e^{-i\omega(t-\tau)} d\tau \right) dt.$$

Again one may interchange the order of integration and therefore

$$\mathcal{F}(f * g)(\omega) = \int_{-\infty}^{\infty} f(\tau)e^{-i\omega\tau} \left(\int_{-\infty}^{\infty} g(t - \tau)e^{-i\omega(t-\tau)} dt \right) d\tau.$$

If we now put $u = t - \tau$, then the inner integral equals $\int_{-\infty}^{\infty} g(u)e^{-i\omega u}\,du = G(\omega)$, and so we finally obtain

$$\mathcal{F}(f * g)(\omega) = G(\omega) \int_{-\infty}^{\infty} f(\tau)e^{-i\omega\tau}\,d\tau = G(\omega)F(\omega).$$

■

An example of this theorem has already been mentioned: for the spectrum of $(p_a * p_a)(t) = aq_a(t)$ one indeed has $(\mathcal{F}p_a)^2(\omega) = a(\mathcal{F}q_a)(\omega)$.

Closely related to convolution are the concepts cross-correlation and autocorrelation. The *cross-correlation* ρ_{fg} of two functions f and g is defined by $\rho_{fg}(t) = (g * \tilde{f})(t)$, where $\tilde{f}$ is the function $\tilde{f}(t) = \overline{f(-t)}$. Taking $g = f$ here one obtains the so-called *autocorrelation* $\rho_{ff}(t) = (f * \tilde{f})(t)$ of the function f.

Cross-correlation

Autocorrelation

EXERCISES

6.23   Calculate the convolution $(\epsilon * \epsilon)(t)$, where $\epsilon(t)$ is the unit step function from example 6.1. Is $(\epsilon * \epsilon)(t)$ absolutely integrable?

6.24   The functions $f(t)$ and $g(t)$ are defined by

$$f(t) = \begin{cases} t^{-1/2} & \text{for } 0 < t < 1, \\ 0 & \text{elsewhere,} \end{cases} \qquad g(t) = \begin{cases} (1-t)^{-1/2} & \text{for } 0 < t < 1, \\ 0 & \text{elsewhere.} \end{cases}$$

a   Show that f and g are both absolutely integrable.
b   Show that $(f * g)(1) = \int_0^1 \tau^{-1}\,d\tau$ and hence that the convolution does not exist for $t = 1$. So in general the absolute integrability of f and g is not sufficient for the existence of the convolution product.

6.25   Let f and g be two functions with $f(t) = g(t) = 0$ for $t < 0$. Functions with this property are called 'causal functions' (see section 1.2.4). The unit step function $\epsilon(t)$ is an example of a causal function. Show that $(f * g)(t) = 0$ for $t < 0$ (so $f * g$ is again causal) and that for $t \geq 0$ one has

$$(f * g)(t) = \int_0^t f(\tau)g(t - \tau)\,d\tau.$$

6.26   **a**   Show that $(e^{-|v|} * e^{-|v|})(t) = (1 + |t|)e^{-|t|}$.
b   Use theorem 6.13 to determine the spectrum of $(1 + |t|)e^{-|t|}$.
c   Determine in a direct way the spectrum of $(1 + |t|)e^{-|t|}$ (use the properties of the Fourier transform).

SUMMARY

The Fourier transform or spectrum $F(\omega)$ of a function $f(t) : \mathbb{R} \to \mathbb{C}$ is defined by

$$F(\omega) = \int_{-\infty}^{\infty} f(t)e^{-i\omega t}\,dt \qquad \text{for } \omega \in \mathbb{R},$$

provided the integral exists as improper Riemann integral. This integral certainly exists as improper Riemann integral when $f(t)$ is absolutely integrable. In that case, $F(\omega)$ is actually a continuous function on $\mathbb{R}$. The mapping $f(t) \to F(\omega)$ is the Fourier transform $\mathcal{F}$.

For a number of frequently used functions (the block function $p_a(t)$, the triangle function $q_a(t)$, the function $e^{-a|t|}$, and the Gauss function e^{-at^2}) we determined the spectrum; these have been included in table 3.

Subsequently, we derived a number of properties. In succession we treated linearity, conjugation, shift in the time domain and the frequency domain, scaling, even and odd functions, selfduality, differentiation in the time and in the frequency domain, and integration in the time domain. These properties can be found in table 4.

As an application of the differentiation rule it was shown that the set $\mathcal{S}$ of rapidly decreasing functions is invariant under the Fourier transform: if $f(t) \in \mathcal{S}$, then $F(\omega) \in \mathcal{S}$.

The convolution product $f * g$ is given by

$$(f * g)(t) = \int_{-\infty}^{\infty} f(\tau)g(t - \tau)\,d\tau,$$

provided the integral exists. One has $f * g = g * f$. If f and g are absolutely integrable, piecewise continuous, and bounded, then the convolution theorem holds: $\mathcal{F}(f * g)(\omega) = F(\omega)G(\omega)$.

SELFTEST

6.27 Consider the function $g(t) = \epsilon(t)e^{-2t}$.
a Sketch the graph of $g(t)$ and determine the spectrum $G(\omega)$. Also give the real and imaginary part of $G(\omega)$.
b Determine the spectrum of $(g * g)(t)$.
c Show that $(g * g)(t) = \epsilon(t)te^{-2t}$ by calculating the convolution and subsequently determine the spectrum of $\epsilon(t)te^{-2t}$.
d Give the relationship between the spectrum of the function $\epsilon(t)te^{-2t}$ and $G(\omega)$ in terms of differentiation. Obtain in this way the spectrum of $\epsilon(t)te^{-2t}$ again and compare the result with part b.

6.28 Consider the Gauss function $g(t) = e^{-t^2/2}$ with Fourier transform $G(\omega) = \sqrt{2\pi}e^{-\omega^2/2}$. Now determine the Fourier transform of the following functions:
a $f_1(t) = \int_{-\infty}^{t} \tau g(\tau)\,d\tau$,
b $f_2(t) = t^2 g(t)$,
c $f_3(t) = (t^2 - 2t + 1)e^{-t^2/2+t-1/2}$,
d $f_4(t) = \sin 4t \int_{-\infty}^{t} \tau g(\tau)\,d\tau$,
e $f_5(t) = e^{-8t^2}$.

6.29 Define the function $T(t)$ by

$$T(t) = \begin{cases} 1 & \text{for } |t| \leq 1, \\ 2 - |t| & \text{for } 1 < |t| \leq 2, \\ 0 & \text{for } |t| > 2. \end{cases}$$

Trapezium function
a Sketch the graph of the function $T(t)$; it is called the *trapezium function*.
b Show that $T(t) = (p_1 * p_3)(t)$, where p_a is the block function on the interval $(-a/2, a/2)$. (Hint: from the definition of $T(t)$ we see that we have to distinguish between the cases $t > 2, t < -2, 1 < t < 2, -2 < t < -1$ and $-1 < t < 1$.)
c Determine the spectrum of $T(t)$.

Contents of Chapter 7

The fundamental theorem of the Fourier integral

The fundamental theorem of the Fourier integral

INTRODUCTION

Now that we have calculated a number of frequently used Fourier transforms and have been introduced to some of the properties of Fourier transforms, it is time to return to the Fourier integral

$$\frac{1}{2\pi} \int_{-\infty}^{\infty} F(\omega)e^{i\omega t}\, d\omega.$$

It is quite reasonable to expect that, analogous to the Fourier series, the Fourier integral will in general be equal to $f(t)$. In section 6.1 this has already been derived intuitively from the fundamental theorem of Fourier series. Therefore, we start this chapter with a proof of this crucial result, which we will call the fundamental theorem of the Fourier integral. It shows that the function $f(t)$ can be recovered from its spectrum $F(\omega)$ through the Fourier integral. We should note, however, that the integral should not be interpreted as an ordinary improper Riemann integral, but as a so-called 'Cauchy principal value'.

Using the fundamental theorem we subsequently prove a number of additional properties of the Fourier transform. One of the most famous is undoubtedly Parseval's identity

$$\int_{-\infty}^{\infty} |f(t)|^2\, dt = \frac{1}{2\pi} \int_{-\infty}^{\infty} |F(\omega)|^2\, d\omega,$$

which has an important interpretation in signal theory: if a signal has a 'finite energy-content' (meaning that $\int_{-\infty}^{\infty} |f(t)|^2\, dt < \infty$), then the spectrum of the signal also has a finite energy-content.

The fundamental theorem from section 7.1, together with its consequences from section 7.2, conclude the Fourier analysis of non-periodic functions. The final section then treats Poisson's summation formula. Although not really a consequence of the fundamental theorem, it forms an appropriate closing subject of this chapter, since this formula gives an elegant relationship between the Fourier series and the Fourier integral. Moreover, we will use Poisson's summation formula in chapter 9 to determine the Fourier transform of the so-called comb distribution.

LEARNING OBJECTIVES
After studying this chapter it is expected that you
- know the Riemann–Lebesgue lemma and know its interpretation for non-periodic functions
- know the concept of Cauchy principal value
- know and can apply the fundamental theorem of the Fourier integral
- know the uniqueness theorem for the Fourier transform
- know and can apply the reciprocity property
- know and can apply the convolution theorem in the frequency domain
- know and can apply Parseval's identities
- know the concepts of energy-content and of energy-signal
- can calculate definite integrals using the fundamental theorem and Parseval's identities
- know and can apply Poisson's summation formula*.

7.1 The fundamental theorem

In this section we give a precise meaning to the Fourier integral in (6.8). We will prove that, under certain conditions on the function f in the time domain, the Fourier integral converges and, just as in the case of Fourier series, will produce the original function f. Hence, through the Fourier integral the function $f(t)$ can be recovered from its Fourier transform $F(\omega)$. The result is crucial for the remainder of the Fourier theory and so we will present a proof of this result in this book. Before we can give this proof, some preparatory results will be derived, although these are also of interest in themselves. A first step is the so-called Riemann–Lebesgue lemma on $\mathbb{R}$. For the concepts absolutely integrable and piecewise continuous we refer to definitions 6.2 and 2.3 respectively.

THEOREM 7.1
Riemann–Lebesgue lemma

Let $f(t)$ be an absolutely integrable and piecewise continuous function on $\mathbb{R}$. Then

$$\lim_{\omega \to \pm\infty} F(\omega) = \lim_{\omega \to \pm\infty} \int_{-\infty}^{\infty} f(t)e^{-i\omega t}\,dt = 0. \tag{7.1}$$

Proof
Let $\epsilon > 0$. Since f is absolutely integrable, there exist $A, B \in \mathbb{R}$ such that

$$\int_{B}^{\infty} |f(t)|\,dt + \int_{-\infty}^{A} |f(t)|\,dt < \epsilon/2.$$

On the remaining bounded interval $[A, B]$ we use the Riemann–Lebesgue lemma from section 4.1 (theorem 4.2): there exists a $G > 0$ such that for $|\omega| > G$ one has

$$\left| \int_{A}^{B} f(t)e^{-i\omega t}\,dt \right| < \epsilon/2.$$

By applying the triangle inequality repeatedly, it follows that

$$\left| \int_{-\infty}^{\infty} f(t)e^{-i\omega t}\,dt \right|$$

$$\leq \left| \int_{-\infty}^{A} f(t)e^{-i\omega t}\,dt \right| + \left| \int_{B}^{\infty} f(t)e^{-i\omega t}\,dt \right| + \left| \int_{A}^{B} f(t)e^{-i\omega t}\,dt \right|$$

$$\leq \int_{-\infty}^{A} |f(t)|\,dt + \int_{B}^{\infty} |f(t)|\,dt + \left| \int_{A}^{B} f(t)e^{-i\omega t}\,dt \right| \leq \epsilon/2 + \epsilon/2 = \epsilon$$

for $|\omega| > G$, where in the second step we used that $\left| \int g(t)\, dt \right| \le \int |g(t)|\, dt$ and $\left| e^{-i\omega t} \right| = 1$. This proves theorem 7.1. ∎

Since $e^{-i\omega t} = \cos \omega t - i \sin \omega t$, it follows immediately that (7.1) is equivalent to

$$\lim_{\omega \to \pm\infty} \int_{-\infty}^{\infty} f(t) \sin \omega t\, dt = 0 \quad \text{and} \quad \lim_{\omega \to \pm\infty} \int_{-\infty}^{\infty} f(t) \cos \omega t\, dt = 0. \qquad (7.2)$$

As a matter of fact, the Riemann–Lebesgue lemma is also valid if we only assume that the function f is absolutely integrable on $\mathbb{R}$. This more general theorem will not be proven here, since we will only need this result for absolutely integrable functions that are piecewise continuous as well; for these functions the proof is easier.

Theorem 7.1 has a nice intuitive interpretation: when we integrate the function f against ever higher frequencies, so for increasing ω, then everything will eventually cancel out. This is because the function f will change little relative to the strong oscillations of the sine and cosine functions; the area of juxtapositioned oscillations of $f(t) \sin \omega t$, for example, will cancel each other better and better for increasing ω.

Besides theorem 7.1 we will use the following identity in the proof of the fundamental theorem:

$$\int_0^{\infty} \frac{\sin t}{t}\, dt = \frac{\pi}{2}. \qquad (7.3)$$

This identity has been proven in section 4.4.1. Using theorem 7.1 and formula (7.3) we can now prove the fundamental theorem. It will turn out, however, that the Fourier integral will not necessarily exist as an improper Riemann integral. Hence, to be able to formulate the fundamental theorem properly, we need the concept of a Cauchy principal value of an integral.

DEFINITION 7.1
Cauchy principal value

The value of $\lim_{A \to \infty} \int_{-A}^{A} f(t)\, dt$ *is called the* Cauchy principal value *of the integral* $\int_{-\infty}^{\infty} f(t)\, dt$, *provided that this limit exists.*

The difference between the Cauchy principal value and the improper Riemann integral is the fact that here the limits tend to ∞ and $-\infty$ *at the same rate*. We recall that in the improper (Riemann) integral we are dealing with *two independent limits* $B \to \infty$ and $A \to -\infty$ (also see the end of section 6.1). When a function has an improper integral, then the Cauchy principal value will certainly exist and it will have the same value as the improper integral (just take '$A = B$'). The converse need not be true, as the next example shows.

EXAMPLE

The improper integral $\int_{-\infty}^{\infty} t\, dt$ does not exist, but the Cauchy principal value is 0 since $\int_{-A}^{A} t\, dt = 0$ for each $A > 0$ and hence also for $A \to \infty$. ◄

Since the Cauchy principal value of the Fourier integral is equal to

$$\lim_{A \to \infty} \frac{1}{2\pi} \int_{-A}^{A} F(\omega) e^{i\omega t}\, d\omega,$$

it seems plausible to investigate for arbitrary $A > 0$ the integral

$$\frac{1}{2\pi} \int_{-A}^{A} F(\omega) e^{i\omega t}\, d\omega = \frac{1}{2\pi} \int_{-A}^{A} \left(\int_{-\infty}^{\infty} f(s) e^{-i\omega s}\, ds \right) e^{i\omega t}\, d\omega \qquad (7.4)$$

more thoroughly. If we may interchange the order of integration (we will return to this in the proof of the fundamental theorem), then it would follow that

$$\frac{1}{2\pi} \int_{-A}^{A} F(\omega) e^{i\omega t}\, d\omega = \frac{1}{2\pi} \int_{-\infty}^{\infty} f(s) \left(\int_{-A}^{A} e^{i\omega (t-s)}\, d\omega \right) ds.$$

The inner integral can be calculated (as a matter of fact, it is precisely the Fourier transform of the block function p_{2A} at $s - t$; see section 6.3.1) and it then follows that

$$\frac{1}{2\pi} \int_{-A}^{A} F(\omega)e^{i\omega t}\, d\omega = \frac{1}{\pi} \int_{-\infty}^{\infty} f(s)\frac{\sin A(t-s)}{t-s}\, ds.$$

By changing to the variable $\tau = t - s$ we finally obtain the important formula

$$\frac{1}{2\pi} \int_{-A}^{A} F(\omega)e^{i\omega t}\, d\omega = \frac{1}{\pi} \int_{-\infty}^{\infty} f(t-\tau)\frac{\sin A\tau}{\tau}\, d\tau. \tag{7.5}$$

This result is important because it shows us *why* the Fourier integral will converge to the value $f(t)$ for $A \to \infty$ (assuming for the moment that f is continuous at t). To that end we take a closer look at the function $D_A(\tau) = \sin A\tau / \tau$ (for $A > 0$). The value at $\tau = 0$ is A, while the zeros of $D_A(\tau)$ are given by $\tau = k\pi/A$ with $k \in \mathbb{Z}$ ($k \neq 0$). For increasing A the zeros are thus getting closer and closer to each other, which means that the function $D_A(\tau)$ has ever increasing oscillations outside $\tau = 0$. This is shown in figure 7.1. This makes it plausible that at a point t where the function f is continuous we will have

$$\lim_{A\to\infty} \frac{1}{\pi} \int_{-\infty}^{\infty} f(t-\tau)\frac{\sin A\tau}{\tau}\, d\tau = f(t), \tag{7.6}$$

under suitable conditions on f. The reason is that for large values of A the strong oscillations of $D_A(\tau)$ outside $\tau = 0$ will cancel everything out (compare this with the interpretation of the Riemann–Lebesgue lemma), and so *only* the value of f at

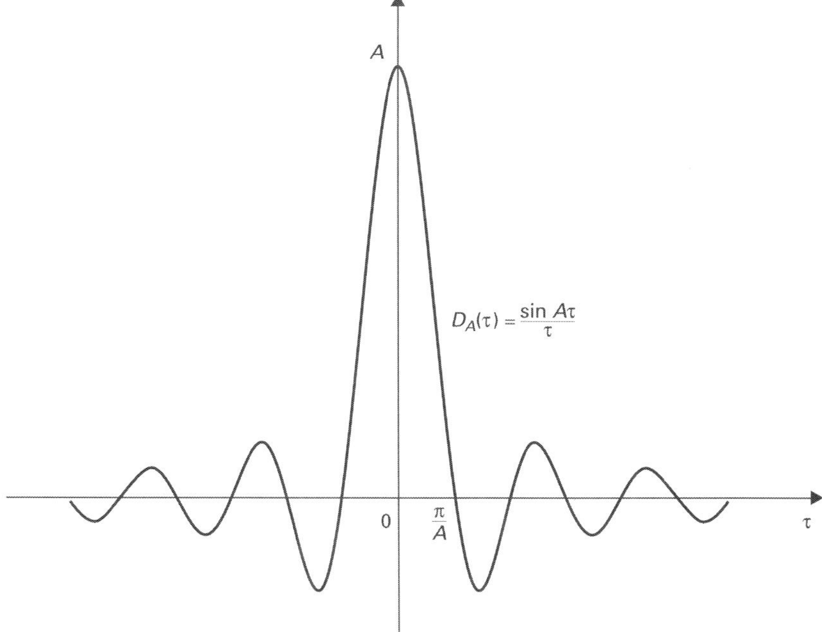

FIGURE 7.1
The function $D_A(\tau)$ for a large value of A.

the point $\tau = 0$ will contribute to the integral (of course, this is *not a proof* of (7.6)). The constant π^{-1} in the left-hand side of (7.6) can easily be 'explained' since (7.3) implies that $\int_{-\infty}^{\infty} \sin A\tau / \tau \, d\tau = \pi$ for every $A > 0$ and hence also for $A \to \infty$; but this is precisely (7.6) if we take for f the constant function 1.

It now becomes clear how we shall proceed with the proof of the fundamental theorem. First we need to prove (7.6); combining this with (7.5) then immediately gives the fundamental theorem of the Fourier integral. As a matter of fact, the assumption that f should be continuous at t was only for our convenience: in the following theorem we will drop this assumption and so we get a slightly different version of (7.6).

THEOREM 7.2

Let $f(t)$ be an absolutely integrable and piecewise smooth function on $\mathbb{R}$. Then one has for each $t \in \mathbb{R}$ that

$$\lim_{A \to \infty} \frac{1}{\pi} \int_{-\infty}^{\infty} f(t - \tau) \frac{\sin A\tau}{\tau} \, d\tau = \frac{1}{2} \left(f(t+) + f(t-) \right).$$

Proof

In order to get a well-organized proof, we divide it into two steps.

Step 1

By splitting the integral in the left-hand side at $\tau = 0$, and changing from τ to $-\tau$ in the resulting integral over $(-\infty, \infty)$, we obtain

$$\lim_{A \to \infty} \frac{1}{\pi} \int_{-\infty}^{\infty} f(t - \tau) \frac{\sin A\tau}{\tau} \, d\tau$$

$$= \lim_{A \to \infty} \frac{1}{\pi} \int_{0}^{\infty} \left(f(t - \tau) + f(t + \tau) \right) \frac{\sin A\tau}{\tau} \, d\tau. \tag{7.7}$$

Now note that by replacing $A\tau$ by v, and so τ by v/A, and applying (7.3), it follows that

$$\lim_{A \to \infty} \int_{0}^{1} \frac{\sin A\tau}{\tau} \, d\tau = \lim_{A \to \infty} \int_{0}^{A} \frac{\sin v}{v} \, dv = \frac{\pi}{2}.$$

If we multiply this equality by $\left(f(t+) + f(t-) \right) / \pi$ we obtain

$$\frac{1}{2} \left(f(t+) + f(t-) \right) = \lim_{A \to \infty} \frac{1}{\pi} \int_{0}^{1} \frac{\sin A\tau}{\tau} \left(f(t+) + f(t-) \right) d\tau \tag{7.8}$$

(in fact, $f(t+) + f(t-)$ does not depend on τ). If we now look at the result to be proven, it is quite natural to study the difference

$$\frac{1}{\pi} \int_{0}^{\infty} \left(f(t - \tau) + f(t + \tau) \right) \frac{\sin A\tau}{\tau} \, d\tau - \frac{1}{\pi} \int_{0}^{1} \left(f(t+) + f(t-) \right) \frac{\sin A\tau}{\tau} \, d\tau$$

of the right-hand sides of (7.7) and (7.8) for $A \to \infty$. If we can show that this difference tends to 0 for $A \to \infty$, then the theorem is proven. We will do this in step 2.

Step 2

If we split the first integral in this difference at $\tau = 1$, then we see that the difference can be rewritten as $I_1 + I_2$ with

$$I_1 = \frac{1}{\pi} \int_{0}^{1} \frac{f(t - \tau) - f(t-) + f(t + \tau) - f(t+)}{\tau} \sin A\tau \, d\tau,$$

$$I_2 = \frac{1}{\pi} \int_{1}^{\infty} \frac{f(t - \tau) + f(t + \tau)}{\tau} \sin A\tau \, d\tau.$$

We will now show that both I_1 and I_2 tend to 0 as $A \to \infty$. To tackle I_1 we define the auxiliary function $g(\tau)$ by

$$
g(\tau) = \begin{cases} \dfrac{f(t-\tau) - f(t-) + f(t+\tau) - f(t+)}{\tau} & \text{for } 0 \leq \tau \leq 1, \\ 0 & \text{otherwise,} \end{cases}
$$

and thus

$$
I_1 = \frac{1}{\pi} \int_{-\infty}^{\infty} g(\tau) \sin A\tau \, d\tau.
$$

Note that outside the point $\tau = 0$, the function $g(\tau)$ is certainly piecewise continuous. Since we assumed that f is piecewise smooth, theorem 2.4 implies that the limits $\lim_{\tau \downarrow 0}(f(t+\tau) - f(t+))/\tau$ and $\lim_{\tau \downarrow 0}(f(t-\tau) - f(t-))/\tau$, occurring in $g(\tau)$, both exist and are equal to $f'(t+)$ and $-f'(t-)$ respectively. The function g is thus piecewise continuous at $t = 0$, and so on $\mathbb{R}$. Moreover, g is certainly absolutely integrable and so it follows from the Riemann–Lebesgue lemma (theorem 7.1) that $\lim_{A \to \infty} I_1 = 0$. We now apply a similar reasoning to I_2. To that end we define the auxiliary function $h(\tau)$ by

$$
h(\tau) = \begin{cases} \dfrac{f(t-\tau) + f(t+\tau)}{\tau} & \text{for } \tau \geq 1, \\ 0 & \text{otherwise,} \end{cases}
$$

and so

$$
I_2 = \frac{1}{\pi} \int_{-\infty}^{\infty} h(\tau) \sin A\tau \, d\tau.
$$

The function h is certainly piecewise continuous on $\mathbb{R}$ and also absolutely integrable since $1/\tau < 1$ if $\tau > 1$ and the function f is absolutely integrable. We can therefore apply the Riemann–Lebesgue lemma again to obtain that $\lim_{A \to \infty} I_2 = 0$. This proves theorem 7.2. ∎

If, in particular, the function f in theorem 7.2 is continuous at a point t, then we obtain (7.6). By combining theorem 7.2 with (7.5) we now immediately obtain the fundamental theorem.

THEOREM 7.3
Fundamental theorem of the Fourier integral

Let $f(t)$ be an absolutely integrable and piecewise smooth function on $\mathbb{R}$ and let $F(\omega)$ be the Fourier transform of f. Then the Fourier integral converges for each $t \in \mathbb{R}$ as a Cauchy principal value and

$$
\frac{1}{2\pi} \int_{-\infty}^{\infty} F(\omega) e^{i\omega t} \, d\omega = \frac{1}{2} \left(f(t+) + f(t-) \right). \tag{7.9}
$$

Here $f(t+) = \lim_{h \downarrow 0} f(t+h)$ and $f(t-) = \lim_{h \uparrow 0} f(t+h)$.

Proof
Without proof we mention that under the conditions of theorem 7.3 interchanging the order of integration, which was used to derive (7.5) from (7.4), is indeed allowed. From (7.5) it then follows that

$$
\lim_{A \to \infty} \frac{1}{2\pi} \int_{-A}^{A} F(\omega) e^{i\omega t} \, d\omega = \lim_{A \to \infty} \frac{1}{\pi} \int_{-\infty}^{\infty} f(t-\tau) \frac{\sin A\tau}{\tau} \, d\tau.
$$

If we now apply theorem 7.2 to the right-hand side, then we obtain

$$
\lim_{A \to \infty} \frac{1}{2\pi} \int_{-A}^{A} F(\omega) e^{i\omega t} \, d\omega = \frac{1}{2} \left(f(t+) + f(t-) \right),
$$

proving theorem 7.3: the Fourier integral converges as a Cauchy principal value to $(f(t+) + f(t-))/2$.　∎

The Fourier integral should in the first instance always be considered as a Cauchy principal value. However, in many cases it is immediately clear that the integral also exists as an improper Riemann integral, for example when $F(\omega)$ is absolutely integrable. Nevertheless, one should not jump to conclusions too easily, and as a warning we present a simple example showing that the Fourier integral can certainly not always be considered as an improper integral.

EXAMPLE 7.1

Let $f(t)$ be the function defined by $f(t) = \epsilon(t)e^{-t}$, where $\epsilon(t)$ is the unit step function (see example 6.1). This function satisfies all the conditions of theorem 7.3 and $F(\omega) = 1/(1+i\omega)$ (see table 3). For $t = 0$ it then follows from the fundamental theorem that

$$\int_{-\infty}^{\infty} \frac{1}{1+i\omega}\, d\omega = \pi,$$

since $(f(t+) + f(t-))/2 = 1/2$ for $t = 0$. This integral must be considered as a Cauchy principal value since it does not exist as an improper Riemann integral. This is because the imaginary part of $1/(1 + i\omega)$ is $-\omega/(1 + \omega^2)$ and

$$\int_{A}^{B} \frac{\omega}{1+\omega^2}\, d\omega = \frac{1}{2}\left(\ln(1 + B^2) - \ln(1 + A^2)\right),$$

which means that the limit does not exist for $A \to -\infty$ (or $B \to \infty$).　◀

For a continuous function $f(t)$, the right-hand side of (7.9) equals $f(t)$ since in this case $f(t+) = f(t-)$ for each $t \in \mathbb{R}$. Let us give another example.

EXAMPLE 7.2

Take $f(t) = e^{-a|t|}$, then $F(\omega) = 2a/(a^2 + \omega^2)$ (see table 3). The function f satisfies all the conditions of the fundamental theorem and it is also continuous (and even). For each $t \in \mathbb{R}$ we thus have

$$e^{-a|t|} = \frac{1}{2\pi} \int_{-\infty}^{\infty} \frac{2a}{a^2 + \omega^2} e^{i\omega t}\, d\omega. \tag{7.10}$$

Since $F(\omega)$ is even, this can also be written as a Fourier cosine transform (see section 6.4.6):

$$e^{-a|t|} = \frac{2}{\pi} \int_{0}^{\infty} \frac{a}{a^2 + \omega^2} \cos \omega t\, d\omega.$$
　◀

Inversion formula

Formula (7.9) is called the *inversion formula* for the Fourier transform on $\mathbb{R}$. The name 'inversion formula' is clear: if the conditions of theorem 7.3 are met, then one can recover the original function $f(t)$ from its spectrum $F(\omega)$ using (7.9). There-

Inverse Fourier transform

fore, the function $f(t)$ is called the *inverse Fourier transform* of $F(\omega)$.

Here we have proven the fundamental theorem under the condition that f is piecewise smooth and absolutely integrable. There are many other conditions for which the theorem remains valid. It is, however, remarkable that up till now there are no known conditions for absolutely integrable functions that are both necessary and sufficient (in other words, a minimal condition) for the inversion formula (7.9) to be valid.

EXAMPLE 7.3

A well-known continuous function which doesn't satisfy the conditions of the fundamental theorem is $f(t) = \sin t/t$. Although this function is piecewise smooth (and even continuously differentiable), it is not absolutely integrable, as was mentioned in section 6.3.1.　◀

If $f(t)$ is an even function, then $F(\omega)$ is also even (see section 6.4.6). Since one then has for any $A > 0$ that

$$\int_{-A}^{A} F(\omega) \cos \omega t \, d\omega = 2 \int_{0}^{A} F(\omega) \cos \omega t \, d\omega \quad \text{and} \quad \int_{-A}^{A} F(\omega) \sin \omega t \, d\omega = 0,$$

it follows that

$$\int_{-\infty}^{\infty} F(\omega) e^{i\omega t} \, d\omega = 2 \int_{0}^{\infty} F(\omega) \cos \omega t \, d\omega$$

as a Cauchy principal value. For even functions we thus obtain the following version of the fundamental theorem:

Fundamental theorem for even functions
$$\frac{2}{\pi} \int_{0}^{\infty} F_c(\omega) \cos \omega t \, d\omega = \frac{1}{2} \left(f(t+) + f(t-) \right), \tag{7.11}$$

where

$$F_c(\omega) = \int_{0}^{\infty} f(t) \cos \omega t \, dt$$

is the Fourier cosine transform of f (see section 6.4.6). Note that the integral in (7.11) is now an ordinary improper integral. When a function $f(t)$ is only defined for $t > 0$, then, just as in section 6.4.6, one can extend the function f to an even function on $\mathbb{R}$ which will be denoted by $f(t)$ again; formula (7.11) then holds for this even extension.

For an odd function $g(t)$ the function $G(\omega)$ is odd and one obtains in a similar fashion a version of the fundamental theorem for odd functions:

Fundamental theorem for odd functions
$$\frac{2}{\pi} \int_{0}^{\infty} G_s(\omega) \sin \omega t \, d\omega = \frac{1}{2} \left(g(t+) + g(t-) \right), \tag{7.12}$$

where

$$G_s(\omega) = \int_{0}^{\infty} g(t) \sin \omega t \, dt$$

is the Fourier sine transform of f (see section 6.4.6). When a function is only defined for $t > 0$ and we extend it to an odd function on $\mathbb{R}$, then (7.12) will hold for this odd extension. Of course, the conditions of the fundamental theorem should be satisfied in all of the preceding statements.

Now that we have done the hard work in proving the fundamental theorem, we will reap the fruits of it in the next section.

EXERCISES

7.1 Verify for the spectra from the exercises in section 6.3 that the Riemann–Lebesgue lemma holds.

7.2 Show that the Fourier transform of the block function p_{2A} at the point $\omega = s - t$ equals $2 \sin(A(t - s))/(t - s)$ (see the derivation of (7.5) from (7.4)).

7.3 Show that for arbitrary $C > 0$ one has

$$\lim_{A \to \infty} \int_{0}^{C} \frac{\sin Au}{u} \, du = \frac{\pi}{2}$$

(see step 1 in the proof of theorem 7.2).

7.4   Calculate in a direct way (so in contrast to example 7.1) the Cauchy principal value
of

$$\int_{-\infty}^{\infty} \frac{1}{1+i\omega}\, d\omega.$$

7.5   **a** Check that the block function $p_a(t)$ satisfies the conditions of the fundamental
theorem and give the inversion formula.
b Use (7.3) to show that the integral in the fundamental theorem of part a exists as
improper integral.

7.6   Let the function $f(t)$ be given by (see exercise 6.9b)

$$f(t) = \begin{cases} \sin t & \text{for } |t| \leq \pi, \\ 0 & \text{elsewhere.} \end{cases}$$

a Check that $f(t)$ satisfies the conditions of the fundamental theorem and that
$F(\omega)$ exists as improper integral.
b Prove that

$$f(t) = \frac{1}{\pi} \int_0^{\infty} \frac{\cos(\pi - t)\omega - \cos(\pi + t)\omega}{1 - \omega^2}\, d\omega.$$

7.7   Show that for an odd function the fundamental theorem can be written as in (7.12).

7.8   Consider the function

$$f(t) = \begin{cases} 1 & \text{for } 0 < t \leq a, \\ 0 & \text{for } t > a. \end{cases}$$

a Use the Fourier sine transform of the function $f(t)$ (see exercise 6.15b) to show
that for $t > 0$ $(t \neq a)$ one has

$$f(t) = \frac{2}{\pi} \int_0^{\infty} \frac{1 - \cos a\omega}{\omega} \sin \omega t\, d\omega.$$

b To which value does the right-hand side of the identity in part a converge for
$t = a$?

7.9   The ordinary multiplication of real numbers has a *unit*, that is to say, there exists a
number e (namely the number 1) such that $ex = x$ for all $x \in \mathbb{R}$. Now take as multiplication of functions the convolution product $f * g$. Use the Riemann–Lebesgue
lemma and the convolution theorem (theorem 6.13) to show that the convolution
product has no unit. In others words, there does not exist a function e such that
$e * f = f$ for all f. (All functions are assumed to be bounded, absolutely integrable
and piecewise smooth.)

7.2   Consequences of the fundamental theorem

In this section we use the fundamental theorem to derive a number of important
additional properties of the Fourier transform.

7.2.1   *Uniqueness*

From the fundamental theorem we immediately obtain the uniqueness of the Fourier
transform.

THEOREM 7.4
Uniqueness theorem

Let $f(t)$ and $g(t)$ be absolutely integrable and piecewise smooth functions on $\mathbb{R}$ with spectra $F(\omega)$ and $G(\omega)$. If $F(\omega) = G(\omega)$ on $\mathbb{R}$, then $f(t) = g(t)$ at all points where f and g are continuous.

Proof
Let $t \in \mathbb{R}$ be a point where f and g are both continuous. Since $F(\omega) = G(\omega)$, it then follows from the fundamental theorem that

$$f(t) = \frac{1}{2\pi} \int_{-\infty}^{\infty} F(\omega)e^{i\omega t}\, d\omega = \frac{1}{2\pi} \int_{-\infty}^{\infty} G(\omega)e^{i\omega t}\, d\omega = g(t).$$

■

The uniqueness theorem is often applied (implicitly) when addressing the following frequently occurring problem: given $F(\omega)$, find a function $f(t)$ with spectrum $F(\omega)$. Let us assume that on the basis of a table, and perhaps in combination with the properties of the Fourier transform, we know a function $f(t)$ indeed having $F(\omega)$ as its spectrum. (In most cases a direct calculation of the Fourier integral is not a very clever method.) Theorem 7.4 then guarantees that the function we have found is the only possibility within the set of absolutely integrable and piecewise smooth functions. This is up to a finite number of points (on an arbitrary bounded interval) where we can give the function f an arbitrary (finite) value (see figure 7.2). This is because if f and g are two functions with the same spectrum, then we can only conclude that $(f(t+) + f(t-))/2 = (g(t+) + g(t-))/2$ at the points where f and/or g are not continuous. These exceptional points are of little importance; however, in order to formulate results like the fundamental theorem accurately, one should keep them in mind. In some of the literature these exceptional points are avoided by assuming that at a point $t \in \mathbb{R}$ where a piecewise smooth function is not continuous, the function value is always defined as $(f(t+) + f(t-))/2$. In that case, theorem 7.4 is thus correct on $\mathbb{R}$. This is the case, for example, for the function h in figure 7.2.

EXAMPLE

We are looking for a function with spectrum $F(\omega) = 1/(1 + i\omega)^2$. This function is the product of $1/(1 + i\omega)$ with itself and from table 3 it follows that $g(t) = \epsilon(t)e^{-t}$ has as its spectrum precisely $1/(1 + i\omega)$. From the convolution theorem (theorem 6.13) it then follows that $f(t) = (g * g)(t)$ has $F(\omega)$ as its spectrum. It is easy to calculate the convolution product (see exercise 6.27) and from this it follows that $f(t) = \epsilon(t)te^{-t}$. According to theorem 7.4 this is the only absolutely integrable and piecewise smooth function with spectrum $F(\omega)$.   ◄

The uniqueness theorem gives rise to a new notation. When $f(t)$ is absolutely integrable and piecewise smooth, and when $F(\omega)$ is the spectrum of $f(t)$, then we will write from now on

$$f(t) \leftrightarrow F(\omega).$$

This expresses the fact that f and F determine each other uniquely according to theorem 7.4. As before, one should adjust the value of f in the exceptional points, where f is not continuous, as before. Often the following equivalent formulation of theorem 7.4 is given (we keep the conditions of the theorem): when $F(\omega) = 0$ on $\mathbb{R}$, then $f(t) = 0$ at all points t where f is continuous. This formulation is equivalent because of the linearity property (see section 6.4.1).

7.2.2 *Fourier pairs*

In section 6.3 the Fourier transforms were calculated for a number of frequently used functions. Using the fundamental theorem it will follow that Fourier transforms

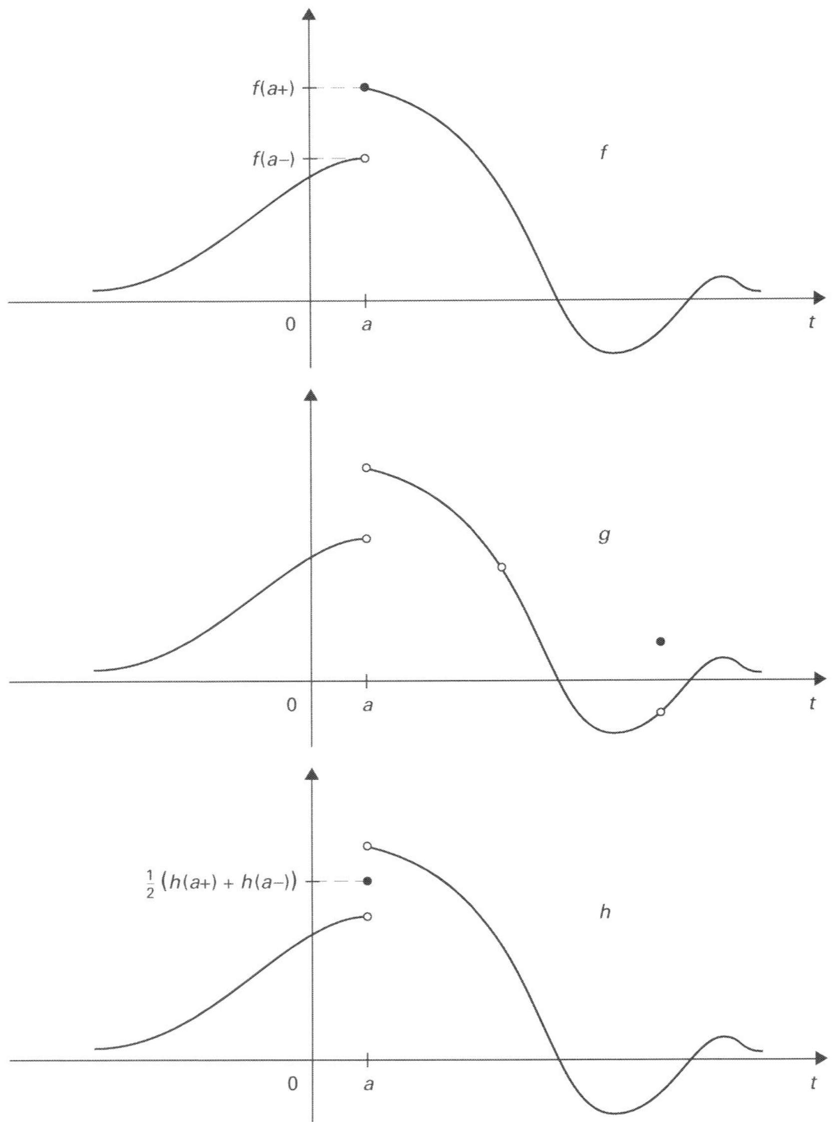

FIGURE 7.2
Three functions f, g and h with the same spectrum.

usually occur *in pairs*, which immediately doubles the number of examples. Let us assume that a function $f(t)$ with spectrum $F(\omega)$ satisfies the conditions of the fundamental theorem. For convenience we also assume that $f(t)$ is continuous, and according to theorem 7.3 we then have for each $t \in \mathbb{R}$ that

$$f(t) = \frac{1}{2\pi} \int_{-\infty}^{\infty} F(\omega)e^{i\omega t}\, d\omega.$$

Let us moreover assume that this integral exists as an improper integral (which usually is the case). Both t and ω are just variables for which one may as well choose another symbol. In particular we may interchange the role of t and ω, and it then follows that

$$f(\omega) = \frac{1}{2\pi} \int_{-\infty}^{\infty} F(t)e^{i\omega t}\, dt.$$

If we now change from the variable t to the variable $-t$ we obtain

$$f(\omega) = \frac{1}{2\pi} \int_{-\infty}^{\infty} F(-t)e^{-i\omega t}\, dt. \tag{7.13}$$

But this last integral is precisely the Fourier transform of the function $F(-t)/2\pi$. Summarizing, we have proven the following theorem.

THEOREM 7.5

Let $f(t)$ be an absolutely integrable and piecewise smooth function on $\mathbb{R}$ with spectrum $F(\omega)$. Assume that f is continuous and that the Fourier integral exists as an improper Riemann integral. Then

$$F(-t) \leftrightarrow 2\pi f(\omega).$$

Duality

Reciprocity

Hence, Fourier transforms almost always occur in pairs. This property is called the *duality* or the *reciprocity* of the Fourier transform and is included as property 11 in table 4. Do not confuse 'duality' with the 'selfduality' from section 6.4.7. If the function $f(t)$ is not continuous, but merely piecewise continuous, then (7.13) and the duality property will still hold, with the exception of the points where f is not continuous (there the left-hand side of (7.13) has to be adjusted). A special case arises when f is an even function. Then F is also an even function, from which it follows that in this case $F(t) \leftrightarrow 2\pi f(\omega)$.

EXAMPLE 7.4

Take $f(t) = e^{-a|t|}$ as in example 7.2. This is an even and continuous function and $F(\omega)$ is absolutely integrable. From theorem 7.5 we then obtain

$$\frac{a}{a^2 + t^2} \leftrightarrow \pi e^{-a|\omega|}, \tag{7.14}$$

and so we have found a new Fourier transform. ◄

We called the mapping assigning Fourier transforms to functions the Fourier transform. For the distribution theory in chapters 8 and 9 it is important to know the image under the Fourier transform of the set $\mathcal{S}$ ($\mathcal{S}$ is the set of rapidly decreasing functions; see section 6.5). In theorem 6.12 it was already shown that $F(\omega) \in \mathcal{S}$ for $f(t) \in \mathcal{S}$. In other words, the Fourier transform maps the space $\mathcal{S}$ into itself. Moreover, functions in $\mathcal{S}$ certainly satisfy the conditions of the uniqueness theorem; the Fourier transform is thus a *one-to-one* mapping in the space $\mathcal{S}$. With the duality property one can now easily show that the image of $\mathcal{S}$ is the whole of $\mathcal{S}$, in other words, the Fourier transform is also a mapping from $\mathcal{S}$ *onto* $\mathcal{S}$.

THEOREM 7.6

The Fourier transform is a one-to-one mapping from $\mathcal{S}$ onto $\mathcal{S}$.

Proof
For a given function $f \in \mathcal{S}$ we must find a function in the time domain having f as its spectrum. Now let $F(\omega)$ be the spectrum of $f(t)$. By theorem 6.12 the function F belongs to $\mathcal{S}$ and so it is certainly absolutely integrable (see the remark preceding theorem 6.12). All conditions of theorem 7.5 are thus satisfied and it follows that the function $F(-t)/2\pi \in \mathcal{S}$ has the function $f(\omega)$ as its spectrum. We have thus found the desired function and this proves theorem 7.6. ■

Sometimes theorem 7.6 is also formulated as follows: for each $F(\omega) \in \mathcal{S}$ there exists a (unique) $f(t) \in \mathcal{S}$ such that $F(\omega)$ is the spectrum of $f(t)$. We recall that $f(t)$ is called the inverse Fourier transform of $F(\omega)$ (see section 7.1). In particular we see that the inverse Fourier transform of a function in $\mathcal{S}$ belongs to $\mathcal{S}$ again (the 'converse' of theorem 6.12). In the next example we apply theorem 7.6 to derive a well-known result from probability theory.

EXAMPLE 7.5

Application of the reciprocity property to the Gauss function e^{-at^2} will not result in a new Fourier transform, since the spectrum of the Gauss function has a similar form. Still, we will have a closer look at the Gauss function and the Fourier pair

$$e^{-at^2} \leftrightarrow \sqrt{\frac{\pi}{a}}e^{-\omega^2/4a} \qquad \text{and} \qquad \frac{1}{2\sqrt{\pi a}}e^{-t^2/4a} \leftrightarrow e^{-a\omega^2}. \qquad (7.15)$$

From (6.15) it follows that

$$\int_{-\infty}^{\infty} \frac{1}{2\sqrt{\pi a}}e^{-t^2/4a}\,dt = 1.$$

Probability distribution

Functions with integral over $\mathbb{R}$ equal to 1, which moreover are positive on $\mathbb{R}$, are called *probability distributions* in probability theory. If we now write

$$W_a(t) = \frac{1}{2\sqrt{\pi a}}e^{-t^2/4a},$$

Normal distribution

then W_a is a positive function with integral over $\mathbb{R}$ equal to 1 and so a probability distribution. It is called the *normal distribution*. As an application of our results, we can now prove a nice and important property of the normal distribution, namely

$$(W_a * W_b)(t) = W_{a+b}(t). \qquad (7.16)$$

(For those familiar with stochastic variables, this result can be reformulated as follows: if X and Y are two independent stochastic variables with normal distributions W_a and W_b, then $X + Y$ has normal distribution W_{a+b}.) To prove this result, we first apply the convolution theorem from section 6.6 and then use (7.15):

$$\begin{aligned}
\mathcal{F}(W_a * W_b)(\omega) &= (\mathcal{F}W_a)(\omega)(\mathcal{F}W_b)(\omega)\\
&= e^{-a\omega^2}e^{-b\omega^2} = e^{-(a+b)\omega^2} = (\mathcal{F}W_{a+b})(\omega).
\end{aligned}$$

Since $\mathcal{F}(W_a * W_b) \in \mathcal{S}$, the inverse Fourier transform $W_a * W_b$ will also belong to $\mathcal{S}$. Hence, both $W_a * W_b$ and W_{a+b} belong to $\mathcal{S}$ and (7.16) then follows from the uniqueness of the Fourier transform on the space $\mathcal{S}$ (theorem 7.6). ◄

7.2.3 Definite integrals

Using the fundamental theorem one can also calculate certain definite integrals. As an example we take the block function $p_a(t)$ with spectrum $2\sin(a\omega/2)/\omega$ (see table 3). Applying the fundamental theorem for $t = 0$, it follows that

$$\int_{-\infty}^{\infty} \frac{\sin(a\omega/2)}{\omega}\,d\omega = \pi.$$

Next we change to the variable $t = a\omega/2$ and use that the integrand is an even function. We then obtain

$$\int_{0}^{\infty} \frac{\sin t}{t}\,dt = \frac{\pi}{2},$$

which proves (7.3) again. Note, however, that in proving the fundamental theorem we have used (7.3), which means that we cannot claim to have found a new proof of (7.3). For other choices for the function $f(t)$, the fundamental theorem can indeed provide new results that may be much harder to prove using other means. Let us give some more examples.

EXAMPLE 7.6

Let $q_a(t)$ be the triangle function with spectrum $4\sin^2(a\omega/2)/(a\omega^2)$ (see table 3 or section 6.3.2). Applying theorem 7.3 for $t = 0$ we obtain

$$\frac{1}{2\pi}\int_{-\infty}^{\infty}\frac{4\sin^2(a\omega/2)}{a\omega^2}\,d\omega = 1.$$

Now change to the variable $t = a\omega/2$ and use that the integrand is an even function. We then find

$$\int_0^{\infty}\frac{\sin^2 t}{t^2}\,dt = \frac{\pi}{2}. \tag{7.17}$$

◀

EXAMPLE 7.7

Take $f(t) = e^{-a|t|}$. All preparations can be found in example 7.2: if we take $t = 0$ and $a = 1$ in (7.10), and change to the variable $t = \omega$, we obtain

$$\int_{-\infty}^{\infty}\frac{1}{1+t^2}\,dt = \pi.$$

(This result can easily be obtained in a direct way since a primitive of the integrand is the well-known function $\arctan t$.) Now write for $a > 0$

$$P_a(t) = \frac{1}{\pi}\frac{a}{a^2+t^2}.$$

Cauchy distribution

Then P_a is a positive function with integral over $\mathbb{R}$ equal to 1 and so P_a is a probability distribution (see example 7.5); it is called the *Cauchy distribution*. ◀

7.2.4 *Convolution in the frequency domain*

In section 6.6 the convolution theorem (theorem 6.13) was proven: if $f(t) \leftrightarrow F(\omega)$ and $g(t) \leftrightarrow G(\omega)$, then $(f*g)(t) \leftrightarrow F(\omega)G(\omega)$. The duality property suggests that a similar result should hold for convolution in the frequency domain. Under certain conditions on the functions f and g this is indeed the case. Since these conditions will return more often, it will be convenient to introduce a new class of functions.

DEFINITION 7.2
Square integrable

A function $f : \mathbb{R} \to \mathbb{C}$ is called square integrable *on $\mathbb{R}$ if $\int_{-\infty}^{\infty}|f(t)|^2\,dt$ exists as improper Riemann integral.*

If a function f is absolutely integrable, then it need not be square integrable. For example, the function $f(t) = p_2(t)\,|t|^{-1/2}$ is absolutely integrable, but not square integrable since $|t|^{-1}$ is not integrable over the interval $-1 \leq t \leq 1$. One now has the following convolution theorem in the frequency domain.

THEOREM 7.7
Convolution theorem in the frequency domain

Let $f(t)$ and $g(t)$ be piecewise smooth functions which, in addition, are absolutely integrable and square integrable on $\mathbb{R}$. Then

$$f(t)g(t) \leftrightarrow (2\pi)^{-1}(F*G)(\omega).$$

Proof
We give the proof under the simplifying assumption that the spectra $F(\omega)$ and $G(\omega)$ exist as improper Riemann integrals. Since $(a+b)^2 = a^2 + 2ab + b^2 \geq 0$, we

see that $| f(t)g(t) | \leq \frac{1}{2}(| f(t) |^2 + | g(t) |^2)$ for all $t \in \mathbb{R}$. Since f and g are square integrable, it now follows that $f(t)g(t)$ is absolutely integrable. Hence, the spectrum of $f(t)g(t)$ will certainly exist. According to the fundamental theorem we may replace $f(t)$ in

$$\mathcal{F}(f(t)g(t))(\omega) = \int_{-\infty}^{\infty} f(t)g(t)e^{-i\omega t}\, dt$$

by

$$\frac{1}{2\pi} \int_{-\infty}^{\infty} F(\tau)e^{i\tau t}\, d\tau,$$

where we may consider this integral as an improper Riemann integral (the exceptional points, where $f(t)$ is not equal to this integral, are not relevant since we integrate over the variable t). It then follows that

$$\mathcal{F}(f(t)g(t))(\omega) = \frac{1}{2\pi} \int_{-\infty}^{\infty} \int_{-\infty}^{\infty} F(\tau)g(t)e^{-i(\omega-\tau)t}\, d\tau\, dt$$

$$= \frac{1}{2\pi} \int_{-\infty}^{\infty} \left(\int_{-\infty}^{\infty} g(t)e^{-i(\omega-\tau)t}\, dt \right) F(\tau)\, d\tau,$$

where we have changed the order of integration. We state without proof that this is allowed under the conditions of theorem 7.7. We recognize the inner integral as $G(\omega - \tau)$. This proves theorem 7.7, since it follows that

$$\mathcal{F}(f(t)g(t))(\omega) = \frac{1}{2\pi} \int_{-\infty}^{\infty} F(\tau)G(\omega - \tau)\, d\tau = \frac{1}{2\pi}(F * G)(\omega).$$

■

For complicated functions one can now still obtain the spectrum, in the form of a convolution product, by using theorem 7.7. In some cases this convolution product can then be calculated explicitly.

EXAMPLE

Suppose we need to find the spectrum of the function $(1+t^2)^{-2}$. In example 7.4 we showed that $1/(1+t^2) \leftrightarrow \pi e^{-|\omega|}$ and since all conditions of theorem 7.7 are met, it follows that $(1+t^2)^{-2} \leftrightarrow (\pi/2)(e^{-|\tau|} * e^{-|\tau|})(\omega)$. By calculating the convolution product (see exercise 6.26a), it then follows that $(1+t^2)^{-2} \leftrightarrow (\pi/2)(1+|\omega|)e^{-|\omega|}$.

◀

7.2.5   Parseval's identities

In theorem 7.7 it was shown that

$$\int_{-\infty}^{\infty} f(t)g(t)e^{-i\omega t}\, dt = \frac{1}{2\pi} \int_{-\infty}^{\infty} F(\tau)G(\omega - \tau)\, d\tau$$

and at the point $\omega = 0$ this gives the identity

$$\int_{-\infty}^{\infty} f(t)g(t)\, dt = \frac{1}{2\pi} \int_{-\infty}^{\infty} F(\tau)G(-\tau)\, d\tau.$$

Now replace $g(t)$ by $\overline{g(t)}$ and use that $\overline{g(t)} \leftrightarrow \overline{G(-\omega)}$ (see table 4), then it follows that

$$\int_{-\infty}^{\infty} f(t)\overline{g(t)}\, dt = \frac{1}{2\pi} \int_{-\infty}^{\infty} F(\omega)\overline{G(\omega)}\, d\omega. \tag{7.18}$$

Parseval's identity

This result is known as the theorem of Parseval or as *Parseval's identity* and it thus holds under the same conditions as theorem 7.7 (taking the complex conjugate has no influence on the conditions). Since $z\bar{z} = |z|^2$ for $z \in \mathbb{C}$, formula (7.18) reduces for $f(t) = g(t)$ to

$$\int_{-\infty}^{\infty} |f(t)|^2 \, dt = \frac{1}{2\pi} \int_{-\infty}^{\infty} |F(\omega)|^2 \, d\omega. \tag{7.19}$$

Plancherel's identity

In order to distinguish between formulas (7.18) and (7.19) one often calls (7.19) *Plancherel's identity*. We shall not make this distinction: both identities will be called Parseval's identity. (Compare (7.18) and (7.19) with Parseval's identities for Fourier series.) Note that in (7.19) it is quite natural to require $f(t)$ to be square integrable.

Identity (7.19) shows that square integrable functions have a Fourier transform that is again square integrable. In signal theory a square integrable function is also called a signal with finite energy-content or *energy-signal* for short. The value $\int_{-\infty}^{\infty} |f(t)|^2 \, dt$ is then called the *energy-content* of the signal $f(t)$ (also see chapter 1). Identity (7.19) shows that the spectrum of an energy-signal has again a finite energy-content.

Energy-signal
Energy-content

Parseval's identities can also be used to calculate certain definite integrals.

EXAMPLE

Again we consider the block function $p_a(t)$ from section 6.3.1 having spectrum $2\sin(a\omega/2)/\omega$. The function p_a satisfies the conditions of theorem 7.7 and from (7.19) it then follows that

$$\int_{-\infty}^{\infty} |p_a(t)|^2 \, dt = \frac{1}{2\pi} \int_{-\infty}^{\infty} \frac{4\sin^2(a\omega/2)}{\omega^2} \, d\omega.$$

The left-hand side is easy to calculate and equals a. Changing to the variable $t = a\omega/2$ in the right-hand side, we again obtain the result (7.17) from example 7.6. ◄

EXAMPLE

In (7.18) we take $f(t) = p_a(t)$ and $g(t) = p_b(t)$ with $0 \leq a \leq b$. It then follows that

$$\int_{-\infty}^{\infty} p_a(t)p_b(t) \, dt = \frac{1}{2\pi} \int_{-\infty}^{\infty} \frac{4\sin(a\omega/2)\sin(b\omega/2)}{\omega^2} \, d\omega.$$

The left-hand side equals a (since $a \leq b$), while the integrand in the right-hand side is an even function. Hence,

$$\int_{0}^{\infty} \frac{\sin(a\omega/2)\sin(b\omega/2)}{\omega^2} \, d\omega = \frac{\pi}{4} a.$$

Now change to the variable $t = a\omega/2$, then it follows for any $c \, (= b/a) \geq 1$ that

$$\int_{0}^{\infty} \frac{\sin t \sin ct}{t^2} \, dt = \frac{\pi}{2}.$$

The previous example is the case $a = b = 1$. ◄

EXERCISES

7.10 Show that theorem 7.4 is equivalent to the following statement: if $F(\omega) = 0$ on $\mathbb{R}$, then $f(t) = 0$ at all points t where f is continuous.

7.11 Use the duality property to determine the Fourier transform of the function $\sin(at/2)/t$. Also see exercise 7.5, where we already verified the conditions.

7.12 Use the duality property to determine the Fourier transform of the function $\sin^2(at/2)/t^2$.

7.13   Use the convolution theorem in the frequency domain to determine the spectrum of the function $\sin^2(at/2)/t^2$ (see section 6.6 for $p_a * p_a$). Check your answer using the result from exercise 7.12 and note that although $\sin(at/2)/t$ is *not* absolutely integrable, theorem 7.7 still produces the correct result.

7.14   According to table 3 one has for $a > 0$ that $\epsilon(t)e^{-at} \leftrightarrow 1/(a + i\omega)$. Can we now use duality to conclude that $1/(a + it) \leftrightarrow 2\pi\epsilon(-\omega)e^{a\omega}$?

7.15   Verify that the duality property applied to the relation

$$e^{-at^2} \leftrightarrow \sqrt{\frac{\pi}{a}}e^{-\omega^2/4a} \qquad \text{leads to the result} \qquad \frac{1}{2\sqrt{\pi a}}e^{-t^2/4a} \leftrightarrow e^{-a\omega^2}.$$

Then show that the second relation is a direct consequence of the first relation by changing from a to $1/(4a)$ in the first relation. Hence, in this case we do not find a new Fourier transform.

7.16   Determine the Fourier transform of the following functions:
a $f(t) = 1/(t^2 - 2t + 2)$,
b $f(t) = \sin 2\pi(t - 3)/(t - 3)$,
c $f(t) = \sin 4t/(t^2 - 4t + 7)$,
d $f(t) = \sin^2 3(t - 1)/(t^2 - 2t + 1)$.

7.17   Determine the function $f(t)$ having the following function $F(\omega)$ as its spectrum:
a $F(\omega) = 1/(\omega^2 + 4)$,
b $F(\omega) = p_{2a}(\omega - \omega_0) + p_{2a}(\omega + \omega_0)$ for $a > 0$,
c $F(\omega) = e^{-3|\omega - 9|}$.

7.18   Let f and g be two functions in $\mathcal{S}$. Use the convolution theorem (theorem 6.13) and theorem 7.6 to show that $f * g$ belongs to $\mathcal{S}$.

7.19   Let $P_a(t)$ be as in example 7.7. Show that $P_a * P_b = P_{a+b}$. Here one may use that $P_a * P_b$ is continuously differentiable (P_a is not an element of $\mathcal{S}$; see exercise 6.20).

7.20   Consider the Fourier transform on the space $\mathcal{S}$. Use the reciprocity property to show that $\mathcal{F}^4$ is the identity on $\mathcal{S}$ (up to a constant): $\mathcal{F}^4(f(t)) = 4\pi^2 f(t)$ for any $f \in \mathcal{S}$.

7.21   For $t > 0$ we define the function $g(t)$ by $g(t) = e^{-t}$. Extend this function to an odd function on $\mathbb{R}$. The spectrum of g has been determined in exercise 6.14.
a   Use the fundamental theorem to show that for $t > 0$ one has

$$\int_0^\infty \frac{x \sin xt}{1 + x^2}\,dx = \frac{\pi}{2}e^{-t}.$$

Why is this identity not correct for $t = 0$?
b   Next use Parseval's identity to show that

$$\int_0^\infty \frac{x^2}{(1 + x^2)^2}\,dx = \frac{\pi}{4}.$$

7.22   Use the function $e^{-a|t|}$ to show that

$$\int_0^\infty \frac{1}{(a^2 + x^2)(b^2 + x^2)}\,dx = \frac{\pi}{2ab(a + b)}.$$

7.23   a   Determine the spectrum of the function $\sin^4 t/t^4$ using the convolution theorem in the frequency domain.
b   Calculate

$$\int_{-\infty}^\infty \frac{\sin^4 x}{x^4}\,dx.$$

7.24   Find the function $f(t)$ with spectrum $1/(1 + \omega^2)^2$ and use this to give a new proof of the identity from exercise 7.22 for the case $a = b = 1$.

7.3   Poisson's summation formula*

The material in this section will only be used to determine the Fourier transform of the so-called *comb distribution* in section 9.1.3 and also to prove the sampling theorem in chapter 15. Sections 7.3 and 9.1.3 and the proof of the sampling theorem can be omitted without any consequences for the remainder of the material.

With the conclusion of section 7.2 one could state that we have finished the theory of the Fourier integral for non-periodic functions. In the next two chapters we extend the Fourier analysis to objects which are no longer functions, but so-called distributions. Before we start with distribution theory, the present section will first examine Poisson's summation formula. It provides an elegant connection between the Fourier series and the Fourier integral. Moreover, we will use Poisson's summation formula in chapter 9 to determine the Fourier transform of the so-called comb distribution, and in chapter 15 to prove the sampling theorem. We note, by the way, that in the proof of Poisson's summation formula we will not use the fundamental theorem of the Fourier integral.

In order to make a connection between the Fourier series and the Fourier integral, we will try to associate a periodic function with period T with an absolutely integrable function $f(t)$. We will do this in two separate ways. First of all we define the periodic function $f_\mathrm{p}(t)$ in the following obvious way:

$$f_\mathrm{p}(t) = \sum_{n=-\infty}^{\infty} f(t + nT). \tag{7.20}$$

Replacing t by $t + T$ in (7.20), it follows from a renumbering of the sum that

$$f_\mathrm{p}(t + T) = \sum_{n=-\infty}^{\infty} f(t + (n + 1)T) = \sum_{n=-\infty}^{\infty} f(t + nT) = f_\mathrm{p}(t).$$

Hence, the function $f_\mathrm{p}(t)$ is indeed periodic with period T. There is, however, yet another way to associate a periodic function with $f(t)$. First take the Fourier transform $F(\omega)$ of $f(t)$ and form a sort of Fourier series associated with f (note again that f is non-periodic):

$$\frac{1}{T} \sum_{n=-\infty}^{\infty} F(2\pi n/T)e^{2\pi int/T}. \tag{7.21}$$

(We will see that this is in fact the Fourier series of $f_\mathrm{p}(t)$.) If we replace t by $t + T$, then (7.21) remains unchanged and (7.21) is thus, as a function of t, also periodic with period T. (We have taken $F(2\pi n/T)/T$ instead of $F(n)$ since a similar connection between Fourier coefficients and the Fourier integral has already been derived in (6.9).) Poisson's summation formula now states that the two methods to obtain a periodic function from a non-periodic function $f(t)$ lead to the same result. Of course we have to require that the resulting series converge, preferably absolutely. In order to give a correct statement of the theorem, we also need to impose some extra conditions on the function $f(t)$.

THEOREM 7.8
Poisson's summation formula

Let $f(t)$ be an absolutely integrable and continuous function on $\mathbb{R}$ with spectrum $F(\omega)$. Let $T > 0$ be a constant. Assume furthermore that there exist constants $p > 1$, $A > 0$ and $M > 0$ such that $|f(t)| < M|t|^{-p}$ for $|t| > A$. Also assume *that $\sum_{n=-\infty}^{\infty} |F(2\pi n/T)|$ converges. Then*

$$\sum_{n=-\infty}^{\infty} f(t + nT) = \frac{1}{T} \sum_{n=-\infty}^{\infty} F(2\pi n/T)e^{2\pi int/T} \tag{7.22}$$

(with absolutely convergent series). In particular

$$\sum_{n=-\infty}^{\infty} f(nT) = \frac{1}{T} \sum_{n=-\infty}^{\infty} F(2\pi n/T). \tag{7.23}$$

Proof
Define $f_{\mathrm{p}}(t)$ as in (7.20). Without proof we mention that, with the conditions on the function $f(t)$, the function $f_{\mathrm{p}}(t)$ exists for every $t \in \mathbb{R}$ and that it is a continuous function. Furthermore, we have already seen that $f_{\mathrm{p}}(t)$ is a periodic function with period T. The proof now consists of the determination of the Fourier series of $f_{\mathrm{p}}(t)$ and subsequently applying some of the results from the theory of Fourier series.

For the nth Fourier coefficient c_n of $f_{\mathrm{p}}(t)$ one has

$$c_n = \frac{1}{T} \int_0^T f_{\mathrm{p}}(t)e^{-2\pi int/T}\, dt = \frac{1}{T} \int_0^T \sum_{k=-\infty}^{\infty} f(t + kT)e^{-2\pi int/T}\, dt.$$

Integrating term-by-term we obtain

$$c_n = \frac{1}{T} \sum_{k=-\infty}^{\infty} \int_0^T f(t + kT)e^{-2\pi int/T}\, dt.$$

From the conditions on the function $f(t)$ it follows that this termwise integration is allowed, but again this will not be proven. Changing to the variable $\tau = t + kT$ in the integral, it then follows that

$$c_n = \frac{1}{T} \sum_{k=-\infty}^{\infty} \int_{kT}^{(k+1)T} f(\tau)e^{-2\pi in\tau/T} e^{2\pi ink}\, d\tau.$$

For $k, n \in \mathbb{Z}$ one has $e^{2\pi ink} = 1$. Furthermore, the intervals $[kT, (k+1)T]$ precisely fill up all of $\mathbb{R}$ when k runs through the set $\mathbb{Z}$, and so

$$c_n = \frac{1}{T} \int_{-\infty}^{\infty} f(\tau)e^{-2\pi in\tau/T}\, d\tau = \frac{1}{T} F(2\pi n/T). \tag{7.24}$$

This determines the Fourier coefficients c_n of $f_{\mathrm{p}}(t)$ and because of the assumption on the convergence of the series $\sum_{n=-\infty}^{\infty} |F(2\pi n/T)|$ we now have that

$$\sum_{n=-\infty}^{\infty} |c_n| \quad \text{converges}.$$

Since $\left| c_n e^{2\pi int/T} \right| = |c_n|$, it then also follows that the Fourier series of $f_{\mathrm{p}}(t)$ converges absolutely (see theorem 4.5). For the moment we call the sum of this series $g(t)$, then $g(t)$ is a continuous function with Fourier coefficients c_n. The two continuous functions $f_{\mathrm{p}}(t)$ and $g(t)$ thus have the same Fourier coefficients and according to the uniqueness theorem 4.4 it then follows that $f_{\mathrm{p}}(t) = g(t)$. Hence,

$$\sum_{n=-\infty}^{\infty} f(t + nT) = f_{\mathrm{p}}(t) = g(t) = \frac{1}{T} \sum_{n=-\infty}^{\infty} F(2\pi n/T)e^{2\pi int/T},$$

which proves (7.22). For $t = 0$ we obtain (7.23). $\blacksquare$

In the proof of theorem 7.8 a number of results were used without proof. All of these results rely on a property of series – the so-called *uniform convergence* – which is not assumed as a prerequisite in this book. The reader familiar with the properties of uniform convergent series can find a more elaborate proof of Poisson's summation formula in, for example, *The theory of Fourier series and integral* by P.L. Walker, Theorem 5.30.

We will call both (7.22) and (7.23) Poisson's summation formula. From the proof we see that the right-hand side of (7.22) is the Fourier series of $f_\mathrm{p}(t)$, that is, of the function $f(t)$ made periodic according to (7.20). The occurring Fourier coefficients are obtained from the spectrum $F(\omega)$ using (7.24). In this manner we have linked the Fourier series to the Fourier integral. It is even possible to derive the fundamental theorem of the Fourier integral from the fundamental theorem of Fourier series using Poisson's summation formula. This gives a new proof of the fundamental theorem of the Fourier integral. We will not go into this any further. In conclusion we present the following two examples.

EXAMPLE 7.8
Take $f(t) = a/(a^2 + t^2)$ with $a > 0$, then $F(\omega) = \pi e^{-a|\omega|}$ (see table 3). We want to apply (7.23) with $T = 1$ and so we have to check the conditions. The assumption about the convergence of the series $\sum_{n=-\infty}^{\infty} |F(2\pi n)|$ is easy since

$$\sum_{n=-\infty}^{\infty} e^{-2\pi a|n|} = 1 + 2\sum_{n=1}^{\infty} e^{-2\pi an},$$

which is a geometric series with ratio $r = e^{-2\pi a}$. Since $a > 0$ it follows that $|r| < 1$, and so the geometric series converges (see section 2.4.1). In this case we even know the sum:

$$\sum_{n=-\infty}^{\infty} e^{-2\pi a|n|} = 1 + 2\frac{e^{-2\pi a}}{1 - e^{-2\pi a}} = \frac{1 + e^{-2\pi a}}{1 - e^{-2\pi a}}.$$

The condition on the function $f(t)$ is also easy to verify: for $t \neq 0$ one has $|f(t)| \leq a/t^2$ and so the condition in theorem 7.8 is met if we take $p = 2$, $M = a$ and $A > 0$ arbitrary. Poisson's summation formula can thus be applied. The right-hand side of (7.23) has just been calculated and hence we obtain

$$\sum_{n=-\infty}^{\infty} \frac{1}{a^2 + n^2} = \frac{\pi}{a}\frac{1 + e^{-2\pi a}}{1 - e^{-2\pi a}}. \tag{7.25}$$

By rewriting (7.25) somewhat, it then follows for any $a > 0$ that

$$\sum_{n=1}^{\infty} \frac{1}{a^2 + n^2} = \frac{\pi}{2a}\frac{1 + e^{-2\pi a}}{1 - e^{-2\pi a}} - \frac{1}{2a^2}.$$

Now take the limit $a \downarrow 0$, then the left-hand side tends to $\sum_{n=1}^{\infty} 1/n^2$, while a little calculation will show that

$$\lim_{a \downarrow 0}\left(\frac{\pi}{2a}\frac{1 + e^{-2\pi a}}{1 - e^{-2\pi a}} - \frac{1}{2a^2}\right) = \lim_{a \downarrow 0}\frac{\pi a(1 + e^{-2\pi a}) - (1 - e^{-2\pi a})}{2a^2(1 - e^{-2\pi a})} = \frac{\pi^2}{6}$$

(apply, for example, De l'Hôpital's rule three times). This gives a new proof of the famous identity (also see exercise 4.8)

$$\sum_{n=1}^{\infty} \frac{1}{n^2} = \frac{\pi^2}{6}.$$

$\blacktriangleleft$

EXAMPLE 7.9

Every rapidly decreasing function $f(t)$ (see section 6.6) satisfies the conditions of theorem 7.8. First of all, it follows straight from the definition of the notion 'rapidly decreasing' that the condition on the function $f(t)$ is met, since for $f(t)$ one has, for example, that there exists a constant $M > 0$ such that $|f(t)| < M/t^2$ for all $t \neq 0$ (so we can choose $p = 2$ and A arbitrary positive). It only remains to be shown that the series $\sum_{n=-\infty}^{\infty} |F(2\pi n/T)|$ converges. Now it has been proven in theorem 6.12 that the spectrum $F(\omega)$ of $f(t)$ is again a rapidly decreasing function. In particular, there again exists a constant $M > 0$ such that $|F(\omega)| < M/\omega^2$ for all $\omega \neq 0$. Hence,

$$\sum_{n=-\infty}^{\infty} |F(2\pi n/T)| = |F(0)| + \sum_{n=1}^{\infty} |F(2\pi n/T)| + \sum_{n=-\infty}^{-1} |F(2\pi n/T)|$$

$$< |F(0)| + 2\frac{MT^2}{4\pi^2} \sum_{n=1}^{\infty} \frac{1}{n^2} < \infty,$$

since $\sum_{n=1}^{\infty} n^{-2}$ converges. All the conditions of theorem 7.8 are thus satisfied and so Poisson's summation formula can be applied to *any* $f \in \mathcal{S}$. This result will be used in section 9.1.3 to determine the Fourier transform of the so-called comb distribution.   ◀

EXERCISES

7.25*

Show that (see example 7.8)

$$\lim_{a \downarrow 0} \frac{\pi a(1 + e^{-2\pi a}) - (1 - e^{-2\pi a})}{2a^2(1 - e^{-2\pi a})} = \frac{\pi^2}{6}.$$

7.26*

Indicate why Poisson's summation formula may be applied to the function $f(t) = e^{-at^2}$ ($a > 0$). Then prove that for every $x > 0$ one has

$$\sum_{n=-\infty}^{\infty} e^{-\pi n^2 x} = x^{-1/2} \sum_{n=-\infty}^{\infty} e^{-\pi n^2/x}.$$

7.27*

Prove the following generalization of (7.25) (here $\sinh x = (e^x - e^{-x})/2$ and $\cosh x = (e^x + e^{-x})/2$):

$$\frac{a}{\pi} \sum_{n=-\infty}^{\infty} \frac{1}{a^2 + (n+t)^2} = \frac{\sinh 2\pi a}{\cosh 2\pi a - \cos 2\pi t}.$$

SUMMARY

For an absolutely integrable and piecewise continuous function $f(t)$ on $\mathbb{R}$ with spectrum $F(\omega)$ one has $\lim_{\omega \to \pm\infty} F(\omega) = 0$ (Riemann–Lebesgue lemma). Using this important property, the fundamental theorem of the Fourier integral was proven:

$$\frac{1}{2\pi} \int_{-\infty}^{\infty} F(\omega)e^{i\omega t}\, d\omega = \frac{1}{2}(f(t+) + f(t-))$$

for any absolutely integrable and piecewise smooth function $f(t)$ on $\mathbb{R}$ (from now on, all functions in the time domain will be assumed to be absolutely integrable and piecewise smooth on $\mathbb{R}$). Here the Fourier integral in the left-hand side converges as a Cauchy principal value.

From the fundamental theorem the uniqueness of the Fourier transform on $\mathbb{R}$ immediately follows: if $F(\omega) = G(\omega)$ on $\mathbb{R}$, then $f(t) = g(t)$ at all points t where f and g are continuous.

In many cases the Fourier integral will exist as an improper integral as well, resulting in the duality or reciprocity property:

$$F(-t) \leftrightarrow 2\pi f(\omega).$$

Because of this, Fourier transforms almost always occur in pairs. The duality property certainly holds for rapidly decreasing functions and this was used to show that the Fourier transform is a one-to-one mapping onto the space $\mathcal{S}$ of rapidly decreasing functions.

For square integrable functions f and g on $\mathbb{R}$, the fundamental theorem was used to prove the convolution theorem in the frequency domain:

$$f(t)g(t) \leftrightarrow (F * G)(\omega)/2\pi.$$

From this, Parseval's identities immediately follow:

$$\int_{-\infty}^{\infty} f(t)\overline{g(t)}\,dt = \frac{1}{2\pi} \int_{-\infty}^{\infty} F(\omega)\overline{G(\omega)}\,d\omega$$

and

$$\int_{-\infty}^{\infty} |f(t)|^2\,dt = \frac{1}{2\pi} \int_{-\infty}^{\infty} |F(\omega)|^2\,d\omega.$$

Both the fundamental theorem and Parseval's identities can be used to determine definite integrals.

Finally, Poisson's summation formula

$$\sum_{n=-\infty}^{\infty} f(t + nT) = \frac{1}{T} \sum_{n=-\infty}^{\infty} F(2\pi n/T)e^{2\pi i n t/T}$$

(where $T > 0$ is a constant) provided a link between Fourier series and the Fourier integral. This formula can, for example, be applied to any $f \in \mathcal{S}$.

SELFTEST

7.28   The function $f(t)$ is defined by

$$f(t) = \begin{cases} \sin t & \text{for } 0 \le t \le \pi, \\ 0 & \text{elsewhere.} \end{cases}$$

(see exercise 6.9 for a similar function).
a   Determine the spectrum $F(\omega)$ of $f(t)$.
b   Show that for each $t \in \mathbb{R}$ one has

$$f(t) = \frac{1}{\pi} \int_0^{\infty} \frac{\cos(\pi - t)\omega + \cos t\omega}{1 - \omega^2}\,d\omega.$$

c   Prove that

$$\int_0^{\infty} \frac{\cos(\pi x/2)}{1 - x^2}\,dx = \frac{\pi}{2}.$$

d   Finally show that

$$\int_0^{\infty} \frac{\cos^2(\pi x/2)}{(1 - x^2)^2}\,dx = \frac{\pi^2}{8}.$$

7.29   Let the function $f(t) = \sin at/(t(1+t^2))$ $(a > 0)$ with spectrum $F(\omega)$ be given.
Find $F(\omega)$ explicitly as follows.
a   Show that $g(t) = (p_{2a}(v) * e^{-|v|})(t)$ has as spectrum $G(\omega) = 4f(\omega)$.
b   Determine $g(t)$ explicitly by calculating the convolution from part a.
c   Verify that the duality property can be applied and then give $F(\omega)$.

7.30   Let $q_a(t)$ be the triangle function and $p_a(t)$ the block function $(a > 0)$, then it is
known that (see table 3)

$$q_a(t) \leftrightarrow \frac{4\sin^2(a\omega/2)}{a\omega^2} \qquad \text{and} \qquad p_a(t) \leftrightarrow \frac{2\sin(a\omega/2)}{\omega}.$$

a   For which values of $t \in \mathbb{R}$ does the Fourier integral corresponding to $q_a(t)$
converge to $q_a(t)$? Does this Fourier integral converge only as Cauchy principal
value or also as improper integral? Justify your answers.
b   Use Parseval's identity to show that

$$\int_0^\infty \frac{\sin^3 x}{x^3}\,dx = \frac{3\pi}{8}.$$

Contents of Chapter 8

Distributions

Distributions

INTRODUCTION

Many new concepts and theories in mathematics arise from the fact that one is confronted with problems that existing theories cannot solve. These problems may originate from mathematics itself, but often they arise elsewhere, such as in physics. Especially fundamental problems, sometimes remaining unsolved for years, decades or even centuries, have a very stimulating effect on the development of mathematics (and science in general). The Greeks, for example, tried to find a construction of a square having the same area as the unit circle. This problem is known as the 'quadrature of the circle' and remained unsolved for some two thousand years. Not until 1882 it was found that such a construction was impossible, and it was discovered that the area of the unit circle, hence the number π, was indeed a very special real number.

Many of the concepts which one day solved a very fundamental problem are now considered obvious. Even the concept of 'function' has one day been heavily debated, in particular relating to questions on the convergence of Fourier series. Problems arising in the context of the solutions of quadratic and cubic equations were solved by introducing the now so familiar complex numbers. As is well-known, the complex numbers form an extension of the set of real numbers.

In this chapter we will introduce new objects, the so-called 'distributions', which form an extension of the concept of function. For twenty years, these distributions were used successfully in physics, prior to the development, in 1946, of a mathematical theory which could handle these problematic objects. It will turn out that these distributions are an important tool, just as the complex numbers. They are indispensable when describing, for example, linear systems in chapter 10.

In section 8.1 we will show how distributions arise in the Fourier analysis of non-periodic functions. We will first concentrate on the so-called delta function – a misleading term by the way, since it is by no means a function. In section 8.2 we then present a mathematically rigorous introduction of distributions, and we treat our first important examples. We will show, among other things, that most functions can be considered as distributions; hence, distributions form an extension of functions (although not *every* function can be considered as a distribution).

It is remarkable that distributions can always be differentiated, as will be established in section 8.3. In this way, one can obtain new distributions by differentiation. In particular one can start with an ordinary function, consider it as a distribution, and then differentiate it (as a distribution). In this manner one can obtain distributions from ordinary functions, which themselves can then no longer be considered as functions. For example, the delta function mentioned above arises as the derivative of the unit step function. In the final section of this chapter two more properties will be developed, which will be useful later on: multiplication and scaling. Fourier analysis will not return until chapter 9.

LEARNING OBJECTIVES
After studying this chapter it is expected that you
- know and can apply the definition of distributions
- know how to consider functions as distributions
- know the definition of a number of specific distributions: the delta function, the principal value $1/t$
- know how to differentiate distributions
- can add distributions and multiply them by a constant
- can multiply distributions by polynomials and some more general functions
- can scale distributions and know the concept of time reversal for distributions
- know the concepts even and odd distributions.

8.1 The problem of the delta function

Without any doubt, the most famous distribution is the 'delta function'. Although the name suggests otherwise, this object is *not* a function. This is because, as we shall see in a moment, a function cannot have the prescribed properties of the delta function. A precise definition of the delta function will be given in section 8.2. First we will show that the notion of the delta function arises naturally in the Fourier analysis of non-periodic functions.

In section 6.2 we already noted that the constant function $f(t) = 1$ has no Fourier transform. However, a good approximation of f is the block function $p_{2a}(t)$ for very large values of a and in fact we would like to take the limit $a \to \infty$. Since the spectrum of $p_{2a}(t)$ is the function $2 \sin a\omega / \omega$, we will be inclined to believe that from this we should get the spectrum of the function $f(t) = 1$ as $a \to \infty$. But what precisely is $\lim_{a\to\infty} 2 \sin a\omega / \omega$? When this is considered as a *pointwise* limit, that is, for each $\omega \in \mathbb{R}$ fixed, then there is no value of ω such that the limit exists, since $\lim_{x\to\infty} \sin x$ does not exist. If we want to obtain a meaningful result, we need to attach a different meaning to the limit $\lim_{a\to\infty} 2 \sin a\omega / \omega$. Now in theorem 7.2 we have shown that for an absolutely integrable and piecewise smooth function one has

$$\lim_{a\to\infty} \frac{1}{\pi} \int_{-\infty}^{\infty} \frac{\sin a\tau}{\tau} f(t - \tau) \, d\tau = f(t), \tag{8.1}$$

where we assume for convenience that f is continuous at t. By substituting $t = 0$ and changing from the variable τ to $-\omega$, we obtain that

$$\lim_{a\to\infty} \frac{1}{\pi} \int_{-\infty}^{\infty} \frac{\sin a\omega}{\omega} f(\omega) \, d\omega = f(0) \tag{8.2}$$

when f is continuous at $t = 0$. This enables us to give a new interpretation for $\lim_{a\to\infty} 2 \sin a\omega / \omega$: for any absolutely integrable and continuously differentiable function $f(\omega)$ formula (8.2) is valid. Only *within this context* will the limit have a meaning. There is no point in asking whether we may interchange the limit and the integral in (8.2). We have come up with this new interpretation *precisely because* the original limit $\lim_{a\to\infty} 2 \sin a\omega / \omega$ had no meaning. Still, one often defines the symbol $\delta(\omega)$ by

$$\delta(\omega) = \frac{1}{2\pi} \lim_{a\to\infty} \frac{2 \sin a\omega}{\omega}, \tag{8.3}$$

and then the limit and the integral are interchanged in (8.2). For the new object $\delta(\omega)$ one then obtains

$$\int_{-\infty}^{\infty} \delta(\omega) f(\omega) \, d\omega = f(0). \tag{8.4}$$

Formulas (8.3) and (8.4) should not be taken literally; the limit in (8.3) does not exist and (8.4) should only be considered as a symbolic way of writing (8.2). Yet, in section 8.2 it will turn out that the object $\delta(\omega)$ can be given a rigorous mathematical meaning that is very close to (8.4). And although $\delta(\omega)$ is called the delta function, it will no longer be a function, but a so-called 'distribution'. The general theory of distributions, of which the delta function is one the most important examples, will be treated in a mathematically correct way in the next section.

We recall that studying the limit $\lim_{a\to\infty} 2\sin a\omega/\omega$ was motivated by the search for the spectrum of the constant function $f(t) = 1$. Since we can write $2\pi\delta(\omega) = \lim_{a\to\infty} 2\sin a\omega/\omega$, taking (8.3) as starting point, it is now plausible that $2\pi\delta(\omega)$ will be the spectrum of the constant function $f(t) = 1$.

Conversely, it can be made plausible in a similar way that the spectrum of the delta function $\delta(t)$ is the constant function $F(\omega) = 1$. To do so, we will take a closer look at (8.4) (which, by the way, is often taken as the defining property in much of the engineering literature). For example, (8.4) should be valid for absolutely integrable and piecewise smooth functions which, moreover, are continuous at $t = 0$. The function

$$f(t) = \begin{cases} 1 & \text{for } a < t < b, \\ 0 & \text{elsewhere,} \end{cases}$$

satisfies all these conditions as long as $a \neq 0$ and $b \neq 0$. From (8.4) it then follows that $\int_a^b \delta(\omega)\,d\omega = f(0)$ for all $a, b \in \mathbb{R}$ with $a \neq 0$, $b \neq 0$ and $a < b$. If we now take $a < 0$ and $b > 0$, then $f(0) = 1$ and so $\int_a^b \delta(\omega)\,d\omega = 1$. This suggests that the integral of the delta function over $\mathbb{R}$ equals 1, that is,

$$\int_{-\infty}^{\infty} \delta(\omega)\,d\omega = 1. \tag{8.5}$$

If, on the other hand, we take $a < b < 0$ or $0 < a < b$, then $f(0) = 0$ and so

$$\int_a^b \delta(\omega)\,d\omega = 0 \qquad \text{for all } a, b \in \mathbb{R} \text{ with } a < b < 0 \text{ or } 0 < a < b. \tag{8.6}$$

A *function* satisfying *both* (8.5) and (8.6) must have a very extraordinary behaviour! (Based on (8.5) and (8.6) one can, for instance, conclude that there is not a single point in $\mathbb{R}$ where $\delta(\omega)$ is continuous.) Sometimes this is solved by describing the delta function as a function *being* 0 *everywhere, except at the point* $\omega = 0$ (in order for (8.6) to hold) and in addition having integral over $\mathbb{R}$ equal to 1 (in order for (8.5) to hold). However, such a function satisfying (8.5) and (8.6) cannot exist since an integral does not change its value if the value of the integrand is changed at one point. Hence, the value of $\delta(\omega)$ at the point $\omega = 0$ is not relevant for the integral as a whole, which means that the integral will be 0 since $\delta(\omega) = 0$ outside the point $\omega = 0$.

The above description of the delta function still has some useful interpretations. Let us consider the block function $a^{-1}p_a(t)$ of height a^{-1} and duration a for ever decreasing a (see figure 8.1). (For small values of a we can interpret $a^{-1}p_a(t)$ physically as an impulse: a big force applied during a short time.) For $a \downarrow 0$ we obtain an object equal to 0 everywhere except at the point $t = 0$ where the limit will be ∞; moreover, the integral over $\mathbb{R}$ of $a^{-1}p_a(t)$ will equal 1 for all $a > 0$, and so in the limit $a \downarrow 0$ the integral over $\mathbb{R}$ will equal 1 as well. We thus obtain an object satisfying precisely the description of the delta function given above. It is then plausible that

$$\lim_{a\downarrow 0} \int_{-\infty}^{\infty} a^{-1}p_a(t)f(t)\,dt = f(0), \tag{8.7}$$

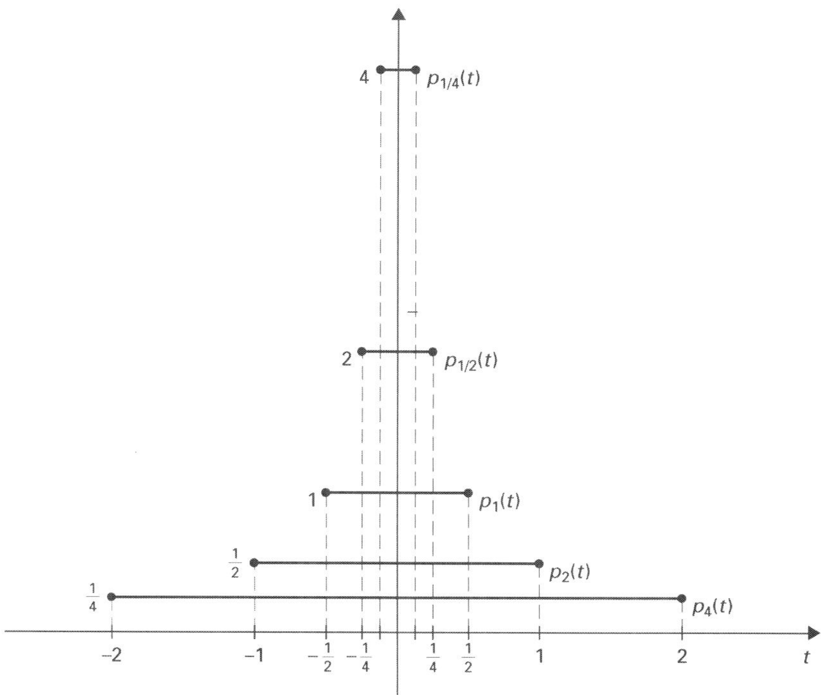

FIGURE 8.1
Block functions approximating the delta function.

and looking at (8.4) we are thus led to believe that

$$\lim_{a\downarrow 0} a^{-1} p_a(t) = \delta(t). \tag{8.8}$$

As a matter of fact, one can prove that (8.7) is indeed correct, under certain conditions on the function $f(t)$. Formulas (8.7) and (8.8) now present us with a situation which is entirely analogous to the situation in (8.2) and (8.3). We can use this to make it plausible that the spectrum of the delta function $\delta(t)$ is the constant function $F(\omega) = 1$. This is because the spectrum of $a^{-1} p_a(t)$ is the function $(2 \sin a\omega/2)/a\omega$ and for arbitrary $\omega \in \mathbb{R}$ one has $\lim_{a\downarrow 0} \sin a\omega/a\omega = 1$. We thus indeed find that the spectrum of $\delta(t) = \lim_{a\downarrow 0} a^{-1} p_a(t)$ will equal the constant function $F(\omega) = 1$ (this also follows if we interchange limit and integral in (8.7) and take $e^{-i\omega t}$ for the function $f(t)$). Also note that the duality or reciprocity property of the Fourier transform seems to hold for the delta function as well: $\delta(t) \leftrightarrow 1$ and $1 \leftrightarrow 2\pi\delta(\omega)$.

Of course, all the conclusions in this section rest upon intuitive derivations. In the next section a mathematically rigorous definition of distributions, and in particular of the delta function, will be given. In chapter 9 all of the results on the Fourier analysis of the delta function described above will be proven and the more general theory of the Fourier transform of distributions will be treated.

EXERCISE

8.1   There are many ways to obtain the delta function as a symbolic limit of functions. We have already seen two of these limits (in (8.3) and (8.8)). Now consider the function $P_a(t)$ from example 7.7:

$$P_a(t) = \frac{a}{\pi(a^2 + t^2)}.$$

a Sketch $\pi P_a(t)$ for $a = 1$ and $a = 1/4$.
b Show that $\lim_{a \downarrow 0} P_a(t) = 0$ for $t \neq 0$, while the limit equals ∞ for $t = 0$. Since the integral of $P_a(t)$ over $\mathbb{R}$ is 1 (see example 7.7), it is plausible that $\delta(t) = \lim_{a \downarrow 0} P_a(t)$.
c Determine the limit of the spectrum of $P_a(t)$ for $a \downarrow 0$ and conclude that it is plausible that the constant function 1 is the spectrum of $\delta(t)$.

8.2   Definition and examples of distributions

8.2.1   Definition of distributions

Loosely speaking, a complex-valued function f is a prescription assigning a value $f(t) \in \mathbb{C}$ to every $t \in \mathbb{R}$. Complex-valued functions are thus *mappings* from $\mathbb{R}$ to $\mathbb{C}$. In section 8.1 we have seen that the expression $\lim_{a \to \infty} 2 \sin a\omega / \omega$ only makes sense in the context of the integral in (8.2). In fact, (8.2) assigns the value $2\pi f(0)$ to every absolutely integrable and, say, continuously differentiable function $f(\omega)$. Here we have discovered an important new principle: the expression $\lim_{a \to \infty} 2 \sin a\omega / \omega$ (written symbolically as $2\pi\delta(\omega)$) can be considered as a *mapping* assigning to every absolutely integrable and continuously differentiable function $f(\omega)$ a certain complex number (namely $2\pi f(0)$). This new principle will be used to give a mathematically rigorous definition of distributions.

Keeping in mind the Fourier analysis of distributions, it turns out that it is not very convenient to work with continuously differentiable functions. In order to get a nice theory it is necessary to use a set of functions which is mapped into itself by the Fourier transform. We already know such a set of functions, namely the set $\mathcal{S}$ of rapidly decreasing functions (see sections 6.5 and 7.2.2). Now distributions will be mappings assigning a complex number to every $f \in \mathcal{S}$. The choice of $\mathcal{S}$ is determined by its usefulness in Fourier analysis. It is quite possible to define certain distributions as mappings from other sets of functions to $\mathbb{C}$, for example from the set of all continuous functions, or the set of all continuously differentiable functions, to $\mathbb{C}$. However, we will mainly confine ourselves to mappings from $\mathcal{S}$ to $\mathbb{C}$.

One additional condition is imposed on these mappings: *linearity*. We will illustrate this using our example from section 8.1. If we replace f in (8.2) by cf, where c is an arbitrary complex constant, then c can be taken outside the integral as well as outside the limit. Hence, we assign to the function $c \cdot f$ the complex number $c \cdot 2\pi f(0)$. So, if we multiply f by c, then the complex number $2\pi f(0)$ assigned to f is also multiplied by c. Next we replace f in (8.2) by a sum $g + h$ of two functions g and h. Then

$$\lim_{a \to \infty} \int_{-\infty}^{\infty} \frac{2\sin a\omega}{\omega}\,(g+h)(\omega)\,d\omega$$

$$= \lim_{a \to \infty} \int_{-\infty}^{\infty} \frac{2\sin a\omega}{\omega}\,g(\omega)\,d\omega + \lim_{a \to \infty} \int_{-\infty}^{\infty} \frac{2\sin a\omega}{\omega}\,h(\omega)\,d\omega$$

$$= 2\pi g(0) + 2\pi h(0).$$

Hence, we assign to the sum $g + h$ the sum $2\pi(g + h)(0)$ of the complex numbers $2\pi g(0)$ and $2\pi h(0)$. Together, these two properties show that the mapping is *linear* (see also chapter 1 on linear systems). This finally brings us to the following definition of the concept of distribution.

DEFINITION 8.1
Distribution

A distribution T is a linear mapping assigning a complex number to every rapidly decreasing function ϕ.

We denote the image of a $\phi \in \mathcal{S}$ under the mapping T by $\langle T, \phi \rangle$; note that $\langle T, \phi \rangle \in \mathbb{C}$. A distribution is thus a mapping $T : \mathcal{S} \to \mathbb{C}$ satisfying
$$\langle T, c\phi \rangle = c \langle T, \phi \rangle,$$
$$\langle T, \phi_1 + \phi_2 \rangle = \langle T, \phi_1 \rangle + \langle T, \phi_2 \rangle,$$
where ϕ, ϕ_1 and ϕ_2 are elements of $\mathcal{S}$ and $c \in \mathbb{C}$.

One uses the notation $\langle T, \phi \rangle$ to prevent confusion with functions. For the same reason it is customary in distribution theory to denote elements in $\mathcal{S}$ with the Greek symbols ϕ, ψ, etc. In section 8.2.3 we will see that many functions can be considered as distributions; it would then be very confusing to use the symbols f, g, etc. for elements in $\mathcal{S}$ as well. Although a distribution T is a linear mapping on $\mathcal{S}$, we will nevertheless often write $T(t)$ to express the fact that T acts on functions that depend on the variable t.

8.2.2   The delta function

Of course, our first example of a distribution should be the delta function $\delta(t)$. In section 8.1 we have argued that (8.4) is the crucial 'property' of $\delta(t)$: to a function $\phi(t)$ the value $\phi(0)$ is assigned. This will be taken as the definition of the delta function.

DEFINITION 8.2
Delta function

The delta function $\delta(t)$ (or δ for short) is the distribution defined by
$$\langle \delta(t), \phi \rangle = \phi(0) \qquad for\ \phi \in \mathcal{S}.$$

Let us verify that δ is indeed a distribution. It is clear that δ is a mapping from $\mathcal{S}$ to $\mathbb{C}$ since $\phi(0) \in \mathbb{C}$. One also has
$$\langle \delta, c\phi \rangle = (c\phi)(0) = c \cdot \phi(0) = c \langle \delta, \phi \rangle$$

and

$$\langle \delta, \phi_1 + \phi_2 \rangle = (\phi_1 + \phi_2)(0) = \phi_1(0) + \phi_2(0) = \langle \delta, \phi_1 \rangle + \langle \delta, \phi_2 \rangle,$$

which proves the linearity of δ. The delta function is thus indeed a distribution. In many books $\langle \delta, \phi \rangle = \phi(0)$ is written as

$$\int_{-\infty}^{\infty} \delta(t)\phi(t)\,dt = \phi(0), \tag{8.9}$$

Sifting property

just as we have done in (8.4) in connection with (8.2). Relation (8.9) is often called the *sifting property* of the delta function; the value of ϕ at the point 0 is 'sifted out' by δ. The graphical representation of the delta function in figure 8.2 is also based on this property: we draw an arrow at the point 0 of height 1. If we agree that the integral in (8.9) is a symbolic representation, then there is no objection. Of course, one cannot prove results about the delta function by applying properties from integral calculus to this integral. This is because it is only a *symbolic* way of writing. However, calculating in an informal way with the integral (8.9) may provide conjectures about possible results for the delta function. As an example one can get an impression of the derivative of the delta function by using integration by parts (see example 8.8).

The delta function was introduced explicitly for the first time by the English physicist P.A.M. Dirac in 1926. He was certainly not the first, nor the only one, to have some notion of a delta function. The classical result (8.2) for example, is already very close to how the delta function operates. Dirac was the first, however, to give an explicit meaning to the delta function and to introduce a separate notation for it. For this reason the delta function is often called the *Dirac function* or *Dirac delta function*. In the years following the introduction of the delta function, its use produced many results, which in physical practice turned out to be correct. Not until 1946 was a rigorous distribution theory developed by the French mathematician L. Schwartz.

Dirac delta function

A slightly more general delta function is obtained by assigning to a function $\phi \in \mathcal{S}$ not the value $\phi(0)$, but the value $\phi(a)$, where $a \in \mathbb{R}$ is fixed. This distribution is denoted by $\delta(t - a)$, so

$$\langle \delta(t - a), \phi \rangle = \phi(a). \tag{8.10}$$

Delta function at a

We will call $\delta(t - a)$ the *delta function at the point a*. If we choose 0 for the point a, then we simply call this the delta function. Symbolically (8.10) is sometimes written as

$$\int_{-\infty}^{\infty} \delta(t - a)\phi(t)\,dt = \phi(a). \tag{8.11}$$

We represent the distribution $\delta(t - a)$ graphically by an arrow of height 1 at the point a. See figure 8.2.

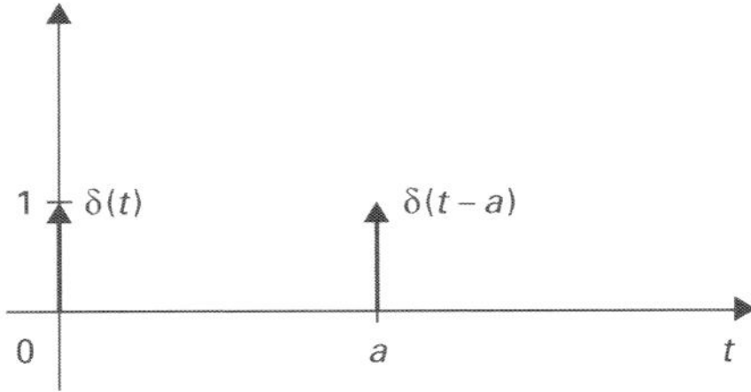

FIGURE 8.2
The delta function at 0 and at a.

At the start of this section it was noted that in order to define a distribution it is not always necessary to confine ourselves to the space $\mathcal{S}$. The definition of $\delta(t - a)$ for example, and in particular of δ, is also meaningful for any continuous function ϕ. Hence, definition 8.2 and (8.10) are often given for all continuous functions ϕ.

8.2.3   Functions as distributions

Generalized functions

Distributions are often called *generalized functions* because they form an extension of the concept of 'function'. Just as any real number can be considered as a complex number, a function apparently can be considered as a distribution. This comparison with $\mathbb{R}$ as a subset of $\mathbb{C}$ is not entirely correct, since *not all* functions can be considered as distributions. If, however, the function $f(t)$ is absolutely integrable, then it

Function as distribution

can certainly be considered as a distribution T_f by defining

$$\langle T_f, \phi \rangle = \int_{-\infty}^{\infty} f(t)\phi(t)\,dt \qquad \text{for } \phi \in \mathcal{S}. \tag{8.12}$$

First, it has to be shown that the integral in (8.12) exists. But for $\phi \in \mathcal{S}$ one has in particular that $\phi(t)$ is bounded on $\mathbb{R}$, say $|\phi(t)| \le M$. It then follows that

$$\left| \int_{-\infty}^{\infty} f(t)\phi(t)\,dt \right| \le \int_{-\infty}^{\infty} |f(t)\phi(t)|\,dt \le M \int_{-\infty}^{\infty} |f(t)|\,dt,$$

and since f is absolutely integrable, the integral exists. Next we have to show that T_f is indeed a distribution. For each $\phi \in \mathcal{S}$ the integral in (8.12) gives a complex number. Hence, T_f is a mapping from $\mathcal{S}$ to $\mathbb{C}$. The linearity of T_f follows immediately from the linearity of integration (see also, for example, section 6.4.1) and so T_f is indeed a distribution. In this way one can consider any absolutely integrable function f as a distribution. But now a problem arises. How do we know for sure that two different functions f and g also lead to two different distributions T_f and T_g? This ought to be true if we consider distributions as an extension of functions. (Two real numbers that are unequal will also be unequal when considered as complex numbers.) But what do we actually mean by 'unequal' or 'equal' distributions?

DEFINITION 8.3
Equality of distributions

Two distributions T_1 and T_2 are called equal *if* $\langle T_1, \phi \rangle = \langle T_2, \phi \rangle$ *for all* $\phi \in \mathcal{S}$. *In this case we write* $T_1 = T_2$.

We now mention without proof that $T_f = T_g$ implies that $f = g$ at the points where both f and g are continuous. When no confusion is possible, the distribution T_f is simply denoted by f. It is then customary to use the phrase 'f as distribution'.

We close this section with some examples of distributions that will often return. Among other things, these examples will show that many functions which are not absolutely integrable still define a distribution T_f through (8.12).

EXAMPLE 8.1
The function $f(t) = 1$

The *constant function* $f(t) = 1$ is not absolutely integrable over $\mathbb{R}$. Still it defines precisely as in (8.12) a distribution, simply denoted by 1:

$$\langle 1, \phi \rangle = \int_{-\infty}^{\infty} \phi(t)\,dt \qquad \text{for } \phi \in \mathcal{S}. \tag{8.13}$$

Since $\phi \in \mathcal{S}$, there exists a constant M such that $(1 + t^2)|\phi(t)| \le M$. Then the integral in (8.13) exists since

$$\left| \int_{-\infty}^{\infty} \phi(t)\,dt \right| \le \int_{-\infty}^{\infty} |\phi(t)|\,dt \le M \int_{-\infty}^{\infty} \frac{1}{1 + t^2}\,dt < \infty.$$

It is now rather easy to show that 1 is indeed a distribution (that is, a linear mapping from $\mathcal{S}$ to $\mathbb{C}$). In chapter 9 it will turn out that 1 is the spectrum of the delta function. This has been already been made plausible in section 8.1. ◄

EXAMPLE 8.2
Unit step function

The *unit step function* $\epsilon(t)$ (see example 6.1) defines a distribution, again denoted by $\epsilon(t)$, or ϵ for short, by

$$\langle \epsilon, \phi \rangle = \int_{-\infty}^{\infty} \epsilon(t)\phi(t)\,dt = \int_{0}^{\infty} \phi(t)\,dt \qquad \text{for } \phi \in \mathcal{S}.$$

Note that (8.12) is again applied here. Further details are almost the same as in example 8.1. ◄

EXAMPLE 8.3
Sign function

The *sign function* sgn t is defined by

$$
\operatorname{sgn} t = \begin{cases} 1 & \text{for } t > 0, \\ 0 & \text{for } t = 0, \\ -1 & \text{for } t < 0. \end{cases}
$$

This function defines a distribution sgn t by

$$
\langle \operatorname{sgn} t, \phi \rangle = \int_{-\infty}^{\infty} \operatorname{sgn} t \, \phi(t) \, dt = \int_{0}^{\infty} (\phi(t) - \phi(-t)) \, dt \qquad \text{for } \phi \in \mathcal{S}.
$$

Here (8.12) is again applied. Further details are the same as in the previous two examples. ◀

EXAMPLE 8.4
The function $f(t) = |t|$

The function $f(t) = |t|$ also defines a distribution using (8.12), and again it will simply be denoted by $|t|$:

$$
\langle |t|, \phi \rangle = \int_{-\infty}^{\infty} |t| \, \phi(t) \, dt = \int_{0}^{\infty} t(\phi(t) + \phi(-t)) \, dt \qquad \text{for } \phi \in \mathcal{S}.
$$

In order to show that the integral exists, we use in this case that there exists a constant $M > 0$ such that $\left| t(1 + t^2)\phi(t) \right| \leq M$ on $\mathbb{R}$. From here on, the proof is exactly the same as in example 8.1. ◀

EXAMPLE 8.5

Principal value $1/t$

The function $1/t$ is not absolutely integrable on $\mathbb{R}$ and even $\phi(t)/t$ with $\phi \in \mathcal{S}$ may not be absolutely integrable on $\mathbb{R}$ since the point $t = 0$ may cause a problem. Hence, for $1/t$ we cannot use (8.12) to define a distribution. This problem can be solved by resorting to a variant of the Cauchy principal value from definition 7.1. Loosely speaking, in this case we let the limits at $t = 0$ tend to zero at the same rate; the distribution arising in this way is denoted by pv$(1/t)$ (pv from 'principal value'). The precise definition of the distribution pv$(1/t)$ is as follows:

$$
\left\langle \operatorname{pv} \frac{1}{t}, \phi \right\rangle = \lim_{\alpha \downarrow 0} \int_{|t| \geq \alpha} \frac{\phi(t)}{t} \, dt = \lim_{\alpha \downarrow 0} \left(\int_{\alpha}^{\infty} \frac{\phi(t)}{t} \, dt + \int_{-\infty}^{-\alpha} \frac{\phi(t)}{t} \, dt \right) \tag{8.14}
$$

for $\phi \in \mathcal{S}$. The existence of the right-hand side is again the most difficult step in proving that pv$(1/t)$ is a distribution. To this end we split the integral in the right-hand side of (8.14) as follows:

$$
\int_{\alpha \leq |t| \leq 1} \frac{\phi(t)}{t} \, dt + \int_{|t| \geq 1} \frac{\phi(t)}{t} \, dt. \tag{8.15}
$$

First look at the second integral in (8.15). Since $|1/t| \leq 1$ for $|t| \geq 1$, it follows that

$$
\left| \int_{|t| \geq 1} \frac{\phi(t)}{t} \, dt \right| \leq \int_{|t| \geq 1} |\phi(t)| \, dt \leq \int_{-\infty}^{\infty} |\phi(t)| \, dt,
$$

and in example 8.1 it was shown that the last integral exists for $\phi \in \mathcal{S}$. Hence, the second integral in (8.15) exists. For the first integral in (8.15) we note that for any $\alpha > 0$ one has

$$
\int_{\alpha \leq |t| \leq 1} \frac{\phi(0)}{t} \, dt = \int_{\alpha}^{1} \frac{\phi(0)}{t} \, dt + \int_{-1}^{-\alpha} \frac{\phi(0)}{t} \, dt = 0,
$$

since $1/t$ is an odd function. Hence,

$$
\int_{\alpha \leq |t| \leq 1} \frac{\phi(t)}{t} \, dt = \int_{\alpha \leq |t| \leq 1} \frac{\phi(t) - \phi(0)}{t} \, dt.
$$

Now define $\psi(t) = (\phi(t) - \phi(0))/t$ for $t \neq 0$ and $\psi(0) = \phi'(0)$. Then ψ is continuous at the point $t = 0$ since

$$\lim_{t \to 0} \psi(t) = \lim_{t \to 0} \frac{\phi(t) - \phi(0)}{t - 0} = \phi'(0) = \psi(0).$$

It then follows that

$$\lim_{\alpha \downarrow 0} \int_{\alpha \leq |t| \leq 1} \frac{\phi(t)}{t} \, dt = \lim_{\alpha \downarrow 0} \int_{\alpha \leq |t| \leq 1} \frac{\phi(t) - \phi(0)}{t} \, dt = \lim_{\alpha \downarrow 0} \int_{\alpha \leq |t| \leq 1} \psi(t) \, dt.$$

Since ψ is continuous at $t = 0$, the limit $\alpha \to 0$ exists and it follows that

$$\lim_{\alpha \downarrow 0} \int_{\alpha \leq |t| \leq 1} \frac{\phi(t)}{t} \, dt = \int_{-1}^{1} \psi(t) \, dt.$$

Since ψ is a continuous function on the closed and bounded interval $[-1, 1]$, it then follows that this final integral, and so the right-hand side of (8.14), exists. ◀

EXERCISES

8.2   Let $\delta(t - a)$ be defined as in (8.10).
a   Show that $\delta(t - a)$ is a distribution, that is, a linear mapping from S to $\mathbb{C}$.
b   Derive the symbolic notation (8.11) by interchanging the limit and the integral in (8.1) (hint: also use that $\delta(-t) = \delta(t)$, which is quite plausible on the basis of (8.3) or (8.8), and which will be proven in section 8.4).

8.3   Show that T_f as defined in (8.12) is a linear mapping.

8.4   Show that 1 as defined in (8.13) is indeed a distribution.

8.5   Prove that the integral in example 8.2 exists and check that ϵ is a distribution.

8.6   Do the same as in exercise 8.5 for the distribution $\mathrm{sgn}\, t$ from example 8.3.

8.7   Prove that $|t|$ from example 8.4 defines a distribution.

8.8   For fixed $a \in \mathbb{R}$ we define T by $\langle T, \phi \rangle = \phi'(a)$ for $\phi \in S$. Show that T is a distribution.

8.9   Let the function $f(t) = |t|^{-1/2}$ be given.
a   Show that f is integrable on the interval $[-1, 1]$. Is f integrable on $\mathbb{R}$?
b   Show that f defines a distribution by means of (8.12). In particular it has to be shown that the defining integral exists (hint: for $|t| \geq 1$ one has $|t|^{-1/2} \leq 1$).

8.10   Prove for the following functions that (8.12) defines a distribution:
a $f(t) = t$,
b $f(t) = t^2$.

8.3   Derivatives of distributions

Switching on a (direct current) apparatus at time $t = 0$ can be described using the unit step function $\epsilon(t)$. This, however, is an ideal description which will not occur in reality of course. More likely there will be a very strong increase in a very short time interval. Let $u(t)$ be the function describing the switching on in a realistic way. The ideal function $\epsilon(t)$ and a typical 'realistic' function $u(t)$ are drawn in figure 8.3. We assume for the moment that $u(t)$ is differentiable and that $u(t)$ increases from the value 0 to the value 1 in the time interval $0 \leq t \leq a$. The derivative $u'(t)$ of $u(t)$ equals 0 for $t < 0$ and $t > a$, while between $t = 0$ and $t = a$ the function

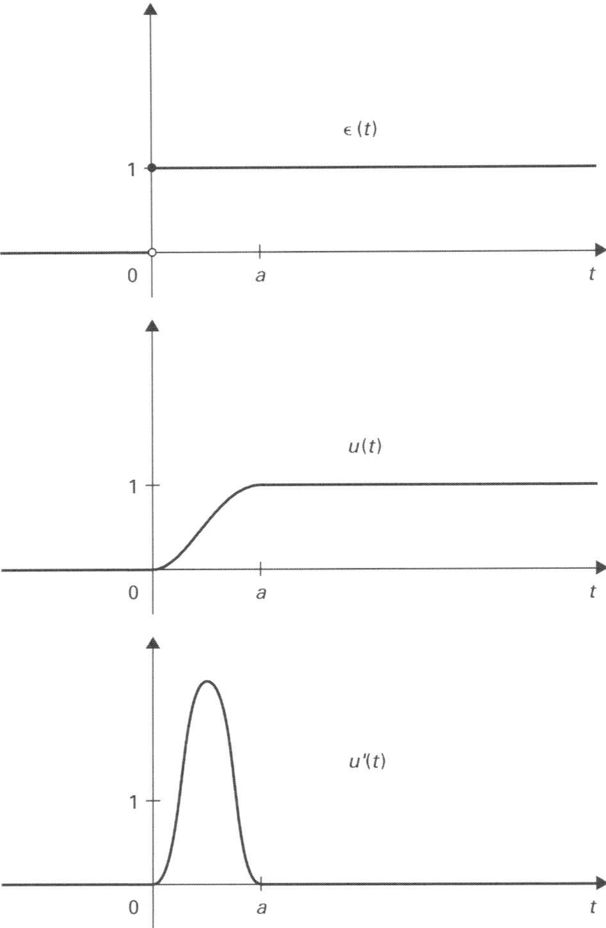

FIGURE 8.3
The ideal $\epsilon(t)$, the realistic $u(t)$ and the derivative $u'(t)$.

$u'(t)$ quickly reaches its maximum and then decreases to 0 rapidly. The graph of a typical $u'(t)$ is also drawn in figure 8.3. If we now take the limit $a \downarrow 0$, then $u(t)$ will transform into the function $\epsilon(t)$, while $u'(t)$ seems to tend towards the delta function. This is because $u(t)$ will have to increase faster and faster over an ever smaller interval; the derivative will then attain ever increasing values in the vicinity of $t = 0$. In the limit $a \downarrow 0$ an object will emerge which is 0 everywhere, except at the point $t = 0$, where the value becomes infinitely large. Since, moreover,

$$\int_{-\infty}^{\infty} u'(t)\,dt = \int_{0}^{a} u'(t)\,dt = [u(t)]_0^a = 1$$

for every $a > 0$, (8.5) is valid as well. We thus obtain an object fitting precisely the description of the delta function from section 8.1. Hence, it is plausible that the derivative of $\epsilon(t)$ will be the delta function.

In order to justify these conclusions mathematically, we should first find a definition for the derivative of a distribution. Of course, this definition should agree with the usual derivative of a function, since distributions are an extension of functions. Now let f be an absolutely integrable function with continuous derivative f' being absolutely integrable as well. Then f' defines a distribution $T_{f'}$ and from an integration by parts it then follows for $\phi \in \mathcal{S}$ that

$$\langle T_{f'}, \phi \rangle = \int_{-\infty}^{\infty} f'(t)\phi(t)\, dt = [f(t)\phi(t)]_{-\infty}^{\infty} - \int_{-\infty}^{\infty} f(t)\phi'(t)\, dt.$$

Since $\phi \in \mathcal{S}$, one has $\lim_{t\to\pm\infty} \phi(t) = 0$ (for $|\phi(t)| \leq M/(1 + t^2)$). Moreover, the final integral can be considered as the distribution T_f applied to the function $\phi'(t)$. We thus have

$$\langle T_{f'}, \phi \rangle = -\langle T_f, \phi' \rangle. \tag{8.16}$$

There is now only one possible definition for the derivative of a distribution.

DEFINITION 8.4
Derivative of distributions

The derivative T' *of a distribution* T *is defined by*

$$\langle T', \phi \rangle = -\langle T, \phi' \rangle \qquad for\ \phi \in \mathcal{S}.$$

Note that $\langle T, \phi' \rangle$ will certainly make sense because T is a distribution and $\phi' \in \mathcal{S}$ whenever $\phi \in \mathcal{S}$. From the linearity of T and of differentiation the linearity of T' immediately follows; hence, T' is indeed a distribution. This means in particular that T' has a derivative as well. As for functions this is called the *second derivative* of T and it is denoted by T''. Applying definition 8.4 twice, it follows that $\langle T'', \phi \rangle = \langle T, \phi'' \rangle$. This process can be repeated over and over again, so that we reach the remarkable conclusion that a distribution can be differentiated an arbitrary number of times. Applying definition 8.4 k times, it follows that the kth derivative $T^{(k)}$ of a distribution T is given by

$$\left\langle T^{(k)}, \phi \right\rangle = (-1)^k \left\langle T, \phi^{(k)} \right\rangle \qquad for\ \phi \in \mathcal{S}. \tag{8.17}$$

In particular it follows that *any* absolutely integrable function, considered as a distribution, is arbitrarily often differentiable. This gives us an obvious way to find new distributions. Start with a function f and consider it as a distribution T_f (if possible). Differentiating several times if necessary, one will in general obtain a distribution which no longer corresponds to a function. In the introduction to this section we have in fact already seen a crucial example of this process. For we have

Derivative of $\epsilon(t)$

argued there that the derivative of the unit step function $\epsilon(t)$ should be the delta function. This can now be proven using definition 8.4. For according to definition 8.4 the distribution ϵ' is given by

$$\langle \epsilon', \phi \rangle = -\langle \epsilon, \phi' \rangle = -\int_0^{\infty} \phi'(t)\, dt = -[\phi(t)]_0^{\infty},$$

where in the second step we used the definition of ϵ (see example 8.2). For $\phi \in \mathcal{S}$ one has $\lim_{t\to\pm\infty} \phi(t) = 0$ and so it follows that

$$\langle \epsilon', \phi \rangle = \phi(0) = \langle \delta, \phi \rangle \qquad for\ \phi \in \mathcal{S}. \tag{8.18}$$

According to definition 8.3 we then have $\epsilon' = \delta$. The informal derivation in the introduction to this section has now been made mathematically sound.

In order to handle distributions more easily, it is convenient to be able to multiply them by a constant and to add them. The definitions are as follows.

DEFINITION 8.5

Let S and T be distributions. Then cT ($c \in \mathbb{C}$) and $S + T$ are defined by
$$\langle cT, \phi \rangle = c \langle T, \phi \rangle \qquad \text{for } \phi \in \mathcal{S},$$
$$\langle S + T, \phi \rangle = \langle S, \phi \rangle + \langle T, \phi \rangle \qquad \text{for } \phi \in \mathcal{S}.$$

EXAMPLE 8.6

The distribution $c\delta(t - a)$ is given by $\langle c\delta(t - a), \phi \rangle = c\phi(a)$. For $c \in \mathbb{R}$ this is graphically represented by an arrow of height c at the point $t = a$. See figure 8.4. ◄

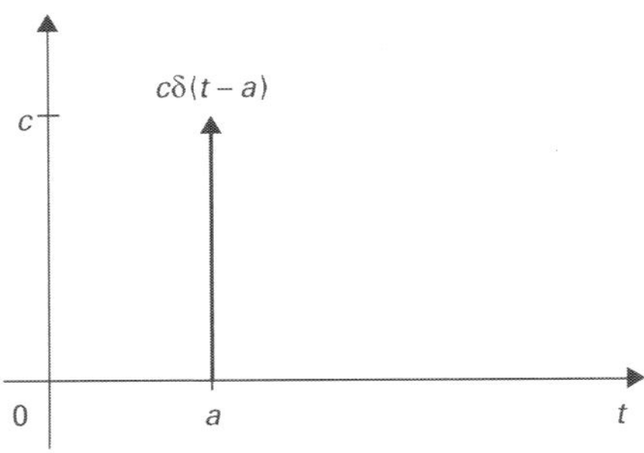

FIGURE 8.4
The distribution $c\delta(t - a)$.

EXAMPLE 8.7

The distribution $2\epsilon(t) + 3i\delta(t)$ is given by
$$\langle 2\epsilon(t) + 3i\delta(t), \phi \rangle = 2 \int_0^\infty \phi(t)\, dt + 3i\phi(0) \qquad \text{for } \phi \in \mathcal{S}.$$
◄

We close this section by determining some derivatives of distributions.

EXAMPLE 8.8
Derivative of $\delta(t)$

We have just shown that $\epsilon' = \delta$, considered as distribution. Differentiating again, we obtain that $\epsilon'' = \delta'$. Just as for δ, there is a simple description for the distribution δ':
$$\langle \delta', \phi \rangle = -\langle \delta, \phi' \rangle = -\phi'(0) \qquad \text{for } \phi \in \mathcal{S}. \tag{8.19}$$

Hence, the distribution δ' assigns the complex number $-\phi'(0)$ to a function ϕ. Note also that δ' is still well-defined when applied to continuously differentiable functions ϕ. Symbolically one writes (8.19) as
$$\int_{-\infty}^\infty \delta'(t)\phi(t)\, dt = -\phi'(0). \tag{8.20}$$

This expression can be derived symbolically from (8.9) by taking the function ϕ' instead of ϕ in (8.9) and performing a formal integration by parts. This example can be extended even further by repeated differentiation. Then the distributions δ'', $\delta^{(3)}$, etc. will arise. ◄

Let f be a function with continuous derivative f' and assume that both f and f' can be considered as a distribution through (8.12). The distribution T_f then has the distribution $(T_f)'$ as derivative; if our definitions have been put together well, then $(T_f)' = T_{f'}$ should be true. For then the two concepts of 'derivative' coincide. Using (8.16) the proof reads as follows: $\langle T_{f'}, \phi \rangle = -\langle T_f, \phi' \rangle = \langle (T_f)', \phi \rangle$ for all $\phi \in \mathcal{S}$. When no confusion is possible, one often simply writes f', when actually T_f' is meant. Most often we then use the phrase f' *as distribution*.

EXAMPLE 8.9

The distribution $|\, t\, |$ has the distribution $\operatorname{sgn} t$ as its derivative (see examples 8.3 and 8.4):

$$
\langle |\, t\, |', \phi \rangle = -\langle |\, t\, |, \phi' \rangle = -\int_0^\infty t\phi'(t)\, dt + \int_{-\infty}^0 t\phi'(t)\, dt
$$

$$
= -[t\phi(t)]_0^\infty + [t\phi(t)]_{-\infty}^0 + \int_0^\infty \phi(t)\, dt - \int_{-\infty}^0 \phi(t)\, dt
$$

$$
= \langle \operatorname{sgn} t, \phi \rangle \qquad \text{for } \phi \in \mathcal{S}.
$$

Here we have used integration by parts and the fact that $\lim_{t\to\pm\infty} t\phi(t) = 0$ (since $\phi \in \mathcal{S}$). ◄

EXAMPLE 8.10

Consider the function $g(t) = \epsilon(t)\cos t$. Since $|\cos t| \leq 1$, one proves precisely as in example 8.2 that $g(t)$ defines a distribution. Note that g is not continuous at the point $t = 0$, since there is a jump of magnitude 1 at this point. We determine the derivative of g as distribution:

$$
\langle (\epsilon(t)\cos t)', \phi \rangle = -\langle \epsilon(t)\cos t, \phi' \rangle = -\int_0^\infty \cos t\, \phi'(t)\, dt \qquad \text{for } \phi \in \mathcal{S}.
$$

From an integration by parts and the fact that $\phi \in \mathcal{S}$ it follows that this equals

$$
[-\phi(t)\cos t]_0^\infty - \int_0^\infty \sin t\, \phi(t)\, dt = \phi(0) - \int_{-\infty}^\infty \epsilon(t)\sin t\, \phi(t)\, dt
$$

$$
= \langle \delta, \phi \rangle - \langle \epsilon(t)\sin t, \phi \rangle .
$$

From definition 8.5 (and definition 8.3) it then follows that

$$
(\epsilon(t)\cos t)' = \delta(t) - \epsilon(t)\sin t.
$$

This identity is graphically represented in figure 8.5. ◄

We now derive a general rule having the result $\epsilon' = \delta$, as well as the examples 8.9 and 8.10, as special cases. Let f be a function, continuously differentiable on $\mathbb{R}$, except at one point a where f has a finite jump. By f' we will mean the derivative of f except at this point a. Assume that both f and f' define distributions T_f and $T_{f'}$ by means of (8.12). This situation occurs for instance in the examples mentioned above. One then has for any $\phi \in \mathcal{S}$ that

$$
\langle T_f', \phi \rangle = -\int_{-\infty}^\infty f(t)\phi'(t)\, dt = -\int_{-\infty}^a f(t)\phi'(t)\, dt - \int_a^\infty f(t)\phi'(t)\, dt,
$$

and from an integration by parts we then obtain

$$
\langle T_f', \phi \rangle = -[f(t)\phi(t)]_{-\infty}^a + \int_{-\infty}^a f'(t)\phi(t)\, dt
$$

$$
-[f(t)\phi(t)]_a^\infty + \int_a^\infty f'(t)\phi(t)\, dt
$$

$$
= -f(a-)\phi(a) + f(a+)\phi(a) + \int_{-\infty}^\infty f'(t)\phi(t)\, dt
$$

$$
= (f(a+) - f(a-))\,\langle \delta(t - a), \phi \rangle + \langle T_{f'}, \phi \rangle .
$$

Jump-formula

This proves the following *jump-formula*:

$$
T_f' = T_{f'} + (f(a+) - f(a-))\,\delta(t - a). \tag{8.21}
$$

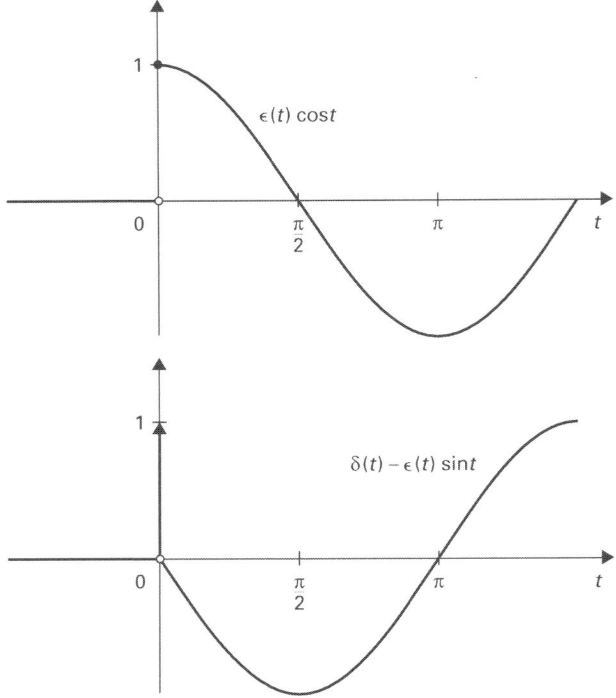

FIGURE 8.5
The function $\epsilon(t)\cos t$ and its derivative, considered as distribution.

If we take for f the function $\epsilon(t)$, then $a = 0$ and $\epsilon(0+) - \epsilon(0-) = 1$. Moreover, $\epsilon'(t) = 0$ for $t \neq 0$, so $T_{f'} = 0$. It then follows from (8.21) that $\epsilon' = \delta$, considered as distribution, in accordance with previous results. In a similar way one obtains examples 8.9 and 8.10 from (8.21).

EXERCISES

8.11   Let T' be defined as in definition 8.4. Show that T' is a linear mapping from $\mathcal{S}$ to $\mathbb{C}$.

8.12   **a**   Which complex number is assigned to $\phi \in \mathcal{S}$ by the distribution $2\delta(t) - i\sqrt{3}\delta'(t) + (1+i)\mathrm{sgn}\,t$?
b   Show that the function $f(t) = at^2 + bt + c$ with a, b and $c \in \mathbb{C}$ defines a distribution through (8.12).

8.13   **a**   Show that for distributions S and T one has $(S+T)' = S'+T'$ and $(cT)' = cT'$. Hence, differentiation of distributions is linear.
b   Show that for the constant function $f(t) = c$ one has $f' = 0$ as distribution.

8.14   **a**   Which complex number is assigned to $\phi \in \mathcal{S}$ by the distribution $\delta^{(3)}$?
b   To which set of functions can one extend the definition of $\delta^{(3)}$?

8.15   **a**   Calculate the derivative of the distribution $\mathrm{sgn}\,t$ in a direct way, using the definition of the derivative of a distribution.

b Verify that $\operatorname{sgn} t = 2\epsilon(t) - 1$ for all $t \neq 0$ and then use exercise 8.13 to determine the derivative of $\operatorname{sgn} t$ again.
c Determine the second derivative of $|t|$.

8.16 Show how examples 8.9 and 8.10 arise as special cases of the jump-formula (8.21).

8.17 Determine the derivative of the following distributions:
a $p_a(t)$,
b $\epsilon(t) \sin t$.

8.18 The (discontinuous) function $f(t)$ is given by

$$
f(t) = \begin{cases} t + 1 & \text{for } t < 1, \\ \pi & \text{for } t = 1, \\ t^2 - 2t + 5 & \text{for } t > 1. \end{cases}
$$

a Verify that $f(t)$ defines a distribution by means of (8.21) (also see exercises 8.10 and 8.12b).
b Determine the derivative of $f(t)$ as distribution.

8.19 Let $a \leq 0$ be fixed and consider $f(t) = \epsilon(t)e^{at}$. Prove that $f'(t) - af(t) = \delta(t)$ (considered as distributions).

8.20 Define for fixed $a \in \mathbb{R}$ ($a \neq 0$) the function $g(t)$ by $g(t) = \epsilon(t)(\sin at)/a$. Prove that $g''(t) + a^2 g(t) = \delta(t)$ (considered as distributions).

8.4 Multiplication and scaling of distributions

In the previous section it was shown that distributions can be added and multiplied by a complex constant. We start this section with a treatment of the multiplication of distributions. Multiplication is important in connection with convolution theorems for the Fourier transform. This is because the convolution product changes into an ordinary product under the Fourier transform. If we want to formulate similar results for distributions, then we ought to be able to multiply distributions. However, in general this is *not* possible (in contrast to functions). The function $f(t) = |t|^{-1/2}$, for example, is integrable on, say, the interval $[-1, 1]$ and thus it defines a distribution through (8.12) (see exercise 8.9). But $f^2(t) = 1/|t|$ is not integrable on an interval containing 0; hence, one cannot define a distribution using (8.12). Still, multiplication is possible in a very limited way: distributions can be multiplied by *polynomials*. As a preparation we will first prove the following theorem.

THEOREM 8.1 *Let $\phi \in \mathcal{S}$ and p be a polynomial. Then $p\phi \in \mathcal{S}$.*

Proof
A polynomial $p(t)$ is of the form $a_n t^n + a_{n-1} t^{n-1} + \cdots + a_1 t + a_0$ with $a_i \in \mathbb{C}$. If $\phi \in \mathcal{S}$, then certainly $c\phi \in \mathcal{S}$ for $c \in \mathbb{C}$. The sum of two elements in $\mathcal{S}$ also belongs to $\mathcal{S}$. Hence, it is sufficient to show that $t^k \phi(t) \in \mathcal{S}$ for $\phi \in \mathcal{S}$ and $k \in \mathbb{N}$. But this has already been observed in section 6.5 (following theorem 6.11). ∎

We can now define the product of a distribution and a polynomial.

DEFINITION 8.6
Product of distribution and polynomial

Let T be a distribution and p a polynomial. The distribution pT is defined by

$$\langle pT, \phi \rangle = \langle T, p\phi \rangle \qquad \text{for } \phi \in \mathcal{S}. \tag{8.22}$$

Theorem 8.1 shows that the right-hand side of (8.22) is meaningful, since T is a distribution and $p\phi \in \mathcal{S}$. Since T is linear, it immediately follows that pT is linear as well:

$$\langle pT, c\phi \rangle = \langle T, p(c\phi) \rangle = \langle T, c(p\phi) \rangle = c \langle T, p\phi \rangle = c \langle pT, \phi \rangle$$

and

$$\langle pT, \phi_1 + \phi_2 \rangle = \langle T, p(\phi_1 + \phi_2) \rangle = \langle T, p\phi_1 + p\phi_2 \rangle$$
$$= \langle T, p\phi_1 \rangle + \langle T, p\phi_2 \rangle = \langle pT, \phi_1 \rangle + \langle pT, \phi_2 \rangle.$$

This proves that pT is indeed a distribution.

EXAMPLE 8.11   For a polynomial p one has

$$p(t)\delta(t) = p(0)\delta(t).$$

This is because according to definitions 8.6 and 8.2 we have $\langle p\delta, \phi \rangle = \langle \delta, p\phi \rangle = (p\phi)(0) = p(0)\phi(0)$ for any $\phi \in \mathcal{S}$. But $p(0)\phi(0) = p(0) \langle \delta, \phi \rangle = \langle p(0)\delta, \phi \rangle$ and according to definition 8.3 the distributions $p\delta$ and $p(0)\delta$ are thus equal. In particular we have for $p(t) = t$:

$$t\delta(t) = 0.$$

Similarly one has for the delta function $\delta(t - a)$ that

$$p(t)\delta(t - a) = p(a)\delta(t - a).$$

◀

Often a distribution can be multiplied by many more functions than just the polynomials. As an example we again look at the delta function $\delta(t)$, which can be defined on the set of all continuous functions (see section 8.2.2). Now if f is a continuous function, then precisely as in (8.22) one can define the product $f\delta$ by

Product of δ and a continuous function

$$\langle f\delta, \phi \rangle = \langle \delta, f\phi \rangle,$$

where ϕ is an arbitrary continuous function. From the definition of $\delta(t)$ it follows that $\langle f\delta, \phi \rangle = f(0)\phi(0) = f(0) \langle \delta, \phi \rangle$ and so one has

$$f(t)\delta(t) = f(0)\delta(t)$$

for any continuous function $f(t)$. For the general delta function $\delta(t - a)$ it follows analogously for any continuous function $f(t)$ that

$$f(t)\delta(t - a) = f(a)\delta(t - a) \qquad \text{for } a \in \mathbb{R}. \tag{8.23}$$

Similarly one can, for example, multiply the distribution δ' by continuously differentiable functions.

The reason why we are constantly working with the space $\mathcal{S}$ lies in the fact that $\mathcal{S}$ is very suitable for Fourier analysis. Moreover, it can be quite tedious to find out exactly for which set of functions the definition of a distribution still makes sense. And finally, it would be very annoying to keep track of all these different sets of functions (the continuous functions for δ, the continuously differentiable functions for δ', etc.). We have made an exception for (8.23) since it is widely used in practical applications and also because in much of the literature the delta function is introduced using continuous functions.

We close this section with a treatment of the scaling of distributions. As in the case of the definition of the derivative of a distribution, we first take a look at the situation for an absolutely integrable function $f(t)$. For $a \in \mathbb{R}$ with $a \neq 0$ one has for the scaled function $f(at)$ that

$$\int_{-\infty}^{\infty} f(at)\phi(t)\,dt = |a|^{-1} \int_{-\infty}^{\infty} f(\tau)\phi(a^{-1}\tau)\,d\tau \qquad \text{for } \phi \in \mathcal{S}.$$

This follows by changing to the variable $\tau = at$, where for $a < 0$ we should pay attention to the fact that the limits of integration are interchanged; this explains the factor $|a|^{-1}$. If we consider this result as an identity for the distributions associated with the functions, then it is clear how scaling of distributions should be defined.

DEFINITION 8.7
Scaling of distributions

Let T be a distribution and $a \in \mathbb{R}$ with $a \neq 0$. Then the scaled distribution $T(at)$ *is defined by*

$$\langle T(at), \phi(t) \rangle = |a|^{-1} \left\langle T(t), \phi(a^{-1}t) \right\rangle \qquad \text{for } \phi \in \mathcal{S}. \tag{8.24}$$

EXAMPLE 8.12

According to definitions 8.7 and 8.2 one has for the scaled delta distribution $\delta(at)$:

$$\langle \delta(at), \phi(t) \rangle = |a|^{-1} \left\langle \delta(t), \phi(a^{-1}t) \right\rangle = |a|^{-1} \phi(0) = |a|^{-1} \langle \delta(t), \phi(t) \rangle$$

for any $\phi \in \mathcal{S}$. Hence,

$$\delta(at) = |a|^{-1} \delta(t).$$

◀

Time reversal of distribution

Even and odd distribution

    A special case of scaling occurs for $a = -1$ and is called *time reversal*. For the delta function one has $\delta(-t) = \delta(t)$, which means that the delta function remains unchanged under time reversal. We recall that a function is called even when $f(-t) = f(t)$ and odd when $f(-t) = -f(t)$. Even and odd distributions are defined in the same way. A distribution T is called *even* when $T(-t) = T(t)$ and odd when $T(-t) = -T(t)$. The delta function $\delta(t)$ is thus an example of an even distribution.

EXERCISES

8.21   In example 8.10 it was shown that $(\epsilon(t)\cos t)' = \delta(t) - \epsilon(t)\sin t$. Derive this result again by formally applying the product rule for differentiation to the product of $\epsilon(t)$ and $\cos t$. (Of course, a product rule for differentiation of distributions cannot exist, since in general the product of distributions does not exist.)

8.22   Show that for the delta function $\delta(t - a)$ one has $p(t)\delta(t - a) = p(a)\delta(t - a)$, where $p(t)$ is a polynomial and $a \in \mathbb{R}$.

8.23   The derivative δ' of the delta function can be defined by $\langle \delta', \phi \rangle = -\phi'(0)$ for the set of all continuously differentiable functions ϕ. Let $f(t)$ be a continuously differentiable function.
  a   Give the definition of the product $f(t)\delta'(t)$.
  b   Show that $f(t)\delta'(t) = f(0)\delta'(t) - f'(0)\delta(t)$.
  c   Prove that $t\delta'(t) = -\delta(t)$ and that $t^2\delta'(t) = 0$.

8.24   Show that for the scaled derivative of the delta function one has $\delta'(at) = a^{-1}|a|^{-1}\delta'(t)$ for $a \neq 0$.

8.25   Show that the product $t \cdot \mathrm{pv}(1/t)$ is equal to the distribution 1.

8.26   Show that a distribution is even if and only if one has $\langle T, \phi(t) \rangle = \langle T, \phi(-t) \rangle$ for all $\phi \in \mathcal{S}$, while T is odd if and only if $\langle T, \phi(t) \rangle = -\langle T, \phi(-t) \rangle$.

8.27   **a**   Show that the distributions $\mathrm{sgn}\,t$ and $\mathrm{pv}(1/t)$ are odd.
  b   Show that the distribution $|t|$ is even.

SUMMARY

Distributions are linear mappings from the space of rapidly decreasing functions $\mathcal{S}$ to $\mathbb{C}$. The delta function $\delta(t - a)$, for example, assigns the number $\phi(a)$ to any $\phi \in \mathcal{S}$ $(a \in \mathbb{R})$. Many ordinary functions f can be considered as distribution by

$$\phi \to \int_{-\infty}^{\infty} f(t)\phi(t)\,dt \qquad \text{for } \phi \in \mathcal{S}.$$

Examples of this are the constant function 1, the unit step function $\epsilon(t)$, the sign function $\operatorname{sgn} t$ and any absolutely integrable function. Like the delta function, the distribution $\operatorname{pv}(1/t)$ is not of this form.

Distributions are arbitrarily often differentiable. The delta function is the derivative of the unit step function. More generally, one has the following. Let f be a function with a jump of magnitude c at the point $t = a$. Then the derivative of f (as distribution) contains the distribution $c\delta(t - a)$ at the point $t = a$.

Distributions can simply be added and multiplied by a constant. In general they cannot be multiplied together. It is possible to multiply a distribution by a polynomial. Sometimes a distribution can also be multiplied by more general functions. For example, for the delta function one has $f(t)\delta(t - a) = f(a)\delta(t - a)$ $(a \in \mathbb{R})$ for any continuous function $f(t)$. Finally, one can scale distributions with a real constant $a \neq 0$. For $a = -1$ this is called time reversal. This also gives rise to the notions even and odd distributions.

SELFTEST

8.28　　Given is the function $f(t) = \ln|t|$ for $t \neq 0$.
a Show that $0 \leq \ln t \leq t$ for $t \geq 1$ and conclude that $0 \leq \ln|t| \leq |t|$ for $|t| \geq 1$.
b Show that f is integrable over $[-1, 1]$. Use part a to show that f defines a distribution by means of (8.12).
c Prove that f defines an even distribution.

8.29　　The continuous function $f(t)$ is given by

$$f(t) = \begin{cases} t^2 & \text{for } t \geq 0, \\ 2t & \text{for } t < 0. \end{cases}$$

a Prove that f can be considered as a distribution.
b Find the derivative of f as distribution.
c Determine the second derivative of f as distribution.

8.30　　Consider the second derivative $\delta''(t)$ of the delta function.
a For which set of functions $f(t)$ can one define the product $f(t)\delta''(t)$?
b Prove that $f(t)\delta''(t) = f''(0)\delta(t) - 2f'(0)\delta'(t) + f(0)\delta''(t)$.
c Show that $t^2\delta''(t) = 2\delta(t)$ and that $t^3\delta''(t) = 0$.
d Show that for $a \neq 0$ one has for the scaled distribution $\delta''(at)$ that $\delta''(at) = a^{-2}|a|^{-1}\delta''(t)$.

Contents of Chapter 9

The Fourier transform of distributions

The Fourier transform of distributions

INTRODUCTION

In the previous chapter we have seen that distributions form an extension of the familiar functions. Moreover, in most cases it is not very hard to imagine a distribution intuitively as a limit of a sequence of functions. Especially when introducing new operations for distributions (such as differentiation), such an intuitive representation can be very useful. In section 8.1 we applied this method to make it plausible that the Fourier transform of the delta function is the constant function 1, and also that the reciprocity property holds in this case.

The purpose of the present chapter is to develop a rigorous Fourier theory for distributions. Of course, the theory has to be set up in such a way that for functions we recover our previous results; this is because distributions are an extension of functions. This is why we will derive the definition of the Fourier transform of a distribution from a property of the Fourier transform of functions in section 9.1. Subsequently, we will determine the spectrum of a number of standard distributions. Of course, the delta function will be treated first.

In section 9.2 we concentrate on the properties of the Fourier transform of distributions. The reciprocity property for distributions is proven. We also treat the correspondence between differentiation and multiplication. Finally, we show that the shift properties also remain valid for distributions.

It is quite problematic to give a rigorous definition of the convolution product or to state (let alone prove) a convolution theorem. Let us recall that the Fourier transform turns the convolution product into an ordinary multiplication (see section 6.6). But in general one cannot multiply two distributions, and so the convolution product of two distributions will not exist in general. In order to study this problem, we start in section 9.3 with an intuitive version of the convolution product of the delta function (and derivatives of the delta function) with an arbitrary distribution. We then look at the case where the distributions are defined by functions. This will tell us in which situations the convolution product of two distributions exists. Finally, the convolution theorem for distributions is formulated. The proof of this theorem will not be given; it would lead us too far into the theory of distributions.

LEARNING OBJECTIVES
After studying this chapter it is expected that you
- know and can apply the definition of the Fourier transform of a distribution
- know the Fourier transform of the delta function and the principal value $1/t$
- can determine the Fourier transform of periodic signals and periodic signals that are switched on
- know and can apply the properties of the Fourier transform of distributions
- know and can apply the convolution product and the convolution theorem for distributions in simple cases
- know the comb distribution and its Fourier transform*.

9.1   The Fourier transform of distributions: definition and examples

It should come as no surprise that we will start section 9.1.1 with the definition of the Fourier transform of a distribution. In section 9.1.2 we then determine the Fourier transform of a number of standard distributions. We also discuss how the Fourier transform of periodic signals (considered as distributions) can be determined. Finally, we treat the so-called comb distribution and its Fourier transform in section 9.1.3.

9.1.1   Definition of the Fourier transform of distributions

In section 8.1 we sketched an intuitive method to obtain the Fourier transform of a distribution. First, a distribution is considered symbolically as a limit of functions f_a whose Fourier transforms F_a are known. Next, the limit of these functions F_a is determined. So when the distribution T is given by $T = \lim f_a$ (for $a \to \infty$ for example) and $f_a \leftrightarrow F_a$, then the Fourier transform of T is given by $\mathcal{F}T = \lim F_a$ (for $a \to \infty$). Mathematically, however, this method has some serious problems. What is meant by the symbolic limits $T = \lim f_a$ and $\mathcal{F}T = \lim F_a$? Is the choice of the sequence of functions f_a (and hence F_a) uniquely determined? And if not, is $\mathcal{F}T$ uniquely determined then? One can solve all of these problems at once using the rigorous definition of distributions given in the previous chapter. The only thing still missing is the rigorous definition of the Fourier transform – or spectrum – of a distribution. Of course, such a definition has to be in agreement with our earlier definition of the Fourier transform of a function, since distributions form an extension of functions. Precisely as in the case of the definition of the derivative of a distribution, we therefore start with a function f that can be considered as a distribution T_f according to (8.12). Is it then possible, using the properties of the Fourier transform of ordinary functions, to come up with a definition of the Fourier transform of the distribution T_f? Well, according to the selfduality property in section 6.4.7 one has for any $\phi \in \mathcal{S}$

$$\int_{-\infty}^{\infty} F(t)\phi(t)\, dt = \int_{-\infty}^{\infty} f(t)\Phi(t)\, dt,$$

where Φ is the spectrum of ϕ and F is the spectrum of f. From theorem 6.12 it follows that $\Phi \in \mathcal{S}$ and so the identity above can also be considered as an identity for distributions: $\langle T_F, \phi \rangle = \langle T_f, \Phi \rangle$. From this, it is obvious how one should define the Fourier transform (or spectrum) of an arbitrary distribution.

DEFINITION 9.1
Fourier transform or
spectrum of distributions

For a distribution T the Fourier transform *or* spectrum *$\mathcal{F}T$ is defined by*

$$\langle \mathcal{F}T, \phi \rangle = \langle T, \Phi \rangle, \tag{9.1}$$

where Φ is the Fourier transform of $\phi \in \mathcal{S}$.

We recall once again theorem 6.12, which established that $\Phi \in \mathcal{S}$ for $\phi \in \mathcal{S}$. Hence, the right-hand side of (9.1) is well-defined and from the linearity of T it follows immediately that $\mathcal{F}T$ is indeed a distribution. The mapping assigning the spectrum $\mathcal{F}T$ to a distribution T is again called the *Fourier transform*. In section 9.1.2 we determine the Fourier transform of a number of distributions.

9.1.2   Examples of Fourier transforms of distributions

First of all, we will use definition 9.1 to determine the Fourier transform of the delta function. We have

$$\langle \mathcal{F}\delta, \phi \rangle = \langle \delta, \Phi \rangle = \Phi(0) = \int_{-\infty}^{\infty} \phi(t)\,dt \qquad \text{for } \phi \in \mathcal{S},$$

Spectrum of $\delta(t)$

where in the last step we used the definition of the ordinary Fourier transform. From example 8.1 it then follows that $\langle \mathcal{F}\delta, \phi \rangle = \langle 1, \phi \rangle$, which shows that the spectrum of δ is indeed the function 1. A short symbolic proof is obtained by taking $e^{-i\omega t}$ for ϕ in (8.9). It then follows that

$$\int_{-\infty}^{\infty} \delta(t) e^{-i\omega t}\,dt = 1, \tag{9.2}$$

which states precisely (but now symbolically) that the spectrum of $\delta(t)$ is the function 1. Figure 9.1 shows $\delta(t)$ and its spectrum.

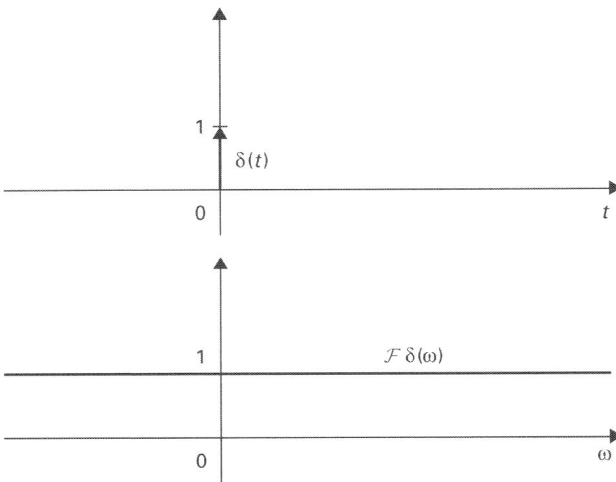

FIGURE 9.1
The delta function $\delta(t)$ and its spectrum.

Conversely, the spectrum of the function 1 is determined as follows:

$$\langle \mathcal{F}1, \phi \rangle = \langle 1, \Phi \rangle = \int_{-\infty}^{\infty} \Phi(\omega)\,d\omega.$$

For $\phi \in \mathcal{S}$ the inversion formula (7.9) certainly holds. Applying it for $t = 0$, it follows that

$$\langle \mathcal{F}1, \phi \rangle = \int_{-\infty}^{\infty} \Phi(\omega)\,d\omega = 2\pi\phi(0) = 2\pi\,\langle \delta, \phi \rangle = \langle 2\pi\delta, \phi \rangle \qquad \text{for } \phi \in \mathcal{S}.$$

Spectrum of 1

Hence, the spectrum of the function 1 (as distribution) is given by the distribution $2\pi\delta(\omega)$. Symbolically, this is sometimes written as

$$\int_{-\infty}^{\infty} e^{-i\omega t}\,dt = 2\pi\delta(\omega). \tag{9.3}$$

(Note that the integral in the left-hand side does not exist as an improper integral; see exercise 6.1.) The results that have been derived intuitively in section 8.1 have now all been proven. Note that $\delta(t) \leftrightarrow 1$ and $1 \leftrightarrow 2\pi\delta(\omega)$ constitute a Fourier pair in the sense of section 7.2.2. In section 9.2.3 we will prove that indeed the reciprocity property for distributions remains valid. In the next example the delta function at the point a is treated (see (8.10) for its definition).

EXAMPLE 9.1
Spectrum of $\delta(t-a)$

The spectrum of $\delta(t-a)$ is the function $e^{-ia\omega}$ (considered as distribution of course). The proof is similar to the case $a = 0$:

$$\langle \mathcal{F}\delta(t-a), \phi \rangle = \langle \delta(t-a), \Phi \rangle = \Phi(a) = \int_{-\infty}^{\infty} \phi(t)e^{-iat}\,dt \qquad \text{for } \phi \in \mathcal{S}.$$

If we now call the variable of integration ω instead of t, then the result follows: $\langle \mathcal{F}\delta(t-a), \phi \rangle = \langle e^{-ia\omega}, \phi \rangle$. Conversely, one has that $2\pi\delta(\omega - a)$ is the spectrum

Spectrum of e^{iat}

of the function e^{iat}:

$$\langle \mathcal{F}e^{iat}, \phi \rangle = \langle e^{iat}, \Phi \rangle = \int_{-\infty}^{\infty} e^{iat}\Phi(t)\,dt$$

and according to the inversion formula (7.9) it then follows that $\langle \mathcal{F}e^{iat}, \phi \rangle = 2\pi\phi(a) = \langle 2\pi\delta(\omega - a), \phi \rangle$ $(\phi \in \mathcal{S})$, which proves the result. We thus again have a Fourier pair $\delta(t-a) \leftrightarrow e^{-ia\omega}$ and $e^{iat} \leftrightarrow 2\pi\delta(\omega - a)$. Also note that the function e^{iat} is a periodic function with period $2\pi/a$. ◄

Just as for the Fourier transform of functions, the Fourier transform of distributions is a linear mapping. For distributions S and T and $a, b \in \mathbb{C}$ one thus has $\mathcal{F}(aS + bT) = a\mathcal{F}S + b\mathcal{F}T$.

EXAMPLE 9.2

Since the Fourier transform is linear, it follows from example 9.1 that the spectrum of $\cos at = (e^{iat} + e^{-iat})/2$ is the distribution $\pi(\delta(\omega - a) + \delta(\omega + a))$. This is represented graphically in figure 9.2. ◄

For the Fourier transform of distributions one has the following result.

THEOREM 9.1

The Fourier transform is a one-to-one mapping on the space of distributions.

Proof
Because of the linearity of the Fourier transform, it is sufficient to show that $\mathcal{F}T = 0$ implies that $T = 0$. So let us assume that $\mathcal{F}T = 0$, then $\langle \mathcal{F}T, \phi \rangle = 0$ for all $\phi \in \mathcal{S}$. From definition 9.1 it then follows that $\langle T, \Phi \rangle = 0$ for all $\phi \in \mathcal{S}$. But according to theorem 7.6 one can write *any* $\psi \in \mathcal{S}$ as the spectrum Φ of some $\phi \in \mathcal{S}$. Hence, $\langle T, \psi \rangle = 0$ for all $\psi \in \mathcal{S}$, which means that $T = 0$ (definition 8.3). ∎

Theorem 9.1 is often used (implicitly) in the following situation. Let a distribution U be given. Suppose that by using some table, perhaps in combination with the properties of the Fourier transform, we have found a distribution whose spectrum is the given distribution U. We may then conclude on the basis of theorem 9.1 that we have found the only possibility. For a given distribution T there is thus only one distribution U which is the spectrum of T. As for functions, $T \leftrightarrow U$ will mean that U is the spectrum of T and that the distributions T and U determine each other uniquely.

EXAMPLE 9.3

Suppose that we are looking for a distribution T whose spectrum is the distribution $U = 4\delta(\omega - 3) - 2\delta(\omega + 2)$. In example 9.1 it was shown that $e^{iat} \leftrightarrow 2\pi\delta(\omega - a)$.

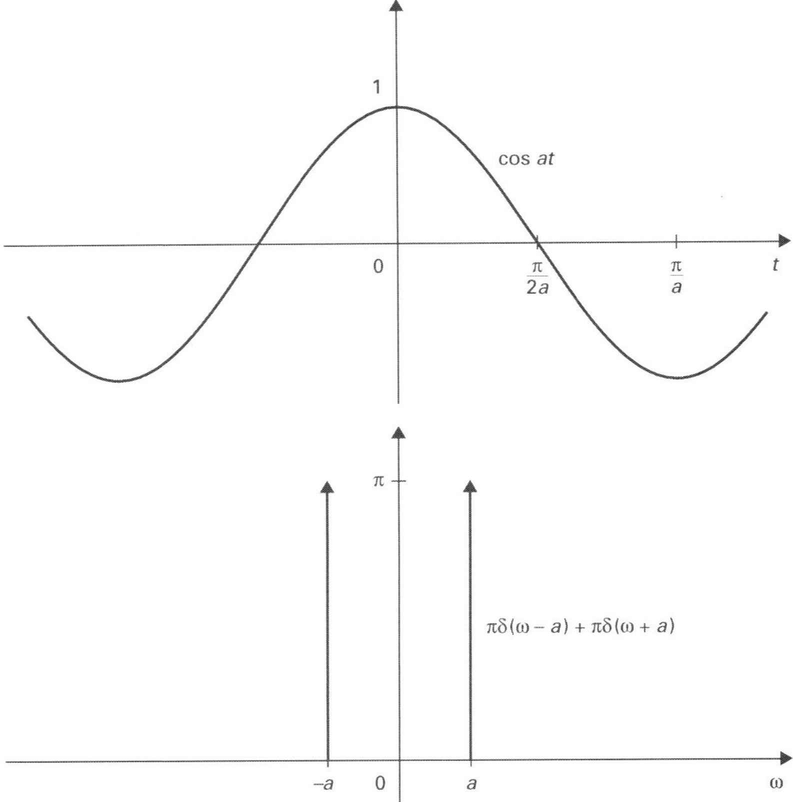

FIGURE 9.2
The function $\cos at$ and its spectrum.

From the linearity of the Fourier transform it then follows that $T = (2/\pi)e^{3it} - (1/\pi)e^{-2it}$ is a distribution with spectrum U. Theorem 9.1 guarantees that it is the only one.   ◀

In the next example the spectrum of the distribution $\mathrm{pv}(1/t)$ will be determined.

EXAMPLE 9.4
Spectrum of $\mathrm{pv}(1/t)$

The spectrum of the distribution $\mathrm{pv}(1/t)$ from example 8.5 is the distribution $-\pi i \operatorname{sgn} \omega$. A mathematically rigorous proof of this result would lead us too far into the theory of distributions. Instead we will only give the following formal proof (using a certain assumption, one can give a rigorous proof; see exercise 9.23).

$$\langle \mathcal{F}\mathrm{pv}(1/t), \phi \rangle = \langle \mathrm{pv}(1/t), \Phi \rangle = \lim_{\alpha \to 0} \int_{|t| \geq \alpha} \frac{\Phi(t)}{t}\, dt \quad \text{for } \phi \in \mathcal{S}.$$

Now apply the definition of spectrum for $\Phi(t)$, then

$$\langle \mathcal{F}\mathrm{pv}(1/t), \phi \rangle = \lim_{\alpha \to 0} \int_{|t| \geq \alpha} \frac{1}{t} \left(\int_{-\infty}^{\infty} \phi(\omega)e^{-i\omega t}\, d\omega \right) dt.$$

Next, we (formally) interchange the order of integration and write the limit inside the integral (again formally). We then obtain

$$\langle \mathcal{F}\mathrm{pv}(1/t), \phi \rangle = \int_{-\infty}^{\infty} \phi(\omega) \left(\lim_{\alpha \to 0} \int_{|t| \geq \alpha} \frac{e^{-i\omega t}}{t}\, dt \right) d\omega. \tag{9.4}$$

The inner integral can easily be calculated since

$$\int_{\alpha}^{\infty} \frac{e^{-i\omega t}}{t}\, dt + \int_{-\infty}^{-\alpha} \frac{e^{-i\omega t}}{t}\, dt = \int_{\alpha}^{\infty} \frac{e^{-i\omega t} - e^{i\omega t}}{t}\, dt = -2i \int_{\alpha}^{\infty} \frac{\sin \omega t}{t}\, dt.$$

Here we changed from the variable t to the variable $-t$ in the second integral of the left-hand side. Now use that

$$\int_{0}^{\infty} \frac{\sin \omega t}{t}\, dt = \frac{\pi}{2}\mathrm{sgn}\,\omega.$$

For $\omega > 0$ this follows from theorem 4.11 (or from (7.3)) by changing from the variable ωt to the variable t, while for $\omega < 0$ an additional minus sign enters because the limits of integration will be reversed; for $\omega = 0$ we have $\sin \omega t = 0$. The limit for $\alpha \to 0$ of the inner integral in (9.4) thus exists and

$$\lim_{\alpha \to 0} \int_{|t| \geq \alpha} \frac{e^{-i\omega t}}{t}\, dt = -\pi i\,\mathrm{sgn}\,\omega.$$

From (9.4) (and example 8.3) it then follows for every $\phi \in \mathcal{S}$ that

$$\langle \mathcal{F}\mathrm{pv}(1/t), \phi \rangle = -\pi i \int_{-\infty}^{\infty} \mathrm{sgn}\,\omega\, \phi(\omega)\, d\omega = \langle -\pi i\,\mathrm{sgn}\,\omega, \phi(\omega) \rangle,$$

which proves that $\mathrm{pv}(1/t) \leftrightarrow -\pi i\,\mathrm{sgn}\,\omega$. This is shown in figure 9.3. Note that $-\pi i\,\mathrm{sgn}\,\omega$ assumes only imaginary values; in figure 9.3 this is emphasized by the dashed lines for the corresponding curves. ◄

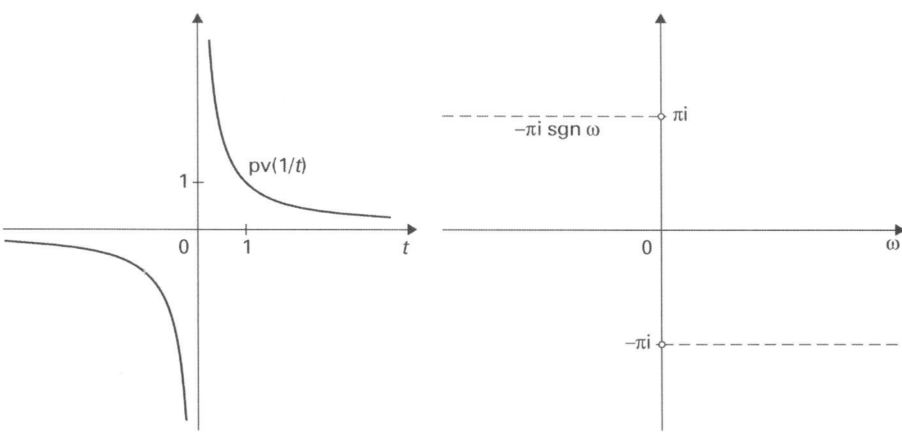

FIGURE 9.3
The distribution $\mathrm{pv}(1/t)$ and its spectrum.

Spectrum of periodic functions

We end section 9.1.2 by determining the spectrum of a periodic signal using the result $(\mathcal{F}e^{iat})(\omega) = 2\pi\delta(\omega - a)$ (see example 9.1). So let f be a periodic function with a converging Fourier series:

$$f(t) = \sum_{k=-\infty}^{\infty} c_k e^{2\pi i k t/T} = \sum_{k=-\infty}^{\infty} c_k e^{i k \omega_0 t},$$

where $\omega_0 = 2\pi/T$. Assume, moreover, that f defines a distribution by means of (8.12). The Fourier series then also defines a distribution and the spectrum of this distribution can be calculated as follows:

$$\mathcal{F}\left(\sum_{k=-\infty}^{\infty} c_k e^{i k \omega_0 t}\right) = \sum_{k=-\infty}^{\infty} c_k \mathcal{F}\left(e^{i k \omega_0 t}\right) = 2\pi \sum_{k=-\infty}^{\infty} c_k \delta\left(\omega - k\omega_0\right). \tag{9.5}$$

Here we assumed that the Fourier transform $\mathcal{F}$ and the summation may be interchanged. We will not prove that this is indeed allowed.

In fact, (9.5) shows once again that a periodic function $f(t)$ has a line spectrum: the spectrum consists of delta functions at the points $\omega = k\omega_0$ ($k \in \mathbb{Z}$) with 'weight' equal to the kth Fourier coefficient c_k (also see section 3.3).

9.1.3   The comb distribution and its spectrum*

The material in this section can be omitted without any consequences for the remainder of the book. Furthermore, we note that Poisson's summation formula from section 7.3* will be used in an essential way.

The main reason that we treat the comb distribution is the fact that it is widely used in the technical literature to represent the *sampling* of a continuous-time signal (see later on in this section).

Comb or shah distribution   The *comb* or *shah distribution* $\text{\textcyrillic{Ш}}$ is defined by

$$\langle \text{\textcyrillic{Ш}}, \phi \rangle = \sum_{k=-\infty}^{\infty} \phi(k) \qquad \text{for } \phi \in \mathcal{S}. \tag{9.6}$$

First we will have to show that $\text{\textcyrillic{Ш}}$ is well-defined, that is, the series in the right-hand side of (9.6) converges. But for $\phi \in \mathcal{S}$ there exists a constant $M > 0$ such that $(1 + t^2)\,|\phi(t)| \leq M$. Hence, it follows that (compare with example 8.1)

$$\sum_{k=-\infty}^{\infty} |\phi(k)| \leq M \sum_{k=-\infty}^{\infty} \frac{1}{1+k^2} = M + 2M \sum_{k=1}^{\infty} \frac{1}{1+k^2}.$$

But $1 + k^2 > k^2$ for $k > 0$ and $\sum_{k=1}^{\infty} k^{-2}$ converges. Hence,

$$\left| \sum_{k=-\infty}^{\infty} \phi(k) \right| \leq \sum_{k=-\infty}^{\infty} |\phi(k)| < M\left(1 + 2\sum_{k=1}^{\infty} k^{-2}\right) < \infty.$$

It now immediately follows that $\text{\textcyrillic{Ш}}$ defines a distribution. Since one has $\langle \delta(t - k), \phi \rangle = \phi(k)$ (see (8.10)), we obtain from (9.6) that

$$\langle \text{\textcyrillic{Ш}}, \phi \rangle = \sum_{k=-\infty}^{\infty} \langle \delta(t - k), \phi \rangle$$

for every $\phi \in \mathcal{S}$. Because of this, one often writes $\text{\textcyrillic{Ш}} = \text{\textcyrillic{Ш}}(t)$ as

$$\text{\textcyrillic{Ш}}(t) = \sum_{k=-\infty}^{\infty} \delta(t - k)$$

Train of delta functions

and calls $\mathcal{W}$ a *train of delta functions*. The distribution $\mathcal{W}$ is then graphically represented as in figure 9.4a. Often, $\mathcal{W}$ is used to represent the sampling of a continuous-

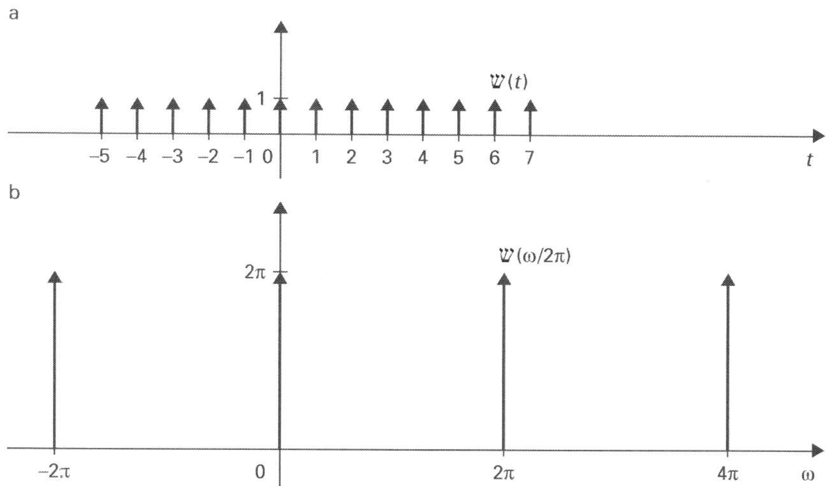

FIGURE 9.4
The impulse train (a) and its spectrum (b).

time signal (for example in Van den Enden and Verhoeckx (1987), chapter 3 (in Dutch)). For when $f(t)$ is a bounded continuous-time signal, then the product $f(t)\mathcal{W}(t)$ exists (we state this without proof) and one has (as in definition 8.6)

$$\langle f(t)\mathcal{W}(t), \phi\rangle = \langle \mathcal{W}(t), f\phi\rangle = \sum_{k=-\infty}^{\infty} f(k)\phi(k) = \sum_{k=-\infty}^{\infty} \langle f(k)\delta(t-k), \phi\rangle$$

for all $\phi \in \mathcal{S}$. Hence we obtain that

$$f(t)\mathcal{W}(t) = \sum_{k=-\infty}^{\infty} f(k)\delta(t-k).$$

Sampling

The right-hand side of this can indeed be interpreted as the *sampling* of the function $f(t)$ at times $t = k$ ($k \in \mathbb{Z}$); see figure 9.5.

After this short intermezzo on sampling, we now determine the spectrum of the comb distribution $\mathcal{W}$. From the definition of $\mathcal{W}$ it follows that

$$\langle \mathcal{F}\mathcal{W}, \phi\rangle = \langle \mathcal{W}, \Phi\rangle = \sum_{k=-\infty}^{\infty} \Phi(k) \qquad \text{for } \phi \in \mathcal{S}.$$

From example 7.9 it follows that Poisson's summation formula (7.23) is valid for functions in $\mathcal{S}$. If we apply it with $T = 2\pi$, then it follows that

$$\langle \mathcal{F}\mathcal{W}, \phi\rangle = \sum_{k=-\infty}^{\infty} \Phi(k) = 2\pi \sum_{k=-\infty}^{\infty} \phi(2\pi k) = \langle \mathcal{W}(\omega/2\pi), \phi\rangle$$

for every $\phi \in \mathcal{S}$ (in the last step we used the scaling property for distributions with scaling factor $1/2\pi$; see definition 8.7). We have thus proven that $\mathcal{F}\mathcal{W} = \mathcal{W}(\omega/2\pi)$, where $\mathcal{W}(\omega/2\pi)$ is the scaled comb distribution $\mathcal{W}$; the spectrum of the

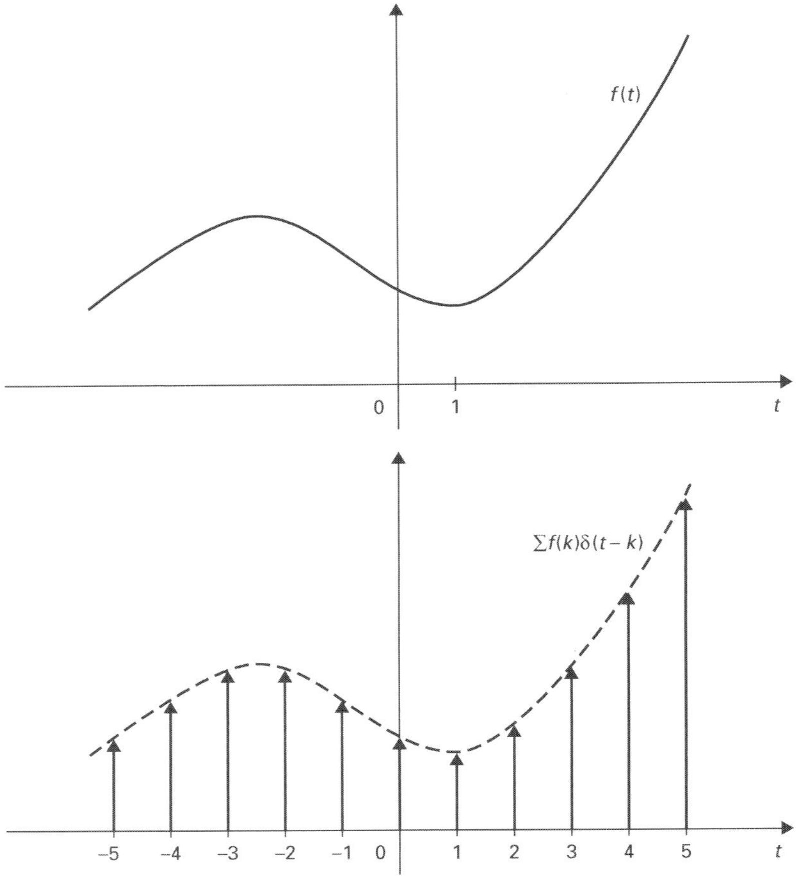

FIGURE 9.5
A continuous-time function $f(t)$ and the sampling $f(t)\ \text{Ш}(t)$.

Spectrum of Ш

comb distribution is thus again a comb distribution:

$$\text{Ш}(t) \leftrightarrow \text{Ш}(\omega/2\pi). \tag{9.7}$$

In the notation of the impulse train, (9.7) becomes

$$\sum_{k=-\infty}^{\infty} \delta(t - k) \leftrightarrow 2\pi \sum_{k=-\infty}^{\infty} \delta(\omega - 2\pi k), \tag{9.8}$$

and one then expresses this by saying that the spectrum of an impulse train is again an impulse train. See figure 9.4b.

EXERCISES

9.1   Let T be a distribution. Show that $\mathcal{F}T$ as defined by (9.1) is a linear mapping from $\mathcal{S}$ to $\mathbb{C}$. Conclude that $\mathcal{F}T$ is a distribution.

9.2   Let S and T be distributions with spectra U and V respectively. Show that for $a, b \in \mathbb{C}$ the spectrum of $aS + bT$ equals $aU + bV$. The Fourier transform of distributions is thus a linear mapping.

9.3   Determine the spectrum of the following distributions:
a $\delta(t - 4)$,
b e^{3it},
c $\sin at$,
d $4\cos 2t + 2i\,\mathrm{pv}(1/t)$.

9.4   Determine the distribution T whose spectrum is the following distributions:
a $\delta(\omega + 5)$,
b $2\pi\delta(\omega + 2) + 2\pi\delta(\omega - 2)$,
c $\mathrm{sgn}\,\omega + 2\cos\omega$.

9.5   Show that the spectrum of an even (odd) distribution is again an even (odd) distribution.

9.6*   **a**   Verify that $\mathcal{W}$ defines a distribution.
b   Define the distribution $\sum_{k=-\infty}^{\infty} e^{ik\omega}$ by

$$\left\langle \sum_{k=-\infty}^{\infty} e^{ik\omega}, \phi \right\rangle = \sum_{k=-\infty}^{\infty} \left\langle e^{ik\omega}, \phi \right\rangle \qquad \text{for } \phi \in \mathcal{S}.$$

Show that $\mathcal{F}\,\mathcal{W} = \sum_{k=-\infty}^{\infty} e^{ik\omega}$.

9.2   Properties of the Fourier transform

Most of the properties of the Fourier transform of functions, as derived in section 6.4, can easily be carried over to distributions. In this section we examine shifting, differentiation and reciprocity (in exercise 9.9 scaling is considered as well).

9.2.1   *Shift in time and frequency domains*

First we treat the shifting property in the time domain. We will show that for a distribution T with spectrum U one has $T(t - a) \leftrightarrow e^{-ia\omega}U(\omega)$ ($a \in \mathbb{R}$), just as for functions (see section 6.4.3). Apart from the proof of this property, there are two problems that we have to address. Does the product of $U(\omega)$ and $e^{-ia\omega}$ exist? And what do we mean by the shifted distribution $T(t - a)$? We start with the first problem. The product $e^{-ia\omega}U(\omega)$ can be defined precisely as in (8.22) by $\left\langle e^{-ia\omega}U(\omega), \phi \right\rangle = \left\langle U(\omega), e^{-ia\omega}\phi \right\rangle$ ($\phi \in \mathcal{S}$). This definition makes sense since it follows immediately from the product rule for differentiation and from $\left| e^{-ia\omega} \right| = 1$ that $e^{-ia\omega}\phi \in \mathcal{S}$. This solves the first problem. We now handle the second problem and define the shifted distribution $T(t - a)$.

DEFINITION 9.2
Shifted distribution

For a distribution $T(t)$ the distribution $T(t - a)$ shifted over $a \in \mathbb{R}$ is defined by $\langle T(t - a), \phi(t) \rangle = \langle T(t), \phi(t + a) \rangle$.

As before (for example, for differentiation, scaling and the Fourier transform), this definition is a direct generalization of the situation that occurs if we take T equal to a distribution T_f, where f is a function (see (8.12)).

EXAMPLE

As is suggested by the notation, the delta function $\delta(t - a)$ at the point a is indeed the shifted delta function $\delta(t)$. This is proven as follows: $\langle \delta(t - a), \phi(t) \rangle = \langle \delta(t), \phi(t + a) \rangle = \phi(a)$, which is in agreement with the earlier definition of $\delta(t - a)$ in (8.10).

$\blacktriangleleft$

We can now prove the shifting property for distributions. When $T(t)$ is a distribution with spectrum $U(\omega)$, then

$$\langle \mathcal{F}(T(t - a)), \phi(t) \rangle = \langle T(t - a), \Phi(t) \rangle = \langle T(t), \Phi(t + a) \rangle \qquad \text{for } \phi \in \mathcal{S}.$$

From the shifting property in theorem 6.4 it follows that

$$\langle T(t), \Phi(t + a) \rangle = \left\langle T(t), \mathcal{F}(e^{-iat}\phi(t)) \right\rangle = \left\langle U(\omega), e^{-ia\omega}\phi(\omega) \right\rangle.$$

Hence, $\langle \mathcal{F}(T(t - a)), \phi(t) \rangle = \left\langle e^{-ia\omega}U(\omega), \phi(\omega) \right\rangle$, proving that

Shift property

$$T(t - a) \leftrightarrow e^{-ia\omega}U(\omega). \tag{9.9}$$

EXAMPLE

Since $\delta(t) \leftrightarrow 1$, it follows from the shifting property that $\delta(t - a) \leftrightarrow e^{-ia\omega}$, in accordance with example 9.1.

$\blacktriangleleft$

In a similar way one can prove the shifting property in the frequency domain (see exercise 9.15):

$$e^{iat}T \leftrightarrow U(\omega - a) \tag{9.10}$$

9.2.2 Differentiation in time and frequency domains

For the ordinary Fourier transform, differentiation in one domain corresponded to multiplication (by $-it$ or $i\omega$) in the other domain (see sections 6.4.8 and 6.4.9). This is also the case for the Fourier transform of distributions. We start by determining

Spectrum of $\delta'(t)$

the spectrum of the derivative $\delta'(t)$ of the delta function. Successively applying definitions 9.1, 8.4 and 8.2, we obtain that

$$\langle \mathcal{F}\delta', \phi \rangle = \langle \delta', \Phi \rangle = -\langle \delta, \Phi' \rangle = -\Phi'(0) \qquad \text{for } \phi \in \mathcal{S}.$$

According to theorem 6.8 one has $\Phi'(\omega) = \mathcal{F}(-it\phi(t))(\omega)$ and so

$$\langle \mathcal{F}\delta', \phi \rangle = \mathcal{F}(it\phi(t))(0) = \int_{-\infty}^{\infty} it\phi(t)\,dt = \langle it, \phi \rangle \qquad \text{for } \phi \in \mathcal{S}.$$

The variable t in the integral – and in the distribution – is irrelevant; by changing to the variable ω we have thus shown that $\delta' \leftrightarrow i\omega$. In general one has for an arbitrary distribution T with spectrum U that $T' \leftrightarrow i\omega U$. In fact, as for the spectrum of the derivative of the delta function, it follows that

$$\langle \mathcal{F}T', \phi \rangle = \langle T', \Phi \rangle = -\langle T, \Phi' \rangle = \langle T, \mathcal{F}(it\phi(t)) \rangle = \langle U, it\phi(t) \rangle.$$

In the last step we use that $T \leftrightarrow U$. Again, the variable t is irrelevant and if we consider the distribution U as acting on functions in the variable ω, then we may also write $\langle \mathcal{F}T', \phi \rangle = \langle U, i\omega\phi(\omega) \rangle$. From definition 8.6 of the multiplication of distributions by polynomials, we then finally obtain that $\langle \mathcal{F}T', \phi \rangle = \langle i\omega U, \phi(\omega) \rangle$, which proves the result. More generally, one has (see exercise 9.8c):

Differentiation in time domain

$$T^{(k)} \leftrightarrow (i\omega)^k U \qquad \text{when} \qquad T \leftrightarrow U. \tag{9.11}$$

Similarly, one can prove the differentiation rule in the frequency domain (see exercise 9.10c):

Differentiation in frequency domain

$$(-it)^k T \leftrightarrow U^{(k)} \qquad \text{when} \qquad T \leftrightarrow U \tag{9.12}$$

(compare with section 6.4.9). This rule is mainly used to determine the spectrum of $t^k T$ when the spectrum of T is known.

EXAMPLE

We know from section 9.1.2 that $1 \leftrightarrow 2\pi\delta(\omega)$. From (9.12) (for $k = 1$ and $k = 2$) it then follows that $t \leftrightarrow 2\pi i\delta'(\omega)$ and $t^2 \leftrightarrow -2\pi\delta''(\omega)$. ◀

9.2.3 Reciprocity

The Fourier pair $\delta \leftrightarrow 1$ and $1 \leftrightarrow 2\pi\delta$ suggests that the reciprocity property also holds for distributions. This is indeed the case: if U is the spectrum of T, then $2\pi T(\omega)$ is the spectrum of $U(-t)$, or

Reciprocity

$$U(-t) \leftrightarrow 2\pi T(\omega) \tag{9.13}$$

(where $U(-t)$ is the scaled distribution from definition 8.7 with $a = -1$). For its proof we first recall the reciprocity property for functions: if $\phi \leftrightarrow \Phi$, then $\Phi(-t) \leftrightarrow 2\pi\phi(\omega)$. Hence, it follows that

$$\langle 2\pi T(\omega), \phi(\omega)\rangle = \langle T(\omega), 2\pi\phi(\omega)\rangle = \langle T(\omega), \mathcal{F}(\Phi(-t))\rangle = \langle U(t), \Phi(-t)\rangle .$$

Formula (9.13) is then proven by applying definition 8.7 of scaling. For then it follows for every $\phi \in \mathcal{S}$ that

$$\langle 2\pi T(\omega), \phi(\omega)\rangle = \langle U(-t), \Phi(t)\rangle = \langle \mathcal{F}U(-t), \phi(\omega)\rangle .$$

EXAMPLE

For T we take the delta function, so U is the function 1. Since 1 is an even distribution, it follows from (9.13) that $1 \leftrightarrow 2\pi\delta$, in accordance with our previous results from section 9.1.2. ◀

EXAMPLE 9.6
Spectrum of sgn t

The distribution $\operatorname{sgn} t$ is odd (see exercise 8.27a). Applying (9.13) to the result of example 9.4, we obtain that the spectrum of $\pi i \operatorname{sgn} t$ is the distribution $2\pi\operatorname{pv}(1/\omega)$, that is,

$$\operatorname{sgn} t \leftrightarrow -2i\operatorname{pv}(1/\omega). \tag{9.14}$$

Spectrum of $\epsilon(t)$

Since $\epsilon(t) = (1 + \operatorname{sgn} t)/2$, where $\epsilon(t)$ is the unit step function, it follows that

$$\epsilon(t) \leftrightarrow \pi\delta(\omega) - i\operatorname{pv}(1/\omega), \tag{9.15}$$

where we used the linearity of the Fourier transform. The function $\epsilon(t)$ and its spectrum are shown in figure 9.6; imaginary values are represented by dashed curves in this figure. ◀

EXAMPLE 9.7

We know from (8.18) that $\epsilon' = \delta$. From (9.11) it then follows that $\mathcal{F}\delta = \mathcal{F}\epsilon' = i\omega\mathcal{F}\epsilon$. Apparently, $i\omega\mathcal{F}\epsilon = 1$. Now we have shown in example 9.6 that $\mathcal{F}\epsilon = \pi\delta(\omega) - i\operatorname{pv}(1/\omega)$ and indeed we have $i\omega(\pi\delta(\omega) - i\operatorname{pv}(1/\omega)) = \pi i\omega\delta(\omega) + \omega\operatorname{pv}(1/\omega) = 0 + 1 = 1$ (see example 8.11 and exercise 8.25). ◀

At the end of section 9.1.2 we determined the spectrum of periodic functions. One can use (9.15) to determine the spectrum of periodic functions that are 'switched on'.

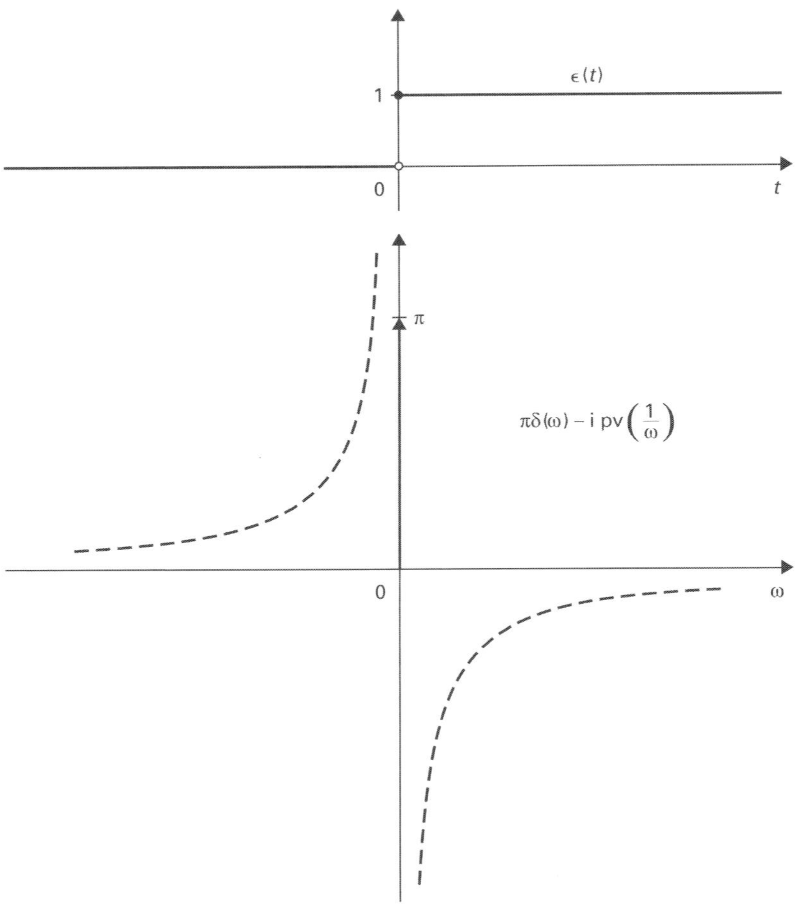

FIGURE 9.6
The unit step and its spectrum.

Let $f(t)$ be a periodic function with period T and converging Fourier series

$$f(t) = \sum_{k=-\infty}^{\infty} c_k e^{2\pi i kt/T}.$$

Switched-on periodic signal   We will call $\epsilon(t)f(t)$ a *switched-on periodic signal*. In (9.15) the spectrum of $\epsilon(t)$ has been determined, and from the shifting property (9.10) it then follows that

$$\epsilon(t)e^{ikt\omega_0} \leftrightarrow \pi\delta(\omega - k\omega_0) - i\,\mathrm{pv}(1/(\omega - k\omega_0)),$$

Spectrum of switched-on periodic signal   where $\omega_0 = 2\pi/T$. If we now assume, as in section 9.1.2, that the Fourier transform and the summation may be interchanged, then it follows that the spectrum of a switched-on periodic signal $\epsilon(t)f(t)$ is given by

$$\sum_{k=-\infty}^{\infty} c_k \left(\pi\delta(\omega - k\omega_0) - i\,\mathrm{pv}\frac{1}{\omega - k\omega_0} \right). \tag{9.16}$$

EXERCISES

9.7 The spectrum of $\operatorname{sgn} t$ is $-2i\operatorname{pv}(1/\omega)$ (see example 9.6). Verify that $\epsilon(t) = (1 + \operatorname{sgn} t)/2$ ($t \neq 0$) and that the spectrum of $\epsilon(t)$ is given by $\pi\delta(\omega) - i\operatorname{pv}(1/\omega)$ (as was stated in example 9.6).

9.8 **a** Determine in a direct way (without using (9.11)) the spectrum of δ''.
b Let T be a distribution with spectrum U. Show that $T'' \leftrightarrow -\omega^2 U$.
c Prove (9.11): $T^{(k)} \leftrightarrow (i\omega)^k U$ when $T \leftrightarrow U$.

9.9 **a** Let T be a distribution with spectrum U. Show that for $a \neq 0$ the scaled distribution $T(at)$ has $|a|^{-1} U(\omega/a)$ as its spectrum.
b Determine the spectrum of $\delta(4t + 3)$.

9.10 **a** Show that $-it\, T \leftrightarrow U'$ when $T \leftrightarrow U$.
b Use part a to determine the spectrum of $t\delta(t)$ and $t\operatorname{pv}(1/t)$; check your answers using example 8.11 and exercise 8.25.
c Show that $(-it)^k T \leftrightarrow U^{(k)}$ when $T \leftrightarrow U$.

9.11 Determine the spectrum of $t\delta'(t)$ and $t\delta''(t)$ using (9.12); check your answers using exercises 8.23 and 8.30.

9.12 From the identity $\epsilon' = \delta$ it follows that $\mathcal{F}\epsilon' = \mathcal{F}\delta = 1$. The differentiation rule (9.11) then leads to $i\omega\mathcal{F}\epsilon = 1$. Hence, $\mathcal{F}\epsilon = 1/i\omega$, where we have to consider the right-hand side as a distribution, in other words, as $\operatorname{pv}(1/\omega)$. Thus we obtain $\mathcal{F}\epsilon = -i\operatorname{pv}(1/\omega)$. Considering the result from example 9.6 (or exercise 9.7), this cannot be true since the term $\pi\delta(\omega)$ is missing. Find the error in the reasoning given above.

9.13 Let T be a distribution. Define for $a \in \mathbb{R}$ the product $e^{iat} T$ by $\left\langle e^{iat} T, \phi \right\rangle = \left\langle T, e^{iat}\phi \right\rangle$. Show that $e^{iat} T$ is a distribution.

9.14 Let T_f be a distribution defined by a function f through (8.12). Show that for this case, definition 9.2 of a shifted distribution reduces to a simple change of variables in an integral.

9.15 Prove (9.10): $e^{iat} T \leftrightarrow U(\omega - a)$ when $T \leftrightarrow U$.

9.16 Determine the spectra of the following distributions:
a $\epsilon(t - 1)$,
b $\epsilon(t)e^{iat}$,
c $\epsilon(t)\cos at$,
d $\delta'(t - 4) + 3i$,
e $\pi i t^3 + \epsilon(t)\operatorname{sgn} t$.

9.17 Determine the distribution T whose spectrum is the following distributions:
a $\operatorname{pv}(1/(\omega - 1))$,
b $(\sin 3\omega)\operatorname{pv}(1/\omega)$,
c $\epsilon(\omega)$,
d $\delta(3\omega - 2) + \delta''(\omega)$.

9.3 Convolution

Turning the convolution product into an ordinary multiplication is an important property of the Fourier transform of functions (see the convolution theorem in section 6.6). In this section we will see to what extent this result remains valid for distributions. We are immediately confronted with a fundamental problem: in general we cannot multiply two distributions S and T. Hence, the desired property $\mathcal{F}(S * T) = \mathcal{F}S \cdot \mathcal{F}T$ will not hold for arbitrary distributions S and T. This is because the right-hand side of this expression will in general not lead to a meaningful expression. If we wish to stick to the important convolution property, then we must conclude that the convolution product of two distributions is not always well-defined. Giving a correct definition of convolution is not a simple matter. Hence, in section 9.3.1 we first present an intuitive derivation of the most important results on convolution. For the remainder of the book it is sufficient to accept these intuitive results as being correct. For completeness (and because of the importance of convolution) we present in section 9.3.2 a mathematically rigorous definition of the convolution of distributions as well as proofs of the intuitive results from section 9.3.1. Section 9.3.2 can be omitted without any consequences for the remainder of the book.

9.3.1   *Intuitive derivation of the convolution of distributions*

As we often did, we first concentrate ourselves on the delta function $\delta(t)$. Let us return to the intuitive definition of the delta function from section 8.1. If we interchange limit and integral in (8.1), then we obtain the symbolic expression

$$\int_{-\infty}^{\infty} \delta(\tau) f(t - \tau) \, d\tau = f(t). \tag{9.17}$$

Changing from the variable $t - \tau$ to τ in the integral leads to

$$\int_{-\infty}^{\infty} f(\tau)\delta(t - \tau) \, d\tau = f(t) \tag{9.18}$$

(also see the similar formula (8.11) and exercise 8.2, although there we also used that $\delta(-t) = \delta(t)$). If we now pretend that the delta function is an ordinary absolutely integrable function, then we recognize in the left-hand sides of (9.17) and (9.18) precisely the convolution of δ and f (see definition 6.4) and so apparently $(\delta * f)(t) = f(t)$ and $(f * \delta)(t) = f(t)$. This is written as $\delta * f = f * \delta = f$ for short. If we now consider δ and f as distributions again, then one should have for an *arbitrary* distribution T that

$$\delta * T = T * \delta = T. \tag{9.19}$$

In a similar way one can derive an intuitive result for the convolution of δ' with a distribution T. For the delta function $\delta'(t - a)$ one has $\langle \delta'(t - a), \phi \rangle = -\phi'(a)$ for each $\phi \in \mathcal{S}$. This can again be symbolically written as

$$\int_{-\infty}^{\infty} \delta'(\tau - a) f(\tau) \, d\tau = -f'(a),$$

where, as in (9.17) and (9.18), we now write f instead of ϕ (for $a = 0$ this is (8.20)). Now $\delta'(\tau - a)$ is an odd distribution (see exercise 9.18), so $\delta'(\tau - a) = -\delta'(a - \tau)$, and if we also write t instead of a, then we see that

$$\int_{-\infty}^{\infty} \delta'(t - \tau) f(\tau) \, d\tau = f'(t).$$

The left-hand side can again be interpreted as the convolution of δ' with f and apparently one has $(\delta' * f)(t) = f'(t)$. By changing from the variable $t - \tau$ to τ it also follows that $(f * \delta')(t) = f'(t)$. If we consider δ' and f as distributions, then one should have for an *arbitrary* distribution T that

$$\delta' * T = T * \delta' = T'. \tag{9.20}$$

Analogously, one can derive for the higher order derivatives $\delta^{(k)}$ of the delta function that

$$\delta^{(k)} * T = T * \delta^{(k)} = T^{(k)}. \tag{9.21}$$

In section 9.3.2 we will define convolution of distributions in a mathematically rigorous way and we will prove formulas (9.19) – (9.21).

We will close this subsection by showing that the convolution theorem (theorem 6.13) is valid in the context of formulas (9.19) and (9.20), that is to say,

$$\mathcal{F}(\delta * T) = \mathcal{F}\delta \cdot \mathcal{F}T \qquad \text{and} \qquad \mathcal{F}(\delta' * T) = \mathcal{F}\delta' \cdot \mathcal{F}T. \tag{9.22}$$

First of all, the first identity in (9.22) follows immediately from $\delta * T = T$ and $\mathcal{F}\delta = 1$ since $\mathcal{F}(\delta * T) = \mathcal{F}T = \mathcal{F}\delta \cdot \mathcal{F}T$. To prove the second identity in (9.22) we first note that (9.11) implies that $\mathcal{F}T' = i\omega\mathcal{F}T$. But $i\omega = \mathcal{F}\delta'$ (see section 9.2.2) and hence it follows from (9.20) that indeed $\mathcal{F}(\delta' * T) = \mathcal{F}T' = i\omega\mathcal{F}T = \mathcal{F}\delta' \cdot \mathcal{F}T$. Similarly, one can show that the convolution theorem also holds for higher order derivatives of the delta function. Finally, in this manner one can also verify that the convolution theorem in the frequency domain, that is, $S \cdot T \leftrightarrow (U * V)/2\pi$ when $S \leftrightarrow U$ and $T \leftrightarrow V$, is valid whenever the delta function or a (higher order) derivative of the delta function occurs in the convolution product. In section 9.3.2 we will discuss in some more detail the convolution theorems for distributions and we will formulate a general convolution theorem (theorem 9.2).

9.3.2 *Mathematical treatment of the convolution of distributions**

In this section we present a mathematically correct treatment of the convolution of distributions. In particular we will prove the results that were derived intuitively in section 9.3.1. As mentioned at the beginning of section 9.3, section 9.3.2 can be omitted without any consequences for the remainder of the book.

In order to find a possible definition for $S * T$, we start, as usual by now, with two functions f and g for which we assume that the convolution product $f * g$ exists and, moreover, defines a distribution by means of (8.12). It then follows for $\phi \in \mathcal{S}$ that

$$\langle T_{f*g}, \phi \rangle = \int_{-\infty}^{\infty} (f * g)(t)\phi(t)\,dt = \int_{-\infty}^{\infty} \left(\int_{-\infty}^{\infty} f(\tau)g(t - \tau)\,d\tau \right) \phi(t)\,dt.$$

Since we are only looking for the right definition, we might as well assume that we may interchange the order of integration. We then see that

$$\begin{aligned} \langle T_{f*g}, \phi \rangle &= \int_{-\infty}^{\infty} f(\tau) \left(\int_{-\infty}^{\infty} g(t - \tau)\phi(t)\,dt \right) d\tau \\ &= \int_{-\infty}^{\infty} f(\tau) \left(\int_{-\infty}^{\infty} g(t)\phi(t + \tau)\,dt \right) d\tau \end{aligned} \tag{9.23}$$

(change the variable from $t - \tau$ to t in the last step). For each fixed $\tau \in \mathbb{R}$ the latter inner integral can be considered as the distribution T_g applied to the function $\phi(t + \tau)$. Here $\phi(t + \tau)$ should be considered as a function of t with τ kept fixed. In order to show this explicitly, we denote this by $\langle T_g(t), \phi(t + \tau) \rangle$; the distribution $T_g(t)$ acts on the variable t in the function $\phi(t + \tau)$. Now $\langle T_g(t), \phi(t + \tau) \rangle$ is a complex number for each $\tau \in \mathbb{R}$ and we will assume that the mapping $\tau \to$

$\langle T_g(t), \phi(t + \tau)\rangle$ from $\mathbb{R}$ into $\mathbb{C}$ is again a function in $\mathcal{S}$; let us denote this function by $\psi(\tau)$. The right-hand side of (9.23) can then be written as the distribution T_f applied to this function $\psi(\tau)$, which means that (9.23) can now be written as

$$\langle T_{f*g}, \phi\rangle = \langle T_f(\tau), \psi(\tau)\rangle = \langle T_f(\tau), \langle T_g(t), \phi(t + \tau)\rangle\rangle.$$

This finally gives us the somewhat complicated definition of the convolution of two distributions.

DEFINITION 9.3
Convolution of distributions

Let S and T be two distributions and define for $\tau \in \mathbb{R}$ the function $\psi(\tau)$ by $\psi(\tau) = \langle T(t), \phi(t + \tau)\rangle$. If for each $\phi \in \mathcal{S}$ the function $\psi(\tau)$ belongs to $\mathcal{S}$, then the convolution product $S * T$ *is defined by*

$$\langle S * T, \phi\rangle = \langle S(\tau), \psi(\tau)\rangle = \langle S(\tau), \langle T(t), \phi(t + \tau)\rangle\rangle. \tag{9.24}$$

The condition that $\psi(\tau)$ should belong to $\mathcal{S}$ is the reason that often $S * T$ cannot be defined.

EXAMPLE 9.8

For T take the delta function $\delta(t)$, then the function $\psi(\tau)$ is given by $\psi(\tau) = \langle T(t), \phi(t + \tau)\rangle = \langle \delta(t), \phi(t + \tau)\rangle = \phi(\tau)$, because of the definition of $\delta(t)$. Hence, we have $\psi(\tau) = \phi(\tau)$ in this case and so we certainly have that $\psi \in \mathcal{S}$. The convolution $S * \delta$ thus exists for all distributions S and $\langle S * \delta, \phi\rangle = \langle S(\tau), \phi(\tau)\rangle$ ($\phi \in \mathcal{S}$). This then proves the identity $S * \delta = S$ in (9.19). ◄

EXAMPLE 9.9

For each distribution T we have that $T * \delta'$ exists and $T * \delta' = T'$. In fact, $\langle T * \delta', \phi\rangle = \langle T(\tau), \langle \delta'(t), \phi(t + \tau)\rangle\rangle = \langle T(\tau), -\phi'(\tau)\rangle$, because of the action of δ' (see (8.19)). Hence $T * \delta'$ exists and $\langle T * \delta', \phi\rangle = \langle T, -\phi'\rangle = \langle T', \phi\rangle$ for all $\phi \in \mathcal{S}$ (in the last step we used definition 8.4). This proves the identity $T * \delta' = T'$ in (9.20). ◄

Convolution of distributions is not easy. For example, for functions f and g we know that $f * g = g * f$, in other words, the convolution product is *commutative* (see section 6.6). This result also holds for distributions. But even this simplest of properties of convolution will not be proven in this book; it would lead us too far into the theory of distributions. Without proof we state the following result: if $S * T$ exists, then $T * S$ also exists and $S * T = T * S$. In particular it follows from examples 9.8 and 9.9 that for an arbitrary distribution T one has $T * \delta = \delta * T = T$ and $T * \delta' = \delta' * T = T'$. In exactly the same way one obtains more generally that for any distribution T and any $k \in \mathbb{Z}^+$ one has

$$T * \delta^{(k)} = \delta^{(k)} * T = T^{(k)}. \tag{9.25}$$

Convolution of distributions is also quite subtle and one should take great care using it. For functions the convolution product is *associative*: $(f * g) * h = f * (g * h)$ for functions f, g and h. For distributions this is no longer true. Both the convolutions $(R * S) * T$ and $R * (S * T)$ may well exist without being equal. An example: one has $(1 * \delta') * \epsilon = 1' * \epsilon = 0 * \epsilon = 0$, while $1 * (\delta' * \epsilon) = 1 * \epsilon' = 1 * \delta = 1$!

Finally, we treat the convolution theorems for distributions. However, even the formulation of these convolution theorems is a problem, since the convolution of two distributions may not exist. Hence, we first have to find a set of distributions for which the convolution exists. To do so, we need a slight extension of the distribution theory. We have already noted a couple of times that distributions can often be defined for more than just the functions in $\mathcal{S}$. The delta function δ, for example, is well-defined for all continuous functions, δ' for all continuously differentiable functions, etc. Now let $\mathcal{E}$ be the space consisting of all functions that are infinitely many times differentiable. All polynomials, for example, belong to the space $\mathcal{E}$. Any function in $\mathcal{S}$ belongs to $\mathcal{E}$ as well, since functions in $\mathcal{S}$ are in particular infinitely many times differentiable (see definition 6.3).

Now let T be a distribution, so T is a linear mapping from $\mathcal{S}$ to $\mathbb{C}$. Then it could be that T is also well-defined for all functions in $\mathcal{E}$, which means that $\langle T, \phi \rangle$ is also meaningful for all $\phi \in \mathcal{E}$ (note that more functions belong to $\mathcal{E}$ than to $\mathcal{S}$: $\mathcal{S}$ is a subset of $\mathcal{E}$). If this is the case, then we will say that T is a distribution that can be defined on $\mathcal{E}$. Such distributions play an important role in our convolution theorems.

Distribution on $\mathcal{E}$

EXAMPLE

The delta function and all the derivatives of the delta function belong to the distributions that can be defined on $\mathcal{E}$. ◄

EXAMPLE

(This example uses the comb distribution from section 9.1.3*.) The comb distribution $\mathcal{W}$ *cannot* be defined on the space $\mathcal{E}$. This is because the function $\phi(t) = 1$ is an element of $\mathcal{E}$ (the function 1 is infinitely many times differentiable with all derivatives equal to 0) and $\langle \mathcal{W}, \phi \rangle = \sum_{k=-\infty}^{\infty} \phi(k) = \sum_{k=-\infty}^{\infty} 1$ diverges. ◄

Distributions that can be defined on $\mathcal{E}$ form a suitable class of distributions for the formulation of the convolution theorem.

THEOREM 9.2
Convolution theorem for distributions

Let S and T be distributions with spectra U and V respectively. Assume that S is a distribution that can be defined on the space $\mathcal{E}$. Then U is a function in $\mathcal{E}$, both $S * T$ and $U \cdot V$ are well-defined, and $S * T \leftrightarrow U \cdot V$.

We will not prove this theorem. It would lead us too far into the distribution theory. However, we can illustrate the theorem using some frequently occurring convolution products.

EXAMPLE 9.10

Take for S in theorem 9.2 the delta function δ. The delta function can certainly be defined on the space $\mathcal{E}$. The spectrum U of δ is the function 1 and this is indeed a function in $\mathcal{E}$. Furthermore, $\mathcal{F}(\delta * T) = \mathcal{F}\delta \cdot \mathcal{F}T$ according to theorem 9.2. But this is obvious since we know that $\delta * T = T$ and $\mathcal{F}\delta = 1$ (also see section 9.3.1). ◄

EXAMPLE 9.11

One can define δ' on the space $\mathcal{E}$ as well. Furthermore, we showed in section 9.2.2 that $\delta' \leftrightarrow i\omega$. Note that the spectrum U of δ' is thus indeed a function in $\mathcal{E}$. Applying theorem 9.2 establishes that $\delta' * T \leftrightarrow i\omega V$ when $T \leftrightarrow V$. This result can again be proven in a direct way since we know that $\delta' * T = T'$ and $T' \leftrightarrow i\omega V$ when $T \leftrightarrow V$ (also see section 9.3.1). ◄

Often, the conditions of theorem 9.2 are *not* satisfied in applications. If one doesn't want to compromise on mathematical rigour, then one has to check case by case whether or not the operations used (multiplication, convolution, Fourier transform) are well-defined. The result $S * T \leftrightarrow U \cdot V$ from theorem 9.2 will then turn out to be valid in many more cases.

Finally, we expect on the basis of the reciprocity property that a convolution theorem in the frequency domain exists as well (compare with section 7.2.4). We will content ourselves with the statement that this is indeed the case: $S \cdot T \leftrightarrow (U * V)/2\pi$, if all operations occurring here are well-defined. This is the case when, for example, U is a distribution that can be defined on the space $\mathcal{E}$.

EXAMPLE 9.12

Take for S the function e^{iat}, considered as a distribution. The spectrum U of S is $2\pi\delta(\omega - a)$ (see example 9.1) and this distribution can indeed be defined on the space $\mathcal{E}$. For an arbitrary distribution T with spectrum V it then follows from the convolution theorem in the frequency domain that the spectrum of $e^{iat}T$ is equal to $(2\pi\delta(\omega - a) * V)/2\pi$. Hence, $e^{iat}T \leftrightarrow \delta(\omega - a) * V$ when $T \leftrightarrow V$. Note that according to the shift property in the frequency domain one has $e^{iat}T \leftrightarrow V(\omega - a)$ (see exercise 9.15). Apparently, $\delta(\omega - a) * V = V(\omega - a)$. ◄

EXERCISES

9.18   Show that $\delta'(t-a)$ is an odd distribution.

9.19   Show that $\delta' * |t| = \operatorname{sgn} t$.

9.20*   Let $T(t)$ be a distribution and $a \in \mathbb{R}$. Show that $T(t) * \delta(t-a)$ exists and that $T(t) * \delta(t-a) = T(t-a)$ (see example 9.12).

9.21*   Prove that $\delta(t-b) * \delta(t-a) = \delta(t-(a+b))$ $(a, b \in \mathbb{R})$. Look what happens when you apply the convolution theorem to this identity!

9.22*   Let the piecewise smooth function f be equal to 0 outside a bounded interval; say $f(t) = 0$ for $|t| \geq A$ (where $A > 0$ is a constant).
a   Prove that the spectrum F of f exists (in the ordinary sense). Hint (which also applies to parts b and c): the function f is bounded.
b   Show that f defines a distribution by means of (8.12).
c   Show that the distribution defined in part b is also well-defined on $\mathcal{E}$.
d   Let T be an arbitrary distribution with spectrum V. We simply write f and F for the distributions defined by the functions f and F. Conclude from part c that $f * T$ exists and that $f * T \leftrightarrow F \cdot V$.
e   Determine the spectrum of $p_2 * \epsilon$ (p_2 is the block function, ϵ the unit step function).
f   Determine the spectrum of $\int_{-\infty}^{t} f(\tau)\,d\tau = (f * \epsilon)(t)$.

9.23*   For functions the only solution to $f'(t) = 0$ (for all $t \in \mathbb{R}$) is the constant function $f(t) = c$. In this exercise we assume that the same result is true for distributions: the only distribution T with $T' = 0$ is the distribution $T = c$ (c a constant). Using this assumption one can determine the spectrum U of the distribution $\operatorname{pv}(1/t)$ in a mathematically rigorous way.
a   Use the result $t \cdot \operatorname{pv}(1/t) = 1$ from exercise 8.25 to show that U satisfies the equation $U' = -2\pi i \delta(\omega)$.
b   Show that all solutions of the equation $S' = \delta$ are necessarily of the form $S = \epsilon + c$, where c is a constant. (Hint: when both S_1 and S_2 are solutions to $S' = \delta$, what then will $S_1 - S_2$ satisfy?) Conclude that $U(\omega) = -2\pi i(\epsilon(\omega) + c)$.
c   Conclude from exercise 8.27a that U is odd. Use this to determine the constant c from part b and finally conclude that $U(\omega) = -\pi i \operatorname{sgn} \omega$.

SUMMARY

The Fourier transform $\mathcal{F}T$ of a distribution T is defined by $\langle \mathcal{F}T, \phi \rangle = \langle T, \Phi \rangle$, where Φ is the Fourier transform of $\phi \in \mathcal{S}$. In this way the Fourier transform is extended from functions to distributions. The most well-known Fourier pair for distributions is $\delta(t) \leftrightarrow 1$ and $1 \leftrightarrow 2\pi\delta(\omega)$. More generally one has $\delta(t-a) \leftrightarrow e^{-ia\omega}$ and $e^{iat} \leftrightarrow 2\pi\delta(\omega - a)$. The latter result can be used to determine the spectrum, in the sense of distributions, of a periodic signal which defines a distribution and, moreover, has a convergent Fourier series. Another important Fourier pair is $\operatorname{pv}(1/t) \leftrightarrow -\pi i \operatorname{sgn} \omega$ and $\operatorname{sgn} t \leftrightarrow -2i\operatorname{pv}(1/\omega)$, from which it follows in particular that $\epsilon(t) \leftrightarrow \pi\delta(\omega) - i\operatorname{pv}(1/\omega)$. Finally, we mention the comb distribution or impulse train $\Sha(t)$, whose spectrum is again a comb distribution, but scaled by $1/2\pi$: $\Sha(t) \leftrightarrow \Sha(\omega/2\pi)$, or $\sum_{k=-\infty}^{\infty} \delta(t-k) \leftrightarrow 2\pi \sum_{k=-\infty}^{\infty} \delta(\omega - 2\pi k)$.
Quite a few of the properties of the ordinary Fourier transform remain valid for distributions. When U is the spectrum of T, then $2\pi T(\omega)$ is the spectrum of $U(-t)$ (reciprocity). Furthermore, one has the differentiation properties $T^{(k)} \leftrightarrow (i\omega)^k U$ and $(-it)^k T \leftrightarrow U^{(k)}$, as well as the shift properties $T(t-a) \leftrightarrow e^{-ia\omega}U(\omega)$ and

$e^{iat}T(t) \leftrightarrow U(\omega - a)$. Since the spectrum of $\epsilon(t)$ is known, the latter property can be used to determine the spectrum of switched-on periodic signals.

Convolution theorems do not hold for arbitrary distributions. In general, multiplication as well as convolution of two distributions are not well-defined. When S and T are distributions with spectra U and V, then the convolution theorems $S * T \leftrightarrow U \cdot V$ and $S \cdot T \leftrightarrow (U * V)/2\pi$ are correct whenever all the operations occurring are well-defined. One has, for example, that $\delta^{(k)} * T$ ($k \in \mathbb{Z}^+$) is well-defined for each distribution T and that $\delta^{(k)} * T = T * \delta^{(k)} = T^{(k)}$ and $\delta^{(k)} * T \leftrightarrow (i\omega)^k U$ when $T \leftrightarrow U$. A list of some standard Fourier transforms of distributions and a list of properties of the Fourier transform of distributions are given in tables 5 and 6.

SELFTEST

9.24   Determine the spectra of the following distributions:
a $\delta(t - 3)$,
b $\cos t \, \delta(t + 4)$,
c $t^2 \epsilon(t)$,
d $(2\epsilon(t) \cos t)'$,
e $(\delta(7t - 1))'$,
f $\frac{\pi}{2} - \frac{2}{\pi} \sum_{k=-\infty}^{\infty} (2k + 1)^{-2} e^{(2k+1)it}$.

9.25   Find the distributions whose spectrum is given by the following distributions:
a $\delta(\omega - 1) - \delta(\omega + 1)$,
b ω^2,
c $e^{i\omega/2}/4$,
d $\omega^3 \sin \omega$,
e $\cos(\omega - 4)$.

9.26   Consider the distribution $U = (\omega - 1)^2$ in the frequency domain.
a   Use the shift property to determine a distribution whose spectrum is U.
b   Use the differentiation property to determine a distribution whose spectrum is the distribution $\omega^2 - 2\omega + 1$.
c   Note that $U = \omega^2 - 2\omega + 1$; the answers in parts a and b should thus be the same. Give a direct proof of this fact.

9.27   Let T be a distribution with spectrum U.
a   Use the convolution theorem to determine the spectrum of $T * \delta''$ in terms of U. Check the result by applying the differentiation property to the distribution T''.
b   The function $|t|$ defines a distribution. Let V be the spectrum of $|t|$. Show that $\delta'' * |t| = 2\delta$ and prove that V satisfies $\omega^2 V = -2$.

Contents of Chapter 10

Applications of the Fourier integral

Introduction *229*

10.1 The impulse response *230*
10.2 The frequency response *234*
10.3 Causal stable systems and differential equations *239*
10.4 Boundary and initial value problems for partial differential equations *243*

Summary *245*

Selftest *246*

Applications of the Fourier integral

INTRODUCTION

The Fourier transform is one of the most important tools in the study of the transfer of signals in control and communication systems. In chapter 1 we have already discussed signals and systems in general terms. Now that we have the Fourier integral available, and are familiar with the delta function and other distributions, we are able to get a better understanding of the transfer of signals in linear time-invariant systems. The Fourier integral plays an important role in continuous-time systems which, moreover, are linear and time-invariant. These have been introduced in chapter 1 and will be denoted here by LTC-systems for short, just as in chapter 5.

Systems can be described by giving the relation between the input $u(t)$ and the corresponding output or response $y(t)$. This can be done in several ways. For example, by a description in the time domain (in such a description the variable t occurs), or by a description in the frequency domain. The latter means that a relation is given between the spectra (the Fourier transforms) $U(\omega)$ and $Y(\omega)$ of, respectively, the input $u(t)$ and the response $y(t)$.

In section 10.1 we will see that for LTC-systems the relation between $u(t)$ and $y(t)$ can be expressed in the time domain by means of a convolution product. Here the response $h(t)$ to the unit pulse, or delta function, $\delta(t)$ plays a central role. In fact, the response $y(t)$ to an input $u(t)$ is equal to the convolution product of $h(t)$, the so-called impulse response, and $u(t)$. Hence, if the impulse response is known, then the system is known in the sense that the response to an arbitrary input can be determined. Properties of LTC-systems, such as stability and causality, can then immediately be derived from the impulse response. Moreover, applying the convolution theorem is now obvious and so the Fourier transform is going to play an important role.

In section 10.2 we will see that the frequency response $H(\omega)$ of a system, which we introduced in chapter 1, is nothing else but the Fourier transform of the impulse response, and that a description of an LTC-system in the frequency domain is simply given by $Y(\omega) = U(\omega)H(\omega)$, where $U(\omega)$ and $Y(\omega)$ are the spectra of, respectively, the input $u(t)$ and the corresponding output $y(t)$. Properties of a continuous-time system can then also be derived from the frequency response. For an all-pass system or a phase shifter, for example, which is an LTC-system with the property that the energy-content of the output is equal to the energy-content of the corresponding input, the modulus of the frequency response is constant. Another example is the ideal low-pass filter, which is characterized by a frequency response being 0 outside a certain frequency band.

Continuous-time systems occurring in practice usually have a rational function of ω as frequency response. Important examples are the electrical RCL-networks having resistors, capacitors and coils as their components. The reason is that for these systems the relation between an input and the corresponding response can be described in the time domain by an ordinary differential equation with constant coefficients. These will be considered in section 10.3.

Applications of the Fourier transform are certainly not limited to just the transfer of signals in systems. The Fourier integral can also be successfully applied to all kinds of physical phenomena, such as heat conduction, which can be described mathematically by a partial differential equation. These kinds of applications are examined in section 10.4.

LEARNING OBJECTIVES
After studying this chapter it is expected that you
- know the concepts of impulse response and step response of an LTC-system and can determine these in simple cases
- know the relation between an input and the corresponding output and can apply it to calculate outputs
- can determine the stability and causality of an LTC-system using the impulse response
- know the concept of frequency response and can determine it in simple cases
- know the relation between the input and the output in the frequency domain and can apply it to calculate outputs
- know what is meant by an all-pass system and an ideal low-pass filter
- can determine the impulse response of a stable and causal LTC-system described by a linear differential equation with constant coefficients
- can apply the Fourier transform in solving partial differential equations with initial and boundary conditions.

10.1 The impulse response

In order to introduce the impulse response, we start, as in chapter 1, with an example of a simple continuous-time system, namely the electrical network from figure 1.1 in chapter 1, consisting of a series connection of a voltage source, a resistor and a coil. The relation between the voltage and the current in this network is described by (1.1), which we will recall here. We denote the voltage by $u(t)$, however, since we will consider it as an input, and the current by $y(t)$, being the corresponding output. Relation (1.1) mentioned above then reads as follows:

$$y(t) = \frac{1}{L} \int_{-\infty}^{t} e^{-(t-\tau)R/L} u(\tau)\, d\tau. \tag{10.1}$$

LTC-system

In chapter 1 we have seen that this relation allows us to consider the network as a continuous-time system which is linear and time-invariant. We recall that we denote this by LTC-system for short. First we will show that relation (10.1) between the input $u(t)$ and the corresponding output $y(t)$ can be written as a convolution product. For this we utilize the unit step function $\epsilon(t)$. Since $\epsilon(t-\tau) = 0$ for $\tau > t$, it follows that

$$\int_{-\infty}^{t} e^{-(t-\tau)R/L} u(\tau)\, d\tau = \int_{-\infty}^{\infty} e^{-(t-\tau)R/L} \epsilon(t-\tau) u(\tau)\, d\tau.$$

From definition 6.4 of the convolution product we then obtain that

$$y(t) = (h * u)(t),$$

where $h(t)$ is the signal

$$h(t) = \frac{1}{L} e^{-tR/L} \epsilon(t).$$

If we now take as input the unit pulse or delta function $\delta(t)$, then it follows from (9.19) that $h(t) = (h * \delta)(t)$ and so we can view $h(t)$ as the output corresponding to the unit pulse. We will show that for LTC-systems in general, the response to the unit pulse plays the same important role. To this end we first use property (9.19) to write an input $u(t)$ in the form

$$u(t) = (u * \delta)(t) = \int_{-\infty}^{\infty} u(\tau)\delta(t - \tau)\, d\tau.$$

Superposition

We say that the signal $u(t)$ is now written as a (continuous) *superposition* of shifted unit pulses $\delta(t - \tau)$. Now if $h(t)$ is the response to $\delta(t)$, then, by the time-invariance of the system, the response to $\delta(t - \tau)$ is $h(t - \tau)$. Next, we tacitly assume that for LTC-systems the linearity property can be extended to the so-called *superposition rule*. By this we mean that the linearity property not only holds for finite sums of inputs, but also for *infinite sums* of inputs, and even for 'continuous sums', that is, for integrals. Applying this superposition rule gives for a system L:

Superposition rule

$$y(t) = \mathrm{L}u(t) = \mathrm{L} \int_{-\infty}^{\infty} u(\tau)\delta(t - \tau)\, d\tau = \int_{-\infty}^{\infty} u(\tau)\mathrm{L}\delta(t - \tau)\, d\tau$$

$$= \int_{-\infty}^{\infty} u(\tau)h(t - \tau)\, d\tau.$$

Impulse response

The signal $h(t)$ is called the *impulse response* of the system L. Hence,

$$\delta(t) \mapsto h(t). \tag{10.2}$$

We have now established the following important property for LTC-systems.

Let $u(t)$ be an input of an LTC-system L with impulse response $h(t)$. Then

$$\mathrm{L}u(t) = (h * u)(t). \tag{10.3}$$

EXAMPLE 10.1
Distortion free system

A continuous-time system is called *distortion free* when the response $y(t)$ to an input $u(t)$ is given by

$$y(t) = K u(t - t_0),$$

where K and t_0 are constants with $K \geq 0$. Compared to the input, the response is of the same form and is shifted over a time t_0. This system is an LTC-system. The linearity can easily be verified. The time-invariance follows from

$$u(t - t_1) \mapsto K u(t - t_1 - t_0) = y(t - t_1) \qquad \text{for all } t_1.$$

Also note that the given system is causal if $t_0 \geq 0$. The impulse response is the response to $\delta(t)$ and thus equal to $h(t) = K\delta(t - t_0)$. According to (10.3) the response $y(t)$ to an input $u(t)$ is then given by

$$y(t) = (h * u)(t) = K \int_{-\infty}^{\infty} u(\tau)\delta(t - t_0 - \tau)\, d\tau = K u(t - t_0).$$

◀

By (10.3), an LTC-system is completely determined by the impulse response. Hence, all properties of an LTC-system can be derived from the impulse response. For instance, an LTC-system is real if and only if the impulse response is real (see exercise 10.1b) and an LTC-system is causal if and only if the impulse response is causal (see exercise 10.1a). Another property, important for the physically realizable systems, is stability as defined in definition 1.3. Using the impulse response one can verify this property in the following way.

THEOREM 10.1

An LTC-system with impulse response $h(t)$ ($h(t)$ being an ordinary function) is stable if and only if $\int_{-\infty}^{\infty} |h(t)|\,dt$ is convergent, in other words, if $h(t)$ is absolutely integrable.

Proof
Let $\int_{-\infty}^{\infty} |h(t)|\,dt$ be convergent and let the input $u(t)$ be bounded, that is to say, there exists a number M such that $|u(t)| \leq M$ for all t. For the corresponding response it then follows from (10.3) that

$$|y(t)| = \left| \int_{-\infty}^{\infty} u(\tau)h(t-\tau)\,d\tau \right| \leq \int_{-\infty}^{\infty} |u(\tau)|\,|h(t-\tau)|\,d\tau$$

$$\leq M \int_{-\infty}^{\infty} |h(\tau)|\,d\tau.$$

The output is thus bounded and so the system is stable.

Now let $\int_{-\infty}^{\infty} |h(t)|\,dt$ be divergent. Take as input

$$u(t) = \begin{cases} \dfrac{\overline{h(-t)}}{|h(-t)|} & \text{for } h(-t) \neq 0, \\[2mm] 0 & \text{for } h(-t) = 0. \end{cases}$$

This signal is bounded: $|u(t)| \leq 1$. Using (10.3) we find for the corresponding response $y(t)$ at $t = 0$:

$$y(0) = \int_{-\infty}^{\infty} u(\tau)h(-\tau)\,d\tau = \int_{-\infty}^{\infty} |h(\tau)|\,d\tau = \infty.$$

The response is not bounded and so the system isn't stable. ∎

Strictly speaking, we cannot apply this theorem when instead of an ordinary function $h(t)$ is a distribution, so for example when $h(t) = \delta(t)$. In fact, for a distribution the notion 'absolute integrability' has no meaning. Consider for example the distortion free system from example 10.1. The impulse response is the distribution $K\delta(t - t_0)$. Still, this system is stable, as follows immediately from definition 1.3 of stability. For if $|u(t)| \leq M$ for all t and a certain M, then one has for the corresponding response that $|y(t)| = K\,|u(t - t_0)| \leq KM$, so the response is bounded as well.

For systems occurring in practice the impulse response will usually consist of an ordinary function with possibly a finite number of shifted delta functions added to this. In order to verify the stability for these cases, one only needs to check the absolute integrability of the ordinary function. We will demonstrate this in the following example.

EXAMPLE 10.2

A system L is given for which the relation between an input $u(t)$ and the response $y(t)$ is given by

$$y(t) = u(t - 1) + \int_{-\infty}^{t} e^{-2(t-\tau)} u(\tau)\,d\tau.$$

In exercise 10.2 the reader is asked to show that L is an LTC-system. The impulse response can be found by substituting the unit pulse $\delta(t)$ for $u(t)$, resulting in $h(t) = \delta(t - 1) + r(t)$, where $r(t) = e^{-2t} \epsilon(t)$. The response $y(t)$ to an input $u(t)$ can be written as $y(t) = u(t - 1) + y_1(t)$, where $y_1(t)$ is the response to $u(t)$ of an LTC-system with impulse response $h_1(t) = r(t)$. Since

$$\int_{-\infty}^{\infty} |h_1(t)|\,dt = \int_{0}^{\infty} e^{-2t}\,dt = \tfrac{1}{2} < \infty,$$

the system is stable according to theorem 10.1. So if $|u(t)| \leq M$, then $|y_1(t)| \leq L$ for some L and we thus have

$$|y(t)| \leq |u(t-1)| + L \leq M + L.$$

This establishes the stability of the LTC-system. ◄

EXAMPLE 10.3
Integrator

An LTC-system for which the relation between an input $u(t)$ and the corresponding output $y(t)$ is given by

$$y(t) = \int_{-\infty}^{t} u(\tau)\,d\tau$$

is called an *integrator*. Since

$$\int_{-\infty}^{t} \delta(\tau)\,d\tau = \epsilon(t),$$

the impulse response of the integrator is equal to the unit step function $\epsilon(t)$. According to property (10.3) we thus have $y(t) = (\epsilon * u)(t)$. The unit step function $\epsilon(t)$ is not absolutely integrable. Hence, it follows from theorem 10.1 that the integrator is unstable. ◄

Step response

We close this section with the introduction of the so-called step response $a(t)$ of an LTC-system. The *step response* is defined as the response to the unit step function $\epsilon(t)$, so

$$\epsilon(t) \mapsto a(t). \tag{10.4}$$

From property (10.3) it follows that $a(t) = (\epsilon * h)(t)$, where $h(t)$ is the impulse response of the system. Convolution with the unit step function is the same as integration (see example 10.3 of the integrator), so

$$a(t) = \int_{-\infty}^{t} h(\tau)\,d\tau. \tag{10.5}$$

This relation implies that $h(t)$ is the derivative of $a(t)$, but not in the ordinary sense. Thanks to the introduction of the distributional derivative, we can say that $h(t)$ is the distributional derivative of $a(t)$: $a'(t) = h(t)$. Apparently, the impulse response follows easily once the step response is known and using (10.3) one can then determine the response to an arbitrary input.

EXERCISES

10.1   Given is an LTC-system L.
a   Show that the system L is causal if and only if the impulse response $h(t)$ of L is a causal signal.
b   Show that the system L is real if and only if the impulse response $h(t)$ of L is a real signal.

10.2   The relation between an input $u(t)$ and the corresponding response $y(t)$ of a system L is given by

$$y(t) = u(t-1) + \int_{-\infty}^{t} e^{-2(t-\tau)} u(\tau)\,d\tau.$$

a   Show that L is a real and causal LTC-system.
b   Determine the step response of the system.

10.3   Given is an LTC-system L with step response $a(t)$. Determine the response to the rectangular pulse $p_2(t)$.

10.4   Given is a stable LTC-system with impulse response $h(t)$.
a   Show that the response to the constant input $u(t) = 1$ is given by the constant output $y(t) = H(0)$ for all t, where

$$H(0) = \int_{-\infty}^{\infty} h(\tau)\,d\tau.$$

b   Let $u(t)$ be a differentiable input with absolutely integrable derivative $u'(t)$ and for which $u(-\infty)$, defined by $\lim_{t \to -\infty} u(t)$, exists. Show that $u(t) = u(-\infty) + (u' * \epsilon)(t)$.
c   Show that for the response $y(t)$ to $u(t)$ one has $y(t) = H(0)u(-\infty) + (u'*a)(t)$.

10.5   For an LTC-system the step response $a(t)$ is given by $a(t) = \cos(2t)e^{-3t}\epsilon(t)$.
a   Determine the impulse response $h(t)$.
b   Show that the system is stable.

10.6   Given are the LTC-systems L_1 and L_2 in a series connection. We denote this so-called *cascade system* by L_2L_1. The response $y(t)$ to an input $u(t)$ is obtained as indicated in figure 10.1.

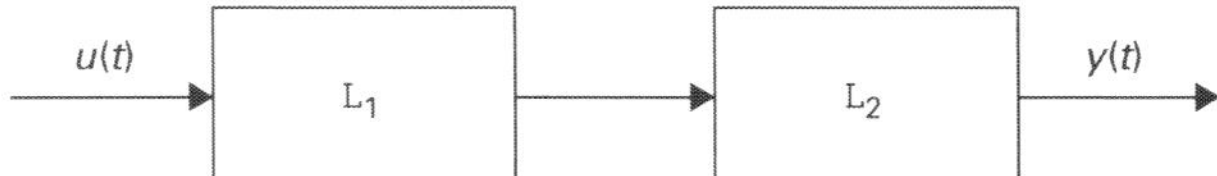

FIGURE 10.1
Cascade system of exercise 10.6.

a   Give an expression for the impulse response $h(t)$ of L_2L_1 in terms of the impulse responses $h_1(t)$ and $h_2(t)$ of, respectively, L_1 and $L_2(t)$.
b   Show that if both L_1 and L_2 are stable, then L_2L_1 is also stable.

10.2   The frequency response

In the previous section we saw that for an LTC-system the relation between an input and the corresponding output is given in the time domain by the convolution product (10.3). In this section we study the relation in the frequency domain. It is quite natural to apply the convolution theorem to (10.3). Since we have to take into account that delta functions may occur in $u(t)$ as well as in $h(t)$, we will need the convolution theorem that is also valid for distributions. We will assume that for the LTC-systems and inputs under consideration, the convolution theorem may always be applied in distributional sense as well. As a result we obtain the following important theorem.

THEOREM 10.2

Let $U(\omega)$, $Y(\omega)$ and $H(\omega)$ denote the Fourier transforms of, respectively, an input $u(t)$, the corresponding output $y(t)$, and the impulse response $h(t)$ of an LTC-system. Then

$$Y(\omega) = H(\omega)U(\omega). \tag{10.6}$$

Property (10.6) describes how an LTC-system operates in the frequency domain. The spectrum of an input is multiplied by the function $H(\omega)$, resulting in the spectrum of the output.

EXAMPLE 10.4

The impulse response of the integrator (see example 10.3) is the unit step function $\epsilon(t)$: $h(t) = \epsilon(t)$. The Fourier transform of $h(t)$ is $1/(i\omega) + \pi\delta(\omega)$ (see table 5; $-i\mathrm{pv}(1/\omega)$ is written as $1/(i\omega)$ for short). Hence, according to (10.6) the integrator is described in the frequency domain by

$$Y(\omega) = (1/(i\omega) + \pi\delta(\omega))\, U(\omega) = U(\omega)/(i\omega) + \pi U(\omega)\delta(\omega)$$
$$= U(\omega)/(i\omega) + \pi U(0)\delta(\omega).$$

In the last equality we assumed that $U(\omega)$ is continuous at $\omega = 0$. ◄

In this example of the integrator we see that substituting $\omega = 0$ is meaningless. The reason for this is the instability of the integrator. For stable systems the impulse response $h(t)$ is absolutely integrable according to theorem 10.1 and $H(\omega)$ will then be an ordinary function, and even a continuous one, defined for all ω (see section 6.4.11).

Now consider in particular the time-harmonic signal $u(t) = e^{i\omega t}$ with frequency ω as input for an LTC-system. According to property (10.3) the response is equal to

$$y(t) = (h * u)(t) = \int_{-\infty}^{\infty} h(\tau)e^{i\omega(t-\tau)}\, d\tau = H(\omega)e^{i\omega t}.$$

Hence,

$$e^{i\omega t} \mapsto H(\omega)e^{i\omega t}. \tag{10.7}$$

This is nothing new. In chapter 1 we have derived that the response of an LTC-system to a time-harmonic signal is again a time-harmonic signal with the same frequency. In this chapter the frequency response $H(\omega)$ has been introduced through property (10.7). We conclude that the following theorem holds.

THEOREM 10.3
Frequency response

The frequency response $H(\omega)$ *of an LTC-system is the spectrum of the impulse response:*

$$h(t) \mapsto H(\omega). \tag{10.8}$$

EXAMPLE 10.5

For an LTC-system the impulse response is given by $h(t) = e^{-t}\epsilon(t)$. The system is stable since

$$\int_{-\infty}^{\infty} |h(t)|\, dt = \int_{0}^{\infty} e^{-t}\, dt = 1 < \infty.$$

The Fourier transform of $h(t)$ can be found in table 3: $H(\omega) = 1/(1 + i\omega)$. The response to the input $e^{i\omega t}$ is thus equal to $e^{i\omega t}/(1 + i\omega)$. ◄

Transfer function
System function

From chapter 1 we know that the frequency response $H(\omega)$ is also called the *transfer function* or the *system function* of an LTC-system. Apparently, an LTC-system can be described in the frequency domain by the frequency response $H(\omega)$.

EXAMPLE 10.6
Ideal low-pass filter

As an example we consider the so-called *ideal low-pass filter* with frequency response $H(\omega)$ given by

$$H(\omega) = \begin{cases} e^{-i\omega t_0} & \text{for } |\omega| \le \omega_c, \\ 0 & \text{for } |\omega| > \omega_c. \end{cases}$$

Hence, only frequencies below the *cut-off frequency* ω_c can pass. By an inverse Fourier transform we find the impulse response of the filter:

$$h(t) = \frac{1}{2\pi} \int_{-\infty}^{\infty} H(\omega)e^{i\omega t}\, d\omega = \frac{1}{2\pi} \int_{-\omega_c}^{\omega_c} e^{i\omega(t-t_0)}\, d\omega = \frac{\sin(\omega_c(t - t_0))}{\pi(t - t_0)}.$$

The function $h(t)$ is shown in figure 10.2. The impulse response $h(t)$ has a maximum value ω_c/π at $t = t_0$; the main pulse of the response is concentrated at $t = t_0$ and has duration $2\pi/\omega_c$. Note that $h(t)$ is not causal, which means that the system is not causal. The step response of the filter follows from integration of $h(t)$:

$$a(t) = \int_{-\infty}^{t} h(\tau)\,d\tau = \int_{-\infty}^{t} \frac{\sin(\omega_c(\tau - t_0))}{\pi(\tau - t_0)}\,d\tau = \frac{1}{\pi}\int_{-\infty}^{\omega_c(t-t_0)} \frac{\sin x}{x}\,dx,$$

where we used the substitution $\omega_c(\tau - t_0) = x$. Using the sine integral (see chapter 4), the step response can be written as

$$a(t) = \frac{1}{2} + \frac{1}{\pi}\mathrm{Si}(\omega_c(t - t_0)).$$

Note that $a(t_0) = \frac{1}{2}$ and that the initial and final values $a(-\infty) = 0$ and $a(\infty) = 1$ are approached in an oscillating way (see figure 10.2b). The maximal overshoot occurs at $t = t_0 + \pi/\omega_c$ and amounts to 9% (compare this with Gibbs' phenomenon in chapter 4). In this example we will also determine the response to a periodic

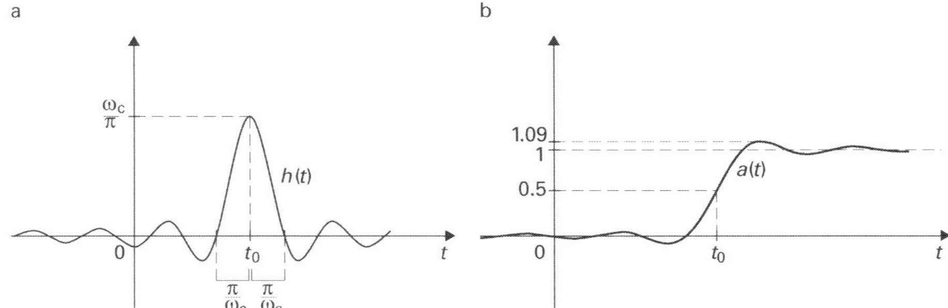

FIGURE 10.2
Impulse response (a) and step response (b) of ideal low-pass filter.

signal $u(t)$ given by the Fourier series

$$u(t) = \sum_{n=-\infty}^{\infty} c_n e^{in\omega_0 t},$$

where $\omega_0 = 2\pi/T$. According to property (10.7), the response to $e^{in\omega_0 t}$ equals

$$H(n\omega_0)e^{in\omega_0 t} = \begin{cases} e^{in\omega_0(t-t_0)} & \text{for } |n\omega_0| \leq \omega_c, \\ 0 & \text{for } |n\omega_0| > \omega_c. \end{cases}$$

Here we assume that ω_c is not an integer multiple of ω_0. Let N be the integer uniquely determined by $N\omega_0 < \omega_c < (N + 1)\omega_0$. For the response $y(t)$ to the periodic input $u(t)$ it then follows that

$$y(t) = \sum_{n=-N}^{N} c_n e^{in\omega_0(t-t_0)} = u_N(t - t_0),$$

where $u_N(t)$ denotes the Nth partial sum of the Fourier series of $u(t)$. ◀

The frequency response can also show us how the energy-content (see section 1.2.3) of an input is effected by an LTC-system. When an energy-signal $u(t)$ is applied to an LTC-system, then by Parseval's identity (see (7.19)) one has for the energy-content of the corresponding output $y(t)$ that

$$\int_{-\infty}^{\infty} |y(t)|^2\, dt = \frac{1}{2\pi} \int_{-\infty}^{\infty} |Y(\omega)|^2\, d\omega = \frac{1}{2\pi} \int_{-\infty}^{\infty} |H(\omega)|^2\, |U(\omega)|^2\, d\omega.$$

Note that for LTC-systems whose amplitude response $|H(\omega)|$ is identical to 1, the energy-content of the output equals the energy-content of the input. Such a system is called an *all-pass system*.

All-pass system

EXAMPLE 10.7

For an LTC-system the frequency response $H(\omega)$ is given by

$$H(\omega) = \frac{\omega - i}{1 + i\omega} e^{-i\omega t_0}.$$

Here t_0 is real. Since

$$\left| \frac{\omega - i}{1 + i\omega} e^{-i\omega t_0} \right| = \left| \frac{\omega - i}{1 + i\omega} \right| = \sqrt{\frac{\omega^2 + 1}{1 + \omega^2}} = 1,$$

the system is an all-pass system. ◀

In this section we have established the importance of the frequency response for LTC-systems. An important class of LTC-systems in practical applications has the property that the frequency response is a rational function of ω. Examples are the electrical networks. In the next section we will examine these in more detail.

EXERCISES

10.7   For an LTC-system L the impulse response is given by $h(t) = \delta(t) + te^{-t}\epsilon(t)$.
a   Determine the frequency response of the system L.
b   Determine for all real ω the response of the LTC-system to the input $u(t) = e^{i\omega t}$.

10.8   For an LTC-system L the frequency response is given by

$$H(\omega) = \frac{\cos \omega}{\omega^2 + 1}.$$

a   Determine the impulse response $h(t)$ of the system L.
b   Determine the response to the input $u(t) = \delta(t - 1)$.

10.9   For a low-pass filter the frequency response $H(\omega)$ is given by the graph of figure 10.3.

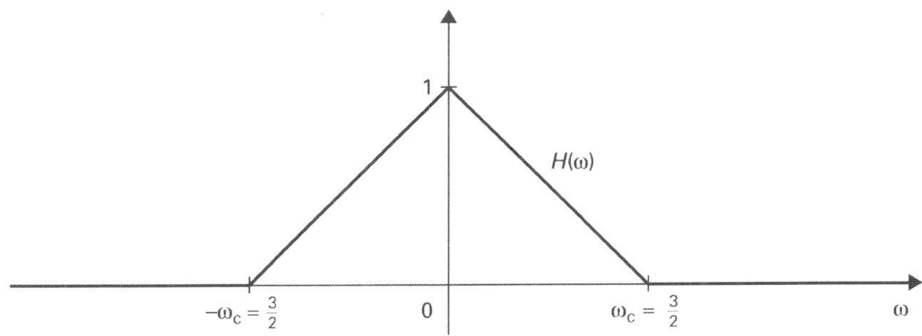

FIGURE 10.3
Frequency response of the low-pass filter of exercise 10.9.

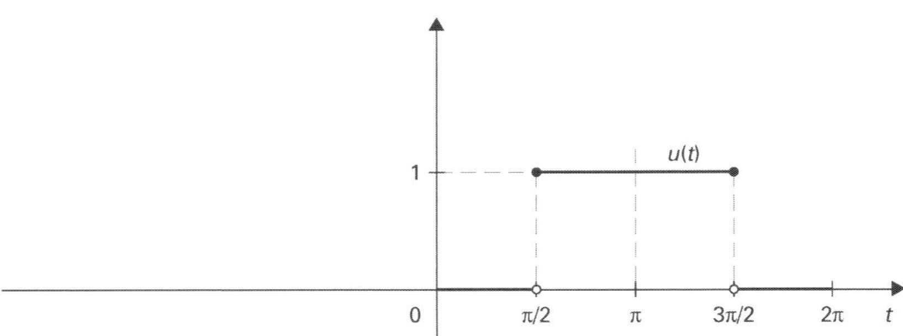

FIGURE 10.4
Periodic signal $u(t)$ of exercise 10.9.

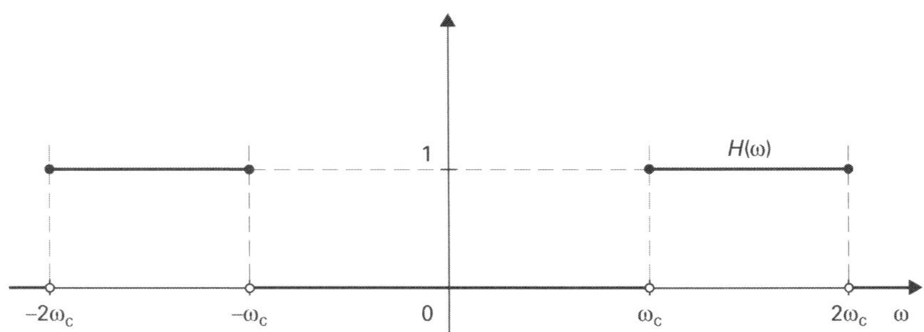

FIGURE 10.5
Band-pass filter of exercise 10.10.

a　Determine the impulse response of the filter.
b　To the filter we apply a periodic signal $u(t)$ with period 2π which, on the time interval $[0, 2\pi)$, is given by the graph of figure 10.4. Find the response to this periodic signal $u(t)$.

10.10
Band-pass filter

For an ideal *band-pass filter* the frequency response is given by the graph of figure 10.5.
a　Determine the impulse response of the filter.
b　Use the sine integral to determine the step response of the filter.

10.11

For an LTC-system the frequency response $H(\omega)$ is given by

$$H(\omega) = \frac{i\omega + 1}{i\omega - 1} \cdot \frac{i\omega - 2}{i\omega + 2}.$$

a　Determine the impulse response of the system.
b　Is the system causal? Justify your answer.
c　To the system we apply a signal $u(t)$ with a finite energy-content. Show that the energy-content of the response $y(t)$ to $u(t)$ is equal to the energy-content of the input $u(t)$.

10.12
Band-limited signal

An LTC-system with frequency response $H(\omega)$ is given. To the system we apply a so-called *band-limited* signal $u(t)$. This means that the spectrum $U(\omega)$ satisfies $U(\omega) = 0$ for $|\omega| \geq \omega_c$ for some $\omega_c > 0$.
a　Show that the output $y(t)$ is also band-limited.

b   For the band-limited signal the values $u(nT)$ for $n \in \mathbb{Z}$ are given, where T is such that $\omega_s = 2\pi/T > 2\omega_c$. In chapter 15 we will derive that this signal can then be written as

$$u(t) = \sum_{n=-\infty}^{\infty} u(nT)\frac{2\sin(\frac{1}{2}\omega_s(t-nT))}{\omega_s(t-nT)}.$$

Show that for the response $y(t)$ one has

$$y(t) = \sum_{n=-\infty}^{\infty} u(nT)h_a(t-nT),$$

where $h_a(t)$ is the signal given by

$$h_a(t) = \frac{1}{\omega_s}\int_{-\omega_s/2}^{\omega_s/2} H(\omega)e^{i\omega t}\,d\omega.$$

10.3   Causal stable systems and differential equations

An example of an LTC-system, occurring quite often in electronics, is an electric network consisting of linear elements: resistors, capacitors and inductors, whose properties should not vary in time (time-invariance). For these systems one can often derive a differential equation, from which the frequency response can then be determined quite easily (see theorem 5.2). From this, the impulse response can be determined by using the inverse Fourier transform, and subsequently one obtains the response to any input by a convolution. Let us start with an example.

EXAMPLE 10.8

In figure 10.6 a series connection is drawn, consisting of a voltage source, a resistor with resistance R and a capacitor with capacitance C. This circuit or network can be considered as a causal LTC-system with input the voltage $u(t)$ drop across the voltage source and with output the voltage drop $y(t)$ across the capacitor. We will

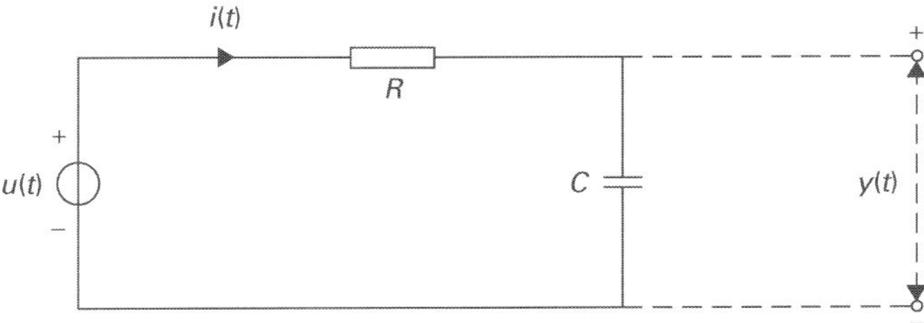

FIGURE 10.6
An RC-network.

now determine the impulse response. The frequency response can be obtained as follows. Let $i(t)$ be the current in the network, then it follows from Kirchhoff's voltage law and Ohm's law that

$$u(t) = Ri(t) + y(t).$$

The voltage–current relation for the capacitor is as follows:

$$y(t) = \frac{1}{C}\int_{-\infty}^{t} i(\tau)\,d\tau.$$

By differentiation it follows that the relation between $u(t)$ and $y(t)$ can be described by the following differential equation:

$$RC\frac{dy}{dt} + y(t) = u(t).$$

From theorem 5.2 it then follows that the frequency response $H(\omega)$ is equal to

$$H(\omega) = \frac{1}{1 + i\omega RC}.$$

Finally, the impulse response is obtained by an inverse Fourier transform. From table 3 we conclude that

$$h(t) = \frac{1}{RC}e^{-t/RC}\epsilon(t).$$

By (10.3), the response to an arbitrary input $u(t)$ then equals

$$y(t) = (u * h)(t) = \frac{1}{RC}\int_{-\infty}^{t} u(\tau)e^{-(t-\tau)/RC}\,d\tau.$$

$\blacktriangleleft$

In this section we will consider *causal* and *stable* LTC-systems described in the time domain by an ordinary differential equation with constant coefficients. Examples are the electrical networks with resistors, capacitors and inductors, the so-called RLC-networks, having one source (a voltage or current source); see also chapter 5. Using Kirchhoff's voltage law and the voltage–current relations for the resistor, capacitor and inductor, one can derive that the relation between the input $u(t)$ and the response $y(t)$ in such networks can always be described by a linear differential equation with constant coefficients of the form

$$a_m\frac{d^m y}{dt^m} + a_{m-1}\frac{d^{m-1}y}{dt^{m-1}} + \cdots + a_1\frac{dy}{dt} + a_0 y$$

$$= b_n\frac{d^n u}{dt^n} + b_{n-1}\frac{d^{n-1}u}{dt^{n-1}} + \cdots + b_1\frac{du}{dt} + b_0 u \tag{10.9}$$

with $n \leq m$. Causal and stable systems described by (10.9) are of practical importance because they can physically be realized, for example as an electric network.

If we have a complete description of a system, then the response $y(t)$ to a given input $u(t)$ must follow uniquely from the given description. For periodic signals this problem has already been discussed in chapter 5, section 5.1. However, if $u(t)$ is known in (10.9) (and so the right-hand side is known), then we know from the theory of ordinary differential equations that a solution $y(t)$ still contains m (the order of the differential equation) unconstrained parameters. Apparently, more data from the output or the system are required to determine $y(t)$ uniquely for a given $u(t)$. If, for example, $y(t)$ and all the derivatives of $y(t)$ up to order $m-1$ are known at $t = 0$, that is to say, if the *initial values* are known, then it follows from the theory of ordinary differential equations that $y(t)$ is uniquely determined for all t.

In this chapter we will assume that the differential equation (10.9) describes a *causal* and *stable* LTC-system. The impulse response is then causal and, after removing delta functions that might occur (see example 10.2), absolutely integrable. Using this, one is able to find a unique solution for the impulse response by substituting $u(t) = \delta(t)$ in (10.9). It is easier, though, to first calculate the frequency response (see theorem 5.2). According to theorem 5.2 one can use the characteristic polynomial $A(s) = a_m s^m + a_{m-1}s^{m-1} + \cdots + a_1 s + a_0$ and the polynomial $B(s) = b_n s^n + b_{n-1}s^{n-1} + \cdots + b_1 s + b_0$ to write the frequency response as

$$A(i\omega)H(\omega)e^{i\omega t} = B(i\omega)e^{i\omega t}.$$

We now impose the condition that $A(i\omega) \neq 0$ for all real ω, which means that the polynomial $A(s)$ has no zeros on the imaginary axis. Dividing by $A(i\omega)e^{i\omega t}$ is then permitted, which leads to the result

$$H(\omega) = \frac{b_n(i\omega)^n + b_{n-1}(i\omega)^{n-1} + \cdots + b_1(i\omega) + b_0}{a_m(i\omega)^m + a_{m-1}(i\omega)^{m-1} + \cdots + a_1(i\omega) + a_0} = \frac{B(i\omega)}{A(i\omega)}. \tag{10.10}$$

The frequency response is thus a rational function of ω. Because we assumed that $n \leq m$, the degree of the denominator is at most equal to the degree of the numerator. The impulse response follows from the inverse Fourier transform of the frequency response. To that end we apply the partial fraction expansion technique, explained in chapter 2, to the rational function $B(s)/A(s)$. Let $s_1, s_2, \ldots, s_m$ be the zeros of the polynomial $A(s)$ and assume, for convenience, that these zeros are simple. Since $n \leq m$, the partial fraction expansion leads to the representation

$$H(\omega) = c_0 + \sum_{k=1}^{m} \frac{c_k}{i\omega - s_k},$$

where $c_0, c_1, \ldots, c_m$ are certain constants. Inverse transformation of c_0 gives the signal $c_0 \delta(t)$. Inverse transformation for $\operatorname{Re} s_k < 0$ gives (see table 3)

$$e^{s_k t}\epsilon(t) \leftrightarrow \frac{1}{i\omega - s_k},$$

while for $\operatorname{Re} s_k > 0$ it gives (use time reversal)

$$-e^{s_k t}\epsilon(-t) \leftrightarrow \frac{1}{i\omega - s_k}.$$

Finally, when $\operatorname{Re} s_k = 0$, so $s_k = i\omega_0$ for some ω_0, then

$$\frac{1}{2}e^{i\omega_0 t}\operatorname{sgn} t \leftrightarrow \frac{1}{i(\omega - \omega_0)}.$$

We have assumed that the system is causal and stable. This then implies that for $k = 1, 2, \ldots, m$ the zeros s_k must satisfy $\operatorname{Re} s_k < 0$. Apparently, the zeros of $A(s)$ lie in the *left-half plane* $\operatorname{Re} s < 0$ of the complex s-plane. The impulse response $h(t)$ of the LTC-system then looks as follows:

$$h(t) = c_0\delta(t) + \sum_{k=1}^{m} c_k e^{s_k t}\epsilon(t).$$

We conclude that for a description of a *causal* and *stable* system by means of a differential equation of type (10.9), the zeros of $A(s)$ must lie in the left-half plane of the complex s-plane. Hence, there should also be no zeros on the imaginary axis. We formulate this result in the following theorem.

THEOREM 10.4

When an LTC-system is described by an ordinary differential equation of type (10.9), then the system is causal *and* stable *if and only if the zeros s of the characteristic polynomial satisfy* $\operatorname{Re} s < 0$.

EXAMPLE 10.9

Consider the RC-network from figure 10.7. The input is the voltage drop $u(t)$ across the voltage source and the output is the voltage drop $y(t)$ between nodes A and B of the network. The network is considered as an LTC-system. The relation between $u(t)$ and $y(t)$ is described by the differential equation

$$RC\frac{dy}{dt} + y = -RC\frac{du}{dt} + u.$$

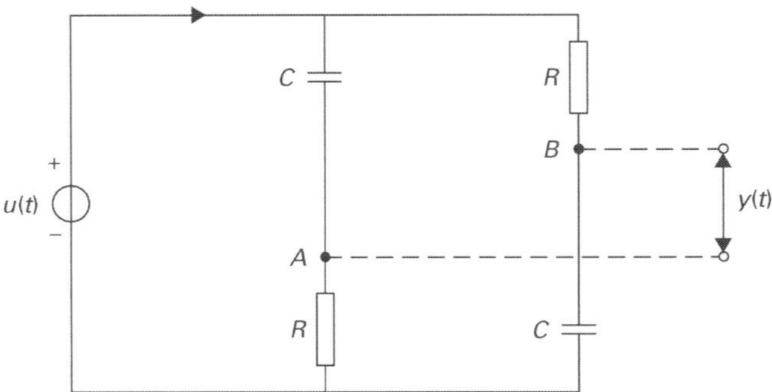

FIGURE 10.7
An RC-network.

Substituting $u(t) = e^{i\omega t}$ gives $y(t) = H(\omega)e^{i\omega t}$ with $H(\omega)$ equal to

$$H(\omega) = \frac{1 - i\omega RC}{1 + i\omega RC}.$$

We see that $|H(\omega)| = 1$. Hence, the amplitude spectrum is identical to 1. Apparently, the network is an example of an all-pass system. Furthermore, we see that $A(s) = 1 + sRC$ has a zero at $s = -1/RC$. This zero lies in the complex left-half plane and so the system is causal and stable. The impulse response follows by an inverse transform of $H(\omega)$, resulting in

$$h(t) = \frac{2}{RC}e^{-t/RC}\epsilon(t) - \delta(t).$$

◀

EXERCISES

10.13　　In the circuit of figure 10.8 we have $L = 2R^2C$. The circuit is considered as an

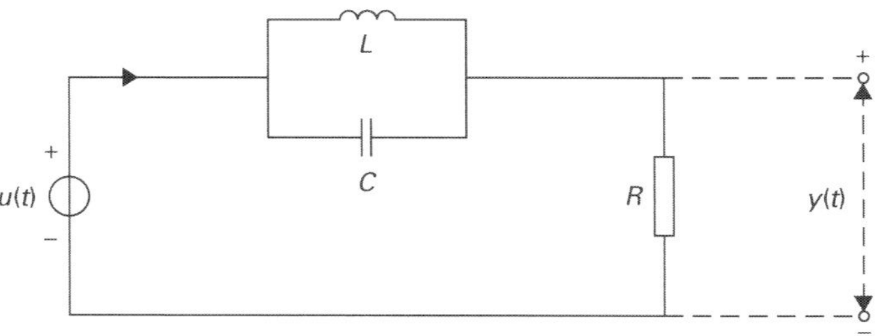

FIGURE 10.8
Circuit with $L = 2R^2C$ of exercise 10.13.

LTC-system with input the voltage drop $u(t)$ across the voltage source and output the voltage drop $y(t)$ across the resistor R. The relation between $u(t)$ and $y(t)$ is given by the differential equation

$$\frac{d^2y}{dt^2} + \omega_0\sqrt{2}\frac{dy}{dt} + \omega_0^2 y = \frac{d^2u}{dt^2} + \omega_0^2 u, \qquad \text{where } \omega_0 = \frac{1}{\sqrt{LC}}.$$

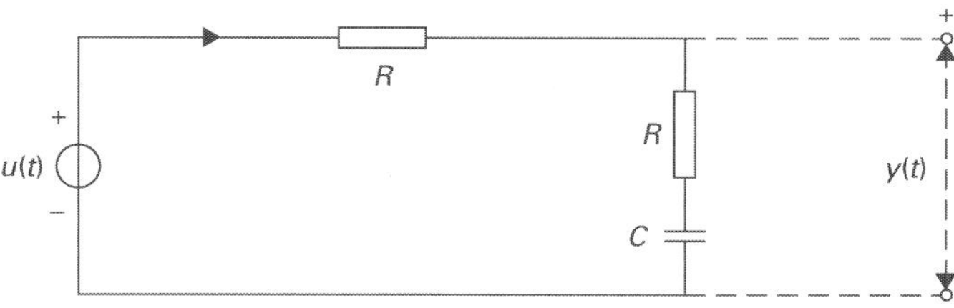

FIGURE 10.9
RC-network of exercise 10.14.

a Determine the frequency response $H(\omega)$.
b Determine the response to $u(t) = \cos(\omega_0 t)$.
c Determine the response to $u(t) = \cos(\omega_0 t)\epsilon(t)$.

10.14 The RC-network from figure 10.9 is considered as an LTC-system with input the voltage drop $u(t)$ across the voltage source and output $y(t)$ the voltage drop across the resistor and the capacitor. The relation between $u(t)$ and $y(t)$ is given by the differential equation

$$2\frac{dy}{dt} + \frac{1}{RC}y = \frac{du}{dt} + \left(1 + \frac{1}{RC}\right)u.$$

a Determine the frequency response and the impulse response of the system.
b Consider the so-called *inverse system*, which takes $y(t)$ as input and $u(t)$ as response. Determine the transfer function $H_1(\omega)$ and the impulse response $h_1(t)$ of the inverse system.
c Determine $(h * h_1)(t)$.

10.15 For an LTC-system the impulse response $h(t)$ is given by $h(t) = t^n e^{-at}\epsilon(t)$. Here n is a non-negative integer and a a complex number with $\operatorname{Re} a > 0$. Show that the system is stable.

10.4 Boundary and initial value problems for partial differential equations

In the previous sections it has been shown that the Fourier transform is an important tool in the study of the transfer of continuous-time signals in LTC-systems. We applied the Fourier transform to signals $f(t)$ being functions of the time t, and the Fourier transform $F(\omega)$ could then be interpreted as a function defined in the frequency domain. However, applications of the Fourier transform are not restricted to continuous-time signals only. For instance, one can sometimes apply the Fourier transform successfully in order to solve boundary and initial value problems for partial differential equations with constant coefficients. In this section an example will be presented. The techniques that we will use are the same as in section 5.2 of chapter 5. By a separation of variables one first determines a class of functions that satisfy the differential equation as well as the linear homogeneous conditions. From this set of functions one subsequently determines by superposition a solution which also satisfies the inhomogeneous conditions.

Again we take the *heat equation* from chapter 5 as an example, and we will also use the notations introduced in that chapter:

$$u_t = k u_{xx} \qquad \text{for } x \in \mathbb{R} \text{ and } t > 0. \tag{10.11}$$

The function $u(x, t)$ describes the heat distribution in a cylinder shaped rod at the point x in the longitudinal direction and at time t. Now consider the conditions

$$u(x, 0) = f(x) \text{ for } x \in \mathbb{R} \text{ and } t > 0,$$
$$u(x, t) \text{ is bounded.}$$

This means that we assume that the rod has infinite length, that the heat distribution at time $t = 0$ is known, and that we are only interested in a bounded solution. A linear homogeneous condition is characterized by the fact that a linear combination of functions that satisfy the condition, also satisfies that condition. In our case this is the boundedness condition. Verify this for yourself.

Separation of variables gives $u(x, t) = X(x)T(t)$, where X is a function of x only and T is a function of t only. In section 5.2 of chapter 5 we derived that this $u(x, t)$ satisfies the given heat equation if for some arbitrary constant c (the separation constant) one has

$$X'' - cX = 0,$$
$$T' - ckT = 0.$$

In order to satisfy the linear homogeneous condition as well, $X(x)T(t)$ has to be bounded, and this implies that both $X(x)$ and $T(t)$ have to be bounded functions.

From the differential equation for $T(t)$ it follows that $T(t) = \alpha e^{ckt}$ for some α. And since $T(t)$ has to be bounded for $t > 0$, the constant c has to satisfy $c \leq 0$. (Unless $\alpha = 0$, but then we obtain the trivial solution $T(t) = 0$, which is of no interest to us.) We therefore put $c = -\omega^2$, where ω is a real number.

For $\omega = 0$ the differential equation for $X(x)$ has as general solution $X(x) = \alpha x + \beta$. The boundedness of $X(x)$ then implies that $\alpha = 0$. For $\omega \neq 0$ the differential equation has as general solution $X(x) = \alpha e^{i\omega x} + \beta e^{-i\omega x}$. This function is bounded for all α and β since

$$|X(x)| \leq |\alpha| \left| e^{i\omega x} \right| + |\beta| \left| e^{-i\omega x} \right| = |\alpha| + |\beta|.$$

From the above it follows that the class of functions we are looking for, that is, satisfying the differential equation and being bounded, can be described by

$$X(x)T(t) = e^{i\omega x} e^{-k\omega^2 t}, \qquad \text{where } \omega \in \mathbb{R}.$$

Now the final step is to construct, by *superposition* of functions from this class, a solution satisfying the inhomogeneous condition as well. This means that for a certain function $F(\omega)$ we will try a solution $u(x, t)$ of the form

$$u(x, t) = \int_{-\infty}^{\infty} F(\omega) e^{-k\omega^2 t} e^{i\omega x} \, d\omega.$$

If we substitute $t = 0$ in this integral representation, then we obtain that

$$f(x) = \int_{-\infty}^{\infty} F(\omega) e^{i\omega x} \, d\omega.$$

We can now apply the theory of the Fourier transform. If we interpret ω as a frequency, then this integral shows that, up to a factor 2π, $f(x)$ equals the inverse

Fourier transform of $F(\omega)$. Hence, $F(\omega)$ is the Fourier transform of $f(x)$, up to a factor 2π:

$$F(\omega) = \frac{1}{2\pi} \int_{-\infty}^{\infty} f(x) e^{-i\omega x}\, dx.$$

Formal solution

We have thus found a solution of the heat conduction problem for the infinite rod, however, without worrying about convergence problems. In fact one should verify afterwards that the $u(x, t)$ we have found is indeed a solution. We will omit this and express this by saying that we have determined a *formal solution*. When, for example, $f(x) = 1/(1 + x^2)$, then it follows from table 3 that $F(\omega) = \frac{1}{2} e^{-|\omega|}$ and so a formal solution is given by

$$u(x, t) = \tfrac{1}{2} \int_{-\infty}^{\infty} e^{-|\omega|} e^{-k\omega^2 t} e^{i\omega x}\, d\omega.$$

EXERCISES

10.16   Let $f(x) = 1/(1 + x^2)$. Determine formally the bounded solution of the following problem from potential theory:

$$u_{xx} + u_{yy} = 0 \qquad \text{for } -\infty < x < \infty \text{ and } y > 0,$$
$$u(x, 0) = f(x) \qquad \text{for } -\infty < x < \infty.$$

10.17   Determine formally the bounded solution $T(x, t)$ of the heat conduction equation

$$T_{xx} = T_t \qquad \text{for } -\infty < x < \infty \text{ and } t > 0$$

with initial condition

$$T(x, 0) = \begin{cases} T_1 & \text{for } x \geq 0, \\ T_2 & \text{for } x < 0. \end{cases}$$

SUMMARY

The Fourier transform is an important tool in the study of linear time-invariant continuous-time systems (LTC-systems). These systems possess the important property that the relation between an input $u(t)$ and the corresponding output $y(t)$ is given in the time domain by means of the convolution product

$$y(t) = (h * u)(t),$$

where $h(t)$ is the response of the LTC-system to the delta function or unit pulse $\delta(t)$. The signal $h(t)$ is called the impulse response. An LTC-system is completely determined when the impulse response is known. Properties of an LTC-system can be derived from the impulse response. For example, an LTC-system is stable if the impulse response, ignoring possible delta functions, is absolutely integrable.

The step response $a(t)$ is defined as the response to the unit step function $\epsilon(t)$. The derivative of the step response as distribution is equal to the impulse response.

The frequency response $H(\omega)$, introduced in chapter 1, turned out to be equal to the Fourier transform of the impulse response $h(t)$. The frequency response has the special property that the LTC-system can be described in the frequency domain by

$$Y(\omega) = H(\omega)U(\omega).$$

Here $U(\omega)$ and $Y(\omega)$ are the spectra of, respectively, the input $u(t)$ and the corresponding output $y(t)$. Hence, an LTC-system is known when $H(\omega)$ is known: in principle the response $y(t)$ to any input $u(t)$ can then be determined.

An ideal low-pass filter is characterized as an LTC-system for which $H(\omega) = 0$ outside the pass-band $-\omega_c < \omega < \omega_c$. An LTC-system for which $H(\omega) = 1$ for all ω is called an all-pass system. An all-pass system has the property that the energy-content of the output equals the energy-content of the corresponding input.

For practical applications the causal and stable systems are important, for which the relationship between $u(t)$ and $y(t)$ can be described by an ordinary differential equation

$$a_m \frac{d^m y}{dt^m} + a_{m-1} \frac{d^{m-1} y}{dt^{m-1}} + \cdots + a_1 \frac{dy}{dt} + a_0 y$$
$$= b_n \frac{d^n u}{dt^n} + b_{n-1} \frac{d^{n-1} u}{dt^{n-1}} + \cdots + b_1 \frac{du}{dt} + b_0 u.$$

Examples are the electrical networks consisting of resistors, capacitors and inductors, whose physical properties should be time-independent. The frequency response is a rational function. Stability and causality can be established by looking at the location of the zeros of the denominator of the frequency response. These should lie in the left-half plane of the complex plane.

The Fourier transform can also successfully be applied to functions depending on a position variable. Particularly for boundary and initial value problems for linear partial differential equations, the Fourier transform can be a valuable tool.

SELFTEST

10.18   A system is described by

$$y(t) = \int_{t-1}^{t} e^{-(t-\tau)} u(\tau) \, d\tau.$$

    a   Determine the frequency response and the impulse response.
    b   Is the system causal? Justify your answer.
    c   Is the system stable? Justify your answer.
    d   Determine the response to the block function $p_2(t)$.

10.19   For an LTC-system the step response $a(t)$ is given by $a(t) = e^{-t} \epsilon(t)$.
    a   Determine the impulse response.
    b   Determine the frequency response.
    c   Determine the response to the input $u(t) = e^{-t} \epsilon(t)$.

10.20   The frequency response of an ideal low-pass filter is given by

$$H(\omega) = \begin{cases} 1 & \text{for } |\omega| \leq \omega_c, \\ 0 & \text{for } |\omega| > \omega_c. \end{cases}$$

Determine the response to the periodic input $u(t)$ with period $T = 5\pi/\omega_c$ given by $u(t) = t$ for $0 \leq t < T$.

10.21   For an LTC-system the frequency response $H(\omega)$ is given by

$$H(\omega) = \begin{cases} 1 - \dfrac{|\omega|}{\omega_c} & \text{for } |\omega| \leq \omega_c, \\[2mm] 0 & \text{for } |\omega| > \omega_c. \end{cases}$$

a Determine the impulse response.

b Show that the step response $a(t)$ is a monotone increasing function.

10.22 The RCL-network of figure 10.10 is considered as an LTC-system with the voltage drop $u(t)$ across the voltage source as input and the voltage drop $y(t)$ across the capacitor as output. The quantities R, L and C are expressed in their respective units ohm, henry and farad. The relation between $u(t)$ and $y(t)$ is given by the differential

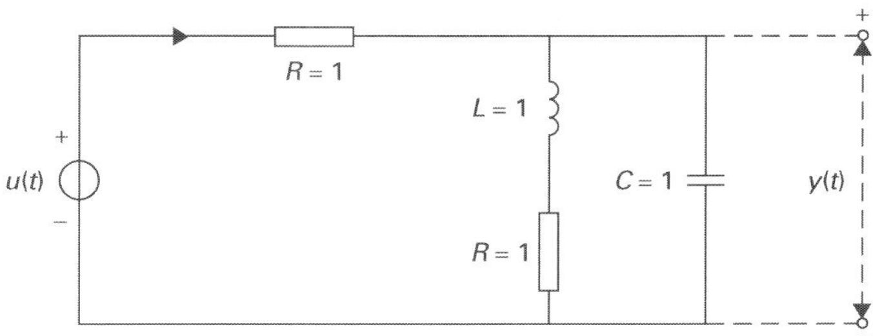

FIGURE 10.10
RCL-network of exercise 10.22.

equation

$$\frac{d^2 y}{dt^2} + 2\frac{dy}{dt} + 2y = \frac{du}{dt} + u.$$

a Determine the impulse response and the step response.

b Determine the response to the input $u(t) = e^{-t}\epsilon(t)$.

10.23 For an LTC-system L the frequency response is given by

$$H(\omega) = \frac{i\omega - 1 - i}{i\omega + 1 - i}\, e^{-i\omega t_0} \qquad \text{where } t_0 > 0.$$

a Show that L is an all-pass system.

b Determine the impulse response.

c Is the system stable? Justify your answer.

d Determine the response to the periodic input $u(t) = 1 + 2\cos t$.

10.24 Determine formally the bounded solution $u(x, y)$ of the following boundary value problem:

$$u_{xx} - 2u_y = 0 \qquad \text{for } -\infty < x < \infty \text{ and } y > 0,$$
$$u(x, 0) = xe^{-x}\epsilon(x) \qquad \text{for } -\infty < x < \infty.$$

Part 4
Laplace transforms

In the previous two parts we considered various forms of Fourier analysis: for periodic functions in part 2 and for non-periodic functions and distributions in part 3. In this part we examine the so-called Laplace transform. On the one hand it is closely related to the Fourier transform of non-periodic functions, but on the other hand it is more suitable in certain applied fields, in particular in signal theory. In physical reality we usually study signals that have been switched on at a certain moment in time. One then chooses this switch-on time as the origin of the time-scale. Hence, in such a situation we are dealing with functions on $\mathbb{R}$ which are zero for $t < 0$, the so-called *causal* functions (see section 1.2.4). The Fourier transform of such a function f is then given by

$$F(\omega) = \int_0^\infty f(t)e^{-i\omega t}\,dt,$$

where $\omega \in \mathbb{R}$. A disadvantage of this integral is the fact that, even for very simple functions, it often does not exist. For the unit step function $\epsilon(t)$ for example, the integral does not exist and in order to determine the spectrum of $\epsilon(t)$ we had to resort to distribution theory. If we multiply $\epsilon(t)$ by a 'damping factor' $e^{-\sigma t}$ for an arbitrary $\sigma > 0$, then the spectrum will exist (see section 6.3.3). It turns out that this is true more generally: when $f(t)$ is a function that is zero for $t < 0$ and whose spectrum does not exist, then there is a fair chance that the spectrum of $g(t) = e^{-\sigma t} f(t)$ does exist (under certain conditions on $\sigma \in \mathbb{R}$). Determining the spectrum of $g(t)$ boils down to the calculation of the integral

$$\int_0^\infty f(t)e^{-(\sigma+i\omega)t}\,dt$$

for arbitrary real σ and ω. The result will be a new function, denoted by F again, which no longer depends on $\omega \in \mathbb{R}$, but on $\sigma + i\omega \in \mathbb{C}$. Hence, if we write $s = \sigma + i\omega$, then this assigns to any causal function $f(t)$ a function $F(s)$ by defining

$$F(s) = \int_0^\infty f(t)e^{-st}\,dt.$$

The function $F(s)$ is called the *Laplace transform* of the causal function $f(t)$ and the mapping assigning the function $F(s)$ to $f(t)$ is called the Laplace transform. When studying phenomena where one has to deal with switched-on signals, the Laplace transform is often given preference over the Fourier transform. In fact, the Laplace transform has a better way 'to deal with the switch-on time $t = 0$'. Another advantage of the Laplace transform is the fact that we do not need distributions very often, since the Laplace transform of 'most' functions exists as an ordinary integral. For most applications it therefore suffices to use only a very limited part of the distribution theory.

Another noticeable difference with the Fourier analysis from parts 2 and 3 is the role of the fundamental theorem. Although the fundamental theorem of the Laplace transform can easily be derived from the one for the Fourier integral, it will play an insignificant role in part 4. In order to recover a function $f(t)$ from its Laplace transform $F(s)$ we will instead use a table, the properties of the Laplace transform and partial fraction expansions.

To really understand the fundamental theorem of the Laplace transform would require an extensive treatment of the theory of functions from $\mathbb{C}$ to $\mathbb{C}$. These functions are called *complex functions* and the Laplace transform is indeed a complex function: to $s \in \mathbb{C}$ the number $F(s) \in \mathbb{C}$ is assigned (we recall that, in contrast, the Fourier transform is a function from $\mathbb{R}$ to $\mathbb{C}$). For a rigorous treatment of the Laplace transform at least some knowledge of complex functions is certainly necessary. We therefore start part 4 with a short introduction to this subject in chapter 11. A thorough study of complex functions, necessary in order to use the fundamental theorem of the Laplace transform in its full strength, lies beyond the scope of this book.

Following the brief introduction to complex functions in chapter 11, we continue in chapter 12 with the definition of the Laplace transform of a causal function. A number of standard Laplace transforms are calculated and some properties, most of which will be familiar by now, are treated.

Chapter 13 starts with a familiar subject as well, namely convolution. However, we subsequently treat a number of properties not seen before in Fourier analysis, such as the so-called initial value and final value theorems. We also consider the Laplace transform of distributions in chapter 13. Finally, the fundamental theorem of the Laplace transform is proven and a method is treated to recover a function $f(t)$ from its Laplace transform $F(s)$ by means of a partial fraction expansion. As in parts 2 and 3, we apply the Laplace transform to the theory of linear systems and (partial) differential equations in the final chapter 14.

Pierre Simon Laplace (1749 – 1827) lived and worked at the end of an epoch that started with Newton, in which the study of the movement of the planets formed an important stimulus for the development of mathematics. Theories developed in this period were recorded by Laplace in the five-part opus *Mécanique Celeste*, which he wrote during the years 1799 to 1825. The shape of the earth, the movements of the planets, and the distortions in their orbits were described in it. Another major work by Laplace, *Théorie analytique des probabilités*, deals with the calculus of probabilities. Both standard works do not only contain his own material, but also that of his predecessors. Laplace, however, made all of this material coherent and moreover wrote extensive introductions in non-technical terms.

Laplace's activities took place in a time where mathematicians were no longer mainly employees of monarchs at courts, but instead were employed by universities and technical institutes. Previously, mathematicians were given the opportunity to work at courts, since enlightened monarchs were on the one hand pleased to have famous scientists associated with their courts, and on the other hand because they realized how useful mathematics and the natural sciences were for the improvement of production processes and warfare. Mathematicians employed by universities and institutes were also given significant teaching tasks. Laplace himself was professor of mathematics at the Paris military school and was also a minister in Napoleon's cabinet for some time. He considered mathematics mainly as a beautiful toolbox which could benefit the progress of the natural sciences.

In Laplace's epoch the idea prevailed that mathematics was developed to such an extent that all could be explained. Based on Newton's laws, numerous different phenomena could be understood. This vision arose from the tendency to identify mathematics mainly with astronomy and mechanics. It led Laplace to the following famous statement: "An intelligence which could oversee all forces acting in nature at a specific moment and, moreover, all relative positions of all parts present in nature, and which would also be sufficiently comprehensive to subject all these data to a mathematical analysis, could in one and the same formula encompass the movements of the largest bodies in the universe as well as that of the lightest atom: nothing would remain uncertain for her, and the future as well as the past would be open to her." Hence, any newly developed mathematics would at best be more of the same. However, in the first decades of the nineteenth century mathematicians, such as Fourier, adopted a new course. In the twentieth century, the view that mathematics could explain everything was thoroughly upset. First by quantum mechanics, which proved that the observer always influences the observed object, and subsequently by chaos theory, which proved that it is impossible to determine the initial state of complex systems, such as the weather, sufficiently accurately to be able to predict all future developments. Notwithstanding, the Laplace transforms remain a very important tool for the analysis and further development of systems and electronic networks.

Contents of Chapter 11

Complex functions

Complex functions

INTRODUCTION

In this chapter we give a brief introduction to the theory of complex functions. In section 11.1 some well-known examples of complex functions are treated, in particular functions that play a role in the Laplace transform. In sections 11.2 and 11.3 continuity and differentiability of complex functions are examined. It will turn out that both the definition and the rules for continuity and differentiability are almost exactly the same as for real functions. Still, complex differentiability is surprisingly different from real differentiability. In the final section we will briefly go into this matter and treat the so-called Cauchy–Riemann equations. The more profound properties of complex functions cannot be treated in the context of this book.

LEARNING OBJECTIVES
After studying this chapter it is expected that you
- know the definition of a complex function and know the standard functions z^n, e^z, $\sin z$ and $\cos z$
- can split complex functions into a real and an imaginary part
- know the concepts of continuity and differentiability for complex functions
- know the concept of analytic function
- can determine the derivative of a complex function.

11.1 Definition and examples

The previous parts of this book dealt almost exclusively with functions that were defined on $\mathbb{R}$ and could have values in $\mathbb{C}$. In this part we will be considering functions that are defined on $\mathbb{C}$ (and can have values in $\mathbb{C}$).

DEFINITION 11.1
Complex function

A function f is called a complex function *when f is defined on a subset of $\mathbb{C}$ and has values in a subset of $\mathbb{C}$.*

Note that in particular $\mathbb{C}$ itself is a subset of $\mathbb{C}$. It is customary to denote the variable of a complex function by the symbol z. The set of all $z \in \mathbb{C}$ for which a complex function is well-defined is called the *domain* of the function. The *range* of a complex function f is the set of values $f(z)$, where z runs through the domain of f.

Domain
Range

EXAMPLE 11.1

The function $f(z) = z$ is a complex function with domain $\mathbb{C}$ and range $\mathbb{C}$. The function assigning the complex conjugate $\overline{z}$ to each $z \in \mathbb{C}$, has domain $\mathbb{C}$ and range $\mathbb{C}$ as well. In fact, since $\overline{\overline{z}} = z$, it follows that $z \in \mathbb{C}$ is the complex conjugate of $\overline{z} \in \mathbb{C}$. In figure 11.1 the function $f(z) = \overline{z}$ is represented: for a point z the image-point $\overline{z}$ is drawn. ◀

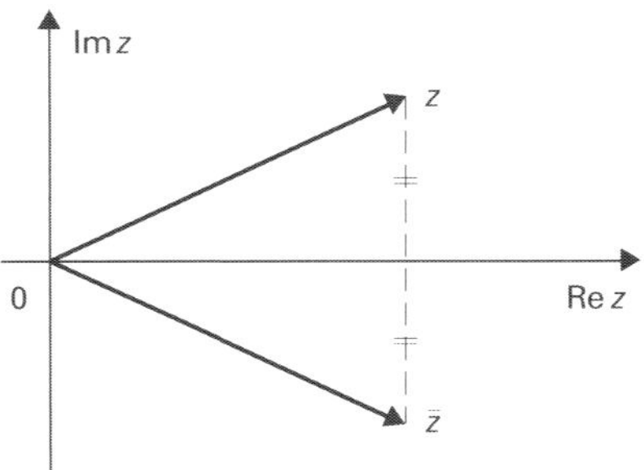

FIGURE 11.1
The function $z \to \bar{z}$.

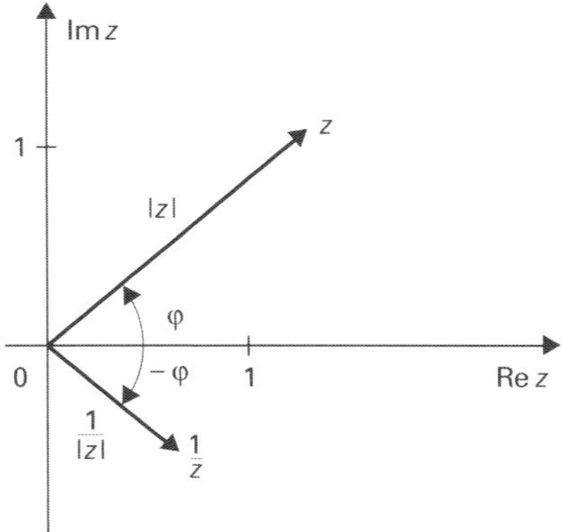

FIGURE 11.2
The function $z \to 1/z$.

EXAMPLE 11.2

Consider the function g assigning the complex number $1/z$ to $z \in \mathbb{C}$, that is, $g(z) = 1/z$. According to section 2.1.1 the number $1/z$ exists for every $z \in \mathbb{C}$ with $z \neq 0$. The domain of g is $\mathbb{C} \setminus \{0\}$. The image of g also is $\mathbb{C} \setminus \{0\}$. In fact, if $z \in \mathbb{C}$ and $z \neq 0$, then one has for $w = 1/z$ that $g(w) = 1/w = z$. In figure 11.2 the function $1/z$ is represented; here $\phi = \arg z$. ◀

Just as for real functions, one can use simple complex functions to build ever more complicated complex functions. The simplest complex functions are of course the constant functions $f(z) = c$, where $c \in \mathbb{C}$. Next we can consider positive integer powers of z, so $f(z) = z^n$ with $n \in \mathbb{N}$. By adding and multiplying by

Polynomial

constants, we obtain the *polynomials* $p(z) = a_n z^n + a_{n-1} z^{n-1} + \cdots + a_1 z + a_0$, where $a_i \in \mathbb{C}$ for each i. Finally, we can divide two of these polynomials to obtain

Rational function

the *rational functions* $p(z)/q(z)$, where $p(z)$ and $q(z)$ are polynomials. As in the real case, these are only defined for those values of z for which $q(z) \neq 0$.

In section 2.1.1 we have seen that $\mathbb{C}$ can also be represented as points in $\mathbb{R}^2$: the complex number $z = x + iy$ is then identified with the point $(x, y) \in \mathbb{R}^2$. If we now write $f(z) = f(x + iy) = u + iv$ with $u, v \in \mathbb{R}$, then u and v will be functions of x and y. Hence,

$$f(z) = u(x, y) + iv(x, y),$$

Real and imaginary part of complex function

where u and v are functions from $\mathbb{R}^2$ to $\mathbb{R}$. The functions u and v are called, respectively, the *real* and *imaginary part* of the complex function f.

EXAMPLE

The function $f(z) = z$ has the function $u(x, y) = x$ as real part and $v(x, y) = y$ as imaginary part. ◄

EXAMPLE 11.3

The function $f(z) = 1/z$ has real part $u(x, y) = x/(x^2 + y^2)$ and imaginary part $v(x, y) = -y/(x^2 + y^2)$. In fact (also see section 2.1.1):

$$\frac{1}{z} = \frac{1}{x + iy} = \frac{x - iy}{(x + iy)(x - iy)} = \frac{x - iy}{x^2 + y^2} = \frac{x}{x^2 + y^2} - i\frac{y}{x^2 + y^2}.$$

◄

EXAMPLE 11.4

Consider the complex function $g(z) = z^2$. One has $(x + iy)^2 = x^2 - y^2 + 2ixy$. Hence, the real part of g is $u(x, y) = x^2 - y^2$, while the imaginary part is given by $v(x, y) = 2xy$. ◄

The most important function in Fourier analysis is without any doubt the complex-valued function $e^{iy} = \cos y + i \sin y$, $y \in \mathbb{R}$ (see definition 2.1). In the Laplace transform the role of this function is taken over by the complex function e^z. In section 2.1.1 we have already defined e^z for $z \in \mathbb{C}$, since we regularly needed the complex-valued function e^{zt}. Here $z \in \mathbb{C}$ is fixed and $t \in \mathbb{R}$ varies. The following definition is, of course, in accordance with definition 2.1, but it emphasizes the fact that we now consider e^z as a complex function.

DEFINITION 11.2
Complex exponential

For $z = x + iy$ we define e^z by

$$e^z = e^x e^{iy} = e^x(\cos y + i \sin y). \tag{11.1}$$

When $z = iy$ with $y \in \mathbb{R}$, then this coincides with definition 2.1 given earlier in section 2.1.1; when $z = x$ with $x \in \mathbb{R}$, then the definition agrees with the usual (real) exponential function e^x. By the way, it immediately follows from the definition that $e^z = u(x, y) + iv(x, y)$ with $u(x, y) = e^x \cos y$ and $v(x, y) = e^x \sin y$. The definition above ensures that the characteristic property $e^{a+b} = e^a e^b$ of the real exponential function remains valid.

THEOREM 11.1

The function e^z has the following properties:
a $\ e^{z+w} = e^z e^w$ *for all $w, z \in \mathbb{C}$;*
b $\ \left| e^{x+iy} \right| = e^x$; *in particular one has $e^z \neq 0$ for all $z \in \mathbb{C}$.*

Proof
Let $z = a + ib$ and $w = c + id$ with $a, b, c, d \in \mathbb{R}$. Then we have $e^{z+w} = e^{(a+c)+i(b+d)} = e^{a+c}e^{i(b+d)}$. In (2.7) it was shown that $e^{i(b+d)} = e^{ib}e^{id}$, while for the real exponential function one has $e^{a+c} = e^a e^c$. Hence, it follows that $e^{z+w} = e^a e^c e^{ib} e^{id} = e^{a+ib}e^{c+id} = e^z e^w$. This proves part a. For part b we use that $|wz| = |w||z|$ (see (2.4)). It then follows that $\left| e^{x+iy} \right| = \left| e^x e^{iy} \right| = \left| e^x \right| \left| e^{iy} \right| = e^x$ since $\left| e^{iy} \right| = 1$ for $y \in \mathbb{R}$ and $e^x > 0$ for $x \in \mathbb{R}$. In particular we see that $\left| e^{x+iy} \right| > 0$ and so $e^z \neq 0$ for all $z = x + iy \in \mathbb{C}$. ∎

In contrast to the real exponential function e^x ($x \in \mathbb{R}$), the complex function e^z is by no means one-to-one. We recall, for example, that $e^{2\pi ik} = 1$ for all $k \in \mathbb{Z}$ (see (2.12)). Using theorem 11.1a it then follows that $e^{z+2\pi ik} = e^z$ for all $k \in \mathbb{Z}$ and $z \in \mathbb{C}$.

The definition of e^z for $z \in \mathbb{C}$ also enables us to extend the sine and cosine functions to complex functions. In (2.11) we already noted that $\sin y = (e^{iy} - e^{-iy})/2i$

and $\cos y = (e^{iy} + e^{-iy})/2$ for $y \in \mathbb{R}$. Since e^{iz} is now defined for $z \in \mathbb{C}$, this suggests the following definition.

DEFINITION 11.3
Complex sine and cosine

For $z \in \mathbb{C}$ we define $\sin z$ and $\cos z$ by

$$\sin z = \frac{e^{iz} - e^{-iz}}{2i}, \qquad \cos z = \frac{e^{iz} + e^{-iz}}{2}.$$

When $z = y \in \mathbb{R}$, these definitions are of course in accordance with the real sine and cosine. Many of the well-known trigonometric identities remain valid for the complex sine and cosine. For example, by expanding the squares it immediately follows that $\sin^2 z + \cos^2 z = 1$. Similarly one obtains for instance the formulas $\cos 2z = \cos^2 z - \sin^2 z$ and $\sin 2z = 2 \sin z \cos z$. However, not all results are the same! For example, it is *not* true that $|\sin z| \leq 1$: for $z = 2i$ we have $|\sin 2i| = (e^2 - e^{-2})/2 > 3$.

EXERCISES

11.1   Determine the domain and range of the following complex functions:
a $f(z) = \overline{z}$,
b $f(z) = z^3$,
c $f(z) = \overline{z - 4 + i}$,
d $f(z) = (3i - 2)/(z + 3)$.

11.2   Determine the real and imaginary part of the complex functions in exercise 11.1.

11.3   Show that $\sin^2 z + \cos^2 z = 1$ and that $\sin 2z = 2 \sin z \cos z$.

11.4   Prove that $\sin(-z) = -\sin z$ and that $\sin(z + w) = \sin z \cos w + \cos z \sin w$.

11.5   For $x \in \mathbb{R}$ the functions *hyperbolic sine* (sinh) and *hyperbolic cosine* (cosh) are
Hyperbolic sine   defined by $\sinh x = (e^x - e^{-x})/2$ and $\cosh x = (e^x + e^{-x})/2$.
Hyperbolic cosine   **a**   Prove that $\sin(iy) = i \sinh y$ and $\cos(iy) = \cosh y$.
b   Use part a and exercise 11.4 to show that the real part of $\sin z$ equals $\sin x \cosh y$ and that the imaginary part of $\sin z$ equals $\cos x \sinh y$.

11.2   Continuity

In this section the concept of continuity is treated for complex functions. Just as for real functions, continuity of a complex function will be defined in terms of limits. However, in order to talk about limits in $\mathbb{C}$, we will first have to specify exactly what will be meant by 'complex numbers lying close to each other'. To do so, we start this section with the notion of a neighbourhood of a complex number z_0.

In section 2.1.1 we noted that the set of complex numbers on the unit circle is given by the equation $|z| = 1$. The set of complex numbers inside the unit circle
Unit disc   will be called the *unit disc*. The complex numbers in the unit disc are thus given by the inequality $|z| < 1$. In the same way all complex numbers inside the circle with centre 0 and radius $\delta > 0$ are given by the inequality $|z| < \delta$. Finally, if we shift the centre to the point $z_0 \in \mathbb{C}$, then all complex numbers inside the circle with centre z_0 and radius $\delta > 0$ are given by the inequality $|z - z_0| < \delta$. We call this a *neighbourhood* of the point z_0. When the point z_0 itself is removed from a neighbourhood of z_0, then we call this a *reduced neighbourhood* of z_0; it is given by the inequalities $0 < |z - z_0| < \delta$. See figure 11.3. We summarize these concepts in definition 11.4.

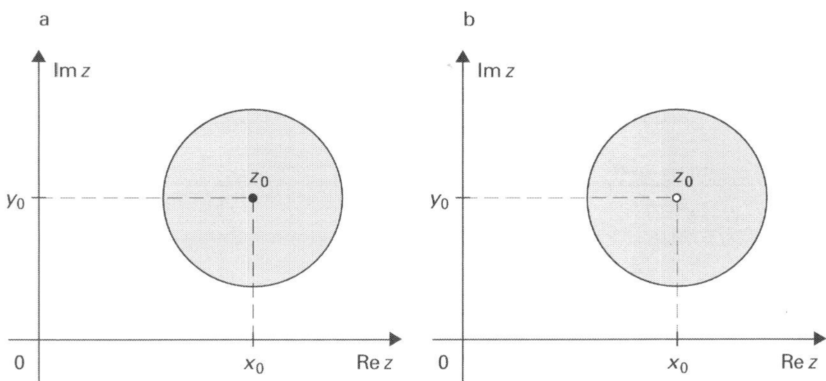

FIGURE 11.3
A neighbourhood (a) and a reduced neighbourhood (b) of z_0.

DEFINITION 11.4
Neighbourhood

Let $\delta > 0$. A neighbourhood of z_0 is defined as the set

$$\left\{ z \in \mathbb{C} \,\Big|\, |z - z_0| < \delta \right\}.$$

Reduced neighbourhood

A reduced neighbourhood of z_0 is defined as the set

$$\left\{ z \in \mathbb{C} \,\Big|\, 0 < |z - z_0| < \delta \right\}.$$

Continuity of a complex function can now be defined precisely as for real functions. First the notion of a limit is introduced and subsequently continuity is defined in terms of limits.

DEFINITION 11.5
Limit in $\mathbb{C}$

Let f be a complex function defined in a reduced neighbourhood of z_0. Then $\lim_{z \to z_0} f(z) = w$ means that for all $\epsilon > 0$ there exists a $\delta > 0$ such that for $0 < |z - z_0| < \delta$ one has $|f(z) - w| < \epsilon$.

Hence, the value $f(z)$ is close to w when z is close to z_0. Geometrically this means that the numbers $f(z)$ will be lying in a disc which is centred around the point w and which is getting smaller and smaller as z tends to z_0.

DEFINITION 11.6
Continuity in $\mathbb{C}$

Let f be a complex function defined in a neighbourhood of $z_0 \in \mathbb{C}$. Then f is called continuous at z_0 if $\lim_{z \to z_0} f(z) = f(z_0)$. The function f is called continuous on a subset G in $\mathbb{C}$ when f is continuous at all points z_0 of G; this subset G should be a set such that every point $z_0 \in G$ has a neighbourhood belonging entirely to G.

Loosely speaking: when f is continuous at z_0, then the value $f(z)$ is close to $f(z_0)$ when z is close to z_0. The condition on the subset G in definition 11.6 is quite natural. This is because continuity at a point z_0 can only be defined if the function is defined in a neighbourhood of the point z_0; hence, together with the point z_0 there should also be a neighbourhood of z_0 belonging entirely to G. This explains the condition on the set G. Definitions 11.5 and 11.6 are completely analogous to the definitions of limit and continuity in the real case. Considering this great similarity with the real case, a warning is justified. In the real case there are only *two* different directions from which a limit can be taken, namely from the right and from the left. Continuity at a point x_0 of a function f defined on $\mathbb{R}$ is thus equivalent to

$$\lim_{x \uparrow x_0} f(x) = f(x_0) = \lim_{x \downarrow x_0} f(x).$$

For a complex function the situation is completely different. The points z may approach z_0 in a completely *arbitrary* fashion in the complex plane, as long as the distance $|z - z_0|$ from z to z_0 decreases. Hence, for a complex function it is often much harder to prove continuity using definition 11.6. However, the rules for limits and continuity of complex functions are indeed precisely the same as for real functions. Using these rules it is in most cases easy to verify the continuity of complex functions. The proofs of the rules are also exactly the same as for real functions. This is why we state the following theorem without proof.

THEOREM 11.2

Let f and g be complex functions defined in a reduced neighbourhood of z_0. If $\lim_{z \to z_0} f(z) = a$ and $\lim_{z \to z_0} g(z) = b$, then

$$\lim_{z \to z_0} (f(z) \pm g(z)) = a \pm b,$$

$$\lim_{z \to z_0} (f(z) \cdot g(z)) = a \cdot b,$$

$$\lim_{z \to z_0} (f(z)/g(z)) = a/b \qquad \text{if } b \neq 0.$$

If $\lim_{w \to a} h(w) = c$ and if the function h is defined in a reduced neighbourhood of the point a, then $\lim_{z \to z_0} h(f(z)) = c$.

From theorem 11.2 one then obtains, as in the real case, the following results. Here the set G is as in definition 11.6.

THEOREM 11.3

When f and g are continuous functions on a subset G of $\mathbb{C}$, then $f + g$ and $f \cdot g$ are continuous on G. Moreover, f/g is continuous on G, provided that $g(z) \neq 0$ for all $z \in G$. If h is a continuous function defined on the range of f, then $(h \circ f)(z) = h(f(z))$ is also a continuous function on G.

As for real functions, theorem 11.3 is used to prove the continuity for ever more complicated functions. The constant function $f(z) = c$ ($c \in \mathbb{C}$) and the function $f(z) = z$ are certainly continuous on $\mathbb{C}$ (the proof is the same as for real functions). According to theorem 11.3, the product $z \cdot z = z^2$ is then also continuous. Repeated application then establishes that z^n is continuous on $\mathbb{C}$ for any $n \in \mathbb{N}$ and hence also that any polynomial is continuous on $\mathbb{C}$. By theorem 11.3 rational functions are then continuous as well, as long as the denominator is unequal to 0; in this case one has to take for the subset G the set $\mathbb{C}$ with the roots of the denominator removed (see section 2.1.2 for the concept of root or zero of a polynomial). Without proof we mention that such a set satisfies the conditions of definition 11.6.

EXERCISES

11.6

a   Show that $\lim_{z \to z_0} (f(z) + g(z)) = a + b$ if $\lim_{z \to z_0} f(z) = a$ and $\lim_{z \to z_0} g(z) = b$.
b   Use part a to prove that $f + g$ is continuous on G if f and g are continuous on G (G is a subset of $\mathbb{C}$ as in definition 11.6).

11.7

Use the definition to show that the following complex functions are continuous on $\mathbb{C}$:
a $f(z) = c$, where $c \in \mathbb{C}$ is a constant,
b $f(z) = z$.

11.8

On which subset G of $\mathbb{C}$ is the following function continuous?

$$g(z) = \frac{3z - 4}{(z - 1)^2 (z + i)(z - 2i)}.$$

11.3 Differentiability

Just as for continuity, the definition of differentiability of a complex function can be copied straight from the real case.

DEFINITION 11.7
Differentiability in $\mathbb{C}$

Let f be a complex function defined in a neighbourhood of $z_0 \in \mathbb{C}$. Then f is called differentiable *at z_0 if*

$$\lim_{z \to z_0} \frac{f(z) - f(z_0)}{z - z_0}$$

exists as a finite number. In this case the limit is denoted by $f'(z_0)$ or by $(df/dz)(z_0)$.

When a complex function is differentiable for every $z \in \mathbb{C}$, then f is not called 'differentiable on $\mathbb{C}$', but *analytic* on $\mathbb{C}$. The following definition is somewhat more general.

DEFINITION 11.8
Analytic function

Derivative

Let f be a complex function, defined on a subset G of $\mathbb{C}$. Then f is called analytic *on G if f is differentiable at every point z_0 of G (here the subset G should again be a set as in definition 11.6). The function f' (now defined on G) is called the* derivative *of f.*

Although these definitions closely resemble the definitions of differentiability and derivative for real functions, there still is a major difference. Existence of the limit in definition 11.7 is much more demanding than in the real case; this is because the limit now has to exist no matter how z approaches z_0. In the real case there are only *two* possible directions, namely from the right or from the left. In the complex case there is much more freedom, since only the distance $|z - z_0|$ from z to z_0 has to decrease and nothing else is assumed about directions (compare this with the remarks following definition 11.6). Yet, here we will again see that for the calculation of derivatives of complex functions one has precisely the same rules as for real functions (see theorem 11.5). As soon as we have calculated a number of standard derivatives, these rules enable us to determine the derivative of more complicated functions. In the following examples we use definition 11.7 to determine our first two standard derivatives.

EXAMPLE 11.5

Consider the constant function $f(z) = c$, where $c \in \mathbb{C}$. Let $z_0 \in \mathbb{C}$. Since $f(z) - f(z_0) = c - c = 0$ for each $z \in \mathbb{C}$, it follows that $f'(z_0) = 0$. As for real constants we thus have that the derivative of a constant equals 0. Put differently, the function $f(z) = c$ is analytic on $\mathbb{C}$ and $f'(z) = 0$. ◀

EXAMPLE 11.6

The function $f(z) = z$ is analytic on $\mathbb{C}$ and has as its derivative the function 1. In fact, for $z_0 \in \mathbb{C}$ one has $f(z) - f(z_0) = z - z_0$ and so $(f(z) - f(z_0))/(z - z_0) = 1$ for each $z \in \mathbb{C}$. This shows that $f'(z_0) = 1$ for each $z_0 \in \mathbb{C}$. ◀

Quite a few of the well-known results for the differentiation of real functions remain valid for complex functions. The proofs of the following two theorems are exactly the same as for the real case and are therefore omitted.

THEOREM 11.4

Let f be a complex function and assume that $f'(z_0)$ exists at the point z_0. Then f is continuous at z_0.

THEOREM 11.5

Linearity

Product rule

Let f and g be analytic on a subset G of $\mathbb{C}$ (G as in definition 11.6). Then the following properties hold:
a *$af + bg$ is analytic on G for arbitrary $a, b \in \mathbb{C}$ and $(af + bg)'(z) = af'(z) + bg'(z)$;*
b *$f \cdot g$ is analytic on G and the* product rule *holds: $(f \cdot g)'(z) = f'(z)g(z) + f(z)g'(z)$;*

c   *if $g(z) \neq 0$ for all $z \in G$, then f/g is analytic on G and the* quotient rule *holds:*

$$\left(\frac{f}{g}\right)'(z) = \frac{f'(z)g(z) - f(z)g'(z)}{(g(z))^2};$$

d   *if h is analytic on the range $f(G)$ of G under f, then the function $(h \circ f)(z) = h(f(z))$ is analytic on G and the* chain rule *holds: $(h \circ f)'(z) = h'(f(z))f'(z)$.*

Using theorem 11.5 and examples 11.5 and 11.6 one can determine the derivative of any polynomial and any rational function. As a first step we calculate the derivative of z^2 using theorem 11.5b and example 11.6: $(z^2)' = (z \cdot z)' = zz' + z'z = z + z = 2z$. By a repeated application of this rule it then follows that $(z^n)' = nz^{n-1}$, just as in the real case. If we now use theorem 11.5a, it follows that any polynomial $a_n z^n + a_{n-1}z^{n-1} + \cdots + a_1 z + a_0$ is analytic on $\mathbb{C}$ and has derivative $na_n z^{n-1} + (n-1)a_{n-1}z^{n-2} + \cdots + a_1$. Subsequently one can apply the quotient rule from theorem 11.5c to conclude that a rational function is analytic on $\mathbb{C}$ if we remove the points where the denominator is equal to zero.

The real exponential function e^x has as its derivative the function e^x again. We will now show that the complex function e^z is analytic on $\mathbb{C}$ and has as its derivative the function e^z again. This gives us an important new standard derivative and although this result will probably come as no surprise, its proof will require quite an effort.

As a preparation we will first consider certain limits of functions of two real variables x and y. Let us consider, for example, the limit

$$\lim_{(x,y)\to(0,0)} \frac{y^3}{x^2 + y^2}.$$

Here $(x, y) \to (0, 0)$ means that the distance from $(x, y) \in \mathbb{R}^2$ to the point $(0, 0)$ keeps decreasing, that is to say, $x^2 + y^2 \to 0$ (if we write $z = x + iy$, then this means precisely that $|z| \to 0$). Now introduce polar coordinates, so $x = r \cos\phi$ and $y = r \sin\phi$ (this means precisely that $z = r \cos\phi + ir \sin\phi = re^{i\phi}$, the well-known expression (2.6)). We then have $x^2 + y^2 = r^2$ and hence

$$\lim_{(x,y)\to(0,0)} \frac{y^3}{x^2 + y^2} = \lim_{r\to 0} \frac{r^3 \sin^3\phi}{r^2} = \lim_{r\to 0} r \sin^3\phi.$$

Since $\left|\sin^3\phi\right| \leq 1$, it follows that

$$\lim_{(x,y)\to(0,0)} \frac{y^3}{x^2 + y^2} = 0.$$

The same method can be applied to the quotient $x^3/(x^2 + y^2)$ or $xy^2/(x^2 + y^2)$. In general one has for $k, l \in \mathbb{N}$ that

$$\lim_{(x,y)\to(0,0)} \frac{x^k y^l}{x^2 + y^2} = 0 \qquad \text{if } k + l \geq 3. \tag{11.2}$$

In fact, by changing to polar coordinates it follows that

$$\lim_{(x,y)\to(0,0)} \frac{x^k y^l}{x^2 + y^2} = \lim_{r\to 0} \frac{r^{k+l} \cos^k\phi \sin^l\phi}{r^2} = \lim_{r\to 0} r^{k+l-2} \cos^k\phi \sin^l\phi = 0,$$

since $k + l - 2 \geq 1$ and $\left|\cos^k\phi \sin^l\phi\right| \leq 1$. We use (11.2) in the proof of the following theorem.

THEOREM 11.6

The function e^z is analytic on $\mathbb{C}$ and

$$(e^z)' = e^z.$$ (11.3)

Proof
Take $z \in \mathbb{C}$ arbitrary, then

$$\lim_{w \to z} \frac{e^w - e^z}{w - z} = \lim_{w \to z} \frac{e^z(e^{w-z} - 1)}{w - z} = e^z \lim_{w-z \to 0} \frac{e^{w-z} - 1}{w - z},$$

where we used the important property in theorem 11.1a. If we now show that $\lim_{u \to 0}(e^u - 1)/u = 1$ for arbitrary $u \in \mathbb{C}$, then it follows that e^z is differentiable at every point $z \in \mathbb{C}$ and that the derivative at that point equals e^z. The function e^z is then analytic on $\mathbb{C}$ and $(e^z)' = e^z$.

In order to show that indeed $\lim_{u \to 0}(e^u - 1)/u = 1$, we note that for $u = x + iy$

$$\frac{e^u - 1}{u} = \frac{e^x(\cos y + i \sin y) - 1}{x + iy} = \frac{\big((e^x \cos y - 1) + ie^x \sin y\big)(x - iy)}{(x + iy)(x - iy)}.$$

Now expand numerator and denominator and note that $u \to 0$ is equivalent to $(x, y) \to (0, 0)$. It then follows that $\lim_{u \to 0}(e^u - 1)/u = 1$ equals

$$\lim_{(x,y) \to (0,0)} \frac{(xe^x \cos y - x + e^x y \sin y) + i(xe^x \sin y - e^x y \cos y + y)}{x^2 + y^2}$$

$$= \lim_{(x,y) \to (0,0)} \frac{x(e^x \cos y - 1) + e^x y \sin y}{x^2 + y^2}$$

$$+ i \lim_{(x,y) \to (0,0)} \frac{xe^x \sin y + y(1 - e^x \cos y)}{x^2 + y^2}.$$

These two limits are now calculated separately by applying (11.2). To this end we develop the functions e^x, $\sin y$ and $\cos y$ in Taylor series around the point $x = 0$ and $y = 0$ respectively (see (2.23) – (2.25)):

$$e^x = 1 + x + x^2/2 + \cdots,$$
$$\sin y = y - y^3/6 + \cdots,$$
$$\cos y = 1 - y^2/2 + \cdots.$$

We now concentrate ourselves on the first limit. From the Taylor series it follows that

$$e^x \cos y - 1 = -1 + (1 + x + x^2/2 + \cdots)(1 - y^2/2 + \cdots)$$
$$= -1 + 1 - y^2/2 + x + x^2/2 + \cdots,$$

from which we see that

$$e^x \cos y - 1 = x + \text{terms of the form } x^k y^l \text{ with } k + l \geq 2.$$

In the same way we obtain

$$e^x \sin y = (1 + x + \cdots)(y - \cdots)$$
$$= y + \text{terms of the form } x^k y^l \text{ with } k + l \geq 2.$$

For the first limit it then follows that

$$\lim_{(x,y)\to(0,0)} \frac{x(e^x \cos y - 1) + e^x y \sin y}{x^2 + y^2}$$

$$= \lim_{(x,y)\to(0,0)} \frac{x^2 + y^2 + \text{terms of the form } x^k y^l \text{ with } k + l \geq 3}{x^2 + y^2}$$

$$= \lim_{(x,y)\to(0,0)} 1 + \frac{\text{terms of the form } x^k y^l \text{ with } k + l \geq 3}{x^2 + y^2} = 1,$$

where we used (11.2) in the final step. The second limit is treated in the same way

$$1 - e^x \cos y = -x + \text{terms of the form } x^k y^l \text{ with } k + l \geq 2,$$

and, with the expression for $e^x \sin y$ as before, we thus find that

$$xe^x \sin y + y(1 - e^x \cos y)$$
$$= xy - yx + \text{terms of the form } x^k y^l \text{ with } k + l \geq 3.$$

Hence, the numerator in the second limit only contains terms of the form $x^k y^l$ with $k + l \geq 3$, since $xy - yx = 0$. From (11.2) it then follows that the second limit equals 0. This proves that $\lim_{u\to0}(e^u - 1)/u = 1$. $\blacksquare$

From theorem 11.6 (and theorem 11.5) it also follows that the functions $\sin z$ and $\cos z$ are analytic on $\mathbb{C}$. This is because $\sin z = (e^{iz} - e^{-iz})/2i$ and $\cos z = (e^{iz} + e^{-iz})/2$ and from the chain rule it then follows that $(\sin z)' = (ie^{iz} + ie^{-iz})/2i = \cos z$ and similarly $(\cos z)' = -\sin z$.

We close this section with an example of a complex function which is *not* differentiable.

EXAMPLE 11.7

Consider the function $f(z) = \overline{z}$ from example 11.1. We will show that f is not differentiable at $z = 0$, in other words, that the limit

$$\lim_{z\to0} \frac{f(z) - f(0)}{z - 0} = \lim_{z\to0} \frac{\overline{z}}{z}$$

from definition 11.7 does not exist. First we take $z = x$ with $x \in \mathbb{R}$; then $\overline{z} = x$ and so $\lim_{z\to0} \overline{z}/z = \lim_{x\to0} x/x = 1$. Next we take $z = iy$ with $y \in \mathbb{R}$; then $\overline{z} = -iy$ and so $\lim_{z\to0} \overline{z}/z = \lim_{y\to0}(-iy)/iy = -1$. This shows that the limit does not exist. $\blacktriangleleft$

EXERCISES

11.9   Use definition 11.7 to prove that the function $f(z) = z^2$ is analytic on $\mathbb{C}$ and determine its derivative.

11.10   Let f and g be analytic on the subset G of $\mathbb{C}$. Show that $f + g$ is analytic on G and that $(f + g)'(z) = f'(z) + g'(z)$.

11.11   Determine on which subset of $\mathbb{C}$ the following functions are analytic and give their derivative:
     a $f(z) = (z - 1)^4$,
     b $f(z) = z + 1/z$,
     c $f(z) = (z^2 - 3z + 2)/(z^3 + 1)$,
     d $f(z) = e^{z^2} + 1$.

11.12   Show that $(\cos z)' = -\sin z$.

11.13   We have seen that many of the rules for complex functions are precisely the same
as for real functions. That not all rules remain the same is shown by the following
Bernoulli paradox: $1 = \sqrt{1} = \sqrt{(-1)(-1)} = \sqrt{(-1)}\sqrt{(-1)} = i \cdot i = i^2 = -1$.
Which step in this argument is apparently not allowed?

11.4   The Cauchy–Riemann equations*

The material in this section will not be used in the remainder of the book and can
thus be omitted without any consequences.

The proof of theorem 11.6 clearly shows that in some cases it may not be easy
to show whether or not a function is differentiable and, in case of differentiability,
to determine the derivative. The reason for this is the fact, mentioned earlier, that
a limit in $\mathbb{C}$ is of a quite different nature from a limit in $\mathbb{R}$. In order to illustrate
this once more, we close with a theorem whose proof cleverly uses the fact that a
limit in $\mathbb{C}$ should not depend on the way in which z approaches z_0 in the expression
$\lim_{z \to z_0}$. The theorem also provides us with a quick and easy way to show that a
function is not analytic.

THEOREM 11.7   *Let $f(z) = u(x, y) + iv(x, y)$ be a complex function. Assume that $f'(z_0)$ exists at
a point $z_0 = x_0 + iy_0$. Then $\partial u/\partial x$, $\partial u/\partial y$, $\partial v/\partial x$ and $\partial v/\partial y$ exist at the point
(x_0, y_0) and at (x_0, y_0) one has*

$$\frac{\partial u}{\partial x} = \frac{\partial v}{\partial y} \qquad and \qquad \frac{\partial u}{\partial y} = -\frac{\partial v}{\partial x}. \tag{11.4}$$

Proof
In the proof we first let $z = x + iy$ tend to $z_0 = x_0 + iy_0$ in the real direction
and subsequently in the imaginary direction. See figure 11.4. It is given that the

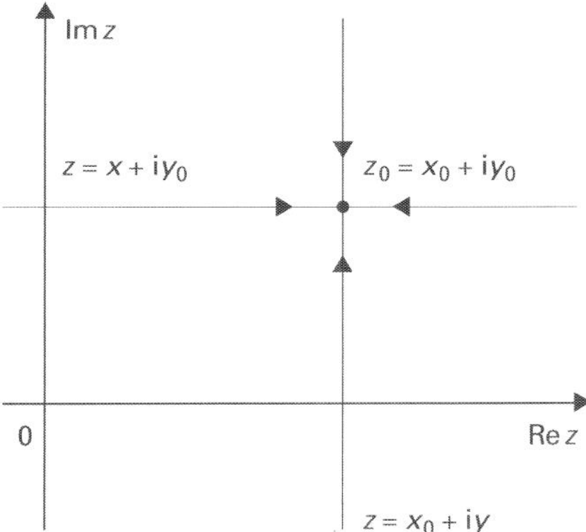

FIGURE 11.4
The limit in the real and in the imaginary direction.

following limit exists:

$$f'(z_0) = \lim_{z \to z_0} \frac{f(z) - f(z_0)}{z - z_0}.$$

We first study the limit in the real direction. We thus take $z = x + iy_0$, which means
that $z \to z_0$ is equivalent to $x \to x_0$ since $z - z_0 = x + iy_0 - x_0 - iy_0 = x - x_0$.

Moreover, we then have

$$
\begin{aligned}
f'(z_0) &= \lim_{z \to z_0} \frac{f(z) - f(z_0)}{z - z_0} \\[2mm]
&= \lim_{x \to x_0} \frac{u(x, y_0) + i v(x, y_0) - u(x_0, y_0) - i v(x_0, y_0)}{x - x_0} \\[2mm]
&= \lim_{x \to x_0} \frac{u(x, y_0) - u(x_0, y_0)}{x - x_0} + i \lim_{x \to x_0} \frac{v(x, y_0) - v(x_0, y_0)}{x - x_0}.
\end{aligned}
$$

Since $f'(z_0)$ exists, both limits in this final expression have to exist as well. But these limits are precisely the partial derivatives $\partial u / \partial x$ and $\partial v / \partial x$ at the point (x_0, y_0), which we denote by $(\partial u / \partial x)(x_0, y_0)$ and $(\partial v / \partial x)(x_0, y_0)$. Hence,

$$
f'(z_0) = \frac{\partial u}{\partial x}(x_0, y_0) + i \frac{\partial v}{\partial x}(x_0, y_0).
$$

Next we study the limit in the imaginary direction. We thus take $z = x_0 + iy$, which means that $z \to z_0$ is equivalent to $y \to y_0$ since $z - z_0 = x_0 + iy - x_0 - iy_0 = i(y - y_0)$. As before, it then follows that

$$
\begin{aligned}
f'(z_0) &= \lim_{y \to y_0} \frac{u(x_0, y) + i v(x_0, y) - u(x_0, y_0) - i v(x_0, y_0)}{i(y - y_0)} \\[2mm]
&= \frac{1}{i} \lim_{y \to y_0} \frac{u(x_0, y) - u(x_0, y_0)}{y - y_0} + \lim_{y \to y_0} \frac{v(x_0, y) - v(x_0, y_0)}{y - y_0}.
\end{aligned}
$$

Hence,

$$
f'(z_0) = \frac{\partial v}{\partial y}(x_0, y_0) - i \frac{\partial u}{\partial y}(x_0, y_0).
$$

From these two expressions for $f'(z_0)$ it follows that at the point (x_0, y_0) we have

$$
\frac{\partial u}{\partial x} + i \frac{\partial v}{\partial x} = \frac{\partial v}{\partial y} - i \frac{\partial u}{\partial y}.
$$

If we compare the real and imaginary parts in this identity, then we see that indeed $\partial u / \partial x = \partial v / \partial y$ and $\partial u / \partial y = -\partial v / \partial x$ at the point (x_0, y_0).   ■

Cauchy–Riemann equations   The equations $\partial u / \partial x = \partial v / \partial y$ and $\partial u / \partial y = -\partial v / \partial x$ from (11.4) are called the *Cauchy–Riemann equations*. They show that for an analytic function the real part $u(x, y)$ and the imaginary part $v(x, y)$ have a very special relationship. One can thus use theorem 11.7 to show in a very simple way that a function is *not* analytic. Indeed, when the real and imaginary part of a complex function f do not satisfy the Cauchy–Riemann equations from theorem 11.7, then f is not analytic.

EXAMPLE   Consider the function $f(z) = \bar{z}$. The real part is $u(x, y) = x$, while the imaginary part is given by $v(x, y) = -y$. Hence $\partial u / \partial x = 1$ and $\partial v / \partial y = -1$. Since the Cauchy–Riemann equations are not satisfied, it follows that f is not analytic on $\mathbb{C}$ (see also example 11.7).   ◀

Even more important than theorem 11.7 is its converse, which we will state without proof: when $\partial u / \partial x$, $\partial u / \partial y$, $\partial v / \partial x$ and $\partial v / \partial y$ exist and are continuous on a subset G of $\mathbb{C}$ and when they satisfy the Cauchy–Riemann equations (11.4) on G, then $f(z) = u(x, y) + i v(x, y)$ $(z = x + iy)$ is analytic on G. This gives us a very simple method to show that a function is analytic!

EXAMPLE   For the function $f(z) = e^z$ one has $u(x, y) = e^x \cos y$ and $v(x, y) = e^x \sin y$. It then follows that $\partial u / \partial x = e^x \cos y$, $\partial u / \partial y = -e^x \sin y$, $\partial v / \partial x = e^x \sin y$ and $\partial v / \partial y = e^x \cos y$. The partial derivatives thus exist and they are continuous on $\mathbb{R}^2$.

Moreover, they satisfy the Cauchy–Riemann equations. Hence, the function e^z is analytic on $\mathbb{C}$ and $f'(z) = \partial u/\partial x + i\,\partial v/\partial x = e^x \cos y + i e^x \sin y = e^z$. Compare the ease of these arguments with the proof of theorem 11.6. ◀

EXERCISES

11.14* We know that the function $f(z) = z^2$ is analytic on $\mathbb{C}$ (see exercise 11.9 or the text following theorem 11.5). Verify the Cauchy–Riemann equations for f (see example 11.4 for the real and imaginary part of f).

11.15* Use the Cauchy–Riemann equations (and the results of example 11.3) to show that $f(z) = 1/z$ is analytic on $\mathbb{C} - \{0\}$.

SUMMARY

Complex functions are functions from (a subset of) $\mathbb{C}$ to $\mathbb{C}$. By identifying $\mathbb{C}$ with $\mathbb{R}^2$, one can split a complex function into a real part and an imaginary part; these are then functions from $\mathbb{R}^2$ to $\mathbb{R}$. A very important complex function is $e^z = e^{x+iy} = e^x(\cos y + i \sin y)$. It has the characteristic property $e^{z+w} = e^z e^w$ for all $w, z \in \mathbb{C}$. Using e^z one can extend the sine and cosine functions from real to complex functions.

Continuity and differentiability of a complex function f can be defined by means of limits, just as for the real case: f is called continuous at $z_0 \in \mathbb{C}$ when $\lim_{z \to z_0} f(z) = f(z_0)$; f is called differentiable when

$$\lim_{z \to z_0} \frac{f(z) - f(z_0)}{z - z_0}$$

exists as a finite number. A complex function f is called analytic on a subset G of $\mathbb{C}$ when f is differentiable at every point of G. The well-known rules from real analysis remain valid for the complex functions treated here. The function e^z, for example, is analytic on $\mathbb{C}$ and has as derivative the function e^z again.

SELFTEST

11.16 Consider the complex function $f(z) = \cos z$.
  a Is it true that $|\cos z| \le 1$? If so, give a proof. If not, give a counter-example.
  b Show that $\cos(w + z) = \cos w \cos z - \sin w \sin z$.
  c Determine the real and imaginary part of $\cos z$.
  d Give the largest subset of $\mathbb{C}$ on which f is analytic. Justify your answer.

11.17 Determine on which subset G of $\mathbb{C}$ the following functions are analytic and give their derivative:
  a $(z^3 + 1)/(z - 1)$,
  b $1/(z^4 + 16)^{10}$,
  c $e^z/(z^2 + 3)$,
  d $\sin(e^z)$.

Contents of Chapter 12

The Laplace transform: definition and properties

The Laplace transform: definition and properties

INTRODUCTION

Signals occurring in practice will always have been switched on at a certain moment in time. Choosing this switch-on moment equal to $t = 0$, we are then dealing with functions that are zero for $t < 0$. If, moreover, such a function is multiplied by a damping factor e^{-at} ($a > 0$), then it is not unreasonable to expect that the Fourier transform of $e^{-at} f(t)$ will exist. As we have seen in the introduction to part 4, this leads to a new transformation, the so-called Laplace transform. In section 12.1 the Laplace transform $F(s)$ of a causal function $f(t)$ will be defined by

$$F(s) = \int_0^\infty f(t)e^{-st}\, dt.$$

Here $s \in \mathbb{C}$ is 'arbitrary' and $F(s)$ thus becomes a complex function. One of the major advantages of the Laplace transform is the fact that the integral is convergent for 'a lot of' functions (which is in contrast to the Fourier transform). For example, the Laplace transform of the unit step function exists, while this is not the case for the Fourier transform.

In section 12.1 we consider in detail the conditions under which the Laplace transform of a function exists. This is illustrated by a number of standard examples of Laplace transforms. Because of the close connection with the Fourier transform, it will hardly be a surprise that for the Laplace transform similar properties hold. A number of elementary properties are treated in section 12.2: linearity, rules for a shift in the time domain as well as in the s-domain, and the rule for scaling.

In section 12.3 the differentiation and integration rules are treated. These are harder to prove, but of great importance in applications. In particular, the rule for differentiation in the time domain proves essential for the application to differential equations in chapter 14. The differentiation rule in the s-domain will in particular show that the Laplace transform $F(s)$ of a causal function $f(t)$ is an analytic function on a certain subset of $\mathbb{C}$.

LEARNING OBJECTIVES
After studying this chapter it is expected that you
- know and can apply the definition of the Laplace transform
- know the concepts of abscissa of convergence and of absolute convergence
- know the concept of exponential order
- know and can apply the standard examples of Laplace transforms
- know and can apply the properties of linearity, shifting and scaling
- know and can apply the rules for differentiation and integration.

12.1   Definition and existence of the Laplace transform

The following definition has been justified in the introduction (and in the introduction to part 4). For the notion 'causal function' or 'causal signal' we refer to section 1.2.4.

DEFINITION 12.1
Laplace transform

Let $f(t)$ be a causal function, so $f(t) = 0$ for $t < 0$. The Laplace transform $F(s)$ of $f(t)$ is the complex function defined for $s \in \mathbb{C}$ by

$$F(s) = \int_0^\infty f(t)e^{-st}\,dt, \tag{12.1}$$

provided the integral exists.

We will see in a moment that for many functions $f(t)$ the Laplace transform $F(s)$ exists (on a certain subset of $\mathbb{C}$). The mapping assigning the Laplace transform $F(s)$ to a function $f(t)$ in the time domain will also be called the *Laplace transform*.

Laplace transform
s-domain

Furthermore, we will say that $F(s)$ is defined in the *s-domain*; one sometimes calls this s-domain the 'complex frequency domain' (although a physical interpretation can hardly be given for arbitrary $s \in \mathbb{C}$). Besides the notation $F(s)$ we will also use $(\mathcal{L}f)(s)$, so $(\mathcal{L}f)(s) = F(s)$. Often the notation $(\mathcal{L}f(t))(s)$, although not very elegant, will be useful in the case of a concrete function.

In this part of the book we will always tacitly assume that the functions are *causal*. The function t, for example, will in this part always stand for $\epsilon(t)t$; it is equal to 0 for $t < 0$ and equal to t for $t \geq 0$. In particular, the constant function 1 is equal to $\epsilon(t)$ in this part. Moreover, for all functions it will always be assumed that they are *piecewise smooth* (see definition 2.4). In particular it then follows that $\int_0^R f(t)e^{-st}\,dt$ will certainly exist for any $R > 0$. The existence of the integral in (12.1) then boils down to the fact that the improper Riemann integral over $\mathbb{R}$, that is, $\lim_{R\to\infty} \int_0^R f(t)e^{-st}\,dt$, has to exist.

Note also that for $s = \sigma + i\omega$ with $\sigma, \omega \in \mathbb{R}$ it immediately follows from the definition of the complex exponential that

$$F(s) = \int_0^\infty f(t)e^{-\sigma t}e^{-i\omega t}\,dt. \tag{12.2}$$

This is an interesting formula, for it shows that the Laplace transform $F(s)$ of $f(t)$ at the point $s = \sigma + i\omega$ is equal to the Fourier transform of $\epsilon(t)f(t)e^{-\sigma t}$ at the point ω, provided all the integrals exist. This is the case, for example, if $\epsilon(t)f(t)e^{-\sigma t}$ is absolutely integrable (see definition 6.2). For the moment, we leave this connection between the Laplace and Fourier transform for what it is, and return to the question of the existence of the integral in (12.1), in other words, the existence of the Laplace transform. As far as the general theory is concerned, we will confine ourselves to absolute convergence.

DEFINITION 12.2
Absolutely convergent

An integral $\int_a^b g(t)\,dt$ with $-\infty \leq a < b \leq \infty$ is called absolutely convergent *if*

$$\int_a^b |g(t)|\,dt < \infty.$$

This concept is entirely analogous to the concept of absolute convergence for series (see section 2.4.2). Since $\left| \int g(t)\,dt \right| \leq \int |g(t)|\,dt$ (as in (2.22)), we see that an absolutely convergent integral is also convergent in the ordinary sense. The converse, however, need not to be true, just as for series. Note that for the absolute convergence of the integral in (12.1) only the value of $\sigma = \mathrm{Re}\,s$ is of importance, since $|e^{i\omega t}| = 1$ for all $\omega \in \mathbb{R}$. In the general theory we will confine ourselves to absolute convergence, since proofs are easier than for the case of ordinary convergence. For concrete cases it is usually quite easy to treat ordinary convergence as well. To get some feeling for the absolute and ordinary convergence of the integral in (12.1), we will first treat two examples.

EXAMPLE 12.1
Laplace transform of 1

Let $\epsilon(t)$ be the unit step function. The Laplace transform of $\epsilon(t)$ (or the function 1) is given by the integral

$$\int_0^\infty e^{-st}\,dt = \lim_{R\to\infty}\int_0^R e^{-\sigma t}e^{-i\omega t}\,dt$$

if $s = \sigma + i\omega$. We first consider absolute convergence. The absolute value of the integrand is equal to $e^{-\sigma t}$ and for $\sigma \neq 0$ one has

$$\int_0^R e^{-\sigma t}\,dt = \left[-\frac{1}{\sigma}e^{-\sigma t}\right]_0^R = \frac{1}{\sigma}(1 - e^{-\sigma R}).$$

The integral thus converges absolutely if $\lim_{R\to\infty} e^{-\sigma R}$ exists and of course this is the case only for $\sigma > 0$. Hence, the Laplace transform of $\epsilon(t)$ certainly exists for $\sigma > 0$, so for $\mathrm{Re}\,s > 0$. We will now show that the integral also converges in the ordinary sense precisely for all $s \in \mathbb{C}$ with $\mathrm{Re}\,s > 0$. For $s = 0$ we have that $\lim_{R\to\infty}\int_0^R 1\,dt$ does not exist. Since $(e^{-st})' = -se^{-st}$ (here we differentiate with respect to the real variable t; see example 2.11), it follows for $s \neq 0$ that

$$\int_0^R e^{-st}\,dt = \left[-\frac{1}{s}e^{-st}\right]_0^R = \frac{1}{s}(1 - e^{-sR}).$$

Hence, the Laplace transform of $\epsilon(t)$ will only exist when $\lim_{R\to\infty} e^{-sR} = \lim_{R\to\infty} e^{-\sigma R}e^{-i\omega R}$ exists. Since $|e^{-i\omega R}| = 1$, the limit will exist precisely for $\sigma > 0$ and in this case the limit will of course be 0 since $\lim_{R\to\infty} e^{-\sigma R} = 0$ for $\sigma > 0$. We conclude that the integral converges in the ordinary sense precisely for all $s \in \mathbb{C}$ with $\mathrm{Re}\,s > 0$ and that the Laplace transform of 1 (or $\epsilon(t)$) is given by $1/s$ for these values of s. Note that in this example the regions of absolute and ordinary convergence are the same. ◀

EXAMPLE 12.2
Laplace transform of e^{at}

Let $a \in \mathbb{R}$. The Laplace transform of the function e^{at} (hence of $\epsilon(t)e^{at}$) is given by

$$\int_0^\infty e^{at}e^{-st}\,dt = \lim_{R\to\infty}\int_0^R e^{-(\sigma-a)t}e^{-i\omega t}\,dt$$

if $s = \sigma + i\omega$. Again we first look at absolute convergence. The absolute value of the integrand is $e^{-(\sigma-a)t}$ and for $\sigma \neq a$ one has

$$\int_0^R e^{-(\sigma-a)t}\,dt = \frac{1}{\sigma-a}\left(1 - e^{-(\sigma-a)R}\right).$$

Hence, the integral converges absolutely when $\lim_{R\to\infty} e^{-(\sigma-a)R}$ exists and this is precisely the case when $\sigma - a > 0$, or $\mathrm{Re}\,s > a$. We will now determine the Laplace transform of e^{at} and moreover show that the integral also converges in the ordinary sense for precisely all $s \in \mathbb{C}$ with $\mathrm{Re}\,s > a$. For $s = a$ the Laplace transform will certainly not exist. Since $(e^{-(s-a)t})' = -(s-a)e^{-(s-a)t}$, it follows for $s \neq a$ that

$$\int_0^\infty e^{at}e^{-st}\,dt = \lim_{R\to\infty}\left[-\frac{1}{s-a}e^{-(s-a)t}\right]_0^R = \frac{1}{s-a} - \frac{1}{s-a}\lim_{R\to\infty} e^{-(s-a)R}.$$

As in example 12.1, we have $\lim_{R\to\infty} e^{-(s-a)R} = \lim_{R\to\infty} e^{-(\sigma-a)R}e^{-i\omega R} = 0$ precisely when $\sigma - a > 0$. Hence, there is ordinary convergence for $\mathrm{Re}\,s > a$ and the Laplace transform of e^{-at} is given by $1/(s-a)$ for these values of s. For $a = 0$ we recover the results of example 12.1 again. ◄

It is not hard to prove the following general result on the absolute convergence of the integral in definition 12.1.

THEOREM 12.1

Let $f(t)$ be a causal function and consider the integral in (12.1).
a *If the integral is absolutely convergent for a certain value $s = \sigma_0 \in \mathbb{R}$, then the integral is absolutely convergent for all $s \in \mathbb{C}$ with $\mathrm{Re}\, s \geq \sigma_0$.*
b *If the integral is not absolutely convergent for a certain value $s = \sigma_1 \in \mathbb{R}$, then the integral is not absolutely convergent for all $s \in \mathbb{C}$ with $\mathrm{Re}\, s \leq \sigma_1$.*

Proof
We first prove part a. Write $s = \sigma + i\omega$. Since $\left| e^{i\omega t} \right| = 1$ and $e^{-\sigma t} > 0$, it follows from 12.2 that

$$\int_0^\infty |f(t)e^{-st}|\, dt = \int_0^\infty |f(t)|\, e^{-\sigma t}\, dt.$$

For $\mathrm{Re}\, s = \sigma \geq \sigma_0$ one has that $e^{-\sigma t} \leq e^{-\sigma_0 t}$ for all $t \geq 0$. Hence,

$$\int_0^\infty |f(t)e^{-st}|\, dt \leq \int_0^\infty |f(t)|\, e^{-\sigma_0 t}\, dt.$$

According to the statement in part a, the integral in the right-hand side of this inequality exists (as a finite number). The integral in (12.1) is thus indeed absolutely convergent for all $s \in \mathbb{C}$ with $\mathrm{Re}\, s \geq \sigma_0$. This proves part a.

Part b immediately follows from part a. Let us assume that there exists an $s_0 \in \mathbb{C}$ with $\mathrm{Re}\, s \leq \sigma_1$ such that the integral is absolutely convergent after all. According to part a, the integral will then be absolutely convergent for all $s \in \mathbb{C}$ with $\mathrm{Re}\, s \geq \mathrm{Re}\, s_0$. But $\mathrm{Re}\, \sigma_1 = \sigma_1 \geq \mathrm{Re}\, s_0$ and hence the integral should in particular be absolutely convergent for $s = \sigma_1$. This contradicts the statement in part b. ∎

From theorem 12.1 we see that for the absolute convergence only the value $\mathrm{Re}\, s = \sigma$ matters. Note that the set $\{\, s \in \mathbb{C} \mid \mathrm{Re}\, s = \sigma \,\}$ is a straight line perpendicular to the real axis. See figure 12.1.

Using theorem 12.1 one can, moreover, show that there are precisely three possibilities regarding the absolute convergence of the integral in (12.1):

a   the integral is absolutely convergent for all $s \in \mathbb{C}$;
b   the integral is absolutely convergent for no $s \in \mathbb{C}$ whatsoever;

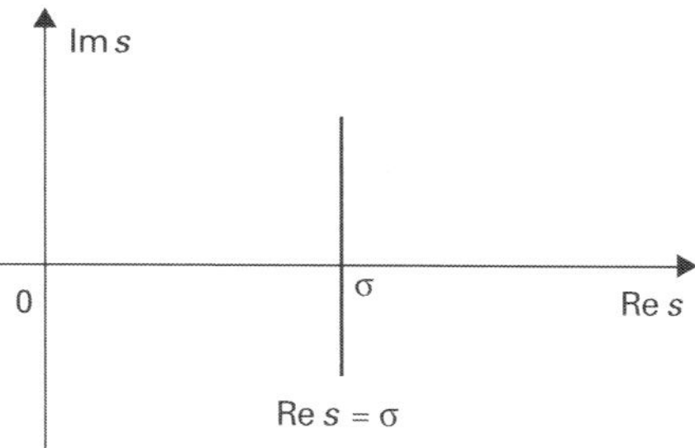

FIGURE 12.1
The straight line $\mathrm{Re}\, s = \sigma$ (for a $\sigma > 0$).

c there exists a number $\sigma_a \in \mathbb{R}$ such that the integral is absolutely convergent for $\operatorname{Re} s > \sigma_a$ and not absolutely convergent for $\operatorname{Re} s < \sigma_a$.

In case c there is no statement about the absolute convergence for $\operatorname{Re} s = \sigma_a$. It is possible that there is absolute convergence on the line $\operatorname{Re} s = \sigma_a$, but it is equally possible that there is no absolute convergence. In example 12.1 we have seen that $\sigma_a = 0$; in this case there is no absolute convergence for $\operatorname{Re} s = \sigma_a = 0$.

Strictly speaking, we have not given a proof that these three possibilities are the only ones. Intuitively, however, this seems quite obvious. Suppose that possibilities a and b do not occur. We then have to show that only possibility c can occur. Now if a does not hold, then there exists a $\sigma_1 \in \mathbb{R}$ such that the integral is not absolutely convergent (we may assume that σ_1 is real, since only the real part is relevant for the absolute convergence). According to theorem 12.1b, the integral is then also not absolutely convergent for all $s \in \mathbb{C}$ with $\operatorname{Re} s \leq \sigma_1$. Since also b does not hold, there similarly exists a $\sigma_2 \in \mathbb{R}$ such that the integral is absolutely convergent for all $s \in \mathbb{C}$ with $\operatorname{Re} s \geq \sigma_2$ (we now use theorem 12.1a). In the region $\sigma_1 < \operatorname{Re} s < \sigma_2$, where nothing yet is known about the absolute convergence, we now choose an arbitrary $\sigma_3 \in \mathbb{R}$. See figure 12.2. If we have absolute convergence for σ_3, then we

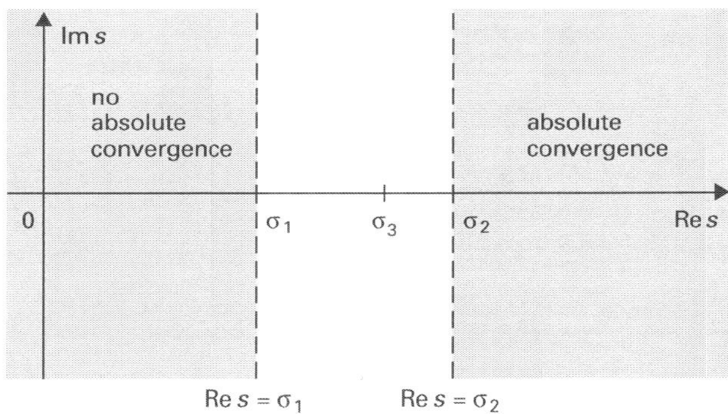

FIGURE 12.2
Regions of absolute convergence.

can extend the region of absolute convergence to $\operatorname{Re} s \geq \sigma_3$. If there is no absolute convergence, then we can extend the region where there is no absolute convergence to $\operatorname{Re} s \leq \sigma_3$. This process can be continued indefinitely and our intuition tells us that at some point the two regions will have to meet, in other words, that possibility c will occur. That this will indeed happen rests upon a fundamental property of the real numbers which we will not go into any further. The above is summarized in the following theorem.

THEOREM 12.2

For a given causal function $f(t)$ there exists a number $\sigma_a \in \mathbb{R}$ with $-\infty \leq \sigma_a \leq \infty$ such that the integral $\int_0^\infty f(t)e^{-st}\,dt$ is absolutely convergent for all $s \in \mathbb{C}$ with $\operatorname{Re} s > \sigma_a$, and not absolutely convergent for all $s \in \mathbb{C}$ with $\operatorname{Re} s < \sigma_a$. By $\sigma_a = -\infty$ we will mean that the integral is absolutely convergent for all $s \in \mathbb{C}$. By $\sigma_a = \infty$ we will mean that the integral is absolutely convergent for no $s \in \mathbb{C}$ whatsoever.

Abscissa of absolute convergence

The number σ_a in theorem 12.2 is called the *abscissa of absolute convergence*. The region of absolute convergence is a half-plane $\operatorname{Re} s > \sigma_a$ in the complex plane. See figure 12.3. The case $\sigma_a = \infty$ (possibility b) almost never occurs in practice.

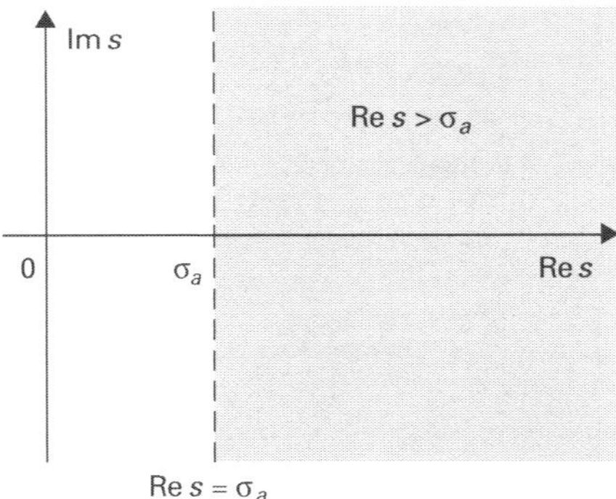

FIGURE 12.3
The half-plane Re $s > \sigma_a$ of absolute convergence.

This is because functions occurring in practice are almost always of 'exponential order'.

DEFINITION 12.3
Exponential order

The causal function $f : \mathbb{R} \to \mathbb{C}$ is called of exponential order *if there are constants $\alpha \in \mathbb{R}$ and $M > 0$ such that $|f(t)| \leq M e^{\alpha t}$ for all $t \geq 0$.*

Functions of exponential order will not assume very large values too quickly. The number α in definition 12.3 is by no means unique, since for any $\beta \geq \alpha$ one has $e^{\alpha t} \leq e^{\beta t}$ for $t \geq 0$.

EXAMPLE 12.3

The unit step function $\epsilon(t)$ is of exponential order with $M = 1$ and $\alpha = 0$ since $|\epsilon(t)| \leq 1$. ◀

EXAMPLE

Let $f(t)$ be a bounded function, so $|f(t)| \leq M$ for some $M > 0$. Then f is of exponential order with $\alpha = 0$. The function $\epsilon(t)$ is a special case and has $M = 1$. ◀

EXAMPLE 12.4

Consider the function $f(t) = t$. From the well-known limit $\lim_{t \to \infty} t e^{-\alpha t} = 0$, for $\alpha > 0$, it follows that $|f(t)| \leq M e^{\alpha t}$ for any $\alpha > 0$ and some constant $M > 0$. Hence, this function is of exponential order with $\alpha > 0$ arbitrary. However, one cannot claim that f is of exponential order with $\alpha = 0$, since it is not true that $|t| \leq M$. ◀

We now recall example 12.2, where it was shown that the Laplace transform of e^{at} exists for Re $s > a$. The following result will now come as no surprise.

THEOREM 12.3

Let f be a function of exponential order as in definition 12.3. Then the integral in (12.1) is absolutely convergent (and so the Laplace transform of f will certainly exist) for Re $s > \alpha$. In particular one has for the abscissa of absolute convergence σ_a that $\sigma_a \leq \alpha$.

Proof
Since $|f(t)| \leq M e^{\alpha t}$ for all $t \geq 0$, it follows that

$$\int_0^\infty |f(t)e^{-st}|\, dt = \int_0^\infty |f(t)|e^{-\sigma t}\, dt \leq M \int_0^\infty e^{-(\sigma - \alpha)t}\, dt,$$

where $s = \sigma + i\omega$. As in example 12.2 it follows that the latter integral exists for $\sigma - \alpha > 0$, so for Re $s > \alpha$. Finally, we then have for σ_a that $\sigma_a \leq \alpha$. For if $\sigma_a > \alpha$, then there will certainly exist a number β with $\alpha < \beta < \sigma_a$. On the one hand it would then follow from $\beta < \sigma_a$ and the definition of σ_a that the integral does not converge absolutely for $s = \beta$, while on the other hand $\beta > \alpha$ would imply that there is absolute convergence. ∎

EXAMPLE

The unit step function $\epsilon(t)$ is of exponential order with $\alpha = 0$. From theorem 12.3 it follows that the Laplace transform will certainly exist for $\mathrm{Re}\, s > 0$ and that $\sigma_a \leq 0$. As was shown in example 12.1, we even have $\sigma_a = 0$. ◄

EXAMPLE 12.5

In example 12.4 it was shown that the function t is of exponential order for arbitrary $\alpha > 0$. From theorem 12.3 it then follows that the Laplace transform of t certainly exists for $\mathrm{Re}\, s > 0$ and that $\sigma_a \leq 0$. In example 12.9 we will see that $\sigma_a = 0$. ◄

In the preceding general discussion of convergence issues, we have confined ourselves to absolute convergence, since the treatment of this type of convergence is relatively easy. Of course one can wonder whether similar results as in theorem 12.3 can also be derived for ordinary convergence. This is indeed the case, but these results are much harder to prove. We will merely state without proof that there

Abscissa of convergence

exists a number σ_c, the so-called *abscissa of convergence*, such that the integral $\int_0^\infty f(t)e^{-st}\,dt$ converges for $\mathrm{Re}\, s > \sigma_c$ and does not converge for $\mathrm{Re}\, s < \sigma_c$. Since absolute convergence certainly implies ordinary convergence, we see that $\sigma_c \leq \sigma_a$. In most concrete examples one can easily obtain σ_c.

EXAMPLE 12.6

In example 12.1 it was shown that the Laplace transform of $\epsilon(t)$ exists precisely when $\mathrm{Re}\, s > 0$. Hence, in this case we have $\sigma_a = \sigma_c = 0$. ◄

EXAMPLE 12.7

The shifted unit step function $\epsilon(t - b)$ in figure 12.4 is defined by

$$\epsilon(t - b) = \begin{cases} 1 & \text{for } t \geq b, \\ 0 & \text{for } t < b. \end{cases}$$

where $b \geq 0$. The Laplace transform of $\epsilon(t - b)$ can be determined as in

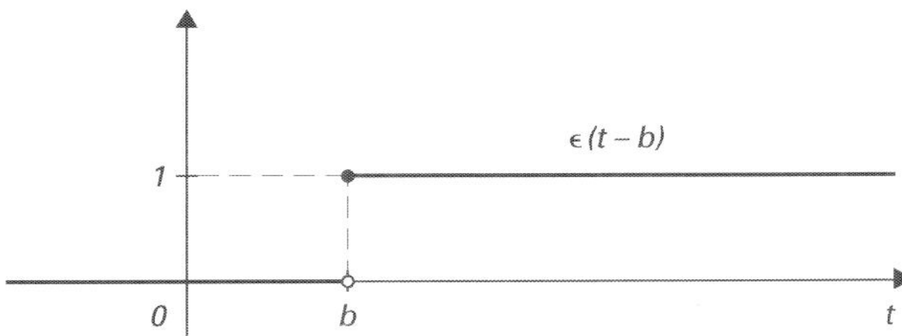

FIGURE 12.4
The shifted unit step function $\epsilon(t - b)$ with $b \geq 0$.

example 12.1. For $s = 0$ the Laplace transform does not exist, while for $s \neq 0$ it follows that

$$\int_0^\infty e^{-st}\epsilon(t - b)\,dt = \int_b^\infty e^{-st}\,dt = \lim_{R \to \infty} \left[-\frac{1}{s}e^{-st} \right]_b^R$$
$$= \frac{e^{-bs}}{s} - \frac{1}{s}\lim_{R \to \infty} e^{-sR}.$$

Again the limit exists precisely for $\mathrm{Re}\, s > 0$ and it equals 0 then. Here we again have $\sigma_a = \sigma_c = 0$. Furthermore, we see that for $\mathrm{Re}\, s > 0$ the Laplace transform of $\epsilon(t - b)$ is given by e^{-bs}/s. ◄

EXAMPLE 12.8

From example 12.2 it follows that for the function e^{bt} we have that $\sigma_a = \sigma_c = b$. For $\operatorname{Re} s > b$ the Laplace transform is given by $1/(s - b)$. ◀

EXAMPLE 12.9

We now determine the Laplace transform of the function $f(t) = t$. For $s = 0$ the Laplace transform does not exist, while for $s \neq 0$ it follows from integration by parts that (from example 12.5 we know that $\sigma_a \leq 0$):

$$
\begin{aligned}
F(s) &= \int_0^\infty t e^{-st}\, dt = \int_0^\infty -\frac{t}{s}(e^{-st})'\, dt \\
&= \lim_{R \to \infty}\left[-\frac{t}{s}e^{-st}\right]_0^R + \frac{1}{s}\int_0^\infty e^{-st}\, dt \\
&= -\frac{1}{s}\lim_{R \to \infty} R e^{-sR} + \lim_{R \to \infty}\left[-\frac{1}{s^2}e^{-st}\right]_0^R \\
&= -\frac{1}{s}\lim_{R \to \infty} R e^{-sR} - \frac{1}{s^2}\lim_{R \to \infty} e^{-sR} + \frac{1}{s^2}.
\end{aligned}
$$

As before, one has for $\operatorname{Re} s = \sigma > 0$ that $\lim_{R \to \infty} e^{-sR} = 0$. Since for $\sigma > 0$ we have $\lim_{R \to \infty} R e^{-\sigma R} = 0$ as well, it also follows that $\lim_{R \to \infty} R e^{-sR} = 0$ for $\operatorname{Re} s > 0$. This shows that $F(s) = 1/s^2$ for $\operatorname{Re} s > 0$. We also see that the limits do not exist if $\operatorname{Re} s \leq 0$; hence, $\sigma_a = \sigma_c = 0$. ◀

The previous example is a prototype of the kind of calculations that are usually necessary in order to calculate the Laplace transform of a function: performing an integration by parts (sometimes more than once) and determining limits. These limits will in general only exist under certain conditions on $\operatorname{Re} s$. Usually this will also immediately give us the abscissa of convergence, as well as the abscissa of absolute convergence. In all the examples we have seen so far, we had $\sigma_a = \sigma_c$. In general, this is certainly not the case; for example, for the function $e^t \sin(e^t)$ one has $\sigma_a = 1$ while $\sigma_c = 0$ (the proof of these facts will be omitted). There are even examples (which are not very easy) of functions for which the integral in (12.1) converges for all $s \in \mathbb{C}$ (so $\sigma_c = -\infty$), but converges absolutely for no $s \in \mathbb{C}$ whatsoever (so $\sigma_a = \infty$)! As a matter of fact, for the application of the Laplace transform it almost always suffices to know that some half-plane of convergence exists; the precise value of σ_a or σ_c is in many cases less important.

We close this section by noting that besides the Laplace transform from definition *Two-sided Laplace transform* 12.1, there also exists a so-called *two-sided Laplace transform*. For functions on $\mathbb{R}$ that are not necessarily causal, this two-sided Laplace transform is defined by

$$
\int_{-\infty}^\infty f(t) e^{-st}\, dt,
$$

for those $s \in \mathbb{C}$ for which the integral exists. Since in most applications it is assumed that the functions involved are causal, we have limited ourselves to the 'one-sided' Laplace transform from definition 12.1. Indeed, the one-sided and two-sided Laplace transforms coincide for causal functions. Also note that

$$
\int_{-\infty}^\infty f(t) e^{-st}\, dt = \int_0^\infty f(t) e^{-st}\, dt + \int_0^\infty f(-t) e^{-(-s)t}\, dt.
$$

Hence, the two-sided Laplace transform of $f(t)$ is equal to the (one-sided) Laplace transform of $f(t)$ plus the (one-sided) Laplace transform at $-s$ of the function $f(-t)$. There is thus a close relationship between the two forms of the Laplace transform.

EXERCISES

12.1   **a**   Indicate why the limit $\lim_{R\to\infty} e^{-i\omega R}$ does not exist for $\omega \in \mathbb{R}$.
       b   For which $\sigma, \omega \in \mathbb{R}$ is it true that $\lim_{R\to\infty} e^{-\sigma R}e^{-i\omega R} = 0$?

12.2   In example 12.2 it was shown that the Laplace transform of e^{at} ($a \in \mathbb{R}$) is given by $1/(s-a)$ for $\operatorname{Re} s > a$. Show that for $a \in \mathbb{C}$ the same result holds for $\operatorname{Re} s > \operatorname{Re} a$.

12.3   Consider the causal function f defined by $f(t) = 1 - \epsilon(t-b)$ for $b \geq 0$.
       a   Sketch the graph of f.
       b   Determine the Laplace transform $F(s)$ of $f(t)$ and give the abscissa of absolute and ordinary convergence.

12.4   Show that the Laplace transform of the function t^2 (hence, of $\epsilon(t)t^2$) is given by $2/s^3$ for $\operatorname{Re} s > 0$.

12.5   Determine for the following functions the Laplace transform and the abscissa of absolute and ordinary convergence:
       a e^{-2t},
       b $\epsilon(t-4)$,
       c $e^{(2+3i)t}$.

12.6   Determine a function $f(t)$ whose Laplace transform is given by the following functions:
       a $1/s$,
       b e^{-3s}/s,
       c $1/(s-7)$,
       d $1/s^3$.

12.2   Linearity, shifting and scaling

From (12.2) it follows that $(\mathcal{L}f)(\sigma + i\omega) = (\mathcal{F}\epsilon(t)f(t)e^{-\sigma t})(\omega)$, provided that all the integrals exist. It will then come as no surprise that for the Laplace transform $\mathcal{L}$ and the Fourier transform $\mathcal{F}$ similar properties hold. In this section we will examine a number of elementary properties: linearity, shifting, and scaling. It will turn out that these properties are quite useful in order to determine the Laplace transform of all kinds of functions.

12.2.1   Linearity

As for the Fourier transform, the linearity of the Laplace transform follows immediately from the linearity of integration (see section 6.4.1). For $\alpha, \beta \in \mathbb{C}$ one thus has

Linearity of $\mathcal{L}$

$$\mathcal{L}(\alpha f + \beta g) = \alpha \mathcal{L}f + \beta \mathcal{L}g$$

in the half-plane where $\mathcal{L}f$ and $\mathcal{L}g$ both exist. Put differently, when F and G are the Laplace transforms of f and g respectively, then $\alpha f + \beta g$ has the Laplace transform $\alpha F + \beta G$; if $F(s)$ exists for $\operatorname{Re} s > \sigma_1$, and $G(s)$ for $\operatorname{Re} s > \sigma_2$, then $(\alpha F + \beta G)(s)$ exists for $\operatorname{Re} s > \max(\sigma_1, \sigma_2)$. This simple rule enables us to find a number of important Laplace transforms.

EXAMPLE 12.10

The Laplace transform of e^{it} is given by $1/(s-i)$ for $\operatorname{Re} s > \operatorname{Re} i$, hence for $\operatorname{Re} s > 0$ (see exercise 12.2). Similarly one has that $(\mathcal{L}e^{-it})(s) = 1/(s+i)$ for

Re $s > 0$. Since $\sin t = (e^{it} - e^{-it})/2i$, it then follows from the linearity property that

$$(\mathcal{L} \sin t)(s) = \frac{1}{2i} \left((\mathcal{L} e^{it})(s) - (\mathcal{L} e^{-it})(s) \right)$$

$$= \frac{1}{2i} \left(\frac{1}{s-i} - \frac{1}{s+i} \right) = \frac{(s+i) - (s-i)}{2i(s-i)(s+i)}$$

$$= \frac{2i}{2i(s-i)(s+i)} = \frac{1}{s^2+1}.$$

For Re $s > 0$ we thus have

$$(\mathcal{L} \sin t)(s) = \frac{1}{s^2+1}. \tag{12.3}$$

◀

EXAMPLE 12.11

For the function $\sinh t$ (see also exercise 11.5) one has

$$(\mathcal{L} \sinh t)(s) = \frac{1}{2} \left((\mathcal{L} e^t)(s) - (\mathcal{L} e^{-t})(s) \right) = \frac{1}{2} \left(\frac{1}{s-1} - \frac{1}{s+1} \right)$$

provided Re $s > 1$ and Re $s > -1$ (see example 12.2). Hence it follows for Re $s > 1$ that

$$(\mathcal{L} \sinh t)(s) = \frac{1}{s^2-1}. \tag{12.4}$$

◀

12.2.2 *Shift in the time domain*

The unit step function is often used to represent the switching on of a signal f at time $t = 0$ (see figure 12.5a). When several signals are switched on at different moments in time, then it is convenient to use the shifted unit step function $\epsilon(t - b)$ (see example 12.7 for its definition). In fact, when the signal f is switched on at time $t = b$ ($b \geq 0$), then this can simply be represented by the function $\epsilon(t - b) f(t - b)$. See figure 12.5b. Using the functions $\epsilon(t - b)$ it is also quite easy to represent

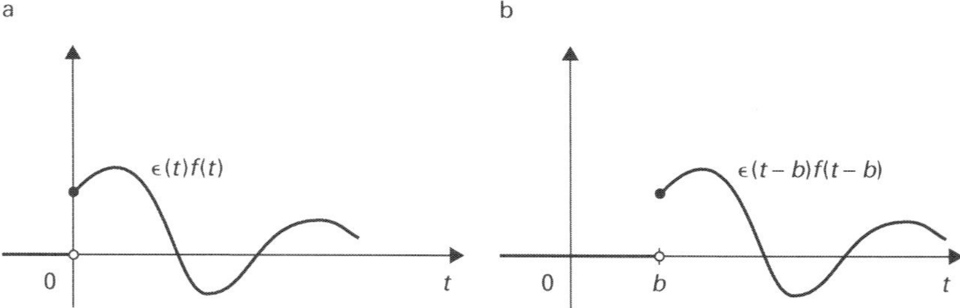

FIGURE 12.5
A signal $f(t)$ switched on at time $t = 0$ (a) and at time $t = b$ (b).

combinations of shifted (switched on) signals. Figure 12.6, for example, shows the graph of the causal function $f(t) = 3 - 2\epsilon(t - 1)(t - 1) + 2\epsilon(t - 3)(t - 3)$. In fact, $f(t) = 3$ for $0 \leq t < 1$, $f(t) = 3 - 2(t - 1) = 5 - 2t$ for $1 \leq t < 3$ and $f(t) = 3 - 2(t - 1) + 2(t - 3) = -1$ for $t \geq 3$.

There is a simple relationship between the Laplace transform $F(s)$ of $f(t)$ (in fact $\epsilon(t) f(t)$) and the Laplace transform of the shifted function $\epsilon(t - b) f(t - b)$.

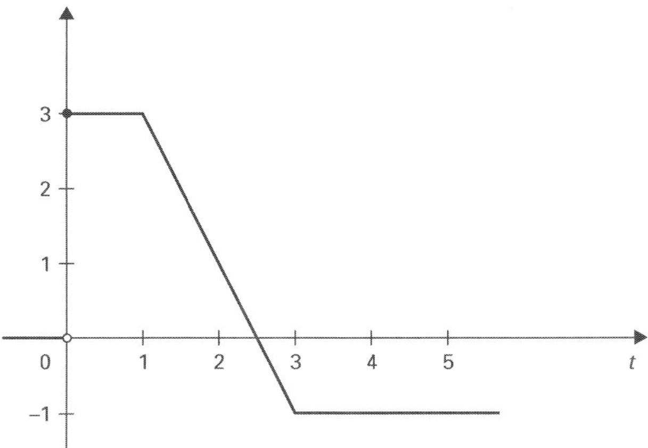

FIGURE 12.6
Graph of the causal function $3 - 2\epsilon(t-1)(t-1) + 2\epsilon(t-3)(t-3)$.

THEOREM 12.4
Shift in the time domain

Let $f(t)$ be a function with Laplace transform $F(s)$ for $\operatorname{Re} s > \rho$ and let $b \geq 0$. Then one has for $\operatorname{Re} s > \rho$ that

$$(\mathcal{L}\epsilon(t-b)f(t-b))(s) = e^{-bs}F(s). \tag{12.5}$$

Proof
By changing to the variable $\tau = t - b$ we obtain that

$$(\mathcal{L}\epsilon(t-b)f(t-b))(s) = \int_b^\infty f(t-b)e^{-st}\,dt = \int_0^\infty f(\tau)e^{-s\tau-bs}\,d\tau$$

$$= e^{-bs}(\mathcal{L}f)(s),$$

where we assumed that $\operatorname{Re} s > \rho$. This proves the theorem. ∎

EXAMPLE

In example 12.1 we have seen for $f(t) = \epsilon(t)$ that $F(s) = 1/s$ for $\operatorname{Re} s > 0$. According to theorem 12.4, the function $\mathcal{L}\epsilon(t-b)$ is then given by e^{-bs}/s for $\operatorname{Re} s > 0$, which is in agreement with the result of example 12.7. ◀

EXAMPLE

Let $g(t) = \epsilon(t-2)\sin(t-2)$; the graph of g is drawn in figure 12.7. From example 12.10 and theorem 12.4 it follows that $(\mathcal{L}g)(s) = e^{-2s}(\mathcal{L}\sin t)(s) = e^{-2s}/(s^2+1)$ for $\operatorname{Re} s > 0$. ◀

EXAMPLE

Let $f(t) = \epsilon(t-3)(t^2 - 6t + 9)$. Since $t^2 - 6t + 9 = (t-3)^2$, it follows from theorem 12.4 and exercise 12.4 that $F(s) = 2e^{-3s}/s^3$. ◀

12.2.3 Shift in the s-domain

We now consider the effect of a shift in the s-domain.

THEOREM 12.5
Shift in the s-domain

Let $f(t)$ be a function with Laplace transform $F(s)$ for $\operatorname{Re} s > \rho$ and let $b \in \mathbb{C}$. Then one has for $\operatorname{Re} s > \rho + \operatorname{Re} b$ that

$$\left(\mathcal{L}e^{bt}f(t)\right)(s) = F(s-b). \tag{12.6}$$

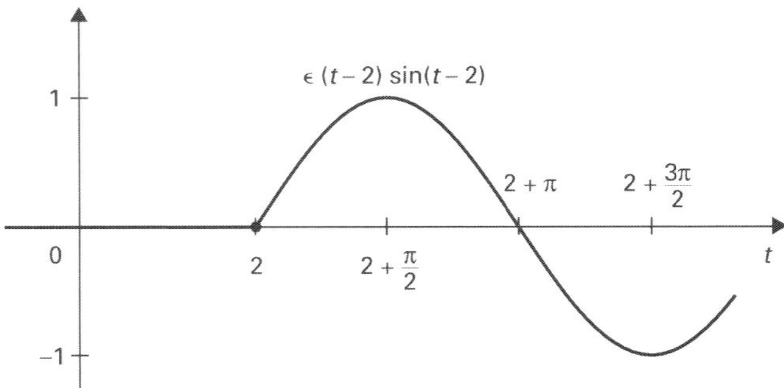

FIGURE 12.7
Graph of $\epsilon(t - 2)\sin(t - 2)$.

Proof
Since $F(s)$ exists for $\operatorname{Re} s > \rho$, it follows that $F(s - b)$ exists for $\operatorname{Re}(s - b) > \rho$, that is, for $\operatorname{Re} s > \rho + \operatorname{Re} b$. For these values of s one has

$$F(s - b) = \int_0^\infty f(t)e^{-(s-b)t}\, dt = \int_0^\infty e^{bt} f(t)e^{-st}\, dt = \left(\mathcal{L}e^{bt} f(t)\right)(s).$$

∎

For $b < 0$ the factor e^{bt} corresponds to a 'damping' of the function $f(t)$; in the s-domain this results in a shift over b. Therefore, theorem 12.5 is sometimes called
Damping theorem   the *damping theorem*.

EXAMPLE   Since $(\mathcal{L}\sin t)(s) = 1/(s^2 + 1)$ (see (12.3)), it follows from theorem 12.5 that $(\mathcal{L}e^{-2t}\sin t)(s) = 1/((s + 2)^2 + 1) = 1/(s^2 + 4s + 5)$. ◀

EXAMPLE   Suppose that we are looking for a function $f(t)$ with $F(s) = 1/(s - 2)^2$. From example 12.9 we know that $(\mathcal{L}t)(s) = 1/s^2$, and theorem 12.5 then implies that $(\mathcal{L}te^{2t})(s) = 1/(s - 2)^2$. Hence, $f(t) = te^{2t}$. At present we do not know whether or not this is the only possible function f having this property. We will return to this matter in section 13.5. ◀

12.2.4   Scaling

The following theorem describes the effect of scaling in the time domain.

THEOREM 12.6   *Let $f(t)$ be a function with Laplace transform $F(s)$ for $\operatorname{Re} s > \rho$ and let $b > 0$.*
Scaling   *Then one has for $\operatorname{Re} s > b\rho$ that*

$$(\mathcal{L}f(bt))(s) = \frac{1}{b}F\left(\frac{s}{b}\right). \tag{12.7}$$

Proof
By changing to the variable $\tau = bt$ for $b > 0$, we obtain that

$$(\mathcal{L}f(bt))(s) = \int_0^\infty f(bt)e^{-st}\, dt = \frac{1}{b}\int_0^\infty f(\tau)e^{-s\tau/b}\, d\tau = \frac{1}{b}F\left(\frac{s}{b}\right).$$

Note that $F(s/b)$ exists for $\operatorname{Re}(s/b) > \rho$, that is, for $\operatorname{Re} s > b\rho$. ∎

EXAMPLE

In example 12.10 it was shown that $(\mathcal{L} \sin t)(s) = 1/(s^2 + 1)$ for $\mathrm{Re}\,s > 0$. From theorem 12.6 it then follows for $a > 0$ that

$$(\mathcal{L} \sin at)(s) = \frac{1}{a} \frac{1}{(s/a)^2 + 1} = \frac{a}{s^2 + a^2} \qquad \text{for } \mathrm{Re}\,s > 0.$$

Since $\sin(-at) = -\sin at$, it follows from the linearity of $\mathcal{L}$ that for any $a \in \mathbb{R}$ we have

$$(\mathcal{L} \sin at)(s) = \frac{a}{s^2 + a^2} \qquad \text{for } \mathrm{Re}\,s > 0. \tag{12.8}$$

This result can also be obtained using the method from example 12.10. ◄

EXERCISES

12.7   **a** Let $F(s)$ for $\mathrm{Re}\,s > \sigma_1$ and $G(s)$ for $\mathrm{Re}\,s > \sigma_2$ be the Laplace transforms of $f(t)$ and $g(t)$ respectively. Show that $(\alpha F + \beta G)(s)$ is the Laplace transform of $(\alpha f + \beta g)(t)$ for $\mathrm{Re}\,s > \max(\sigma_1, \sigma_2)$.
b Determine the Laplace transform of $f(t) = 3t^2 - it + 4$.

12.8   **a** Use the formula $\cos t = (e^{it} + e^{-it})/2$, exercise 12.2, and the linearity of $\mathcal{L}$ to determine the Laplace transform of $\cos t$.
b Determine also, as in part a, $\mathcal{L} \cos at$ for $a \in \mathbb{R}$.
c Similarly, determine $\mathcal{L} \cosh at$ for $a \in \mathbb{R}$.

12.9   Determine $\mathcal{L} \cos(at + b)$ and $\mathcal{L} \sin(at + b)$ for $a, b \in \mathbb{R}$. (Suggestion: use the addition formulas for the sine and cosine functions.)

12.10   Determine the Laplace transform $F(s)$ of the following functions:
a $f(t) = 10t^2 - 5t + 8i - 3$,
b $f(t) = \sin 4t$,
c $f(t) = \cosh 5t$,
d $f(t) = t + 1 - \cos t$,
e $f(t) = e^{2t} + e^{-3t}$,
f $f(t) = \sin^2 t$,
g $f(t) = \sin(t - 2)$,
h $f(t) = 3^t$.

12.11   The function f is given by $f(t) = 1 - \epsilon(t - 1)(2t - 2) + \epsilon(t - 2)(t - 2)$.
a Sketch the graph of f.
b Determine $F(s) = (\mathcal{L} f(t))(s)$.

12.12   Draw the graph of the function $g(t) = \cos t - \epsilon(t - 2\pi) \cos(t - 2\pi)$ and determine $G(s) = (\mathcal{L} g(t))(s)$.

12.13   For $\mathrm{Re}\,s > 0$ one has $(\mathcal{L} 1)(s) = 1/s$. Use the shift property in the s-domain to determine the Laplace transform of e^{bt} for $b \in \mathbb{C}$. Compare the result with exercise 12.2.

12.14   Use the scaling property and $\mathcal{L} \cos t$ to determine $\mathcal{L} \cos at$ $(a \in \mathbb{R})$ again, and compare the result with exercise 12.8b.

12.15   Let $f(t)$ be a function with Laplace transform

$$F(s) = \frac{s^2 - s + 1}{(2s + 1)(s - 1)} \qquad \text{for } \mathrm{Re}\,s > 1.$$

Determine the Laplace transform $G(s)$ of $g(t) = f(2t)$.

12.16   Determine the Laplace transform $F(s)$ of the following functions:
a $f(t) = te^{2t}$,
b $f(t) = \epsilon(t-1)(t-1)^2$,
c $f(t) = e^{-3t}\sin 5t$,
d $f(t) = e^{bt}\cos at$ for $a \in \mathbb{R}$ and $b \in \mathbb{C}$,
e $f(t) = \epsilon(t-3)\cosh(t-3)$,
f $f(t) = t^2 e^{t-3}$.

12.17   Draw the graph of the following functions and determine their Laplace transform $F(s)$:
a $f(t) = \epsilon(t-1)(t-1)$,
b $f(t) = \epsilon(t)(t-1)$,
c $f(t) = \epsilon(t-1)t$.

12.18   Consider the (causal) function

$$f(t) = \begin{cases} t & \text{for } 0 \leq t < 1, \\ 0 & \text{for } t \geq 1. \end{cases}$$

Write f as a combination of shifted unit step functions and determine the Laplace transform $F(s)$.

12.19   Determine a function $f(t)$ whose Laplace transform $F(s)$ is given by:
a $F(s) = 2/(s-3)$,
b $F(s) = 3/(s^2+1)$,
c $F(s) = 4s/(s^2+4)$,
d $F(s) = 1/(s^2-4)$,
e $F(s) = e^{-2s}/s^2$,
f $F(s) = se^{-3s}/(s^2+1)$,
g $F(s) = 1/((s-1)^2+16)$,
h $F(s) = (3s+2)/((s+1)^2+1)$,
i $F(s) = -6/(s-3)^3$,
j $F(s) = (s-2)/(s^2-4s+8)$,
k $F(s) = se^{-s}/(4s^2+9)$.

12.3   Differentiation and integration

In this section we continue our investigation into the properties of the Laplace transform with the treatment of the differentiation and integration rules. We will examine differentiation both in the time domain and in the s-domain. For the application of the Laplace transform to differential equations (see chapter 14) it is especially important to know how the Laplace transform behaves with respect to differentiation in the time domain. Differentiation in the s-domain is complex differentiation. In particular we will show that $F(s)$ is analytic in a certain half-plane in $\mathbb{C}$. And finally an integration rule in the time domain will be derived from the differentiation rule in the time domain. The rule for integration in the s-domain will not be treated. This rule isn't used very often and a proper treatment would moreover require a thorough understanding of integration of complex functions over curves in $\mathbb{C}$.

12.3.1   Differentiation in the time domain

In section 6.4.8 we have seen that differentiation in the time domain and multiplication by a factor $i\omega$ in the frequency domain correspond with each other under the

Fourier transform. A similar correspondence, but now involving the factor s, exists for the Laplace transform.

THEOREM 12.7
Differentiation in the time domain

Let f be a causal function which, in addition, is differentiable on $\mathbb{R}$. In a half-plane where $\mathcal{L}f$ and $\mathcal{L}f'$ both exist one has

$$(\mathcal{L}f')(s) = s(\mathcal{L}f)(s). \tag{12.9}$$

Proof
Let $s \in \mathbb{C}$ be such that $\mathcal{L}f$ and $\mathcal{L}f'$ both exist at s. By applying integration by parts we obtain that

$$(\mathcal{L}f')(s) = \int_0^\infty f'(t)e^{-st}\,dt = \left[f(t)e^{-st}\right]_0^\infty + s\int_0^\infty f(t)e^{-st}\,dt.$$

Since f is differentiable on $\mathbb{R}$, f is certainly continuous on $\mathbb{R}$. But f is also causal and so we must have $f(0) = \lim_{t\uparrow 0} f(t) = 0$. From this it follows that

$$(\mathcal{L}f')(s) = \lim_{R\to\infty} f(R)e^{-sR} + s(\mathcal{L}f)(s).$$

Since $(\mathcal{L}f')(s)$ and $s(\mathcal{L}f)(s)$ exist, the limit $\lim_{R\to\infty} f(R)e^{-sR}$ must also exist. But then this limit has to equal 0, for $(\mathcal{L}f)(s) = \int_0^\infty f(t)e^{-st}\,dt$ exists (here we use the simple fact that for a continuous function $g(t)$ with $\lim_{R\to\infty} g(R) = a$, where $a \in \mathbb{R}$ and $a \neq 0$, the integral $\int_0^\infty g(t)\,dt$ does not exist; see exercise 12.20). This proves the theorem. ∎

Using the concept of a 'function of exponential order' (see definition 12.3), one is able to specify the half-planes where $\mathcal{L}f$ and $\mathcal{L}f'$ both exist. If we assume that the function $f(t)$ from theorem 12.7 is of exponential order for a certain value $\alpha \in \mathbb{R}$, then $\mathcal{L}f$ exists for $\operatorname{Re} s > \alpha$ (see theorem 12.3). One can show that in this case $\mathcal{L}f'$ also exists for $\operatorname{Re} s > \alpha$. We will not go into this any further.

By repeatedly applying theorem 12.7, one can obtain the Laplace transform of the higher derivatives of a function. Of course, the conditions of theorem 12.7 should then be satisfied throughout. Suppose, for example, that a causal function $f(t)$ is continuously differentiable on $\mathbb{R}$ (so f' exists and is continuous on $\mathbb{R}$) and that f' is differentiable on $\mathbb{R}$. By applying theorem 12.7 twice in a half-plane where all Laplace transforms exist, it then follows that

$$(\mathcal{L}f'')(s) = s(\mathcal{L}f')(s) = s^2(\mathcal{L}f)(s).$$

Now, more generally, let $f(t)$ be a causal function which is $n-1$ times continuously differentiable on $\mathbb{R}$ (so the $(n-1)$th derivative $f^{(n-1)}(t)$ of $f(t)$ exists and is continuous on $\mathbb{R}$) and let $f^{(n-1)}(t)$ be differentiable on $\mathbb{R}$ (in the case $n = 1$ we have $f^{(0)}(t) = f(t)$ and by a '0 times continuously differentiable function' we simply mean a continuous function). In a half-plane where all Laplace transforms exist, we then have the following differentiation rule in the time domain:

Differentiation in the time domain

$$(\mathcal{L}f^{(n)})(s) = s^n(\mathcal{L}f)(s). \tag{12.10}$$

EXAMPLE

For $f(t) = t^2$ we have $f'(t) = 2t$. The function $\epsilon(t)t^2$ is indeed differentiable on $\mathbb{R}$ and according to (12.9) we thus have $s(\mathcal{L}t^2)(s) = 2(\mathcal{L}t)(s)$. From example 12.9 we know that $(\mathcal{L}t)(s) = 1/s^2$ for $\operatorname{Re} s > 0$, and so $(\mathcal{L}t^2)(s) = 2/s^3$ for $\operatorname{Re} s > 0$. Compare this method with exercise 12.4. ◀

The method from the example above can be used to determine $\mathcal{L}t^n$ for every $n \in \mathbb{N}$ with $n \geq 2$. In fact, the function $\epsilon(t)t^n$ satisfies the conditions of theorem 12.7 for $n \in \mathbb{N}$ with $n \geq 2$ and so it follows from (12.9) that

$$(\mathcal{L}nt^{n-1})(s) = s(\mathcal{L}t^n)(s).$$

Since $\lim_{t \to \infty} t^n e^{-\alpha t} = 0$ for every $\alpha > 0$, it follows just as in the examples 12.4 and 12.5 that $\mathcal{L}t^n$ exists for $\operatorname{Re} s > 0$ ($n \in \mathbb{N}$). Hence,

$$(\mathcal{L}t^n)(s) = \frac{n}{s}(\mathcal{L}t^{n-1})(s) \qquad \text{for } \operatorname{Re} s > 0$$

(this result can also easily be derived by a direct calculation). Applying this result repeatedly, we find that

$$(\mathcal{L}t^n)(s) = \frac{n}{s} \cdot \frac{n-1}{s} \cdot \ldots \cdot \frac{1}{s}(\mathcal{L}1)(s) \qquad \text{for } \operatorname{Re} s > 0.$$

Now finally use that $(\mathcal{L}1)(s) = 1/s$ for $\operatorname{Re} s > 0$ to establish the following important result:

$$(\mathcal{L}t^n)(s) = \frac{n!}{s^{n+1}} \qquad \text{for } \operatorname{Re} s > 0, \tag{12.11}$$

where $n! = 1 \cdot 2 \cdot 3 \cdot \ldots \cdot (n-1) \cdot n$. Note that (12.11) is also valid for $n = 1$ and $n = 0$ ($0! = 1$ by convention).

Theorem 12.7 cannot be applied to an arbitrary piecewise smooth function. This is because the function in theorem 12.7 has to be differentiable on $\mathbb{R}$ and so it can certainly have no jump discontinuities. In section 13.4 we will derive a differentiation rule for distributions, which in particular can then be applied to piecewise smooth functions.

12.3.2 Differentiation in the s-domain

On the basis of the properties of the Fourier transform (see section 6.4.9) we expect that here differentiation in the s-domain will again correspond to a multiplication by a factor in the time domain. It will turn out that this is indeed the case. Still, for the Laplace transform this result is of a quite different nature, since we are dealing with a complex function $F(s)$ here and so with complex differentiation. As in the case of the Fourier transform we will thus have to show that $F(s)$ is in fact differentiable. Put differently, we will first have to show that $F(s)$ is an analytic function on a certain subset of $\mathbb{C}$. One has the following result.

THEOREM 12.8
Differentiation in the s-domain

Let f be a function with Laplace transform $F(s)$ and let σ_a be the abscissa of absolute convergence. Then $F(s)$ is an analytic function of s for $\operatorname{Re} s > \sigma_a$ and

$$\frac{d}{ds}F(s) = -(\mathcal{L}tf(t))(s). \tag{12.12}$$

Proof
We have to show that $\lim_{h \to 0}(F(s+h) - F(s))/h$ (with $h \to 0$ in $\mathbb{C}$; see section 11.2) exists for $s \in \mathbb{C}$ with $\operatorname{Re} s > \sigma_a$. Now

$$F(s+h) - F(s) = \int_0^\infty f(t)e^{-(s+h)t}\, dt - \int_0^\infty f(t)e^{-st}\, dt$$

$$= \int_0^\infty f(t)(e^{-ht} - 1)e^{-st}\, dt,$$

which means that we have to show that

$$\lim_{h \to 0} \int_0^\infty f(t)\left(\frac{e^{-ht} - 1}{h}\right)e^{-st}\, dt$$

exists. We now assume that under the condition mentioned in the theorem we may interchange the limit and the integral. Note that we thus assume in particular that the integral resulting from the interchange will again exist. From definition 11.7, theorem 11.6, and the chain rule it follows that

$$\lim_{h \to 0} \frac{e^{-ht} - 1}{h} = \left(\frac{d}{ds} e^{-st}\right)(0) = -t.$$

This shows that

$$\lim_{h \to 0} \frac{F(s+h) - F(s)}{h} = \int_0^\infty (-tf(t))e^{-st}\,dt,$$

or

$$\frac{d}{ds} F(s) = -(\mathcal{L}tf(t))(s).$$

$\blacksquare$

The difficult step in this theorem is precisely interchanging the limit and the integral, which in fact proves the existence of the derivative (and thus proves that $F(s)$ is analytic). It takes quite some effort to actually show that the interchanging is allowed (see e.g. Körner, *Fourier analysis*, 1990, Theorem 7.5.2). If we compare theorem 12.8 with theorem 6.8 for the Fourier transform, then it is remarkable that in theorem 12.8 we do not require in advance that the Laplace transform of $tf(t)$ exists, but that this fact *follows* as a side result of the theorem. It also means that the theorem can again be applied to the function $-tf(t)$, resulting in

$$\frac{d^2}{ds^2} F(s) = (\mathcal{L}t^2 f(t))(s)$$

for $\operatorname{Re} s > \sigma_a$. Repeated application then leads to the remarkable result that $F(s)$ is arbitrarily often differentiable and that for $\operatorname{Re} s > \sigma_a$

Differentiation in the s-domain

$$\frac{d^k}{ds^k} F(s) = (-1)^k (\mathcal{L}t^k f(t))(s) \qquad \text{for } k \in \mathbb{N}. \tag{12.13}$$

We will call this the *differentiation rule in the s-domain*. Usually, (12.13) is applied in the following way: let $F(s) = (\mathcal{L}f(t))(s)$, then

$$(\mathcal{L}t^k f(t))(s) = (-1)^k \frac{d^k}{ds^k} F(s) \qquad \text{for } k \in \mathbb{N}.$$

Multiplication by t^k

Hence, this rule is sometimes referred to as *multiplication by t^k*.

EXAMPLE 12.12

Since $(\mathcal{L}\sin t)(s) = 1/(s^2 + 1)$ for $\operatorname{Re} s > 0$, it follows that

$$(\mathcal{L}t \sin t)(s) = -\frac{d}{ds}\left(\frac{1}{s^2 + 1}\right) = \frac{2s}{(s^2 + 1)^2} \qquad \text{for } \operatorname{Re} s > 0.$$

$\blacktriangleleft$

EXAMPLE

We know that $(\mathcal{L}e^{-3t})(s) = 1/(s + 3)$, so

$$(\mathcal{L}t^2 e^{-3t})(s) = \frac{d^2}{ds^2}\left(\frac{1}{s + 3}\right) = \frac{2}{(s + 3)^3}.$$

This result also follows by noting that $(\mathcal{L}t^2)(s) = 2/s^3$ and subsequently using the shift property from theorem 12.5.

$\blacktriangleleft$

12.3.3 Integration in the time domain

From the differentiation rule in the time domain one quickly obtains the following result.

THEOREM 12.9
Integration in the time domain

Let f be a causal function which is continuous on $\mathbb{R}$ and has Laplace transform $F(s)$. Then one has in a half-plane contained in the region $\mathrm{Re}\,s > 0$

$$\left(\mathcal{L}\int_0^t f(\tau)\,d\tau\right)(s) = \frac{1}{s}F(s). \tag{12.14}$$

Proof
Define the function $g(t)$ by $g(t) = \int_{-\infty}^t f(\tau)\,d\tau = \int_0^t f(\tau)\,d\tau$, then g is the primitive of f with $g(0) = 0$ and $g(t) = 0$ for $t < 0$. Since f is continuous, it follows that g is differentiable on $\mathbb{R}$. According to theorem 12.7 one then has, in a half-plane where both Laplace transforms exist, that $(\mathcal{L}g')(s) = s(\mathcal{L}g)(s)$. But $g' = f$ and so

$$s\left(\mathcal{L}\int_0^t f(\tau)\,d\tau\right)(s) = F(s).$$

In a half-plane where both Laplace transforms exist and which lies to the right of $\mathrm{Re}\,s = 0$, one may divide by s and so the result follows. ∎

EXAMPLE 12.13

The causal function $\sin t$ is continuous on $\mathbb{R}$ and since $\int_0^t \sin\tau\,d\tau = 1 - \cos t$, it then follows from theorem 12.9 that

$$(\mathcal{L}(1 - \cos t))(s) = \frac{1}{s}(\mathcal{L}\sin t)(s) = \frac{1}{s(s^2 + 1)}.$$

This result is easy to verify since we know from table 7 that $(\mathcal{L}\cos t)(s) = s/(s^2+1)$ and $(\mathcal{L}1)(s) = 1/s$. Hence,

$$(\mathcal{L}(1 - \cos t))(s) = \frac{1}{s} - \frac{s}{s^2 + 1} = \frac{(s^2 + 1) - s^2}{s(s^2 + 1)} = \frac{1}{s(s^2 + 1)}. \quad \blacktriangleleft$$

As in the case of theorem 12.7 one can use the concept 'function of exponential order' to specify the half-planes where the Laplace transforms exist. If a function $f(t)$ is of exponential order for a certain $\alpha \in \mathbb{R}$, then one can show that the result from theorem 12.9 is correct for $\mathrm{Re}\,s > \max(0, \alpha)$. We will not go into this matter any further.

Theorem 12.9 can also be applied in the 'opposite direction'. When we are looking for a function $f(t)$ whose Laplace transform is $F(s)$, then we could start by first ignoring factors $1/s$ that might occur in $F(s)$. In fact, such factors can afterwards be re-introduced by an integration.

EXAMPLE 12.14

Let $g(t)$ be a function with Laplace transform $G(s) = 4/(s(s^2 + 4))$. If we ignore the factor $1/s$, then we are looking for a function $h(t)$ having Laplace transform $H(s) = 4/(s^2 + 4)$. This is easy: $h(t) = 2\sin 2t$. Integrating $h(t)$ we find $g(t)$:

$$g(t) = \int_0^t 2\sin 2\tau\,d\tau = [-\cos 2\tau]_0^t = 1 - \cos 2t. \quad \blacktriangleleft$$

EXERCISES

12.20

Let $g(t)$ be a continuous function on $\mathbb{R}$. Show that $\int_0^\infty g(t)\,dt$ does not exist if $\lim_{R\to\infty} g(R) = a$ where $a \in \mathbb{R}$ and $a \neq 0$.

12.21 Show that $\lim_{t\to\infty} t^n e^{-\alpha t} = 0$ for any $n \in \mathbb{N}$ and $\alpha > 0$ and use this to show that the function t^n is of exponential order with $\alpha > 0$ arbitrary. Conclude that $(\mathcal{L}t^n)(s)$ exists for $\mathrm{Re}\, s > 0$.

12.22 Use the definition to show that for any $n \in \mathbb{N}$ one has

$$(\mathcal{L}t^n)(s) = \frac{n}{s}(\mathcal{L}t^{n-1})(s) \qquad \text{for } \mathrm{Re}\, s > 0.$$

12.23 One has that $(\mathcal{L}1)(s) = 1/s$ for $\mathrm{Re}\, s > 0$. Use the differentiation rule in the s-domain to show that $(\mathcal{L}t^n)(s) = n!/s^{n+1}$ for $\mathrm{Re}\, s > 0$.

12.24 In example 12.12 we used the differentiation rule in the s-domain to show that $(\mathcal{L}t \sin t)(s) = 2s/(s^2 + 1)^2$ for $\mathrm{Re}\, s > 0$. Since $t \sin t = (te^{it} - te^{-it})/2i$, one can also derive this result quite easily using the shift property. Give this derivation.

12.25 Consider the function $f(t) = t^n e^{at}$ for $a \in \mathbb{C}$ and let $F(s) = (\mathcal{L}f(t))(s)$.
   a Determine $F(s)$ using a shift property.
   b Determine $F(s)$ using a differentiation rule.

12.26 Determine the Laplace transform $G(s)$ of $g(t) = \int_0^t \tau \cos 2\tau \, d\tau$.

12.27 One has that $(\mathcal{L}\sinh at)(s) = a/(s^2 - a^2)$ for $\mathrm{Re}\, s > a$ $(a \in \mathbb{R})$. Which function $f(t)$ has $a/(s(s^2 - a^2))$ as its Laplace transform?

12.28 Determine the Laplace transform $F(s)$ of the following functions:
   a $f(t) = t^2 \cos at$,
   b $f(t) = (t^2 - 3t + 2) \sinh 3t$.

12.29 Determine a function $f(t)$ whose Laplace transform $F(s)$ is given by:

   a $F(s) = \dfrac{d^2}{ds^2}\left(\dfrac{1}{s^2 + 1}\right),$

   b $F(s) = \dfrac{1}{s^2(s^2 - 1)}.$

SUMMARY

The Laplace transform $F(s)$ of a causal function $f(t)$ is defined for $s \in \mathbb{C}$ by

$$F(s) = \int_0^\infty f(t)e^{-st}\, dt.$$

There exists a number $\sigma_a \in \mathbb{R}$ with $-\infty \leq \sigma_a \leq \infty$, such that the integral is absolutely convergent for all $s \in \mathbb{C}$ with $\mathrm{Re}\, s > \sigma_a$ and is not absolutely convergent for all $s \in \mathbb{C}$ with $\mathrm{Re}\, s < \sigma_a$. The number σ_a is called the abscissa of absolute convergence. The case $\sigma_a = \infty$ almost never occurs in practice, since most functions are of exponential order, so $|f(t)| \leq Me^{\alpha t}$ for certain $M > 0$ and $\alpha \in \mathbb{R}$. For ordinary convergence there are similar results; in this case we have a abscissa of convergence σ_c. For the unit step function $\epsilon(t)$ one has, for example, $\sigma_a = \sigma_c = 0$; for $\mathrm{Re}\, s > 0$ the Laplace transform of $\epsilon(t)$ is given by $1/s$. A number of standard Laplace transforms, together with their abscissa of convergence, are given in table 7.

There is a simple relationship between the Laplace and the Fourier transform: $(\mathcal{L}f)(\sigma + i\omega) = (\mathcal{F}\epsilon(t)f(t)e^{-\sigma t})(\omega)$. Therefore, the properties of the Laplace transform are very similar to the properties of the Fourier transform. In this chapter the following properties were treated: linearity, shifting in the time and the s-domain, scaling in the time domain, differentiation in the time and the s-domain, and integration in the time domain. These properties are summarized in table 8. In particular it was shown that the Laplace transform is an analytic function in a certain half-plane in $\mathbb{C}$.

SELFTEST

12.30   Consider the function

$$f(t) = \begin{cases} 2t & \text{for } 0 \leq t < 1, \\ t & \text{for } t \geq 1. \end{cases}$$

a   Sketch the graph of f and write f as a linear combination of functions of the form $\epsilon(t-a)$ and $\epsilon(t-b)(t-b)$.
b   Determine the Laplace transform $F(s)$ of $f(t)$.
c   Determine f' (at all points where f' exists) and determine $\mathcal{L}f'$.
d   Is the differentiation rule in the time domain valid in this case? Explain your answer.

12.31   In this exercise $F(s)$ is the Laplace transform of a certain piecewise smooth causal function $f(t)$.
a   Determine $\mathcal{L}f(t)\sin at$, where $a \in \mathbb{R}$.
b   Determine $\mathcal{L}e^{-2t}f(3t)$ when $F(s) = e^{-s}/s$.
c   Determine $\mathcal{L}\int_0^t \tau^3 f(\tau)\,d\tau$ when it is also given that $f(t)$ is continuous on $\mathbb{R}$.

12.32   Determine the Laplace transform $F(s)$ of the following functions:
a $f(t) = 3e^{t-2} + \epsilon(t-2)$,
b $f(t) = (t-1)^2$,
c $f(t) = e^{2t}\epsilon(t-4)$,
d $f(t) = e^{-t}(\cos 2t + i\sin 2t)$,
e $f(t) = e^{t+3}\epsilon(t-2)\sin(t-2)$,
f $f(t) = 3^t\cos 2t$,

$$\textbf{g} \quad f(t) = \begin{cases} 1 & \text{for } 0 \leq t < 1, \\ 0 & \text{for } 1 \leq t < 2, \\ 1 & \text{for } 2 \leq t < 3, \\ 0 & \text{for } t \geq 3. \end{cases}$$

12.33   Determine a function $f(t)$ whose Laplace transform $F(s)$ is given by the following functions:
a $F(s) = (1 - e^{-s})/s$,
b $F(s) = (s^3 + 3)/s^4$,
c $F(s) = 1/(s+1)^2 + 1/(s^2 - 4) + (1 + e^{-\pi s})/(s^2 + 1)$,
d $F(s) = (3s - 2)/(s^2 - 4s + 20)$,
e $F(s) = (s + 3)/(s^2 + 8s + 16)$,
f $F(s) = e^{-4s}/(s-2)^3$,
g $F(s) = e^{-s}/(s(s^2 + 9))$.

Contents of Chapter 13

Further properties, distributions, and the fundamental theorem

Further properties, distributions, and the fundamental theorem

INTRODUCTION

In the first three sections of this chapter the number of properties of the Laplace transform will be extended even further. We start in section 13.1 with the treatment of the by now well-known convolution product. As for the Fourier transform, the convolution product is transformed into an ordinary product by the Laplace transform.

In section 13.2 we treat two theorems that have not been encountered earlier in the Fourier transform: the so-called *initial* and *final value* theorems for the Laplace transform. The initial value theorem relates the 'initial value' $f(0+)$ of a function $f(t)$ to the behaviour of the Laplace transform $F(s)$ for $s \to \infty$. Similarly, the final value theorem relates the 'final value' $\lim_{t \to \infty} f(t)$ to the behaviour of $F(s)$ for $s \to 0$. Hence, the function $F(s)$ can provide information about the behaviour of the original function $f(t)$ shortly after switching on (the value $f(0+)$) and 'after a considerable amount of time' (the value $\lim_{t \to \infty} f(t)$).

In section 13.3 we will see how the Laplace transform of a periodic function can be determined. It will turn out that this is closely related to the Laplace transform of the function which arises when we limit the periodic function to one period.

In order to determine the Laplace transform of a periodic function, it is not necessary to turn to the theory of distributions. This is in contrast to the Fourier transform (see section 9.1.2). Still, a limited theory of the Laplace transform of distributions will be needed. The delta function, for example, remains an important tool as a model for a strong signal with a short duration (a 'pulse'). Moreover, the response to the delta function is essential in the theory of linear systems (see chapter 14). The theory of the Laplace transform of distributions will be developed in section 13.4. In particular it will be shown that the Laplace transform of the delta function is the constant function 1, just as for the Fourier transform. We will also go into the relationship between the Laplace transform of distributions and differentiation, and we will treat some simple results on the Laplace transform and convolution of distributions.

In section 13.5 the fundamental theorem of the Laplace transform is proven. In the theory of the Laplace transform this is an important theorem; it implies, for example, that the Laplace transform is one-to-one. However, in order to apply the fundamental theorem in practice (and so recover the function $f(t)$ from $F(s)$), a fair amount of knowledge of the theory of complex integration is needed. This theory is beyond the scope of this book. Therefore, if we want to recover $f(t)$ from $F(s)$, we will confine ourselves to the use of tables, the properties of the Laplace transform, and partial fraction expansions. This method will be illustrated by means of examples.

LEARNING OBJECTIVES
After studying this chapter it is expected that you
- know and can apply the convolution of causal functions and the convolution theorem of the Laplace transform
- know and can apply the initial and final value theorems
- can determine the Laplace transform of a periodic function
- know and can apply the Laplace transform of some simple distributions
- know and can apply the differentiation rule for the Laplace transform of distributions
- can apply the convolution theorem for distributions in simple cases
- know the uniqueness theorem for the Laplace transform
- can find the inverse Laplace transform of complex functions by using a table, applying the properties of the Laplace transform, and applying partial fraction expansions.

13.1 Convolution

We have already encountered the convolution product of two functions f and g : $\mathbb{R} \to \mathbb{C}$ in definition 6.4. When, moreover, f and g are both causal (which is assumed throughout part 4), then one has for $t > 0$ that

Convolution of causal functions

$$(f * g)(t) = \int_{-\infty}^{\infty} f(\tau)g(t - \tau)\,d\tau = \int_0^t f(\tau)g(t - \tau)\,d\tau,$$

since the integrand is zero for both $\tau < 0$ and $t - \tau < 0$. For the same reason one has $(f * g)(t) = 0$ if $t \leq 0$ (also see exercise 6.25). If, moreover, we assume that the causal functions f and g are piecewise smooth, then the existence of the convolution product is easy to prove. In fact, for fixed $t > 0$ the function $\tau \to f(\tau)g(t - \tau)$ is then again piecewise smooth as a function of τ and such a function is always integrable over the bounded interval $[0, t]$. Hence, for two piecewise smooth causal functions f and g, the convolution product exists for every $t \in \mathbb{R}$ and $f * g$ is again a causal function. One now has the following convolution theorem (compare with theorem 6.13).

THEOREM 13.1
Convolution theorem for $\mathcal{L}$

Let f and g be piecewise smooth and causal functions. Let the Laplace transforms $F = \mathcal{L}f$ and $G = \mathcal{L}g$ exist as absolutely convergent integrals in a half-plane $\operatorname{Re} s > \rho$. Then $\mathcal{L}(f * g)$ exists for $\operatorname{Re} s > \rho$ and

$$\mathcal{L}(f * g)(s) = F(s)G(s). \tag{13.1}$$

Proof
Since $f(t) = g(t) = 0$ for $t < 0$, it follows for $\operatorname{Re} s > \rho$ that

$$F(s)G(s) = \int_{-\infty}^{\infty} f(t)e^{-st}\,dt \int_{-\infty}^{\infty} g(u)e^{-su}\,du.$$

Since the second integral does not depend on t, we can write

$$F(s)G(s) = \int_{-\infty}^{\infty} \left(\int_{-\infty}^{\infty} f(t)g(u)e^{-s(t+u)}\,du \right) dt.$$

Now change to the new variable $t + u = \tau$, then

$$F(s)G(s) = \int_{-\infty}^{\infty} \left(\int_{-\infty}^{\infty} f(t)g(\tau - t)e^{-s\tau}\,d\tau \right) dt.$$

Without proof we mention that under the conditions of theorem 13.1 we may change the order of integration. We then obtain:

$$F(s)G(s) = \int_{-\infty}^{\infty} e^{-s\tau} \left(\int_{-\infty}^{\infty} f(t)g(\tau - t)\, dt \right) d\tau.$$

As indicated above, the inner integral is the convolution product of the two causal functions f and g. Hence, $\mathcal{L}(f * g)$ exists and $\mathcal{L}(f * g)(s) = F(s)G(s)$. ∎

In theorem 13.1 a half-plane is mentioned for which the Laplace transforms F and G exist as absolutely convergent integrals. If $f(t)$ and $g(t)$ are of exponential order for $\alpha \in \mathbb{R}$ and $\beta \in \mathbb{R}$ respectively, then more can be said. For we then know from theorem 12.3 that the integrals for $F(s)$ and $G(s)$ are absolutely convergent for $\operatorname{Re} s > \alpha$ and $\operatorname{Re} s > \beta$. Hence, in this case both Laplace transforms exist as absolutely convergent integrals in the half-plane $\operatorname{Re} s > \rho$, where $\rho = \max(\alpha, \beta)$.

EXAMPLE 13.1

Let $f(t) = e^t$ and $g(t) = t$, then $F(s) = 1/(s - 1)$ and $G(s) = 1/s^2$ (see table 7), so $F(s)G(s) = 1/(s^2(s - 1))$. From the convolution theorem it then follows that $\mathcal{L}(e^v * v)(s) = 1/(s^2(s - 1))$. The convolution theorem can easily be verified in this case by calculating the convolution product and then determining its the Laplace transform. We calculate the convolution product using integration by parts:

$$(e^v * v)(t) = \int_0^t e^\tau (t - \tau)\, d\tau = \left[e^\tau (t - \tau) \right]_0^t + \int_0^t e^\tau\, d\tau$$

$$= -t + \left[e^\tau \right]_0^t = -t + e^t - 1.$$

Furthermore, we have $\mathcal{L}(e^t - t - 1)(s) = 1/(s - 1) - 1/s^2 - 1/s$ and a simple calculation will show that this result is indeed equal to $1/(s^2(s - 1))$. ◀

EXAMPLE

Let $f(t)$ be an arbitrary (causal) function and $g(t) = \epsilon(t)$. Then

$$(f * g)(t) = \int_0^t f(\tau)\epsilon(t - \tau)\, d\tau = \int_0^t f(\tau)\, d\tau,$$

while $F(s)G(s) = F(s)/s$. Hence, in this case the convolution theorem reduces to the integration rule (12.14) from section 12.3.3. ◀

EXAMPLE

Suppose one is asked to determine a function f with $(\mathcal{L}f)(s) = F(s) = 1/(s^2 + 1)^2$. From table 7 we know that $(\mathcal{L}\sin t)(s) = 1/(s^2 + 1)$ and by virtue of the convolution theorem we thus have $f(t) = (\sin v * \sin v)(t)$. It is not so hard to calculate this convolution using the trigonometric identity $2 \sin \alpha \sin \beta = \cos(\alpha - \beta) - \cos(\alpha + \beta)$:

$$(\sin v * \sin v)(t) = \int_0^t \sin \tau \sin(t - \tau)\, d\tau = \tfrac{1}{2} \int_0^t (\cos(2\tau - t) - \cos t)\, d\tau$$

$$= \tfrac{1}{2} \left[\tfrac{1}{2} \sin(2\tau - t) \right]_0^t - \tfrac{1}{2} \cos t \, [\tau]_0^t = \tfrac{1}{2} \sin t - \tfrac{1}{2} \cos t.$$

Check for yourself that indeed $(\mathcal{L}(\sin t - t \cos t)/2)(s) = 1/(s^2 + 1)^2$. ◀

EXERCISES

13.1

Show that $(\cos v * \cos v)(t) = (\sin t + t \cos t)/2$. Use this to verify the convolution theorem for the functions $f(t) = g(t) = \cos t$. (Suggestion: use the trigonometric identity $2 \cos \alpha \cos \beta = \cos(\alpha + \beta) + \cos(\alpha - \beta)$.)

13.2

a Determine a function f with $(\mathcal{L}f)(s) = F(s) = 1/(s - a)$, where $a \in \mathbb{C}$.

b Use the convolution theorem to determine a function $g(t)$ such that

$(\mathcal{L}g)(s) = 1/((s-a)(s-b))$. Calculate the obtained convolution product explicitly and verify the convolution theorem for this case.

13.3 Verify the convolution theorem for the functions $f(t) = t^2$ and $g(t) = e^t$.

13.4 Determine a convolution product $(f * g)(t)$ whose Laplace transform is given by the following complex function:
a $1/(s^2(s + 1))$,
b $s/((s + 2)(s^2 + 4))$,
c $s/(s^2 - 1)^2$,
d $1/(s^2 - 16)^2$.

13.2 Initial and final value theorems

In this section we treat two theorems that can give us information about a function $f(t)$ straight from its Laplace transform $F(s)$, without the need to determine $f(t)$ explicitly. The issues at stake are the limiting value of f at the point $t = 0$, so immediately after the 'switching on', and the limiting value of f for $t \to \infty$, that is, the final value after 'a long period of time'. These two results are therefore called the initial and final value theorems. As a matter of fact, they can also be used in the *opposite* direction: given $f(t)$, one obtains from these theorems information about the behaviour of $F(s)$ for $s \to 0$ and for $s \to \infty$, without having to determine $F(s)$ explicitly.

A brief explanation of the notation '$s \to \infty$' is appropriate here, since s is complex. In general it will mean that $|s| \to \infty$ (so the modulus of s keeps increasing). In most cases this cannot be allowed for a Laplace transform $F(s)$ since we might end up outside the half-plane of convergence. By $\lim_{s \to \infty} F(s)$ we will therefore always mean that $|s| \to \infty$ and that simultaneously $\operatorname{Re} s \to \infty$. In particular, s will lie in the half-plane of convergence for sufficiently large values of $\operatorname{Re} s$ (specifically, for $\operatorname{Re} s > \sigma_c$).

Similar remarks apply to the limit $s \to 0$, which will again mean that $|s| \to 0$ (as in section 11.2). If the limit for $s \to 0$ of a Laplace transform $F(s)$ is to exist, then $F(s)$ will certainly have to exist in the half-plane $\operatorname{Re} s > 0$. When the half-plane of convergence is precisely $\operatorname{Re} s > 0$, then the limit for $s \to 0$ has to be taken in such a way that $\operatorname{Re} s > 0$ as well. By $\lim_{s \to 0} F(s)$ we will therefore always mean that $|s| \to 0$ and that simultaneously $\operatorname{Re} s \downarrow 0$.

Before we start our treatment of the initial value theorem, we will first derive the following result, which, for that matter, is also useful in other situations and will therefore be formulated for a somewhat larger class of functions.

THEOREM 13.2

For the Laplace transform $F(s)$ of a piecewise continuous function $f(t)$ we have

$$\lim_{s \to \infty} F(s) = 0,$$

where the limit $s \to \infty$ has to be taken in such a way that $\operatorname{Re} s \to \infty$ *as well.*

Proof
If $s \to \infty$ such that $\operatorname{Re} s \to \infty$ as well, then $\lim_{s \to \infty} e^{-st} = 0$ for any fixed $t > 0$. In fact, for $s = \sigma + i\omega$ we have $e^{-st} = e^{-\sigma t} e^{-i\omega t}$ and $\lim_{\sigma \to \infty} e^{-\sigma t} = 0$ for any $t > 0$. Hence we obtain that

$$\lim_{s \to \infty} F(s) = \lim_{s \to \infty} \int_0^\infty f(t) e^{-st} \, dt = 0,$$

if we assume that we may interchange the integral and the limit. When the function $f(t)$ is of exponential order, then the problem of the interchanging of the limit and the integral can be avoided. For if $|f(t)| \leq Me^{\alpha t}$, then

$$|F(s)| \leq \int_0^\infty |f(t)|e^{-\sigma t}\,dt \leq M\int_0^\infty e^{(\alpha-\sigma)t}\,dt = \frac{M}{\alpha-\sigma}\left[e^{(\alpha-\sigma)t}\right]_0^\infty$$

for all $\sigma \neq \alpha$. We also agreed that $\operatorname{Re} s = \sigma \to \infty$ and for sufficiently large σ one will have that $\sigma > \alpha$, so $\alpha - \sigma < 0$. It thus follows that

$$\left[e^{(\alpha-\sigma)t}\right]_0^\infty = \lim_{R\to\infty} e^{(\alpha-\sigma)R} - 1 = -1,$$

which proves that $|F(s)| \leq M/(\sigma - \alpha)$. In the limit $s \to \infty$ with $\operatorname{Re} s = \sigma \to \infty$, the right-hand side of this inequality tends to zero, from which it follows that $\lim_{s\to\infty} F(s) = 0$ as well. $\blacksquare$

EXAMPLE

For $f(t) = \epsilon(t)$ we have $F(s) = 1/s$. Indeed, $\lim_{s\to\infty} F(s) = 0$. ◄

EXAMPLE 13.2

The constant function $F(s) = 1$ cannot be the Laplace transform of a piecewise continuous function $f(t)$. This is because $\lim_{s\to\infty} F(s) = 1$. ◄

We recall that for a piecewise smooth function $f(t)$ the limit $f(0+) = \lim_{t\downarrow 0} f(t)$ will always exist. The *initial value theorem* is a stronger version of theorem 13.2 and reads as follows.

THEOREM 13.3
Initial value theorem

Let $f(t)$ be a piecewise smooth function with Laplace transform $F(s)$. Then

$$\lim_{s\to\infty} sF(s) = f(0+), \tag{13.2}$$

where the limit $s \to \infty$ has to be taken in such a way that $\operatorname{Re} s \to \infty$ as well.

Proof
We will not prove the theorem in its full generality. However, if we impose an additional condition on the function $f(t)$, then a simpler proof of the initial value theorem can be given using theorem 13.2. We therefore assume that in addition $f(t)$ is continuous for $t > 0$. Let f' be the derivative of f at all points where f' exists. As in the proof of theorem 12.7, it then follows from an integration by parts that

$$(\mathcal{L}f')(s) = \lim_{R\to\infty} f(R)e^{-sR} - f(0+) + sF(s) = sF(s) - f(0+),$$

since $\lim_{R\to\infty} e^{-sR} = 0$ (see the proof of theorem 12.7). The difference with theorem 12.7 is the appearance of the value $f(0+)$ because f is not necessarily continuous at $t = 0$. If we now apply theorem 13.2 to $f'(t)$, then it follows that $\lim_{s\to\infty}(\mathcal{L}f')(s) = 0$. Hence we obtain that $\lim_{s\to\infty}(sF(s) - f(0+)) = 0$, which proves the theorem, under the additional condition mentioned earlier. $\blacksquare$

Theorem 13.3 can be used in both directions. When $F(s)$ is known and $f(t)$ is hard to determine explicitly, then one can still determine $f(0+)$, provided that we know that f is piecewise smooth. When on the other hand $f(t)$ is known and $F(s)$ is hard to determine, then theorem 13.3 reveals information about the behaviour of $F(s)$ for $s \to \infty$.

EXAMPLE

The function $\epsilon(t)$ has the function $F(s) = 1/s$ as Laplace transform. Indeed, $1 = \epsilon(0+) = \lim_{s\to\infty} sF(s)$. ◄

EXAMPLE

Consider the function $f(t) = e^{-bt}\cosh at$. Then $f(0+) = 1$ and the Laplace transform $F(s)$ exists, so $\lim_{s\to\infty} sF(s) = 1$. This can easily be verified since $F(s) = (s+b)/((s+b)^2 - a^2)$ (see table 7). ◄

We now move on to the *final value theorem*, which relates the final value $f(\infty) = \lim_{t \to \infty} f(t)$ (a notation which will be used henceforth) to the behaviour of $F(s)$ for $s \to 0$.

THEOREM 13.4
Final value theorem

Let $f(t)$ be a piecewise smooth function with Laplace transform $F(s)$. When $f(\infty) = \lim_{t \to \infty} f(t)$ exists, then

$$\lim_{s \to 0} s F(s) = f(\infty), \tag{13.3}$$

where the limit $s \to 0$ has to be taken in such a way that $\operatorname{Re} s \downarrow 0$ as well.

Proof
Again, theorem 13.4 will not be proven in full generality. If we impose a number of additional conditions on the function, then a simpler proof can be given, as was the case for theorem 13.3. We first of all assume that in addition $f(t)$ is continuous for $t > 0$. As in the proof of theorem 13.3, it then follows that $(\mathcal{L}f')(s) = s F(s) - f(0+)$. Next we will assume that this result is valid in the half-plane $\operatorname{Re} s > 0$. For the limit $s \to 0$ (with $\operatorname{Re} s \downarrow 0$) we then have

$$\lim_{s \to 0} s F(s) = f(0+) + \lim_{s \to 0} \int_0^\infty f'(t) e^{-st} \, dt.$$

Now, finally, assume that the limit and the integral may be interchanged, then we obtain that

$$\lim_{s \to 0} s F(s) = f(0+) + \int_0^\infty f'(t) \, dt.$$

Since f is piecewise smooth and continuous for $t > 0$ and since, moreover, $\lim_{t \to \infty} f(t)$ exists, we have finally established that

$$\lim_{s \to 0} s F(s) = f(0+) + [f(t)]_0^\infty = \lim_{t \to \infty} f(t) = f(\infty).$$

This proves theorem 13.4, using quite a few additional conditions. ∎

We note once again that $F(s)$ in theorem 13.4 must surely exist for $\operatorname{Re} s > 0$, because otherwise one cannot take the limit $s \to 0$. When $F(s)$ is a rational function, then this means in particular that the denominator cannot have any zero for $\operatorname{Re} s > 0$ (see example 13.3).

One cannot omit the condition that $\lim_{t \to \infty} f(t)$ should exist. This can be shown using a simple example. The function $f(t) = \sin t$ has Laplace transform $F(s) = 1/(s^2 + 1)$. We have $\lim_{s \to 0} s F(s) = 0$, but $\lim_{t \to \infty} f(t)$ does not exist.

Theorem 13.4 can again be applied in two directions. When $F(s)$ is known, $f(\infty)$ can be determined, provided that we know that $f(\infty)$ exists. If, on the other hand, $f(t)$ is known and $f(\infty)$ exists, then theorem 13.4 reveals information about the behaviour of $F(s)$ for $s \to 0$.

EXAMPLE

The function $\epsilon(t)$ has Laplace transform $F(s) = 1/s$ and indeed we have $1 = \lim_{t \to \infty} \epsilon(t) = \lim_{s \to 0} s F(s)$. ◄

EXAMPLE 13.3

Consider the function $f(t) = e^{-at}$ with $a > 0$. Then $\lim_{t \to \infty} f(t) = 0$ and so $\lim_{s \to 0} s F(s) = 0$. This can easily be verified since $F(s) = 1/(s + a)$. Note that for $a < 0$ the denominator of $F(s)$ has a zero for $s = -a > 0$, which means that in this case $F(s)$ does not exist in the half-plane $\operatorname{Re} s > 0$. This is in agreement with the fact that $\lim_{t \to \infty} f(t)$ does not exist for $a < 0$. Of course, the complex function $F(s) = 1/(s + a)$ remains well-defined for all $s \neq -a$ and in particular one has for $a \neq 0$ that $\lim_{s \to 0} s F(s) = 0$. However, the function $F(s)$ is *not* the Laplace transform of the function $f(t)$ for $\operatorname{Re} s \leq (-a)$. ◄

EXERCISES

13.5 Can the function $F(s) = s^n$ ($n \in \mathbb{N}$) be the Laplace transform of a piecewise continuous function $f(t)$? Justify your answer.

13.6 Determine the Laplace transform $F(s)$ of the following functions $f(t)$ and verify the initial value theorem:
a $f(t) = \cosh 3t$,
b $f(t) = 2 + t \sin t$,
c $f(t) = \int_0^t g(\tau)\, d\tau$, where g is a continuous function on $\mathbb{R}$.

13.7 In the proof of theorem 13.3 we used the property $(\mathcal{L}f')(s) = s(\mathcal{L}f)(s) - f(0+)$. Now consider the function $f(t) = \epsilon(t-1)\cos(t-1)$.
a Verify that the stated property does not hold for f.
b Show that the initial value theorem does apply.

13.8 Determine the Laplace transform $F(s)$ of the following functions $f(t)$ and verify the final value theorem:
a $f(t) = e^{-3t}$,
b $f(t) = e^{-t} \sin 2t$,
c $f(t) = 1 - \epsilon(t-1)$.

13.9 Determine whether the final value theorem can be applied to the functions $\cos t$ and $\sinh t$.

13.10 For the complex function $F(s) = 1/(s(s-1))$ one has $\lim_{s \to 0} sF(s) = -1$. Let $f(t)$ be the function with Laplace transform $F(s)$ in a certain half-plane in $\mathbb{C}$.
a Explain, without determining $f(t)$, why the final value theorem cannot be applied.
b Verify that $F(s) = 1/(s-1) - 1/s$. Subsequently determine $f(t)$ and check that $f(\infty) = \lim_{t \to \infty} f(t)$ does not exist.

13.3 Periodic functions

In general, the Laplace transform is more comfortable to use than the Fourier transform since many of the elementary functions possess a Laplace transform, while on the contrary the Fourier transform often only exists when the function is considered as a distribution. For a periodic function the Fourier transform also exists only if it is considered as a distribution (see section 9.1.2). In this section we will see that the Laplace transform of a periodic function can easily be determined without distribution theory. Since we are only working with causal functions in the Laplace transform, a periodic function $f(t)$ with period $T > 0$ will from now on be a function on $[0, \infty)$ for which $f(t + T) = f(t)$ for all $t \geq 0$. In figure 13.1a a periodic function with period T is drawn. Now consider the function $\phi(t)$ obtained from $f(t)$ by restricting $f(t)$ to one period T, so

Causal periodic function

$$\phi(t) = \begin{cases} f(t) & \text{for } 0 \leq t < T, \\ 0 & \text{elsewhere.} \end{cases}$$

See figure 13.1b. Using the shifted unit step function, the function $\phi(t)$ can be written as

$$\phi(t) = f(t) - \epsilon(t - T)f(t - T).$$

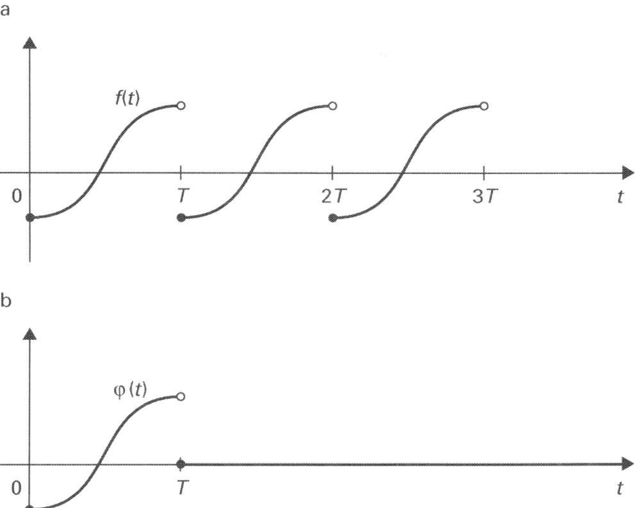

FIGURE 13.1
A periodic function f (a) and its restriction ϕ (b).

If we now apply the shift property in the time domain (theorem 12.4), then it follows that

$$\Phi(s) = F(s) - e^{-sT} F(s) = (1 - e^{-sT}) F(s),$$

where $\Phi(s)$ and $F(s)$ are the Laplace transforms of ϕ and f respectively. If $s = \sigma + i\omega$ and $\sigma > 0$, then $\left| e^{-sT} \right| = e^{-\sigma T} < 1$, and hence $1 - e^{-sT} \neq 0$. For $\operatorname{Re} s > 0$ we may thus divide by $1 - e^{-sT}$ and it then follows that

$$F(s) = \frac{\Phi(s)}{1 - e^{-sT}}, \qquad \text{where} \qquad \Phi(s) = \int_0^{\infty} \phi(t) e^{-st}\, dt = \int_0^{T} f(t) e^{-st}\, dt.$$

Note that for a piecewise continuous function the preceding integral over the bounded interval $[0, T]$ exists for every $s \in \mathbb{C}$; for the function ϕ the abscissa of convergence is thus equal to $-\infty$. For the periodic function f the abscissa of convergence is equal to 0 and hence the Laplace transform $F(s)$ exists for $\operatorname{Re} s > 0$. We see here that the Laplace transform of a periodic function can be expressed in a simple way in terms of the Laplace transform of the function restricted to one period. These results are summarized in the following theorem.

THEOREM 13.5
Laplace transform of periodic functions

Let f be a piecewise smooth and periodic function with period T and let $\Phi(s)$ be the Laplace transform of $\phi(t) = f(t) - \epsilon(t - T) f(t - T)$. Then the Laplace transform $F(s)$ of $f(t)$ is for $\operatorname{Re} s > 0$ given by

$$F(s) = \frac{\Phi(s)}{1 - e^{-sT}}, \qquad \text{where} \qquad \Phi(s) = \int_0^{T} f(t) e^{-st}\, dt. \tag{13.4}$$

EXAMPLE 13.4

Consider the periodic block function $f(t)$ with period 2 defined by $f(t) = 1 - \epsilon(t - 1)$ for $0 \leq t < 2$; so $f(t) = 1$ for $0 \leq t < 1$ and $f(t) = 0$ for $1 \leq t < 2$. See figure 13.2. We then have $\phi(t) = 1 - \epsilon(t - 1)$ and from tables 7 and 8 we see

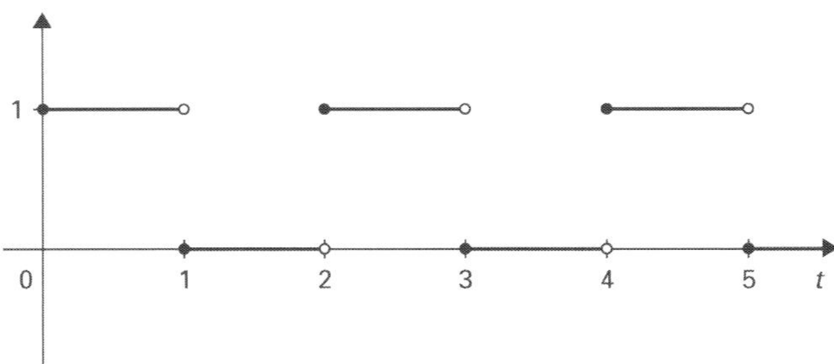

FIGURE 13.2
Periodic block function.

that $(\mathcal{L}\phi(t))(s) = \Phi(s) = 1/s - e^{-s}/s$. Using theorem 13.5 we thus find for the Laplace transform $F(s)$ of $f(t)$ that

$$F(s) = \frac{1 - e^{-s}}{s(1 - e^{-2s})} = \frac{1}{s(1 + e^{-s})}.$$

◀

EXERCISES

13.11 Let $f(t)$ for $t \geq 0$ be a periodic function with period T.
a Can we apply the final value theorem to the periodic function f? Justify your answer.
b Show that

$$\lim_{s \to 0} s F(s) = \frac{1}{T} \int_0^T f(t)\, dt.$$

In other words, in this case the limit $\lim_{s \to 0} s F(s)$ equals the average value of f over one period. (Suggestion: use the definition of the derivative of the complex function e^{-zT} at the point $z = 0$.)
c Verify the result from part b for the function $f(t)$ from example 13.4.

13.12 In figure 13.3 the graph is drawn of a periodic block function $f(t)$ with period $2a$, defined for $0 \leq t < 2a$ by

$$f(t) = \begin{cases} 1 & \text{for } 0 \leq t < a, \\ -1 & \text{for } a \leq t < 2a. \end{cases}$$

a Define $\phi(t)$ as the restriction of f to the period $[0, 2a)$. Show that $\phi(t) = \epsilon(t) - 2\epsilon(t - a) + \epsilon(t - 2a)$. Determine the Laplace transform $F(s)$ of $f(t)$.
b Show that $F(s) = (\tanh(as/2))/s$, where $\tanh z = \sinh z / \cosh z = (e^z - e^{-z})/(e^z + e^{-z})$.

13.13 The periodic sawtooth function f with period 2 is given by $f(t) = t(\epsilon(t) - \epsilon(t - 2)) - 2(\epsilon(t - 1) - \epsilon(t - 2))$ for $0 \leq t < 2$.
a Sketch the graph of f.
b Determine $F(s) = (\mathcal{L}f)(s)$.

13.14 Let f be the periodic function with period 2 given by $f(t) = t$ for $0 \leq t < 2$. Show that $(\mathcal{L}f)(s) = F(s) = (1 + 2s)/s^2 - 2/(s(1 - e^{-2s}))$.

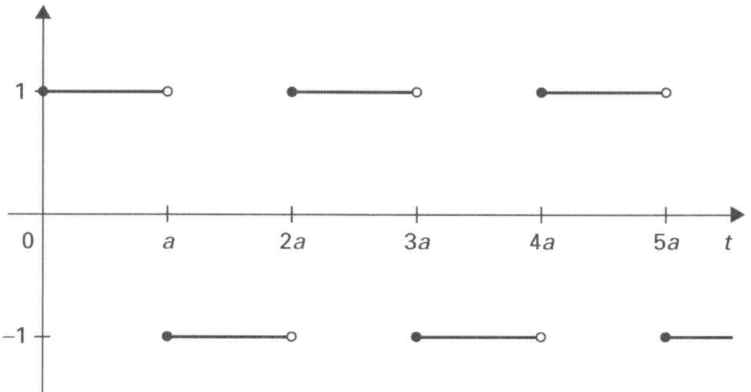

FIGURE 13.3
Periodic block function of exercise 13.12.

13.4 Laplace transform of distributions

Up till now we have been able to avoid the use of the theory of distributions in the Laplace transform. For every function we could always calculate the Laplace transform using the defining integral (12.1). Still, we will need a limited theory of the Laplace transform of distributions. This is because the delta function will remain an important tool in the theory of linear systems: in the application of the Laplace transform the impulse response again plays an essential role (see chapter 14).

In section 13.4.1 the main results will be derived in an intuitive way. For the remainder of this book it will suffice to accept these results as being correct. In section 13.4.2 we treat the mathematical background necessary to give a rigorous definition of the Laplace transform of a distribution. This will enable us to prove the results from section 13.4.1. Section 13.4.2 may be omitted without any consequences for the remainder of the book.

13.4.1 Intuitive derivation

To get an intuitive idea of the Laplace transform of the delta function $\delta(t)$, we consider the causal rectangular pulse $r_b(t)$ of height $1/b$ and duration $b > 0$. Hence, $r_b(t) = (\epsilon(t) - \epsilon(t - b))/b$. See figure 13.4. Note that $\int_{-\infty}^{\infty} r_b(t)\,dt = 1$ for every $b > 0$. For $b \downarrow 0$ we thus obtain an object which, intuitively, will be an approximation for the delta function (see section 8.1). Since $(\mathcal{L}r_b)(s) = (1 - e^{-bs})/sb$, we expect that for $b \downarrow 0$ this will give us the Laplace transform of the delta function. When $b \downarrow 0$, then also $-sb \to 0$ for any $s \in \mathbb{C}$. Now write $z = -sb$, then we have to determine the limit $\lim_{z \to 0}(e^z - 1)/z$. But this is precisely the derivative of the analytic function e^z at $z = 0$. Since $(e^z)' = e^z$, we obtain for $z = 0$ that $(e^z)'(0) = 1$ and hence $\lim_{b \downarrow 0}(\mathcal{L}r_b)(s) = 1$. As for the Fourier transform, we thus expect that the Laplace transform of the delta function will equal the constant function 1.

We will now try to find a possible definition for the Laplace transform of a distribution. First we recall that a function $f(t)$ can be considered as a distribution T_f by means of the rule

$$\langle T_f, \phi \rangle = \int_{-\infty}^{\infty} f(t)\phi(t)\,dt \qquad \text{for } \phi \in \mathcal{S}. \tag{13.5}$$

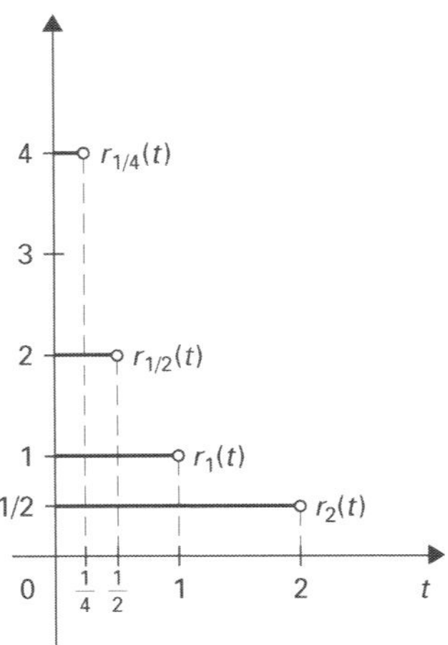

FIGURE 13.4
The rectangular pulse function $r_b(t)$ for some values of b.

(See (8.12).) If we now take for ϕ the function e^{-st}, then it follows for a causal function $f(t)$ that

$$\langle T_f, e^{-st} \rangle = \int_0^\infty f(t)e^{-st}\, dt = F(s). \tag{13.6}$$

The definition of the Laplace transform $U = \mathcal{L}T$ of a distribution T now seems quite obvious, namely as the complex function

$$U(s) = \langle T(t), e^{-st} \rangle. \tag{13.7}$$

Note that $T(t)$ acts on the variable t and that s is always an (arbitrary) fixed complex number. Furthermore, the Laplace transform of a distribution is no longer a distribution, but just a complex function. When $f(t)$ is a causal function defining a distribution T_f, then it follows from (13.6) that the definition in (13.7) results in the ordinary Laplace transform $F(s)$ of $f(t)$ again:

$$(\mathcal{L}T_f)(s) = F(s) \tag{13.8}$$

(assuming that $F(s)$ exists in a certain half-plane of convergence).

Two problems arise from definition (13.7). First of all it is easy to see that the function e^{-st} is *not* an element of the space $\mathcal{S}(\mathbb{R})$ of rapidly decreasing functions; hence, $\langle T, e^{-st} \rangle$ is not well-defined for an arbitrary distribution T. A second problem concerns the analogue of the notion 'causal function' for a distribution, since (13.6) is only valid for causal functions. Of course we would like to call the distribution T_f 'causal' if f is a causal function. But what shall we mean in general by a 'causal distribution'? This will have to be a distribution being 'zero for $t < 0$'. In section 13.4.2 we will return to these problems and turn definition (13.7) into a rigorous one. In this section we use (13.7) for all distributions that are 'zero for $t < 0$' according to our intuition. Examples of such distributions are the delta function $\delta(t)$, the derivatives $\delta^{(n)}(t)$ ($n \in \mathbb{N}$), the delta function $\delta(t-a)$ with $a > 0$, and the derivatives $\delta^{(n)}(t-a)$ ($a > 0$ and $n \in \mathbb{N}$).

EXAMPLE 13.5   From formula (13.7) and definition 8.2 of $\delta(t)$ it follows that $(\mathcal{L}\delta(t))(s) = \langle \delta(t), e^{-st} \rangle = 1$ since $e^{-st} = 1$ for $t = 0$. Hence, $\mathcal{L}\delta = 1$.   ◀

EXAMPLE 13.6

For the delta function $\delta(t-a)$ with $a > 0$ it follows from formulas (13.7) and (8.10) that $(\mathcal{L}\delta(t-a))(s) = \langle \delta(t-a), e^{-st} \rangle = e^{-as}$. Hence, $(\mathcal{L}\delta(t-a))(s) = e^{-as}$. ◄

EXAMPLE 13.7

For the derivative $\delta^{(n)}(t)$ it follows from formulas (13.7) and (8.17) that $(\mathcal{L}\delta^{(n)}(t))(s) = \langle \delta^{(n)}(t), e^{-st} \rangle = (-1)^n \langle \delta(t), (e^{-st})^{(n)} \rangle$. But $(e^{-st})^{(n)} = (-s)^n e^{-st}$ and so $(\mathcal{L}\delta^{(n)}(t))(s) = s^n \langle \delta(t), e^{-st} \rangle = s^n$. This proves that $(\mathcal{L}\delta^{(n)}(t))(s) = s^n$. ◄

The properties of the Laplace transform of distributions are similar to the properties of the Laplace transform of functions (and to the properties of the Fourier transform). The simplest property is *linearity*. It follows immediately from the definition in (13.7) and definition 8.5 of the addition of distributions and the multiplication of a distribution by a complex constant.

Linearity

EXAMPLE

The Laplace transform of $3i\delta(t-4) + 5\sin t$ is given by the complex function $3ie^{-4s} + 5/(s^2 + 1)$. ◄

Differentiation in the time domain

Besides linearity, the most important property will be the *differentiation rule in the time domain*: when T is a distribution with Laplace transform $\mathcal{L}T$, then

$$(\mathcal{L}T^{(n)})(s) = s^n(\mathcal{L}T)(s) \tag{13.9}$$

for $n \in \mathbb{N}$. (Compare this with the differentiation rule in the time domain in (12.10).) The proof of (13.9) is easy and follows just as in example 13.7 from formulas (13.7) and (8.17):

$$(\mathcal{L}T^{(n)})(s) = \langle T^{(n)}, e^{-st} \rangle = (-1)^n \langle T, (e^{-st})^{(n)} \rangle = s^n \langle T, e^{-st} \rangle$$
$$= s^n(\mathcal{L}T)(s).$$

EXAMPLE

We know from example 13.5 that $\mathcal{L}\delta = 1$. From (13.9) it then follows that $(\mathcal{L}\delta^{(n)}(t))(s) = s^n$, in agreement with example 13.7. ◄

Formula (13.9) can in particular be applied to a causal function f defining a distribution T_f, and so it is much more general than the differentiation rule in the time domain from theorem 12.7. We will give some examples.

EXAMPLE

The Laplace transform $F(s)$ of $f(t) = \epsilon(t-1)$ is given by $F(s) = e^{-s}/s$. Of course, the function $\epsilon(t-1)$ is not differentiable on $\mathbb{R}$, but considered as a distribution we have that $\epsilon'(t-1) = \delta(t-1)$. According to (13.9) with $n = 1$ one then obtains that $(\mathcal{L}\delta(t-1))(s) = s(\mathcal{L}\epsilon(t-1))(s) = e^{-s}$. This is in accordance with example 13.6. Applying (13.9) for $n \in \mathbb{N}$ it follows that $(\mathcal{L}\delta^{(n-1)}(t-1))(s) = s^n(\mathcal{L}\epsilon(t-1))(s) = s^{n-1}e^{-s}$. It is not hard to obtain this result in a direct way (see exercise 13.16). ◄

EXAMPLE 13.8

For the (causal) function $f(t) = \cos t$ one has $(\cos t)' = \delta(t) - \sin t$, considered as a distribution (see example 8.10). From formula (13.9) with $n = 1$ it then follows that $(\mathcal{L}(\delta(t) - \sin t))(s) = s(\mathcal{L}\cos t)(s)$. This identity can easily be verified since $\mathcal{L}\delta = 1$, $(\mathcal{L}\sin t)(s) = 1/(s^2 + 1)$ and $(\mathcal{L}\cos t)(s) = s/(s^2 + 1)$. ◄

In example 13.8 we encounter a situation that occurs quite often: a causal function $f(t)$ having a jump at the point $t = 0$, being continuously differentiable otherwise, and defining a distribution T_f. Let us assume for convenience that $f(0) = f(0+)$, in other words, let us take the function value at $t = 0$ equal to the limiting value $f(0+)$. The magnitude of the jump at $t = 0$ is then given by the value $f(0)$. We moreover assume that the Laplace transforms of f and f' exist (in the ordinary sense), and according to (13.8) we thus have $\mathcal{L}T_f = \mathcal{L}f$ and $\mathcal{L}T_{f'} = \mathcal{L}f'$. For the derivative T'_f of f, considered as a distribution, one has, according to the jump formula (8.21), that $T'_f = f'(t) + f(0)\delta(t)$, where $f'(t)$ is the derivative of $f(t)$ for $t \neq 0$. Hence,

$$(\mathcal{L}T'_f)(s) = (\mathcal{L}f')(s) + f(0)(\mathcal{L}\delta)(s).$$

Since $\mathcal{L}\delta = 1$ it then follows from (13.9) that

$$(\mathcal{L}f')(s) = (\mathcal{L}T'_f)(s) - f(0) = s(\mathcal{L}f)(s) - f(0),$$

where we used $\mathcal{L}T_f = \mathcal{L}f$. Applying this rule repeatedly, we obtain for a causal function $f(t)$ being n times continuously differentiable for $t \geq 0$ that

$$(\mathcal{L}f^{(n)})(s) = s^n(\mathcal{L}f)(s) - \sum_{k=1}^{n} s^{n-k} f^{(k-1)}(0). \tag{13.10}$$

Here $f^{(k)}$ is the kth derivative of $f(t)$ for $t \neq 0$ and it is assumed that all Laplace transforms exist in the ordinary sense. Formula (13.10) is used especially for solving differential equations by means of the Laplace transform. In addition to an unknown function $f(t)$ satisfying a differential equation, only the values $f^{(k)}(0)$ ($k = 0, 1, \ldots, n - 1$) are given (see chapter 14).

Laplace transform and convolution

We close with some elementary results on the Laplace transform of a *convolution product of distributions*. From section 9.3 it is known that the convolution product $\delta * T$ exists for any distribution T and that $\delta * T = T$ (see (9.19)). If the Laplace transform $\mathcal{L}T$ of T exists, then this implies that $\mathcal{L}(\delta * T) = \mathcal{L}T$, and since $\mathcal{L}\delta = 1$ we thus see that $\mathcal{L}(\delta * T) = \mathcal{L}\delta \cdot \mathcal{L}T$. This shows that in this particular case the convolution theorem for the Laplace transform also holds for distributions. Using the same method one can verify in a direct way the convolution theorem for distributions for a limited number of other cases as well. Let us give a second example. In (9.20) we saw that $\delta' * T = T'$ for a distribution T. Since $(\mathcal{L}T')(s) = s(\mathcal{L}T)(s)$, it follows that $\mathcal{L}(\delta' * T)(s) = s(\mathcal{L}T)(s)$. But $(\mathcal{L}\delta')(s) = s$ and so we indeed have $\mathcal{L}(\delta' * T) = \mathcal{L}\delta' \cdot \mathcal{L}T$.

In most cases these simple results on the Laplace transform of convolution products will suffice in the applications.

13.4.2   Mathematical treatment*

In section 13.4.1 it was pointed out that the definition in (13.7) of the Laplace transform of a distribution gives rise to two problems. First of all it was noted that the function e^{-st} is not an element of the space S. For $\operatorname{Re} s < 0$, for example, we have that $e^{-st} \to \infty$ for $t \to \infty$. This problem is solved by simply allowing a larger class of functions ϕ in (13.5). To this end we replace S by the space $\mathcal{E}$ defined as the set of all C^∞-functions on $\mathbb{R}$ (this space of arbitrarily often differentiable functions has previously been used towards the end of section 9.3). As a consequence of this change, the number of distributions for which we can define the Laplace transform is reduced considerably. However, this is unimportant to us, since we will only need a very limited theory of the Laplace transform of distributions (in fact, only the delta function and its derivatives are needed). Note that the complex-valued function e^{-st} indeed belongs to $\mathcal{E}$.

The second problem involved finding the analogue of the notion of causality for a distribution. This should be a distribution being 'zero for $t < 0$'. If the function

f in (13.5) is causal and we choose a $\phi \in \mathcal{S}$ such that $\phi(t) = 0$ for $t \geq 0$, then it follows that

$$\langle T_f, \phi \rangle = \int_{-\infty}^{\infty} f(t)\phi(t)\,dt = \int_{0}^{\infty} f(t)\phi(t)\,dt = 0.$$

We can now see that the definition of a causal distribution should be as follows.

DEFINITION 13.1
Causal distribution

Let T be a distribution. We say that $T = 0$ on the interval $(-\infty, 0)$ when $\langle T, \phi \rangle = 0$ for every $\phi \in \mathcal{S}$ with $\phi(t) = 0$ for $t \geq 0$. Such a distribution is called a causal distribution.

EXAMPLE

If f is a causal function defining a distribution T_f, then T_f is a causal distribution. This has been shown earlier. ◀

EXAMPLE

The delta function δ is a causal distribution since $\langle \delta, \phi \rangle = \phi(0) = 0$ for $\phi \in \mathcal{S}$ with $\phi(t) = 0$ for $t \geq 0$. More generally we have that the delta function $\delta(t - a)$ with $a \geq 0$ is a causal distribution. In fact, it follows for $\phi \in \mathcal{S}$ with $\phi(t) = 0$ for $t \geq 0$ that $\langle \delta(t - a), \phi \rangle = \phi(a) = 0$ since $a \geq 0$. ◀

The causal distributions, which moreover can be defined on the space $\mathcal{E}$, will now form the set of distributions for which the definition of the Laplace transform of a distribution in (13.7) makes sense.

DEFINITION 13.2
Laplace transform of a distribution

Let T be a distribution which can be defined on the space $\mathcal{E}$ of all C^{∞}-functions on $\mathbb{R}$. Assume moreover that T is causal. Then the Laplace transform $U = \mathcal{L}T$ of T is defined as the complex function $U(s) = \langle T(t), e^{-st} \rangle$.

As was noted following (13.7), the Laplace transform $U(s)$ of a distribution is a complex function. From definition 13.2 we see that $U(s)$ is defined on the whole of $\mathbb{C}$. One even has that $U(s)$ is an *analytic* function on $\mathbb{C}$! The proof of this result is outside the scope of this book; we will not need it anyway. Also, in concrete examples this result will follow from the calculations. We will give some examples.

EXAMPLE

The delta function is a causal distribution which can be defined on the space $\mathcal{E}$, since $\langle \delta, \phi \rangle = \phi(0)$ has a meaning for every *continuous* function ϕ (see section 8.2.2). Hence, the Laplace transform of δ is well-defined and as in example 13.5 it follows that $\mathcal{L}\delta = 1$. Note that the constant function 1 is an analytic function on $\mathbb{C}$. ◀

EXAMPLE

Consider the delta function $\delta(t - a)$ at the point a for $a > 0$. Then $\mathcal{L}\delta(t - a)$ is again well-defined and $(\mathcal{L}\delta(t - a))(s) = e^{-as}$ (see example 13.6). This is again an analytic function on $\mathbb{C}$. ◀

EXAMPLE

For all derivatives of the delta function $\delta(t)$ the Laplace transform is also well-defined and as in example 13.7 it follows that $(\mathcal{L}\delta^{(n)}(t))(s) = s^n$. The function s^n is again an analytic function on $\mathbb{C}$. ◀

Many of the properties that hold for the Laplace transform of functions, can be translated into properties of the Laplace transform of distributions. The linearity and the differentiation rule for the Laplace transform of distributions have already been treated in section 13.4.1. The shift property in the s-domain and the scaling rule also remain valid for distributions, but because of the limited applicability of these rules, we will not prove them. As an illustration we will prove the shift property in the time domain here.

Let $T(t)$ be a distribution whose Laplace transform $U(s)$ exists (so T is causal and defined on $\mathcal{E}$). Then one has for $a \geq 0$ that

Shift in the time domain

$$(\mathcal{L}T(t - a))(s) = e^{-as}U(s), \tag{13.11}$$

where $T(t - a)$ is the distribution shifted over a (see definition 9.2). In order to prove this rule, we first show that $T(t - a)$ is causal. So let $\phi \in S$ with $\phi(t) = 0$ for $t \geq 0$. Then $\langle T(t - a), \phi(t) \rangle = \langle T(t), \phi(t + a) \rangle = 0$ for $t \geq 0$ since T is causal and $t + a \geq 0$ (because $a \geq 0$ and $t \geq 0$). Hence, $T(t - a)$ is causal. It also follows immediately that $T(t - a)$ is defined on $\mathcal{E}$, since T is defined on $\mathcal{E}$ and the function $\psi(t) = \phi(t + a)$ belongs to $\mathcal{E}$ when $\phi(t) \in \mathcal{E}$. Hence, the Laplace transform of $T(t - a)$ exists and from definition 13.2 it then follows that

$$(\mathcal{L}T(t - a))(s) = \langle T(t - a), e^{-st} \rangle = \langle T(t), e^{-s(t+a)} \rangle$$
$$= e^{-as} \langle T(t), e^{-st} \rangle = e^{-as} U(s),$$

proving (13.11).

EXAMPLE

We know that $\mathcal{L}\delta = 1$. From (13.11) it then follows that $(\mathcal{L}\delta(t - a))(s) = e^{-as}$, which is in accordance with example 13.6. ◀

Some simple results on the convolution in relation to the Laplace transform of distributions have already been treated in section 13.4.1. As for the Fourier transform, there are of course general convolution theorems for the Laplace transform of distributions. A theorem comprising all the examples we have treated earlier reads as follows.

THEOREM 13.6
Convolution theorem

Let S and T be causal distributions which can be defined on the space $\mathcal{E}$. Then $S * T$ is a causal distribution which can again be defined on the space $\mathcal{E}$ and $\mathcal{L}(S * T) = \mathcal{L}S \cdot \mathcal{L}T$.

Proof
The proof that $S * T$ is a causal distribution which can be defined on $\mathcal{E}$ is beyond the scope of this book. Assuming this result, it is not hard to prove that $\mathcal{L}(S * T) = \mathcal{L}S \cdot \mathcal{L}T$. For $s \in \mathbb{C}$ one has $\mathcal{L}(S * T)(s) = \langle (S * T)(t), e^{-st} \rangle = \langle S(\tau), \langle T(t), e^{-s(\tau+t)} \rangle \rangle$, where we used definition 9.3 of convolution. It then follows that $\mathcal{L}(S * T)(s) = \langle S(\tau), \langle T(t), e^{-st} e^{-s\tau} \rangle \rangle$. For fixed τ, the complex number $e^{-s\tau}$ does not depend on t. The number $e^{-s\tau}$ can thus be taken outside of the action of the distribution $T(t)$. This results in $\mathcal{L}(S * T)(s) = \langle S(\tau), \langle T(t), e^{-st} \rangle e^{-s\tau} \rangle$. But now $\langle T(t), e^{-st} \rangle$ is, for fixed t, a complex number which does not depend on τ and so it can be taken outside of the action of the distribution $S(\tau)$. This gives the desired result: $\mathcal{L}(S * T)(s) = \langle S(\tau), e^{-s\tau} \rangle \langle T(t), e^{-st} \rangle = (\mathcal{L}S)(s) \cdot (\mathcal{L}T)(s)$. ■

All the examples in section 13.4.1 satisfy the conditions of theorem 13.6. This is because the delta function and all of its derivatives are causal distributions which can be defined on the space $\mathcal{E}$.

EXERCISES

13.15

Consider the function $f_a(t) = ae^{-at} \epsilon(t)$ for $a > 0$ and let $F_a(s) = (\mathcal{L}f_a)(s)$.
a   Sketch the graph of $f_a(t)$ and show that $\int_{-\infty}^{\infty} f_a(t)\, dt = 1$.
b   Show that $\lim_{a \to \infty} f_a(t) = 0$ for every $t > 0$ and that $\lim_{a \to \infty} f_a(t) = \infty$ for $t = 0$. Conclude from parts a and b that the function $f_a(t)$ is an approximation of the delta function $\delta(t)$ for $a \to \infty$.
c   Determine $F_a(s)$ and calculate $\lim_{a \to \infty} F_a(s)$. Explain your answer.

13.16

Use formula (13.7) to determine the Laplace transform of the nth derivative $\delta^{(n)}(t - a)$ of the delta function at the point a for $a \geq 0$.

13.17

Show that the Laplace transform of distributions is linear.

13.18   Verify the convolution theorem of the Laplace transform for $\delta^{(m)} * T$ ($m \in \mathbb{N}$), where T is a distribution whose Laplace transform exists. Hence, show in a direct way that $\mathcal{L}(\delta^{(m)} * T) = \mathcal{L}\delta^{(m)} \cdot \mathcal{L}T$.

13.19   Determine the Laplace transform of the following distributions (and combinations of functions and distributions):

a $\delta(t) + \sin t$,
b $\delta'(t) + 3\delta''(t)$,
c $\epsilon(t) + \delta(t-2) + 2i\delta''(t-4)$,
d $\delta^{(3)}(t) * \delta(t-a)$.

13.20   Determine a combination of functions and/or distributions whose Laplace transform is given by the following complex functions:

a $F(s) = s + 3 - e^{-2s}$,
b $F(s) = (s-2)^2 + 1/(s-2)$,
c $F(s) = e^{-2s}/(s^2+1) + e^{-2s}s^3$,
d $F(s) = s^2/(s^2+1)$.

13.21*   Let T be a distribution with Laplace transform U and consider for $a \geq 0$ the shifted distribution $T(t-a)$ (see definition 9.2). Verify the convolution theorem for $T(t) * \delta(t-a)$. (Hint: see exercise 9.20.)

13.5   The inverse Laplace transform

In this final section on the theory of the Laplace transform we consider the problem of the inverse of the Laplace transform. We start with the proof of the fundamental theorem of the Laplace transform, which describes how a function $f(t)$ in the time domain can be recovered from its Laplace transform $F(s)$. By using the connection between the Fourier and the Laplace transform, the proof of the fundamental theorem is quite easy.

THEOREM 13.7
Fundamental theorem of the Laplace transform

Let $f(t)$ be a piecewise smooth (and causal) function of exponential order $\alpha \in \mathbb{R}$. Let $F(s)$ be the Laplace transform of $f(t)$. Then one has for $t \geq 0$ and $s = \sigma + i\omega$ with $\sigma > \alpha$ that

$$\lim_{A \to \infty} \frac{1}{2\pi} \int_{-A}^{A} F(s)e^{st}\, d\omega = \frac{1}{2}\left(f(t+) + f(t-)\right). \tag{13.12}$$

Proof
Write $s = \sigma + i\omega$ and define $g(t) = \epsilon(t)f(t)e^{-\sigma t}$. Note that $g(t)$ is absolutely integrable for $\sigma > \alpha$ since $f(t)$ is of exponential order $\alpha \in \mathbb{R}$ (see the proof of theorem 12.3). The Fourier transform of $g(t)$ thus exists for $\sigma > \alpha$ and according to (12.2) we then have $F(s) = (\mathcal{F}g)(\omega)$. Since f is piecewise smooth, g is also piecewise smooth and, moreover, absolutely integrable. The fundamental theorem of the Fourier integral (theorem 7.3) can thus be applied to the function g and since $(\mathcal{F}g)(\omega) = F(\sigma + i\omega)$, it then follows from (7.9) that

$$\frac{1}{2\pi} \int_{-\infty}^{\infty} F(\sigma + i\omega)e^{i\omega t}\, d\omega = \frac{1}{2}\left(g(t+) + g(t-)\right).$$

Here the integral should be interpreted as a Cauchy principal value, hence as $\lim_{A \to \infty} \int_{-A}^{A} \dots$. For $t \geq 0$ we have $g(t+) = \epsilon(t+)f(t+)e^{-\sigma t+} = f(t+)e^{-\sigma t}$ and similarly $g(t-) = f(t-)e^{-\sigma t}$, which leads to

$$\lim_{A \to \infty} \frac{1}{2\pi} \int_{-A}^{A} F(\sigma + i\omega)e^{i\omega t}\, d\omega = \frac{1}{2}\left(f(t+) + f(t-)\right)e^{-\sigma t}.$$

If we now multiply the left- and right-hand sides by $e^{\sigma t}$, then (13.12) indeed follows, valid for $\sigma > \alpha$ and $t \geq 0$.   ■

Note that in (13.12) we only integrate over ω and that the value of σ is irrelevant (as long as $\sigma > \alpha$). As for the Fourier transform (see section 7.2.1), the fundamental theorem immediately implies that the Laplace transform is *one-to-one*.

THEOREM 13.8
The Laplace transform is one-to-one

Let $f(t)$ and $g(t)$ be two piecewise smooth functions of exponential order and let $F(s)$ and $G(s)$ be the Laplace transforms of $f(t)$ and $g(t)$. When $F(s) = G(s)$ in a half-plane $\mathrm{Re}\, s > \rho$, then $f(t) = g(t)$ at all points where f and g are continuous.

Proof
Let $t \in \mathbb{R}$ be a point where both f and g are continuous. Since $F(s) = G(s)$ for $\mathrm{Re}\, s > \rho$, it follows from the fundamental theorem that (in the following integrals we have $s = \sigma + i\omega$ with $\sigma > \rho$)

$$f(t) = \lim_{A \to \infty} \frac{1}{2\pi} \int_{-A}^{A} F(s)e^{st}\, d\omega = \lim_{A \to \infty} \frac{1}{2\pi} \int_{-A}^{A} G(s)e^{st}\, d\omega = g(t).$$

                                    ■

This theorem is often used implicitly if we are asked to determine the function $f(t)$ whose Laplace transform $F(s)$ is given. Suppose that an $f(t)$ is found within the class of piecewise smooth functions of exponential order. Then we know by theorem 13.8 that this is the only possible function within this class, except for a finite number of points on a bounded interval (also see the similar remarks on the uniqueness of the Fourier transform in section 7.2.1). Without proof we also mention that the Laplace transform of distributions is one-to-one as well.

Theorem 13.7, and the resulting theorem 13.8, are important results in the theory of the Laplace transform. As for the Fourier transform, theorem 13.7 tells us precisely how we can recover the function $f(t)$ from $F(s)$. Obtaining f from F is called the inverse problem and therefore theorem 13.7 is also known as the *inversion theorem* and (13.12) as the *inversion formula*. We will call the function f the *inverse Laplace transform* of F. Still, (13.12) will *not* be used for this purpose. In fact, calculating the integral in (13.12) requires a thorough knowledge of the integration of complex functions over lines in $\mathbb{C}$, an extensive subject which is outside the scope of this book. Hence, the fundamental theorem of the Laplace transform will not be used in the remainder of this book, except in the form of the frequent (implicit) application of the fact that the Laplace transform is one-to-one. Moreover, in practice it is often a lot easier to determine the inverse Laplace transform of a function $F(s)$ by using tables, applying the properties of the Laplace transform, and using partial fraction expansions.

Partial fraction expansions have been treated in detail in section 2.2 and will be used to obtain the inverse Laplace transform of a *rational function $F(s)$*. It will be assumed that $F(s)$ has real coefficients; in practice this is usually the case. We will now describe in a number of steps how the inverse Laplace transform of such a rational function $F(s)$ can be determined.

Inversion theorem
Inversion formula
Inverse Laplace transform

Step 1

If the degree of the numerator is greater than or equal to the degree of the denominator, then we perform a division. The function $F(s)$ is then the sum of a polynomial and a rational function for which the degree of the numerator is smaller than the degree of the denominator. The polynomial gives rise to distributions in the inverse Laplace transform since $s^n = (\mathcal{L}\delta^{(n)}(t))(s)$.

EXAMPLE

We want to determine the function/distribution $f(t)$ having Laplace transform $F(s) = (s^3 - s^2 + s)/(s^2 + 1)$. Since the degree of the numerator is greater than the degree of the denominator, we first divide: $F(s) = s - 1 + 1/(s^2 + 1)$. Now $\mathcal{L}\delta = 1$, $\mathcal{L}\delta' = s$ and $(\mathcal{L}\sin t)(s) = 1/(s^2 + 1)$, so $f(t) = \delta'(t) - \delta(t) + \sin t$. ◀

Step 2

From step 1 it follows that henceforth we may assume that $F(s)$ is a rational function for which the degree of the numerator is smaller than the degree of the denominator. From the results of section 2.2 it then follows that $F(s)$ can be written as a sum of fractions of the form

$$\frac{A}{(s + a)^k} \qquad \text{and} \qquad \frac{Bs + C}{(s^2 + 2bs + c)^l},$$

with $k, l \in \mathbb{N}$ and where all constants are real and $s^2 + 2bs + c$ cannot be factorized into factors with real coefficients. This latter fact means that the discriminant of $s^2 + 2bs + c$ is negative. We will now determine the inverse Laplace transform for each of these fractions separately.

Step 3

From table 7 we immediately obtain the inverse Laplace transform of $A/(s + a)^k$:

$$(\mathcal{L}t^{k-1}e^{-at})(s) = \frac{(k-1)!}{(s + a)^k}. \tag{13.13}$$

EXAMPLE

Determine the function $f(t)$ with $(\mathcal{L}f(t))(s) = F(s) = 1/(s^2 + 3s + 2)$. The discriminant of the denominator is positive and so it can be factorized into two real linear factors: $s^2 + 3s + 2 = (s + 1)(s + 2)$. From a partial fraction expansion it follows that $F(s) = 1/(s + 1) - 1/(s + 2)$. Hence, $f(t) = e^{-t} - e^{-2t}$. Compare this with the use of the convolution theorem in exercise 13.2. ◀

EXAMPLE

Determine the inverse Laplace transform $f(t)$ of $F(s) = 1/(s^3 + 4s^2 + 3s)$. Note that $s^3 + 4s^2 + 3s = s(s^2 + 4s + 3)$ and that the quadratic form has a positive discriminant; it follows that $s^3 + 4s^2 + 3s = s(s + 3)(s + 1)$ and from a partial fraction expansion it follows that

$$F(s) = \frac{1}{3s} + \frac{1}{6(s + 3)} - \frac{1}{2(s + 1)}.$$

From (13.13) we then obtain that $f(t) = (2 + e^{-3t} - 3e^{-t})/6$. ◀

EXAMPLE

The denominator of the function $F(s) = 1/((s + 1)^3(s - 2)^2)$ has two multiple zeros. The partial fraction expansion of this function will take some effort, but will eventually result in

$$F(s) = \frac{1}{27}\left(\frac{3}{(s + 1)^3} + \frac{2}{(s + 1)^2} + \frac{1}{s + 1} + \frac{1}{(s - 2)^2} - \frac{1}{s - 2}\right).$$

From (13.13) we then see that the inverse Laplace transform of $F(s)$ is given by $f(t) = e^{-t}(\frac{3}{2}t^2 + 2t + 1)/27 + e^{2t}(t - 1)/27$. ◀

Step 4

In order to determine the inverse Laplace transform of $(Bs + C)/(s^2 + 2bs + c)^l$, we complete the square in the denominator: $s^2 + 2bs + c = (s + b)^2 + (c - b^2)$. For convenience we write the positive constant $c - b^2$ simply as c^2 (for some new constant c), which means that we want to determine the inverse Laplace transform of the function

$$\frac{Bs + c}{((s + b)^2 + c^2)^l} \qquad \text{for } l \in \mathbb{N}. \tag{13.14}$$

For $l = 1$ we obtain the inverse Laplace transform of this function from (13.14) by taking a suitable linear combination of the following results from table 7:

$$(\mathcal{L}e^{-bt}\sin ct)(s) = \frac{c}{(s+b)^2 + c^2}, \quad (\mathcal{L}e^{-bt}\cos ct)(s) = \frac{s+b}{(s+b)^2 + c^2}. \quad (13.15)$$

EXAMPLE

Determine the function $f(t)$ with Laplace transform $F(s) = 1/(s^2 - 2s + 17)$. The discriminant of the denominator is negative and so we complete the square: $s^2 - 2s + 17 = (s-1)^2 + 16$. From (13.15) it then follows that $f(t) = (e^t \sin 4t)/4$ (also see exercise 12.19g). ◀

EXAMPLE

Determine the function $f(t)$ with Laplace transform $F(s) = 1/(s(s^2+1))$. A partial fraction expansion leads to $F(s) = 1/s - s/(s^2+1) = (\mathcal{L}1)(s) - (\mathcal{L}\cos t)(s)$. From a very simple case of (13.15) it then follows that $f(t) = 1 - \cos t$. In example 12.13 we used the integration rule to prove this. ◀

EXAMPLE

Determine the function $f(t)$ with Laplace transform $F(s) = 1/((s^2+4)(s^2+16))$. Since $F(s)$ is a function of s^2, we put $y = s^2$ and apply partial fraction expansion to the function $1/((y+4)(y+16))$, resulting in $1/(12(y+4)) - 1/(12(y+16))$. Hence, $F(s) = 1/(12(s^2+4)) - 1/(12(s^2+16))$ and then it again follows from a simple case of (13.15) that $f(t) = (\sin 2t)/24 - (\sin 4t)/48$. ◀

EXAMPLE

Determine the inverse Laplace transform $f(t)$ of the function $F(s) = (2s+5)/(s^2+9)$. Since

$$\frac{2s+5}{s^2+9} = 2\frac{s}{s^2+9} + \frac{5}{3}\frac{3}{s^2+9},$$

it follows from (13.15) that $f(t) = (6\cos 3t + 5\sin 3t)/3$. ◀

EXAMPLE

Determine the inverse Laplace transform $f(t)$ of the function

$$F(s) = \frac{e^{-4s}(s+1)}{s^2(s+5)} + \frac{s+2}{s^2+2s+5}.$$

Since the discriminant of $s^2 + 2s + 5$ is negative, we complete the square: $s^2 + 2s + 5 = (s+1)^2 + 4$. To the first term we apply partial fraction expansion:

$$F(s) = e^{-4s}\left(\frac{1}{5s^2} + \frac{4}{25s} - \frac{4}{25(s+5)}\right) + \frac{s+1}{(s+1)^2+4} + \frac{1}{(s+1)^2+4}.$$

Using (13.13), (13.15) and the shift property in the time domain we obtain $f(t) = \epsilon(t-4)(5(t-4) + 4 - 4e^{-5(t-4)})/25 + e^{-t}(2\cos 2t + \sin 2t)/2$. ◀

Larger values of l in (13.14) can be obtained by a repeated application of the differentiation rule in the s-domain, although the calculations become increasingly tedious. For a simple case we refer to example 12.12. The final example of this section will show that even for the case $l = 2$ calculations can be quite complicated.

EXAMPLE

Determine the inverse Laplace transform $f(t)$ of the function $F(s) = (6s-4)/(s^2+4s+8)^2$. First complete the square in the denominator: $s^2 + 4s + 8 = (s+2)^2 + 4$. Starting from (13.15) with $b = 2$ and $c = 2$ we want to apply the differentiation rule in the s-domain. A bit of calculation will show that

$$\frac{d}{ds}\frac{1}{(s+2)^2+4} = \frac{-2s-4}{((s+2)^2+4)^2}, \quad \frac{d}{ds}\frac{s+2}{(s+2)^2+4} = \frac{-(s+2)^2+4}{((s+2)^2+4)^2}.$$

Writing $-(s+2)^2 + 4 = -((s+2)^2 + 4) + 8$ will reveal that

$$\frac{d}{ds}\frac{s+2}{(s+2)^2+4} = -\frac{1}{(s+2)^2+4} + \frac{8}{((s+2)^2+4)^2}$$

and thus

$$a\frac{d}{ds}\frac{1}{(s+2)^2+4} + b\left(\frac{d}{ds}\frac{s+2}{(s+2)^2+4} + \frac{1}{(s+2)^2+4}\right) = \frac{-2as-4a+8b}{((s+2)^2+4)^2}.$$

The right-hand side is equal to $F(s)$ for $a = -3$ and $b = -2$ and hence

$$F(s) = -3\frac{d}{ds}\frac{1}{(s+2)^2+4} - 2\frac{d}{ds}\frac{s+2}{(s+2)^2+4} - \frac{2}{(s+2)^2+4}.$$

Using (13.15) and the differentiation rule in the s-domain we can then finally determine the inverse Laplace transform: $f(t) = (3te^{-2t}\sin 2t)/2 + 2te^{-2t}\cos 2t - e^{-2t}\sin 2t$. ◀

EXERCISES

13.22 Determine the inverse Laplace transform $f(t)$ of the following complex functions:
    a $s/(s^2 + 5s + 6)$,
    b $1/(s^2 + 6s + 10)$,
    c $s/((s-1)(s+2)(s+3))$,
    d $(s^2 + 1)/((s^2 - 4)(s^2 - 1))$,
    e $(s^2 + 7)/((s+1)^2(s-1))$,
    f $1/(s^2 - 1)^2$,
    g $(s^3 + 4)/((s^2 + 4)(s-1))$,
    h $e^{-2s}(s^6 + s^2 - 1)/(s^2(s^2 - 1))$.

13.23 For integer k we define $F_k(s) = 1/(s^k - 1)$. Determine the inverse Laplace transform $f_k(t)$ of $F_k(s)$ for $k = 1, 2, 3, 4$ and -1.

SUMMARY

First, a number of additional properties of the Laplace transform were treated in this chapter.

It was shown that the convolution product of two causal functions f and g is again a causal function and that, under certain conditions, the convolution theorem holds: $\mathcal{L}(f * g)(s) = (\mathcal{L}f)(s) \cdot (\mathcal{L}g)(s)$.

In general one has for the Laplace transform $F(s)$ that $\lim_{s\to\infty} F(s) = 0$. If $f(t)$ is a piecewise smooth function, then $\lim_{s\to\infty} sF(s) = f(0+)$ according to the initial value theorem. If the final value $\lim_{t\to\infty} f(t) = f(\infty)$ exists, then the final value theorem states that $\lim_{s\to 0} sF(s) = f(\infty)$.

For a periodic function $f(t)$ with period T there is a simple relationship between $F(s)$ and the Laplace transform of the function restricted to one period. When $\phi(t) = f(t) - \epsilon(t - T)f(t - T)$ has Laplace transform $\Phi(s)$, then

$$F(s) = \frac{\Phi(s)}{1 - e^{-sT}} \quad \text{with} \quad \Phi(s) = \int_0^T f(t)e^{-st}\,dt.$$

For distributions T the Laplace transform $\mathcal{L}T$ is defined as the complex function $(\mathcal{L}T)(s) = \langle T(t), e^{-st}\rangle$. One then has, for example, that $\mathcal{L}\delta = 1$ and $(\mathcal{L}\delta(t - a))(s) = e^{-as}$. The differentiation rule for distributions reads as follows:

$(\mathcal{L}T^{(m)})(s) = s^m(\mathcal{L}T)(s)$ and so one has in particular $(\mathcal{L}\delta^{(m)})(s) = s^m$. For certain classes of distributions the convolution theorem remains valid, so $\mathcal{L}(S * T) = \mathcal{L}S \cdot \mathcal{L}T$. One has, for example, that $\mathcal{L}(\delta * T) = \mathcal{L}T$ and $\mathcal{L}(\delta' * T) = s(\mathcal{L}T)(s)$.

From the fundamental theorem of the Fourier transform one immediately obtains the fundamental theorem of the Laplace transform. From this fundamental theorem the uniqueness of the Laplace transform follows: when $f(t)$ and $g(t)$ are two piecewise smooth functions of exponential order and $F(s) = G(s)$ in a certain half-plane in $\mathbb{C}$, then $f(t) = g(t)$ at all points where f and g are continuous. For distributions the uniqueness of the Laplace transform remains valid. In practical problems the inverse Laplace transform is usually determined by using tables, properties of the Laplace transform and partial fraction expansions.

SELFTEST

13.24   Let $f(t)$ be the function with Laplace transform $F(s) = s/(s^2 + 4)^2$.
a   Determine a convolution product $(g * h)(t)$ with Laplace transform $F(s)$.
b   Determine $f(t)$ by calculating the convolution from part a.
c   Calculate the derivative of $1/(s^2 + 4)$ and use the differentiation rule in the s-domain to determine $f(t)$ once again.

13.25   Consider the function $f(t) = t - \epsilon(t - 2)(t - 2)$ having Laplace transform $F(s)$.
a   Determine $\lim_{s \to 0} sF(s)$ and $\lim_{s \to \infty} sF(s)$ without calculating $F(s)$.
b   Calculate $F(s)$ and verify the results from part a.

13.26   Let $f(t)$ be a function with Laplace transform $F(s)$. Determine the function $f(t)$ for each of the functions $F(s)$ given below and verify the final value theorem, or explain why the final value theorem cannot be applied.
a $F(s) = (s + 5)/(s(s^2 + 2s + 5))$.
b $F(s) = s^2/((s^2 - 1)(s^2 + 4))$.

13.27   Let $f(t)$ be the (causal) periodic function with period 2 defined for $0 \leq t < 2$ by

$$f(t) = \begin{cases} t^2 & \text{for } 0 \leq t < 1, \\ 0 & \text{for } 1 \leq t < 2. \end{cases}$$

a   Sketch the graph of f and show that $f(t) = t^2(\epsilon(t) - \epsilon(t - 1))$ for $0 \leq t < 2$.
b   Determine $(\mathcal{L}f)(s) = F(s)$.

13.28   Determine the inverse Laplace transform $f(t)$ of the following complex functions $F(s)$:
a $F(s) = (1 - e^{-3s})/(s(s + 1))$,
b $F(s) = (s^2 + 1)/(s^2(s - 1)^2)$,
c $F(s) = (1 + e^{-\pi s}(s^5 - 4s^4 - 8s + 64))/(s^2(s^2 + 4))$.

Contents of Chapter 14

Applications of the Laplace transform

Applications of the Laplace transform

INTRODUCTION

The greater part of this chapter consists of section 14.1 on linear time-invariant continuous-time systems (LTC-systems). The Laplace transform is very well suited for the study of causal LTC-systems where switch-on phenomena occur as well: at time $t = 0$ 'a switch is thrown' and a process starts, while prior to time $t = 0$ the system was at rest. The input $u(t)$ will thus be a causal signal and since the system is causal, the output $y(t)$ will be causal as well. Applying the Laplace transform is then quite natural, especially since the Laplace transform exists for a large class of inputs $u(t)$ as an ordinary integral in a certain half-plane $\mathrm{Re}\,s > \rho$. This is in contrast to the Fourier transform, where distributions are needed more often. For the Laplace transform we can usually restrict the distribution theory to the delta functions $\delta(t - a)$ with $a \geq 0$ (and their derivatives). As in chapter 10, the response $h(t)$ to the delta function $\delta(t)$ again plays an important role. The Laplace transform $H(s)$ of the impulse response is called the *transfer function* or *system function*. An LTC-system is then described in the s-domain by the simple relationship $Y(s) = H(s)U(s)$, where $Y(s)$ and $U(s)$ are the Laplace transforms of, respectively, the output $y(t)$ and the input $u(t)$ (compare this with (10.6)).

In this chapter we will mainly limit ourselves to systems described by ordinary linear differential equations with constant coefficients and with initial conditions all equal to zero (since the system is at rest at $t = 0$). The transfer function $H(s)$ is then a rational function of s and the impulse response can thus be determined by a partial fraction expansion and then transforming this back to the time domain. As we know, the response $y(t)$ of the system to an arbitrary input $u(t)$ is given by the convolution of $h(t)$ with $u(t)$ (see section 10.1). In order to find the response $y(t)$ for a given input $u(t)$, it is often easier first to determine the Laplace transform $U(s)$ of $u(t)$ and subsequently to transform $H(s)U(s)$ back to the time domain. This is because $U(s)$, and hence $Y(s) = H(s)U(s)$ as well, is a rational function for a large class of inputs. The inverse Laplace transform $y(t)$ of $Y(s)$ can then immediately be determined by a partial fraction expansion. This simple standard solution method is yet another advantage of the Laplace transform over the Fourier transform.

If we compare this with the classical method for solving ordinary linear differential equations with constant coefficients (using the homogeneous and the particular solution), then the Laplace transform again has the advantage. This is because it will turn out that the Laplace transform takes the initial conditions immediately into account in the calculations. This reduces the amount of calculation considerably, especially for higher order differential equations. As a disadvantage of the Laplace transform, we mention that in general one cannot give a straightforward interpretation in terms of spectra, as is the case for the Fourier transform.

For the differential equations treated in section 14.1, all the initial conditions will always be zero. In section 14.2 we will show that the Laplace transform can equally well be applied to ordinary linear differential equations with constant coefficients and with *arbitrary* initial conditions. In essence, nothing will change in the solution method from section 14.1. Even more general are the *systems* of several *coupled* ordinary linear differential equations with constant coefficients from section 14.3. Again these can be solved using the same method, although in the *s*-domain we have not one equation, but a system of several equations. For convenience we confine ourselves to systems of two differential equations. Finally, we briefly describe in section 14.4 how the Laplace transform can be used to solve partial differential equations with initial and boundary conditions. By applying the Laplace transform to one of the variables, the partial differential equation becomes an ordinary differential equation, which is much easier to solve.

LEARNING OBJECTIVES
After studying this chapter it is expected that you
- know the concept of transfer function or system function of an LTC-system
- can determine the transfer function and the impulse and step response of a causal LTC-system described by an ordinary linear differential equation with constant coefficients
- know the relation between the input and the output in the *s*-domain using the transfer function and can use it to calculate outputs
- can verify the stability using the transfer function
- can apply the Laplace transform in solving ordinary linear differential equations with constant coefficients and arbitrary initial conditions
- can apply the Laplace transform in solving systems of two coupled ordinary linear differential equations with constant coefficients
- can apply the Laplace transform in solving partial differential equations with initial and boundary conditions.

14.1   Linear systems

14.1.1   The transfer function

The basic concepts from the theory of LTC-systems (linear time-invariant continuous-time systems) have been treated extensively in chapter 1 and section 10.1. Let us summarize the most important concepts.

An LTC-system L associates with any input $u(t)$ an output $y(t)$. One also calls $y(t)$ the response to $u(t)$. When $h(t)$ is the impulse response, that is, $h(t)$ is the response to $\delta(t)$, then it follows for an arbitrary input $u(t)$ that

$$y(t) = \mathrm{L}u(t) = (h * u)(t) \tag{14.1}$$

(see (10.3)). An LTC-system is thus completely determined by the impulse response $h(t)$. Besides the impulse response we also introduced in section 10.1 the step response $a(t)$, that is to say, the response of the system to the unit step function $\epsilon(t)$. We recall that $h(t)$ is the derivative of $a(t)$ (considered as a distribution, if necessary). Using the convolution theorem of the Laplace transform one can translate relation (14.1) to the *s*-domain. To this end we assume, as in section 10.2, that for the LTC-systems under consideration we may apply the convolution theorem, in the distribution sense if necessary. If $U(s)$, $Y(s)$ and $H(s)$ are the Laplace transforms of $u(t)$, $y(t)$ and $h(t)$ respectively, then it follows from the convolution theorem that

$$Y(s) = H(s)U(s). \tag{14.2}$$

The function $H(s)$ plays the same important role as the frequency response from chapter 10.

DEFINITION 14.1
System function
Transfer function

Let $h(t)$ be the impulse response of an LTC-system. Then the system function or transfer function $H(s)$ of the LTC-system is the Laplace transform of $h(t)$ (in the distribution sense, if necessary).

EXAMPLE 14.1

Consider the integrator from example 10.3 with impulse response $h(t) = \epsilon(t)$. The Laplace transform of $\epsilon(t)$ is $1/s$ (see table 7) and so $H(s) = 1/s$. Hence, the response $y(t)$ of the integrator to an input $u(t)$ is described in the s-domain by $Y(s) = U(s)/s$. ◄

Switched-on system

In practical situations we are usually dealing with systems where switch-on phenomena may occur: at time $t = 0$ a system, being at rest, is switched on. The inputs are then causal and when the LTC-system is causal, then the output will be causal as well (theorem 1.2). In this chapter we will limit ourselves to causal LTC-systems and, moreover, we will always assume that all inputs $u(t)$, and thus all outputs $y(t)$ as well, are causal: $u(t) = y(t) = 0$ for $t < 0$. Here we will also admit distributions 'which are zero for $t < 0$': the delta functions $\delta(t - a)$ with $a \geq 0$ and their derivatives (see section 13.4). In particular the impulse response will also be a causal signal. If now the Laplace transform $H(s)$ exists in a certain half-plane $\operatorname{Re} s > \rho$ with $\rho < 0$, then we can take $s = i\omega$ and since $h(t)$ is causal, it then follows that

$$H(i\omega) = (\mathcal{L}h)(i\omega) = \int_0^\infty h(t)e^{-i\omega t}\, dt = \int_{-\infty}^\infty h(t)e^{-i\omega t}\, dt = (\mathcal{F}h)(\omega).$$

Hence, in this case we see that the frequency response from section 10.2 is given by $H(i\omega)$. Many Laplace transforms only exist in a half-plane contained in $\operatorname{Re} s > 0$. Of course, substituting $s = i\omega$ is then *not allowed*. Because of this, the term 'frequency response' for the function $H(s)$ is misplaced: in general $H(s)$ is a complex function having no interpretation in terms of frequencies.

EXAMPLE 14.2

Consider the integrator from example 10.3 with system function $H(s) = 1/s$ (for $\operatorname{Re} s > 0$). The frequency response is *not* given by the function $1/i\omega$, but by $1/i\omega + \pi\delta(\omega)$ (see example 10.4). ◄

In section 10.3 we already noted that a large and important class of LTC-systems occurring in practice (such as RLC-networks) can be described by ordinary linear differential equations with constant coefficients of the form (10.9) (also see (14.5)). From now on we confine ourselves to such systems. In section 14.1.2 we will first explain the fundamental principle underlying the application of the Laplace transform to linear differential equations with constant coefficients. In essence, this principle remains unchanged throughout the remainder of this chapter.

14.1.2　The method of Laplace transforming

Using an elementary example we will illustrate how the Laplace transform can be used to obtain solutions to linear differential equations with constant coefficients. Moreover, we will show the difference between the method using the Laplace transform and the 'classical' solution method using the homogeneous and particular solutions. This classical solution method has already been explained in section 5.1 and can also be found in many introductions to this subject. In the following example we first solve a certain initial value problem using the classical method.

EXAMPLE 14.3

Consider for the unknown function $y = y(t)$ the initial value problem

$$y'' - y = 2t, \tag{14.3}$$

$$y(0) = y'(0) = 0. \tag{14.4}$$

The corresponding homogeneous equation $y'' - y = 0$ has characteristic equation $\lambda^2 - 1 = 0$ and so $\lambda = \pm 1$. Hence, the solution y_h of the homogeneous equation is $y_h(t) = \alpha e^t + \beta e^{-t}$, where α and β are arbitrary constants. To find a particular solution y_p, we try $y_p(t) = bt + c$. Substituting this into (14.3) gives $c = 0$ and $b = -2$. The general solution is thus

$$y(t) = y_h(t) + y_p(t) = \alpha e^t + \beta e^{-t} - 2t.$$

Substituting the initial values (14.4) gives $\alpha + \beta = 0$ and $\alpha - \beta - 2 = 0$. Solving this system of two equations in the two unknowns α and β we obtain that $\alpha = 1$ and $\beta = -1$. The solution to the initial value problem is thus given by

$$y(t) = e^t - e^{-t} - 2t.$$

◄

Characteristic for this classical method is the fact that we *first* determine the general solution to the differential equation ($y(t)$ in example 14.3) and *then* determine the unknown constants in the general solution from the initial values (α and β follow from $y(0)$ and $y'(0)$ in example 14.3). For an mth order linear differential equation with constant coefficients it is known that a solution is uniquely determined by specifying (for example) m initial conditions (say $y(0), \ldots, y^{(m-1)}(0)$). With the m initial conditions one can determine the m unknown constants in the general solution, by solving a system of m linear equations in m unknowns. For $m \geq 3$ this is a tedious calculation. Another disadvantage of the classical method is finding a particular solution. This is not always easy and may again require some tedious calculations. We will now use the Laplace transform to solve the initial value problem from example 14.3.

EXAMPLE 14.4

Apply the Laplace transform to both sides of the differential equation (14.3). Assume that $Y(s) = (\mathcal{L}y)(s)$ exists in a certain half-plane and that moreover the differentiation rule in the time domain ((12.10) or table 8) can be applied. Then $(\mathcal{L}(y'' - y))(s) = s^2 Y(s) - Y(s)$ and since $(\mathcal{L}t)(s) = 1/s^2$, the initial value problem transforms into

$$(s^2 - 1)Y(s) = \frac{2}{s^2}.$$

Note that instead of a differential equation for $y(t)$ we now have an algebraic equation for $Y(s)$. Solving for $Y(s)$ and applying a partial fraction expansion we obtain that

$$Y(s) = \frac{2}{s^2(s^2 - 1)} = \frac{2}{s^2 - 1} - \frac{2}{s^2}.$$

The inverse Laplace transform $y(t)$ of $Y(s)$ follows from table 7:

$$y(t) = 2 \sinh t - 2t = e^t - e^{-t} - 2t.$$

This is in accordance with the result from example 14.3. One can easily verify that the (causal) function $y(t)$ satisfies the differential equation (14.3) and the initial conditions (14.4) (in general it is necessary to verify the result since a number of assumptions have been made in order to find the solution). ◄

                    14 Applications of the Laplace transform

Before we pass on to a comparison of the classical method with the method from example 14.4, we first summarize the most important steps of the solution method in example 14.4.

Step 1   The Laplace transform is applied to the differential equation for $y(t)$. Here we assume that the Laplace transform $Y(s)$ of the unknown function $y(t)$ exists and that the differentiation rule in the time domain may be applied (either in the ordinary sense or in the sense of distributions). From the differential equation for $y(t)$ we obtain an algebraic equation for $Y(s)$ which is much easier to solve.

Step 2   The algebraic equation in the s-domain is solved for $Y(s)$.

Step 3   The solution we have found in the s-domain is then transformed back into the t-domain. For this we use tables, the properties of the Laplace transform and partial fraction expansions (see section 13.3). For the solution $y(t)$ found in this way, one can verify whether it satisfies the differential equation and the initial conditions.

This procedure is represented once more in figure 14.1.

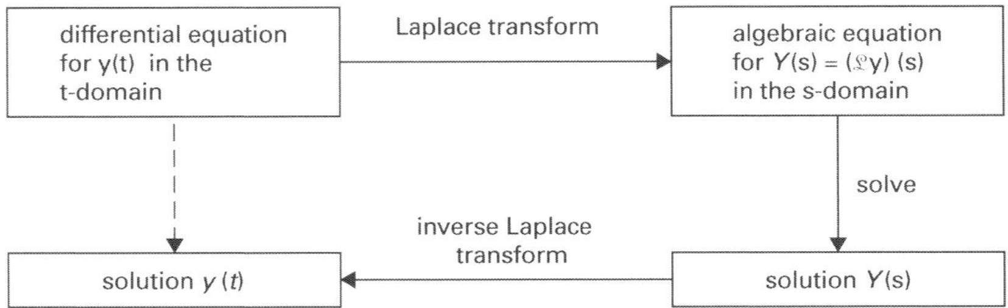

FIGURE 14.1
Global procedure for the application of the Laplace transform in solving differential equations.

If we now compare the method from example 14.4 with the classical method from example 14.3, then the advantage of the new method is not obvious yet. In general the advantages increase as the order of the differential equation increases and/or the right-hand side of the differential equation gets more complex. This is because the Laplace transform immediately leads to the desired solution, without the need to determine the *general solution* first. Solving a system of m linear equations in m unknowns afterwards (for an mth order differential equation), is no longer necessary. This is caused by the fact that the Laplace transform immediately takes the initial conditions into account in the calculations (which is not very visible in example 14.4 since we have initial conditions $y(0) = y'(0) = 0$). Besides this, finding a particular solution is not necessary since again the right-hand side of the differential equation is immediately taken into account in the calculations by the Laplace transform.

14.1.3   *Systems described by differential equations*

We now return to the general theory of the causal LTC-systems described by an ordinary linear differential equation with constant coefficients of the form

$$a_m \frac{d^m y}{dt^m} + a_{m-1}\frac{d^{m-1} y}{dt^{m-1}} + \cdots + a_1\frac{dy}{dt} + a_0 y$$

$$= b_n \frac{d^n u}{dt^n} + b_{n-1}\frac{d^{n-1} u}{dt^{n-1}} + \cdots + b_1\frac{du}{dt} + b_0 u, \tag{14.5}$$

where $n \leq m$ (also see (10.9)). Here $u(t)$ is the input and $y(t)$ the output or response.

At first sight one could think that any differential equation of the form (14.5) describes an LTC-system. This, however, is not the case since a differential equation on its own admits a large class of solutions. If sufficient additional conditions are supplied, for example initial conditions, then we may have reduced the number of solutions to just one, but in general the linearity property of linear systems (see definition 1.1) will not be fulfilled then. We illustrate this with an example.

EXAMPLE 14.5

Consider the system described by the differential equation $y'' - y = u(t)$ (also see example 14.3) with two *different* initial conditions:

$$y(0) = 0, y'(0) = -2 \qquad \text{and} \qquad y(0) = 0, y'(0) = 0.$$

The system having $y(0)$ and $y'(0)$ both equal to zero (the 'right-hand system') will be called a system at 'initial rest'. Now take as input $u(t) = u_1(t) = 2t$ (as in example 14.3). The response $y_1(t)$ of these two systems is given by, respectively,

$$y_1(t) = -2t \qquad \text{and} \qquad y_1(t) = e^t - e^{-t} - 2t.$$

For both situations we now take another input, namely $u(t) = u_2(t) = t^2$. The response $y_2(t)$ of these two systems is then given by, respectively,

$$y_2(t) = 2e^{-t} - t^2 - 2 \qquad \text{and} \qquad y_2(t) = e^t + e^{-t} - t^2 - 2.$$

If we add the inputs, then the linearity of the system should guarantee that we obtain as output the sum of the separate outputs. We will verify this for both situations. The response $y_3(t)$ to the input $u_1(t) + u_2(t) = 2t + t^2$ is given by, respectively,

$$y_3(t) = e^t + e^{-t} - t^2 - 2t - 2 \qquad \text{and} \qquad y_3(t) = 2e^{-t} - t^2 - 2t - 2.$$

We see that for the left-hand and right-hand system we have, respectively,

$$y_3(t) \neq y_1(t) + y_2(t) \qquad \text{and} \qquad y_3(t) = y_1(t) + y_2(t).$$

◀

If we now look at how these differences in example 14.5 arise, then we observe that the right-hand system is at initial rest, that is to say, $y(0) = y'(0) = 0$. This is in contrast to the left-hand system, where $y'(0) = -2$. For the solutions y_1 and y_2 one then has $y_1'(0) = y_2'(0) = -2$, so $(y_1 + y_2)'(0) = -4$, while $y_3'(0) = -2$. From this we immediately see that $y_3 \neq y_1 + y_2$.

For a system described by a differential equation of the form (14.5) we now define in general the condition of *initial rest* by

Initial rest

$$y(0) = y'(0) = \ldots = y^{(m-1)}(0) = 0. \tag{14.6}$$

Example 14.5 suggests that the condition of initial rest is sufficient to ensure that a system described by a differential equation of the form (14.5) is linear. This is indeed the case. In fact, if an input $u_1(t)$ with response $y_1(t)$ satisfies differential equation (14.5), and an input $u_2(t)$ with response $y_2(t)$ also satisfies differential equation (14.5), then $(ay_1 + by_2)(t)$ will satisfy differential equation (14.5) for the input $(au_1 + bu_2)(t)$ because differentiating is linear. When, in addition, both y_1 and y_2 satisfy the condition of initial rest, then $(ay_1 + by_2)^{(k)}(0) = ay_1^{(k)}(0) + by_2^{(k)}(0) = 0$ for $k = 0, 1, \ldots, m - 1$. Hence, $ay_1 + by_2$ also satisfies the condition of initial rest. Moreover, since solutions are uniquely determined by the m initial conditions in (14.6), it follows that $(ay_1 + by_2)(t)$ must be the response to the input $(au_1 + bu_2)(t)$. Thus, the system is linear.

For a causal LTC-system described by differential equation (14.5) and condition of initial rest (14.6) we now apply the Laplace transforming method from section 14.1.2.

Step 1

Assume that the Laplace transform $Y(s)$ of the unknown function $y(t)$ exists and apply the Laplace transform to all terms in the differential equation (14.5). We do this under the assumption that we may apply the differentiation rule (12.10) in the time domain. The (causal) function $y(t)$ should then be $m-1$ times continuously differentiable on $\mathbb{R}$ and $y^{(m-1)}$ should be differentiable on $\mathbb{R}$. Note that in this case the conditions of initial rest are automatically satisfied since for $k = 0, 1, \ldots, m-1$ it follows from the continuity of $y^{(k)}$ that $y^{(k)}(0) = y^{(k)}(0-) = 0$ (for $y^{(k)}$ is causal). Similar remarks can be made for the function $u(t)$. From (12.10) it then follows that, after Laplace transforming, the differential equation (14.5) is transformed into the algebraic equation

$$a_m s^m Y(s) + \cdots + a_1 s Y(s) + a_0 Y(s) = b_n s^n U(s) + \cdots + b_0 U(s),$$

where $U(s)$ is the Laplace transform of $u(t)$.

Step 2

The algebraic equation in the s-domain is solved for $Y(s)$. Using the polynomials

$$A(s) = a_m s^m + a_{m-1} s^{m-1} + \cdots + a_1 s + a_0,$$
$$B(s) = b_n s^n + b_{n-1} s^{n-1} + \cdots + b_1 s + b_0, \tag{14.7}$$

we can write the algebraic equation for $Y(s)$ as $A(s)Y(s) = B(s)U(s)$. Solving for $Y(s)$ gives

$$Y(s) = \frac{B(s)}{A(s)} U(s). \tag{14.8}$$

Step 3

The function $y(t)$ we are looking for is the inverse Laplace transform of $Y(s)$. In many cases $U(s)$ will be a rational function with real coefficients and the differential equation (14.5) will have real coefficients as well. It then follows from (14.8) that $Y(s)$ is also a rational function with real coefficients. The partial fraction expansion method from section 13.3 can then be applied. One can then verify whether the solution $y(t)$ we have found satisfies both the differential equation and the initial conditions.

EXAMPLE 14.6

A system satisfying the condition of initial rest is described by the differential equation $y'' - 3y' + 2y = u(t)$. We want to determine the response to the signal $u(t) = t$. Let $Y(s)$ be the Laplace transform of $y(t)$. Since $(\mathcal{L}t)(s) = 1/s^2$, it follows that $(s^2 - 3s + 2)Y(s) = 1/s^2$, or

$$Y(s) = \frac{1}{s^2(s^2 - 3s + 2)}.$$

But $s^2 - 3s + 2 = (s-1)(s-2)$ and a partial fraction expansion gives

$$Y(s) = \frac{1}{4(s-2)} - \frac{1}{s-1} + \frac{3}{4s} + \frac{1}{2s^2}.$$

From table 7 we obtain $y(t) = (e^{2t} - 4e^t + 3 + 2t)/4$. It is easy to check that $y(t)$ satisfies the differential equation and the initial conditions.   ◀

By combining (14.8) and (14.2) we obtain the following important result for the transfer function:

$$H(s) = \frac{B(s)}{A(s)}, \tag{14.9}$$

where $B(s)$ and $A(s)$ are the polynomials from (14.7). These two polynomials follow immediately from the *form* of the differential equation: to get the transfer function $H(s)$ we do *not* have to calculate the impulse response $h(t)$ explicitly! Equation (14.9) can also be obtained by taking the delta function $\delta(t)$ as input in (14.8). Since in this case $U(s) = (\mathcal{L}\delta)(s) = 1$ and $Y(s) = H(s)$ (definition 14.1), it then indeed follows from (14.8) that $H(s) = B(s)/A(s)$.

EXAMPLE 14.7

For the system in example 14.6 we take as input $u(t) = \delta(t)$. Since $(\mathcal{L}\delta)(s) = 1$, it follows from $y'' - 3y' + 2y = \delta(t)$ that $(s^2 - 3s + 2)H(s) = 1$. The transfer function $H(s)$ is thus given by $H(s) = 1/(s^2 - 3s + 2)$. Let us also determine the impulse response $h(t)$. From the partial fraction expansion it follows that $H(s) = 1/(s - 2) - 1/(s - 1)$ and so $h(t) = e^{2t} - e^t$. It is easy to check that $h(t)$ satisfies the differential equation, provided that we differentiate in the distribution sense! ◄

In most of the engineering literature one is quite happy with the derivation of (14.9) based on the input $\delta(t)$. However, using example 14.7 it is easy to show that this derivation is not quite rigorous. For this we note that the impulse response $h(t)$ from example 14.7 does satisfy $h(0) = 0$, but that $h'(0)$ *does not exist*. There is even a jump of $h'(t)$ at $t = 0$! This is because $h'(0-) = 0$, since h' is causal, while $h'(0+) = 1$, since $h'(t) = 2e^{2t} - e^t$ for $t > 0$. Indeed, the function $h'(t)$ must have a jump at $t = 0$, since otherwise no delta function would occur in the input. The conclusion must be that $h(t)$ does *not* satisfy the condition of initial rest. This problem can be solved by applying the differentiation rule for *distributions* in (13.9) (or table 10). In fact, as we have already mentioned in example 14.7, differentiation should be taken in the distribution sense. The condition of initial rest is then irrelevant, since in general distributions have no meaning at all at $t = 0$. If we take as input $u(t) = \delta(t)$ in differential equation (14.5), then it indeed follows from $(\mathcal{L}\delta^{(k)})(s) = s^k$ and the differentiation rule for distributions in (13.9) that $H(s) = B(s)/A(s)$, where $B(s)$ and $A(s)$ are the polynomials from (14.7).

Besides the impulse response one often uses the step response as well. Here similar phenomena occur as for the impulse response: although strictly speaking the differentiation rule in the time domain cannot be applied, one can again justify the result with the differentiation rule for distributions. We will not go into this any further and instead present another example.

EXAMPLE 14.8

For the system from example 14.6 we take as input $u(t) = \epsilon(t)$. Since $(\mathcal{L}\epsilon)(s) = 1/s$, it follows from $y'' - 3y' + 2y = \epsilon(t)$ that $(s^2 - 3s + 2)Y(s) = 1/s$, so $Y(s) = 1/(s(s^2 - 3s + 2))$ (of course, this also follows straight from (14.2) since in this case $U(s) = (\mathcal{L}u)(s) = 1/s$). A partial fraction expansion leads to

$$Y(s) = \frac{1}{2s} + \frac{1}{2(s - 2)} - \frac{1}{s - 1},$$

and so the step response is given by $y(t) = (1 + e^{2t} - 2e^t)/2$. Note that $y(0) = y'(0) = 0$ and that indeed $y'(t) = h(t)$, with $h(t)$ the impulse response from example 14.7. Since in this case the impulse response had already been determined, the step response could of course also have been calculated using (14.1):

$$y(t) = (h * \epsilon)(t) = \int_0^t h(\tau)\, d\tau = \int_0^t (e^{2\tau} - e^\tau)\, d\tau = \tfrac{1}{2}(1 + e^{2t} - 2e^t).$$

◄

Let us now assume in general that for a certain system the impulse response $h(t)$ is known. Using (14.1) one can then calculate the output $y(t)$ for any input $u(t)$. In example 14.8, for instance, it is clear that the step response follows quite easily from the impulse response using (14.1). In general the convolution in (14.1) is not so easy to calculate and so the Laplace transforming method is applied for each input $u(t)$ separately (as we have done in example 14.8 in order to determine the step response). Here the input $u(t)$ may be any arbitrary piecewise smooth (causal) function, or even a distribution (being zero for $t < 0$). Assuming the condition of initial rest, we state without proof that the method outlined above always leads to the desired result. When the input contains the delta function $\delta(t)$, or derivatives $\delta^{(k)}(t)$, then we should keep in mind that a number of initial conditions will be irrelevant, since we are then dealing with an output for which one or several of the derivatives has a jump at $t = 0$. If we want to verify afterwards that the solution $y(t)$ we have found satisfies the differential equation, then we should interpret the derivatives in the sense of distributions.

EXAMPLE 14.9

For the system from example 14.6 we take as input $u(t) = 2\delta(t - 1)$. Since $(\mathcal{L}\delta(t - 1))(s) = e^{-s}$, it follows from (14.2) and the expression for $H(s)$ in example 14.7 that

$$Y(s) = 2e^{-s}H(s) = \frac{2e^{-s}}{s - 2} - \frac{2e^{-s}}{s - 1}.$$

Since $(\mathcal{L}e^{2t})(s) = 1/(s - 2)$ and $(\mathcal{L}e^t)(s) = 1/(s - 1)$, it follows from the shift property in the time domain that $y(t) = 2\epsilon(t - 1)(e^{2t-2} - e^{t-1})$.   ◄

The advantages of the Laplace transform, compared to both the classical method and the Fourier transform, now become clear. Towards the end of section 14.1.2 it was already noted that the Laplace transform immediately takes the initial conditions into account. Afterwards we do not have to use the initial conditions to solve a number of constants from a system of linear equations. A second advantage is the fact that the Laplace transform exists as an ordinary integral for a large class of functions. We only need a limited distribution theory. Finally we note that for a large class of inputs the Laplace transform $Y(s)$ of the response $y(t)$ is given by a rational function with real coefficients. The response $y(t)$ can then be found by a partial fraction expansion followed by an inverse transform. A disadvantage of the Laplace transform is the fact that there is no obvious interpretation of the Laplace transform in terms of the spectrum of the input (also see example 14.2 and the remarks preceding it).

14.1.4   Stability

Again it will be assumed that all systems are causal LTC-systems described by a differential equation of the form (14.5) and with condition of initial rest (14.6). In section 14.1.3 it was shown that the transfer function $H(s)$ of such a system is given by the rational function $B(s)/A(s)$, where $B(s)$ and $A(s)$ are the polynomials in (14.7). We will now show that the *stability* of the system (see definition 1.3) immediately follows from the location of the zeros of $A(s)$.

First note that the degree n of the numerator $B(s)$ is less than or equal to the degree m of the denominator $A(s)$ since $n \leq m$ in (14.5). From the theory of partial fraction expansions (see chapter 2) it then follows that $H(s)$ is a linear combination of the constant function 1 and fractions of the form

$$\frac{1}{(s - \alpha)^n},$$

where $n \in \mathbb{N}$ and $\alpha \in \mathbb{C}$ is a zero of the denominator of $H(s)$, so $A(\alpha) = 0$. We recall that the zeros of the denominator of a rational function are called the *poles* of the function (see chapter 2). Note that the constant function occurs when $n = m$, that is to say, when the degree of the numerator equals the degree of the denominator. The inverse Laplace transform of the function 1 is the delta function $\delta(t)$, and the inverse Laplace transform of $1/(s - \alpha)^n$ is given by the function $t^{n-1} e^{\alpha t}/(n-1)!$ (see table 7), and hence we conclude that the impulse response $h(t)$ is a linear combination of the delta function $\delta(t)$ and terms of the form $t^{n-1} e^{\alpha t}$, where $n \in \mathbb{N}$ and $\alpha \in \mathbb{C}$. According to theorem 10.1, the system is stable if and only if $h(t)$ is absolutely integrable. Since for a causal system the function $h(t)$ is causal, this means that we have to check whether $\int_0^\infty |h(t)|\,dt$ is convergent. Here we may ignore possible delta functions (see the remarks following the proof of theorem 10.1). Now take a term of the form $t^{n-1} e^{\alpha t}$ with $n \in \mathbb{N}$ and $\alpha \in \mathbb{C}$, then we thus have to check whether $\int_0^\infty t^{n-1} e^{\alpha t}\,dt$ is absolutely convergent. But

$$\int_0^\infty t^{n-1} e^{\alpha t}\,dt = (\mathcal{L}t^{n-1})(-\alpha),$$

and we know that the abscissa of absolute convergence of the function t^{n-1} is equal to 0 (see examples 12.4 and 12.5 for the case $n = 2$ and section 12.3.1 for the general case). Since the function $t^{n-1} e^{\alpha t}$ is *not* absolutely integrable for $\operatorname{Re} \alpha = 0$ (for then $\left| t^{n-1} e^{\alpha t} \right| = t^{n-1}$ for $t > 0$ and t^{n-1} is not integrable), it thus follows that the function $t^{n-1} e^{\alpha t}$ is absolutely integrable precisely for $\operatorname{Re}(-\alpha) > 0$, so for $\operatorname{Re} \alpha < 0$. The impulse response $h(t)$ is thus absolutely integrable when in each of the terms of the form $t^{n-1} e^{\alpha t}$ one has $\operatorname{Re} \alpha < 0$. We summarize the above in the following theorem.

THEOREM 14.1
Stability

Let $H(s)$ be the transfer function of a causal LTC-system described by a differential equation of the form (14.5) and satisfying the condition of initial rest (14.6). Then the system is stable if and only if the poles of $H(s)$ lie in the half-plane $\operatorname{Re} s < 0$.

EXAMPLE 14.10

Consider the system from example 14.6. From example 14.7 it follows that the poles of the transfer function are given by $s = 1$ and $s = 2$. They do not lie in the half-plane $\operatorname{Re} s < 0$. Hence, according to theorem 14.1 the system is not stable. In example 14.7 it was shown that $h(t) = e^{2t} - e^t$ and indeed this function is not absolutely integrable (we even have $h(t) \to \infty$ if $t \to \infty$). ◄

We close this section on the application of the Laplace transform to linear systems with the treatment of an example which is known as the 'harmonic oscillator'.

14.1.5 The harmonic oscillator

In figure 14.2 an RLC-network is drawn consisting of a series connection of a resistor of resistance R, an inductor of inductance L, a capacitor of capacity C and a voltage source generating a voltage $v(t)$. Assume that $L > 0$, $C > 0$ and $R \geq 0$. As a consequence of the voltage $v(t)$ there will be a current $i(t)$ in the network after closing the switch S at time $t = 0$. We assume that the network is at rest at the moment of switching on. For such systems one can use Kirchhoff's laws (see section 5.1) to obtain the following relationship between the voltage $v(t)$ and the current $i(t)$:

$$Li'(t) + Ri(t) + \frac{1}{C} \int_{-\infty}^t i(\tau)\,d\tau = v(t).$$

Now let $q(t)$ be the charge on the capacitor, then $q(t) = \int_{-\infty}^t i(\tau)\,d\tau$ and so $i(t) = q'(t)$. From this we obtain for $q(t)$ the differential equation

$$Lq''(t) + Rq'(t) + \frac{1}{C} q(t) = v(t). \tag{14.10}$$

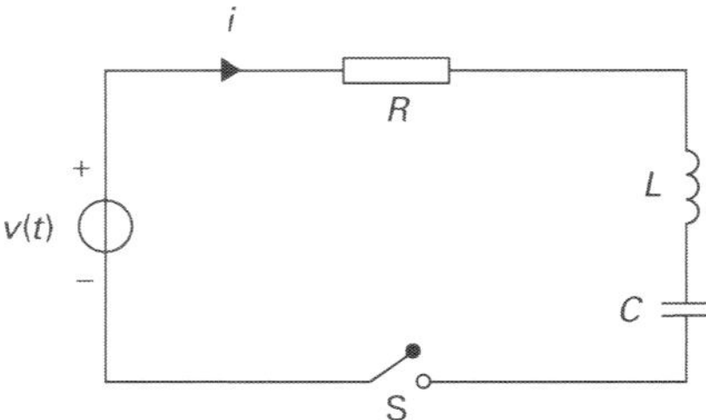

FIGURE 14.2
RLC-network.

Harmonic oscillator

The system described by differential equation (14.10) is known as the *harmonic oscillator* (exactly the same differential equation occurs in the study of so-called mass–dashpot systems; see the end of section 5.1). Since we assumed that the network was at rest, the condition of initial rest holds: $q(0) = q'(0) = 0$. Applying the Laplace transform to (14.10) we obtain that

$$(Ls^2 + Rs + 1/C)Q(s) = V(s),$$

where $Q(s)$ and $V(s)$ are the Laplace transforms of $q(t)$ and $v(t)$ respectively. From this it follows in particular that the transfer function $H(s)$ is given by $H(s) = 1/(Ls^2 + Rs + 1/C)$. The denominator of $H(s)$ is a quadratic polynomial with real coefficients and discriminant $R^2 - 4L/C$. Let s_1 and s_2 be the roots of $Ls^2 + Rs + 1/C$, then $Ls^2 + Rs + 1/C = L(s - s_1)(s - s_2)$ with

$$s_1 = \frac{-R + \sqrt{R^2 - 4L/C}}{2L} \quad \text{and} \quad s_2 = \frac{-R - \sqrt{R^2 - 4L/C}}{2L}.$$

We assume from now on that $R > 0$ and determine the impulse response of the system in the following three cases.
Case 1: $R^2 - 4L/C < 0$; then s_1 and s_2 are complex and $s_1 = \overline{s_2}$.
Case 2: $R^2 - 4L/C = 0$; then s_1 and s_2 are real and $s_1 = s_2$.
Case 3: $R^2 - 4L/C > 0$; then s_1 and s_2 are real and $s_1 \neq s_2$.

Case 1

The partial fraction expansion of $H(s)$ gives

$$H(s) = \frac{1}{L(s - s_1)(s - s_2)} = \frac{1}{L(s_1 - s_2)} \left(\frac{1}{s - s_1} - \frac{1}{s - s_2} \right), \tag{14.11}$$

and by an inverse transform it then follows from table 7 that

$$h(t) = \frac{1}{L(s_1 - s_2)} \left(e^{s_1 t} - e^{s_2 t} \right). \tag{14.12}$$

Now put

$$\omega_0 = \frac{1}{2L} \sqrt{\frac{4L}{C} - R^2} \quad \text{and} \quad \sigma = -\frac{R}{2L},$$

then it follows that $s_1 = \overline{s_2} = \sigma + i\omega_0$, and using this we can write (14.12) as

$$h(t) = \frac{1}{2L\omega_0 i} \left(e^{(\sigma + i\omega_0)t} - e^{(\sigma - i\omega_0)t} \right) = \frac{1}{L\omega_0} e^{\sigma t} \frac{e^{i\omega_0 t} - e^{-i\omega_0 t}}{2i}.$$

Hence,

$$h(t) = \frac{1}{L\omega_0} e^{\sigma t} \sin \omega_0 t \tag{14.13}$$

with $\sigma < 0$. The impulse response is a damped sinusoidal with frequency ω_0. See figure 14.3a. This case is called a 'damped vibration' and is also called 'undercritical damping'.

Case 2

Put $\sigma = -R/2L = s_1$. Since $H(s) = 1/(L(s - \sigma)^2)$ it immediately follows from table 7 that $h(t) = t e^{\sigma t}/L$ with $\sigma < 0$. In figure 14.3b $h(t)$ is sketched. This case is called 'critical damping'.

Case 3

Since $s_1 \neq s_2$, the partial fraction expansion of $H(s)$ is exactly the same as in (14.11) and so $h(t)$ is again given by (14.12). Since $L > 0$ and $C > 0$ it follows that $R > \sqrt{R^2 - 4L/C}$ and so $s_2 < s_1 < 0$. We thus see that $h(t)$ is the sum of two exponentially damped functions. Now put

$$\omega_0 = \frac{1}{2L}\sqrt{R^2 - \frac{4L}{C}} \qquad \text{and} \qquad \sigma = -\frac{R}{2L},$$

then $s_1 = \sigma + \omega_0$ and $s_2 = \sigma - \omega_0$. As in case 1 we can write $h(t)$ as

$$h(t) = \frac{1}{L\omega_0} e^{\sigma t} \sinh \omega_0 t = \frac{1}{2L\omega_0} e^{(\sigma + \omega_0)t} \left(1 - e^{-2\omega_0 t}\right)$$

with $\sigma < 0$. The impulse response is a damped hyperbolic sine (it is damped since $\sigma + \omega_0 = s_1 < 0$). See figure 14.3c. This case is called 'overdamped' or 'overcritical damping'.

EXERCISES

14.1   The impulse response $h(t)$ of an LTC-system is given by $h(t) = \delta(t) + t e^{-t}$ (also see exercise 10.7).
a   Determine the transfer function $H(s)$ of the system.
b   Can we obtain the frequency response of the system by substituting $s = i\omega$ in $H(s)$? Justify your answer.
c   Determine the response to the input $u(t) = e^{-t} \sin t$ in two ways: first by applying (14.1), then by determining $U(s) = (\mathcal{L}u)(s)$ and calculating the inverse Laplace transform of $H(s)U(s)$ (see (14.2)).

14.2   For a system at rest it is known that the response to the input $u(t) = t$ is given by $y(t) = t - \cos 2t$. Determine the impulse response.

14.3   **a**   Verify that the function $y(t) = e^t - e^{-t} - 2t$ obtained in example 14.4 is indeed a solution of the initial value problem given by (14.3) and (14.4). Check in particular that $y(t)$ satisfies the differential equation (14.3), whether we consider the differentiations in the ordinary sense or in the sense of distributions.
b   Verify that the impulse response $h(t)$ from example 14.7 satisfies the differential equation if we consider the differentiations in the sense of distributions.
c   Do the same for the response $y(t)$ from example 14.9.

14.4   Use the Laplace transform to determine the solution of the initial value problem $y'' + 4y = 12 \sin 4t$ with the condition of initial rest.

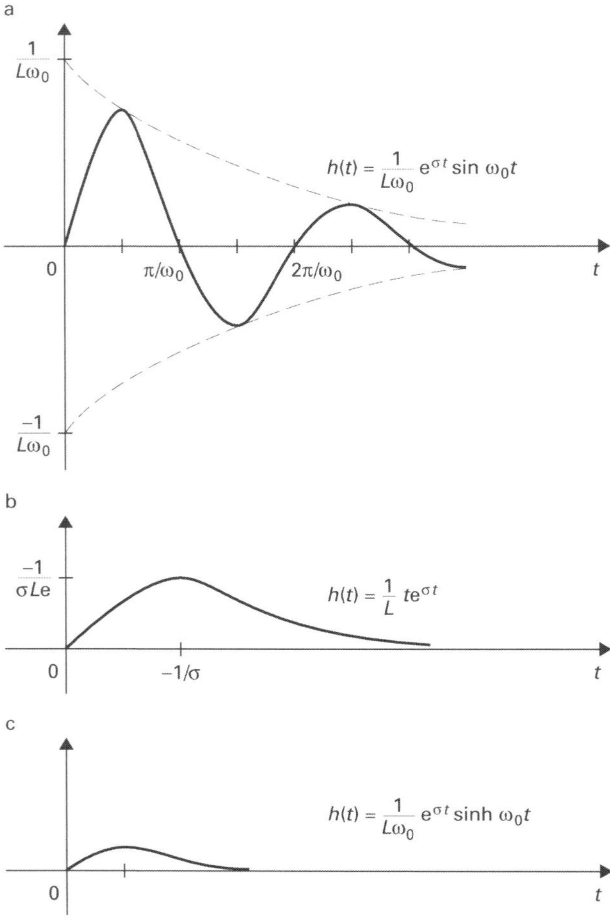

FIGURE 14.3
The impulse response in case of undercritical damping (a), critical damping (b), and overcritical damping (c).

14.5   Consider the causal LTC-system described by the differential equation $y'' - 5y' + 4y = u(t)$.

a   Determine the transfer function $H(s)$ and determine whether the system is stable.

b   Determine the impulse response $h(t)$. Is $h(t)$ absolutely integrable?

c   Determine the step response $a(t)$.

d   Determine the response to the input $u(t) = e^{2t}$.

e   Determine the response to the input $u(t) = 3\delta(t - 1)$.

14.6   We consider the harmonic oscillator with the condition of initial rest and with $L > 0$, $C > 0$ and $R > 0$. Verify for the three different cases from section 14.1.5 whether the system is stable or not.

14.7   The RC-network in figure 14.4 with resistance $R > 0$ and capacitance $C > 0$ is considered as an LTC-system with input the voltage $v(t)$ and output the charge $q(t)$

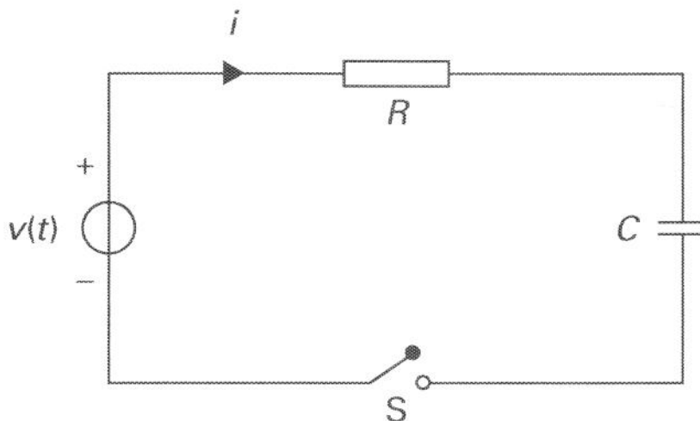

FIGURE 14.4
RC-network from exercise 14.7.

of the capacitor. At time $t = 0$, while the system is at rest, we close the switch. One has that $Rq'(t) + C^{-1}q(t) = v(t)$. Furthermore, let $E > 0$ be a constant.

a Determine the transfer function and verify whether the system is stable.

b Determine the charge $q(t)$ of the capacitor when $v(t) = E$. Also determine the current $i(t)$.

c Determine the response to the input $v(t) = E \sin at$.

d Determine the response to the input $v(t) = E\delta(t - 3)$.

14.8   The RL-network in figure 14.5 with resistance $R > 0$ and inductance $L > 0$ is considered as an LTC-system with input the voltage $v(t)$ and output the current $i(t)$.

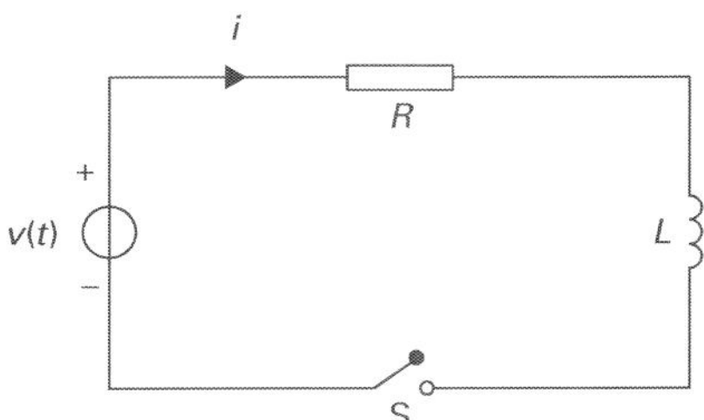

FIGURE 14.5
RL-network from exercise 14.8.

At time $t = 0$, while the system is at rest, we close the switch. One has that $Li'(t) + Ri(t) = v(t)$.

a Determine the transfer function and the impulse response. Is the system stable?

b Determine the step response.

c Determine the response to the input $v(t) = \epsilon(t - a)$ with $a > 0$.

14.2   Linear differential equations with constant coefficients

In section 14.1 the Laplace transform was applied to linear differential equations with constant coefficients and with the condition of initial rest. The method described in section 14.1.2 can also be used when *arbitrary* initial conditions are allowed. We confine ourselves in this section to the case $n = 0$ of (14.5) (this is not an essential limitation since by taking linear combinations of several functions $u(t)$ we can always obtain the right-hand side of (14.5)). We will still call the functions $u(t)$

and $y(t)$ in (14.5) the input and the output or response respectively, and as always we will assume that $u(t)$ and $y(t)$ are causal functions. The initial conditions will be certain given values for $y(0)$, $y'(0)$, $\dots$, $y^{(m-1)}(0)$. If $y^{(k)}(0) \neq 0$ then this implies that the function $y^{(k)}(t)$ has a jump at $t = 0$. The condition for $y^{(k)}(0)$ should then be interpreted as $y^{(k)}(0+)$, and when applying the differentiation rule one should now use (13.10) instead of (12.10). We start with a simple example.

EXAMPLE 14.11

Consider the differential equation $y'' - y = 2t$ with initial conditions $y(0) = 0$ and $y'(0) = -2$ (also see example 14.5). Apply the Laplace transform to both sides of the differential equation, then (13.10), applied to the function y for $n = 2$, implies that

$$\left(s^2 Y(s) - s y(0) - y'(0)\right) - Y(s) = \frac{2}{s^2},$$

where $Y(s)$ is the Laplace transform of $y(t)$. Now substitute the initial conditions $y(0) = 0$ and $y'(0) = -2$ and solve the equation for $Y(s)$. We then obtain

$$Y(s) = \left(\frac{2}{s^2} - 2\right) \frac{1}{s^2 - 1} = \frac{2 - 2s^2}{s^2(s^2 - 1)} = -\frac{2}{s^2}.$$

The inverse Laplace transform of this is given by $y(t) = -2t$. Let us check that this solution satisfies all the conditions that were imposed. It is obvious that $y(0) = 0$. Interpreting $y'(0) = -2$ as $y'(0+) = -2$, we see that all the initial conditions are satisfied. Since for $t > 0$ we have $y'(t) = -2$ and $y''(t) = 0$, we also see that $y(t)$ satisfies the differential equation for $t > 0$. Here it is *essential* that we do *not* differentiate in the sense of distributions at the point $t = 0$. For if we consider $y(t)$ as a distribution, then $y''(t) = -2\delta(t)$. This is certainly not what we want, since $y(t)$ will then no longer satisfy the original differential equation, but instead it will satisfy $y'' - y = 2t - 2\delta(t)$. ◄

From example 14.11 it is immediately clear that in general the initial conditions $y(0)$, $y'(0)$, $\dots$, $y^{(m-1)}(0)$ should indeed be considered as the *limiting values* $y(0+)$, $y'(0+)$, $\dots$, $y^{(m-1)}(0+)$ and that we are looking for a function $y(t)$ satisfying the differential equation for $t > 0$ and having the right limiting values. One should *not* interpret the differentiations at $t = 0$ in the sense of distributions. For then the initial conditions $y(0+)$, $y'(0+)$, $\dots$, $y^{(m-1)}(0+)$ would in general have no meaning and the input would have to contain distributions at $t = 0$.

We finally mention without proof that (13.10) can also be used to obtain the desired solution when the input $u(t)$ is an arbitrary piecewise smooth function. As a second example we therefore choose an input which is not continuous.

EXAMPLE 14.12

Consider the differential equation $y' + 2y = 1 - \epsilon(t - 1)$ with initial condition $y(0) = 2$. Applying the Laplace transform and using (13.10) leads to the following equation for $Y(s)$:

$$(s Y(s) - y(0)) + 2Y(s) = \frac{1}{s} - \frac{e^{-s}}{s}.$$

We then substitute the initial condition and find, after some calculations, that

$$Y(s) = \frac{1}{s(s + 2)} - \frac{e^{-s}}{s(s + 2)} + \frac{2}{s + 2}. \tag{14.14}$$

Now apply a partial fraction expansion to the first term in (14.14):

$$\frac{1}{s(s + 2)} = \frac{1}{2s} - \frac{1}{2(s + 2)}.$$

Since $(\mathcal{L}1)(s) = 1/s$ and $(\mathcal{L}e^{-2t})(s) = 1/(s+2)$, it follows from the shift property in the time domain (theorem 12.4 or table 8) that

$$\mathcal{L}(\epsilon(t-1)(1 - e^{-2(t-1)}))(s) = e^{-s}\left(\frac{1}{s} - \frac{1}{s+2}\right) = \frac{2e^{-s}}{s(s+2)}.$$

The inverse Laplace transform $y(t)$ is thus given by $y(t) = \frac{1}{2} - \frac{1}{2}e^{-2t} - \epsilon(t-1)$ $(\frac{1}{2} - \frac{1}{2}e^{-2(t-1)}) + 2e^{-2t}$. This can also be written as

$$y(t) = \begin{cases} \frac{1}{2} + \frac{3}{2}e^{-2t} & \text{for } 0 \leq t < 1, \\ \frac{3}{2}e^{-2t} + \frac{1}{2}e^{-2(t-1)} & \text{for } t \geq 1. \end{cases}$$

The result is shown in figure 14.6. Note that $y(t)$ is not differentiable at $t = 1$ and that at $t = 0$ we do not differentiate in the sense of distributions, since otherwise delta functions at $t = 0$ would occur and $y(t)$ would then no longer satisfy the differential equation. Moreover, the initial condition would no longer make sense. ◀

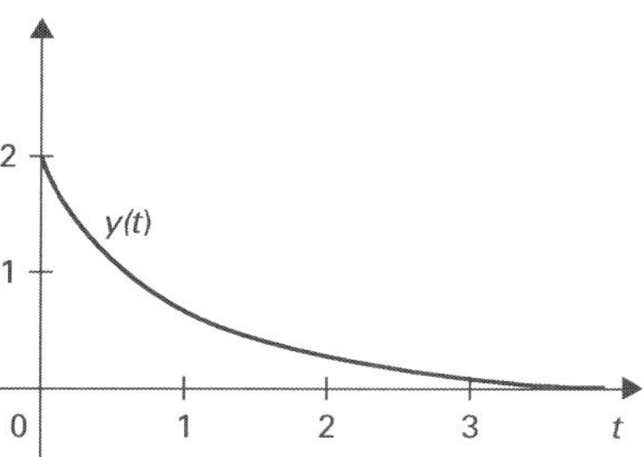

FIGURE 14.6
Response to the input $1 - \epsilon(t-1)$.

Even if the input contains distributions of the form $\delta(t-a)$ or $\delta^{(k)}(t-a)$ with $a > 0$ (!), one can prescribe arbitrary initial conditions and still find the solution using (13.10).

EXAMPLE 14.13

Consider the differential equation $y'' - 3y' + 2y = \delta(t-2)$ with initial conditions $y(0) = 0$ and $y'(0) = 1$. When $Y(s)$ is the Laplace transform of $y(t)$, then it follows from (13.10) and table 9 that $(s^2 Y(s) - 1) - 3sY(s) + 2Y(s) = e^{-2s}$. Solving for $Y(s)$ we obtain

$$Y(s) = \frac{1 + e^{-2s}}{s^2 - 3s + 2}.$$

In example 14.7 we have seen that the inverse Laplace transform of $1/(s^2 - 3s + 2)$ is $e^{2t} - e^t$. From the shift property in the time domain it then follows that

$$\mathcal{L}\left(\epsilon(t-2)(e^{2(t-2)} - e^{t-2})\right)(s) = \frac{e^{-2s}}{s^2 - 3s + 2}.$$

Hence, $y(t) = e^{2t} - e^t + \epsilon(t-2)(e^{2t-4} - e^{t-2})$. Let us verify that $y(t)$ satisfies all the conditions. Note that $y(0) = 0$ and that y has no jump at $t = 2$, which implies that $y'(t) = 2e^{2t} - e^t + \epsilon(t-2)(2e^{2t-4} - e^{t-2})$ for $t > 0$ in the sense of distributions. We now see that $y'(0) = 1$ and that y' has a jump at $t = 2$ of magnitude 1. Hence,

$$y''(t) = 4e^{2t} - e^t + \epsilon(t-2)(4e^{2t-4} - e^{t-2}) + \delta(t-2) \text{ for } t > 0 \text{ in the sense}$$

of distributions. Using this it is now easy to check that $y(t)$ satisfies the differential equation, provided that we differentiate in the sense of distributions for $t > 0$. Note that again we do not differentiate in the sense of distributions at $t = 0$.   ◄

If we do not have the condition of initial rest, then we will not allow delta functions $\delta(t)$ and/or derivatives of $\delta(t)$ in the input $u(t)$. For then it is no longer clear whether or not we should differentiate the solution $y(t)$ in the sense of distributions at $t = 0$. One can resolve this issue by dropping the causality of $y(t)$. We will not go into this any further, since situations like these will not be considered in this book (but see exercise 14.15).

In summary one can use (13.10) to solve linear differential equations with constant coefficients and arbitrary initial conditions $y(0) = y(0+)$, $y'(0) = y'(0+)$, $\ldots$, $y^{(m-1)}(0) = y^{(m-1)}(0+)$ whenever the (causal) input $u(t)$ consists of a piecewise smooth function and/or delta functions $\delta(t-a)$ or $\delta^{(k)}(t-a)$ with $a > 0$. The solutions then satisfy the differential equation for $t > 0$ if we differentiate in the sense of distributions. At $t = 0$ we should not differentiate in the sense of distributions.

EXERCISES

14.9   **a** Verify that the solution $y(t)$ from example 14.12 satisfies the differential equation $y' + 2y = 2\delta(t) + 1 - \epsilon(t-1)$ (instead of the original differential equation $y' + 2y = 1 - \epsilon(t-1)$) if we differentiate in the sense of distributions at $t = 0$.
b Give the differential equation that is satisfied by the solution $y(t)$ from example 14.13 if we differentiate in the sense of distributions at $t = 0$.

14.10   Use the Laplace transform to solve the following initial value problem:
$y'' + y = t$ with $y(0) = 0$ and $y'(0) = 1$.

14.11   Use the Laplace transform to solve the following initial value problem:
$y'' + 4y = 0$ with $y(0) = 1$ and $y'(0) = 2$.

14.12   Use the Laplace transform to solve the following initial value problem:
$y'' - 4y' - 5y = 3e^t$ with $y(0) = 3$ and $y'(0) = 1$.

14.13   Use the Laplace transform to solve the following initial value problem:
$y'' + y = u(t)$ with $y(0) = 1$ and $y'(0) = 0$ and

$$u(t) = \begin{cases} t & \text{for } 0 \leq t < 2, \\ 2 & \text{for } t \geq 2. \end{cases}$$

14.14   Use the Laplace transform to solve the following initial value problem:
$y'' + 2y' + 5y = 2\delta(t-2) + 1$ with $y(0) = 2$ and $y'(0) = -2$.

14.15   In this exercise we show for a simple differential equation which kinds of phenomena may occur if we take $\delta(t)$ as input without having the condition of initial rest. Consider the initial value problem $y'' - y = \delta(t)$ with $y(0) = 1$ and $y'(0) = 0$.
a Show that the well-known solution method using the Laplace transform leads to the function $y(t) = \cosh t + \sinh t$.
b Verify that $y(t)$ from part a does not satisfy the original differential equation, but instead satisfies the differential equation $y'' - y = 0$ if we *do not* differentiate in the sense of distributions at $t = 0$.
c Verify that $y(t)$ from part a does not satisfy the original differential equation, but instead satisfies the differential equation $y'' - y = \delta(t) + \delta'(t)$ if we *do* differentiate in the sense of distributions at $t = 0$.

d Consider the function $y(t) = \cosh t + \epsilon(t) \sinh t$ on $\mathbb{R}$ (so this is *not* a causal function!). Show that this function does satisfy the original initial value problem, if indeed we consider $y'(0)$ as $y'(0-)$ (and hence *not* as $y'(0+)$).

14.3 Systems of linear differential equations with constant coefficients

Using the Laplace transform one can also solve *systems* of several coupled ordinary linear differential equations with constant coefficients. We confine ourselves here to systems of *two* such coupled differential equations, since these can still be solved relatively easy without using techniques from matrix theory. Systems of more than two coupled differential equations will not be considered in this book. We merely note that they can be solved entirely analogously, although matrix theory becomes indispensable.

In general, a system of two coupled ordinary linear differential equations with constant coefficients and of first order has the following form:

$$\begin{cases} a_{11}x' + a_{12}y' + b_{11}x + b_{12}y = u_1(t), \\ a_{21}x' + a_{22}y' + b_{21}x + b_{22}y = u_2(t), \end{cases}$$

with initial conditions certain given values for $x(0)$ and $y(0)$. Similarly, one can describe the general system of second order with initial conditions $x(0)$, $x'(0)$, $y(0)$ and $y'(0)$ (for higher order and/or more differential equations the vector and matrix notation is much more convenient). The solution method based on the Laplace transform again consists of Laplace transforming all the functions that occur, and then solving the resulting system of linear equations with polynomials in s as coefficients. As far as the initial conditions are concerned, the same phenomena may occur as in sections 14.1 and 14.2. We will not go into this any further and confine ourselves to two simple examples. For a more extensive treatment of this subject we refer to the literature (see for example *Guide to the applications of Laplace transforms* by G. Doetsch, sections 15–19).

EXAMPLE 14.14

Consider the system

$$\begin{cases} 7x' + y' + 2x = 0, \\ x' + 3y' + y = 0, \end{cases}$$

with initial conditions $x(0) = 1$ and $y(0) = 0$. Let $X(s)$ and $Y(s)$ be the Laplace transforms of $x(t)$ and $y(t)$. According to (13.10) one then has

$$\mathcal{L}(7x' + y' + 2x)(s) = 7(sX - x(0)) + (sY - y(0)) + 2X,$$

and by substituting the initial conditions one obtains that $7(sX - 1) + sY + 2X = 0$, or $(7s + 2)X + sY = 7$. Transforming the second differential equation of the system in a similar way, we see that the Laplace transform turns the system into the algebraic system

$$\begin{cases} (7s + 2)X + sY = 7, \\ sX + (3s + 1)Y = 1. \end{cases}$$

Solving this system of two linear equations in the unknowns $X = X(s)$ and $Y = Y(s)$, we find that

$$X(s) = \frac{7(3s + 1) - s}{(3s + 1)(7s + 2) - s^2} = \frac{20s + 7}{(4s + 1)(5s + 2)},$$

$$Y(s) = \frac{7s - (7s + 2)}{s^2 - (7s + 2)(3s + 1)} = \frac{2}{(4s + 1)(5s + 2)}$$

($X(s)$ is found by multiplying the first and the second equation by, respectively, $3s + 1$ and s and subtracting; $Y(s)$ follows similarly). Using partial fraction expansions we obtain that

$$X(s) = \frac{8}{3(4s+1)} + \frac{5}{3(5s+2)} = \frac{2}{3(s+1/4)} + \frac{1}{3(s+2/5)},$$

$$Y(s) = \frac{8}{3(4s+1)} - \frac{10}{3(5s+2)} = \frac{2}{3(s+1/4)} - \frac{2}{3(s+2/5)},$$

and by inverse transforming this we see that the solution to the system is given by

$$\begin{cases} x(t) = \frac{2}{3}e^{-t/4} + \frac{1}{3}e^{-2t/5}, \\ y(t) = \frac{2}{3}e^{-t/4} - \frac{2}{3}e^{-2t/5}. \end{cases}$$

It is easy to verify that $x(t)$ and $y(t)$ satisfy the system of differential equations for $t > 0$ and that $x(0+) = 1$ and $y(0+) = 0$ (note that at $t = 0$ we should not differentiate in the sense of distributions). ◀

Systems of coupled differential equations occur, for example, in RCL-networks having more than one closed loop. This is because, according to Kirchhoff's laws, there is then a differential equation for every closed loop. We close this section with an example of such a situation.

EXAMPLE 14.15　　Consider the RL-network from figure 14.7. At the node K, the current i splits into two currents i_1 and i_2. In the closed loop containing R and L we have $Ri + Li_1' =$

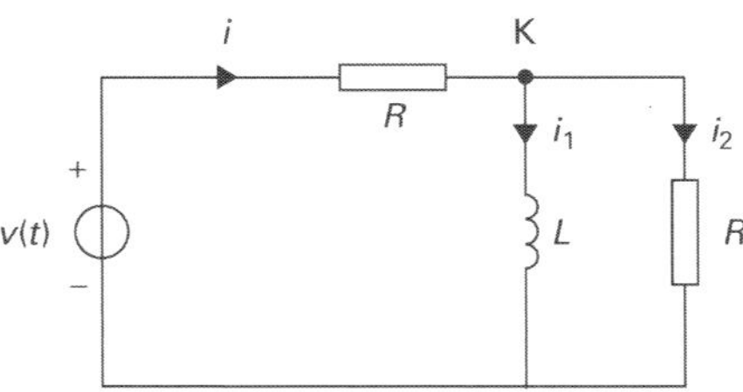

FIGURE 14.7
RL-network described by a system of two differential equations.

$v(t)$, while in the closed loop containing the two resistors R we have $Ri + Ri_2 = v(t)$. We want to determine the step response, that is, $v(t) = \epsilon(t)$, with the system being at initial rest. Since $i = i_1 + i_2$, hence $Ri_2 = Ri - Ri_1$, we have to solve the following system in the unknown functions $i(t)$ and $i_1(t)$:

$$\begin{cases} Ri + Li_1' = \epsilon(t), \\ 2Ri - Ri_1 = \epsilon(t), \end{cases}$$

with initial conditions $i(0) = i_1(0) = 0$. Laplace transforming leads to

$$\begin{cases} RI + sLI_1 = 1/s, \\ 2RI - RI_1 = 1/s, \end{cases}$$

where $I = \mathcal{L}i$ and $I_1 = \mathcal{L}i_1$. It is not hard to solve this for I (multiply the first equation by R, the second by sL, and add) and then to apply a partial fraction expansion:

$$I = \frac{1}{R} \frac{R+sL}{s(R+2sL)} = \frac{1}{R}\left(\frac{1}{s} - \frac{1}{2(s+R/2L)}\right).$$

After inverse transforming this we see that the step response is given by the function
$a(t) = (2 - e^{-Rt/2L})/2R$. ◄

EXERCISES

14.16   Use the Laplace transform to solve the following system with initial conditions
$x(0) = -1$ and $y(0) = 0$:

$$\begin{cases} x' + y = 2\cos 2t, \\ y' + x = \sin 2t. \end{cases}$$

14.17   Use the Laplace transform to solve the following system with initial conditions
$x(0) = 0$, $x'(0) = 2$, $y(0) = -1$ and $y'(0) = 0$:

$$\begin{cases} x'' + y' = 0, \\ y'' - x' = 0. \end{cases}$$

14.18   Use the Laplace transform to solve the following system with initial conditions
$x(0) = 1$ and $y(0) = -1$:

$$\begin{cases} 2y' + y + 5x' - 2x = 2e^{-t}, \\ -y' - 2x' + x = \sin t. \end{cases}$$

14.19   Consider the RCL-network from figure 14.8 with resistors R, inductor L and capacitor C. At time $t = 0$ we close the switch S while the network is at rest. In the

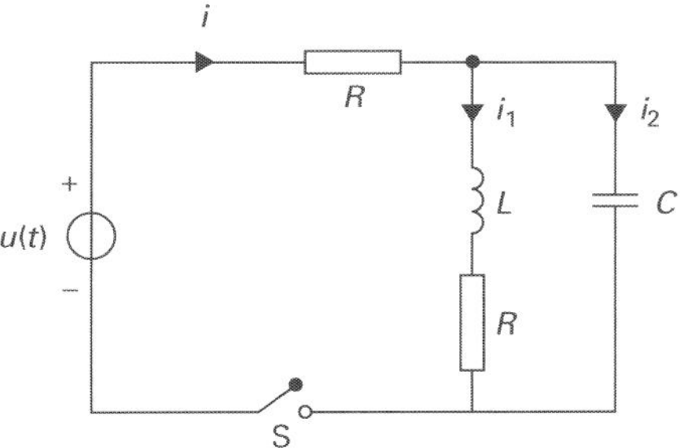

FIGURE 14.8
RCL-network from exercise 14.19.

closed loop containing the two resistors one has $Ri + Li_1' + Ri_1 = u(t)$ and since
$i = i_1 + i_2$ it follows that $Ri_2 + Li_1' + 2Ri_1 = u(t)$. In the closed loop containing the
capacitor and the inductor one has $C^{-1} \int_{-\infty}^{t} i_2(\tau)\,d\tau - Ri_1 - Li_1' = 0$. We now
take $R = 1$ ohm, $L = 1$ henry, $C = 1$ farad and $u(t) = 1$.
a   Use the Laplace transform to solve the resulting system

$$\begin{cases} i_2 + 2i_1 + i_1' = 1, \\ \int_{-\infty}^{t} i_2(\tau)\,d\tau - i_1 - i_1' = 0 \end{cases}$$

in the unknown functions i_1 and i_2 (the initial conditions are $i_1(0) = i_2(0) = 0$).
(Hint: since $i_2(0) = 0$, the function i_2 will be continuous on $\mathbb{R}$ and so the integration
rule in the time domain can be applied.)
b   Calculate the voltage drop $y(t) = \int_{-\infty}^{t} i_2(\tau)\,d\tau$ across the capacitor. Now note
that here we are studying the same network as in exercise 10.22 and so compare the
answer with the step response found in exercise 10.22a.

14.4   Partial differential equations

Up till now we have only studied *ordinary* differential equations in this chapter: the unknown functions depended only on one variable, which was mostly associated with time. In *partial* differential equations the unknown function depends on several variables. As in chapters 5 and 10, we only consider partial differential equations for functions $u(x, t)$ of *two* variables. Just as in chapter 5, we will denote the partial derivatives of the function $u(x, t)$ with respect to x and t by u_x, u_t, u_{xx}, etc. The t-variable will again be interpreted as *time*, while the x-variable will be associated with *position*. The switch-on moment is chosen as being $t = 0$ in the time domain, and the conditions at $t = 0$ (for example $u(x, 0)$ or $u_t(x, 0)$) will be called the *initial conditions*. Conditions in the position domain x (such as $u(0, t)$ or $u_x(0, t)$) will be called *boundary conditions* (although sometimes this term is also used for all conditions taken together). In sections 5.2 and 10.4 we have seen how the Fourier transform can be used to solve partial differential equations. In this section we will show how this can be done using the Laplace transform. Again, the method is essentially the same as in the previous sections, having the same advantages. However, in partial differential equations we are dealing with functions depending on two variables. Since the Laplace transform can only be applied to *one* variable, we will have to make a choice. In most cases one chooses a transform with respect to the t-variable. From the function $u(x, t)$ one then obtains for each x a Laplace transform depending on s. This will be denoted by $U(x, s)$, so

Initial condition

Boundary condition

$$(\mathcal{L}u(x, t))(s) = \int_0^\infty u(x, t)e^{-st}\, dt = U(x, s). \tag{14.15}$$

Similarly, we have initial conditions for each x, like $u(x, 0)$ or $u_t(x, 0)$, which again should be interpreted as limiting values $u(x, 0+)$ or $u_t(x, 0+)$ if necessary. All previous results, in particular the differentiation rules in the time domain from (12.10) and (13.10), can now be applied in the t-variable for fixed x. Concerning the differentiations in the x-variable, we will assume that differentiation and Laplace transform may be interchanged, for example:

$$\begin{aligned}(\mathcal{L}u_x(x, t))(s) &= \int_0^\infty \frac{\partial u}{\partial x}(x, t)e^{-st}\, dt = \frac{\partial}{\partial x}\left(\int_0^\infty u(x, t)e^{-st}\, dt\right) \\ &= \frac{\partial}{\partial x}(\mathcal{L}u(x, t))(s) = U_x(x, s).\end{aligned}$$

We will now show how the Laplace transform can be used to solve partial differential equations with initial and boundary conditions.

EXAMPLE 14.16

Half-infinite string

Consider a *half-infinite string* being at rest at time $t = 0$ along the positive x-axis. We let the string vibrate by moving the left end at $x = 0$ once up and down in the following way:

$$u(0, t) = f(t) = \begin{cases} \sin t & \text{for } 0 \le t \le 2\pi, \\ 0 & \text{elsewhere.} \end{cases}$$

Of course, an infinitely long string does not exist in reality, but we use it as a mathematical model for a long string or a long piece of rope. From experience we know that if we move a piece of rope once up and down, a travelling wave in the rope will arise. See figure 14.9. It is known that the equation of motion of the string is given by

$$u_{tt} = c^2 u_{xx} \qquad \text{for } x > 0 \text{ and } t > 0, \tag{14.16}$$

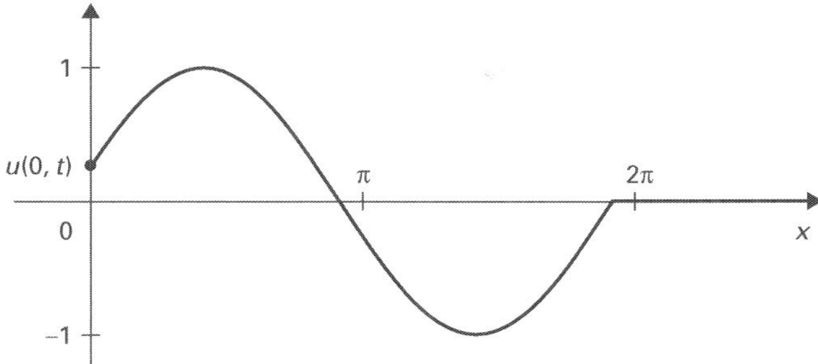

FIGURE 14.9
Travelling wave in a half-infinite string.

where $c > 0$ is a physical constant. Because of the state of rest at $t = 0$, we have the initial conditions $u(x, 0) = 0$ and $u_t(x, 0) = 0$ for $x \geq 0$. The second initial condition indicates that the initial velocity of the string at $t = 0$ is zero. Since the string has infinite length in the positive x-direction, we have $\lim_{x \to \infty} u(x, t) = 0$ for $t \geq 0$ (such a condition will be called a boundary condition as well). Summarizing, the initial and boundary conditions are given by

$$
\begin{cases}
u(x, 0) = u_t(x, 0) = 0 & \text{for } x \geq 0, \\
\lim_{x \to \infty} u(x, t) = 0 & \text{for } t \geq 0, \\
u(0, t) = f(t) & \text{for } t \geq 0.
\end{cases}
$$

We now assume that the Laplace transform $U(x, s)$ of $u(x, t)$ exists and that all the operations we will be performing are allowed. Once a solution $u(x, t)$ has been found, it has to be verified afterwards that it indeed satisfies the partial differential equation and all the initial and boundary conditions. We will not go into this any further and omit this verification.

If we apply the Laplace transform with respect to t to the partial differential equation (14.16), then it follows from (13.10) that

$$
s^2 U(x, s) - su(x, 0) - u_t(x, 0) = c^2 (\mathcal{L}u_{xx})(s).
$$

Since we assumed that differentiations with respect to x and the Laplace transform may be interchanged, it follows that $(\mathcal{L}u_{xx})(s) = U_{xx}(x, s)$. Substituting the initial conditions $u(x, 0) = u_t(x, 0) = 0$, we obtain that $s^2 U(x, s) = c^2 U_{xx}(x, s)$, or $U_{xx} - (s^2/c^2)U = 0$. We have thus transformed the partial differential equation for $u(x, t)$ into the *ordinary* differential equation

$$
U'' - \frac{s^2}{c^2} U = 0 \tag{14.17}
$$

for $U(x, s)$ as function of x (s being constant), which is a considerable simplification. The general solution of differential equation (14.17) is

$$
U(x, s) = Ae^{sx/c} + Be^{-sx/c},
$$

where A and B can still be functions of s. To determine A and B, we translate the remaining boundary conditions to the s-domain by Laplace transforming them.

From the condition $\lim_{x\to\infty} u(x,t) = 0$ it follows that

$$\lim_{x\to\infty} U(x,s) = \lim_{x\to\infty} \int_0^\infty u(x,t)e^{-st}\,dt = \int_0^\infty \left(\lim_{x\to\infty} u(x,t)\right) e^{-st}\,dt = 0,$$

where we assume that the limit and the integral may be interchanged. Since $U(x,s)$ exists in a certain half-plane, it will in particular be defined for values of $s \in \mathbb{C}$ with $\operatorname{Re} s > 0$, and so having $\operatorname{Re} s/c > 0$ as well (because $c > 0$). Using this we see that $\lim_{x\to\infty} e^{sx/c}$ does not exist and so $A = 0$.

If we now apply the Laplace transform to the remaining boundary condition $u(0,t) = f(t)$, then it follows that $B = U(0,s) = F(s)$ with $F(s) = (\mathcal{L}f)(s)$. Hence, the solution in the s-domain is now completely determined:

$$U(x,s) = F(s)e^{-sx/c}.$$

The shift property in the time domain (table 8) finally gives us the function $u(x,t) = \epsilon(t - x/c)f(t - x/c)$ as the inverse Laplace transform. With our choice for the function f this means that the solution is given by

$$u(x,t) = \begin{cases} \sin(t - x/c) & \text{for } x/c < t < x/c + 2\pi, \\ 0 & \text{elsewhere.} \end{cases}$$

This is a sinusoidal wave, travelling to the right with speed c. A point x remains at rest until $t = x/c$, the time necessary to reach this position. The motion of the point x is then identical to the motion of the left end $x = 0$.   ◀

In summary, the most important steps in solving partial differential equations using the Laplace transform are as follows:

Step 1

Each term of the partial differential equation for $u(x,t)$ is Laplace transformed with respect to one of the variables, mostly the time variable t. For the Laplace transform $U(x,s)$ we obtain an ordinary differential equation, since only derivatives of $U(x,s)$ with respect to x occur in the transformed equation. Moreover, this equation contains all the initial conditions.

Step 2

The ordinary differential equation is solved using all known means. To do so, we must also determine the boundary conditions for $U(x,s)$. These can be found by transforming the boundary conditions for $u(x,t)$.

Step 3

The solution obtained in the s-domain is transformed back into the t-domain. For the solution $u(x,t)$ thus found, one can verify whether it satisfies the partial differential equation and all the conditions.

EXERCISES

14.20

A string attached at $x = 0$ and $x = 1$ is given the initial position $u(x,0) = 2\sin 2\pi x$ at time $t = 0$ and is then released. Let $u(x,t)$ be the position of the string at point x and time t. The equation of motion of the string is given by $u_{tt} = 4u_{xx}$ ($0 < x < 1$, $t > 0$). The initial and boundary conditions are given by

$$\begin{cases} u(0,t) = u(1,t) = 0 & \text{for } t \geq 0, \\ u_t(x,0) = 0 & \text{for } 0 < x < 1, \\ u(x,0) = 2\sin 2\pi x & \text{for } 0 \leq x \leq 1. \end{cases}$$

Solve this problem using the Laplace transform.

14.21

An insulated rod is fixed at $x = 0$ and $x = \pi$ and has initial temperature $u(x,0) = 4\sin x$. Let $u(x,t)$ be the temperature at position x and time t. The ends of the rod are cooled and kept at a temperature of 0 degrees, so $u(0,t) = 0$ and $u(\pi,t) = 0$.

The temperature distribution is described by the heat equation $u_t = u_{xx}$ ($0 < x < \pi$, $t > 0$). Find the temperature distribution $u(x, t)$ using the Laplace transform (in the calculations for this exercise you may use the complex function $\sqrt{s}$ in a formal way; this function has not been treated in chapter 11 since the 'multivaluedness' problem arises here: for one value of s there are two values $w \in \mathbb{C}$ with $w = \sqrt{s}$).

14.22   Consider for $u(x, t)$ the partial differential equation $u_t = u_{xx} - 6u$ ($0 < x < \pi$, $t > 0$) with initial condition $u(x, 0) = \cos(x/2)$ ($0 \leq x \leq \pi$) and boundary conditions $u(\pi, t) = 0$ ($t \geq 0$) and $u_x(0, t) = 0$ ($t > 0$). Solve this problem using the Laplace transform.

SUMMARY

The Laplace transform is very well suited for the study of causal LTC-systems for which switch-on phenomena occur. If $h(t)$ is the impulse response of such a system, then the transfer function or system function is given by the Laplace transform $H(s)$ of $h(t)$. If $u(t)$ is an input with $(\mathcal{L}u)(s) = U(s)$ and $y(t)$ the response with $(\mathcal{L}y)(s) = Y(s)$, then $Y(s) = H(s)U(s)$.

An important class of causal LTC-systems is described by an ordinary linear differential equation with constant coefficients (in the form of (14.5)) and with the condition of initial rest (as in (14.6)). In this case the system function is given by the rational function

$$H(s) = \frac{b_n s^n + \cdots + b_1 s + b_0}{a_m s^m + \cdots + a_1 s + a_0},$$

where $n \leq m$. After a partial fraction expansion, one can obtain the impulse response $h(t)$ from this by an inverse transform. The response $y(t)$ to an arbitrary input is then given by $y(t) = (h * u)(t)$. Calculating this convolution is in general not very easy. Instead, one can also immediately apply the Laplace transform for a given input $u(t)$. The differential equation, including the initial conditions, is then transformed into an algebraic equation for $Y(s)$. This equation is easy to solve and for a large class of inputs, $Y(s)$ is again a rational function of s with real coefficients. Using a partial fraction expansion and an inverse transform, one then obtains $y(t)$.

The stability of the system follows from the location of the poles of $H(s)$: these should lie in the half-plane $\operatorname{Re} s < 0$.

In exactly the same way one can apply the Laplace transform to ordinary linear differential equations with constant coefficients and having arbitrary initial conditions. The input may be any combination of piecewise smooth functions and distributions $\delta(t - a)$ or $\delta^{(k)}(t - a)$ with $a > 0$. The obtained solution should not be differentiated in the sense of distributions at $t = 0$.

We can also apply the Laplace transform to a system of coupled ordinary linear differential equations with constant coefficients. Transforming it will turn it into a system of linear equations having polynomials in s as coefficients. From this, the unknowns in the s-domain can be solved. An inverse transform then leads to the solution of the system of differential equations.

Finally, the Laplace transform can also be used to solve certain partial differential equations with initial and boundary conditions. Transforming this will lead to an ordinary differential equation containing the initial conditions. By also transforming the remaining boundary conditions, one can solve the ordinary differential equation. The solution in the s-domain thus obtained can be transformed back to the t-domain.

SELFTEST

14.23   For an LTC-system it is known that the step response $a(t)$ is given by $a(t) = \cosh 2t - 2\cos t + e^{-t}$.
a   Determine the impulse response and the transfer function.
b   Determine the response to the input $u(t) = 2\delta(t - 1)$.
c   Determine the response to the input $u(t) = t$.

14.24   Consider the harmonic oscillator with the condition of initial rest and with $L > 0$, $C > 0$ but $R = 0$. We are then dealing with an LC-network as in figure 14.10. At $t = 0$ we close the switch S. The network is considered as an LTC-system with

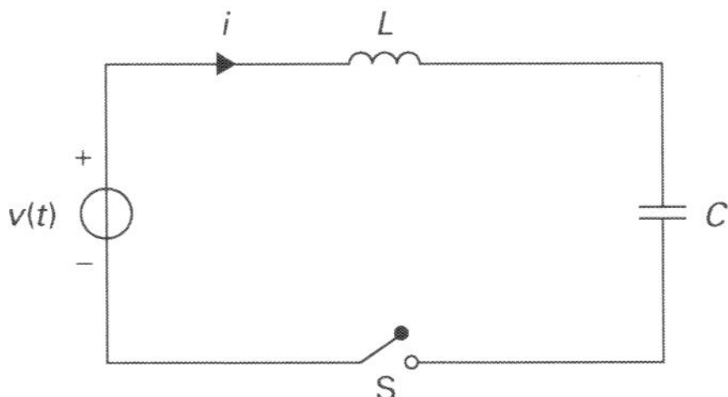

FIGURE 14.10
LC-network from exercise 14.24.

input the voltage $v(t)$ and output the charge $q(t)$ of the capacitor. One has that $Lq''(t) + C^{-1}q(t) = v(t)$ with $q(0) = q'(0) = 0$.
a   Determine the transfer function and the impulse response.
b   Is the system stable or not? Justify your answer.
c   Determine the charge $q(t)$ of the capacitor if $v(t) = e^{-at}$ with $a > 0$.
d   Determine the response $q(t)$ to $v(t) = \cos at$ with $a \neq (LC)^{-1/2}$.
e   Put $\omega_0 = (LC)^{-1/2}$. Determine the response $q(t)$ to $v(t) = 2\cos \omega_0 t$. Verify that $\lim_{t \to \infty} |q(t)| = \infty$ and sketch $q(t)$.

14.25   Use the Laplace transform to solve the following initial value problem:
$y'' + y' - 2y = u(t)$ with $y(0) = 1$ and $y'(0) = 1$ and with $u(t)$ given by:
a $u(t) = \begin{cases} \cos t & \text{for } 0 \leq t < \pi, \\ 0 & \text{for } t \geq \pi. \end{cases}$

b $u(t) = 3\delta(t - 2) + 6\delta'(t - 3)$.

14.26   Consider the RL-network from figure 14.11 with the three resistors R, $2R$ and $2R$, the inductors L and $2L$, and the voltage source $v(t)$. At time $t = 0$ we close the switch S, while the network is at rest, so $i(0) = i_1(0) = i_2(0) = 0$. By considering the closed loops ABEF and BCDE, one obtains the following system of differential equations in the unknown functions i_1 and i_2 (use that $i = i_1 + i_2$):

$$\begin{cases} 2Ri_2 + 3Ri_1 + Li'_1 = v(t), \\ 2Li'_2 + 2Ri_2 - Ri_1 - Li'_1 = 0. \end{cases}$$

Use the Laplace transform to determine the currents $i(t)$, $i_1(t)$ and $i_2(t)$ when the voltage $v(t)$ is given by:
a $v(t) = E$, with E a constant,
b $v(t) = \sin 2t$.

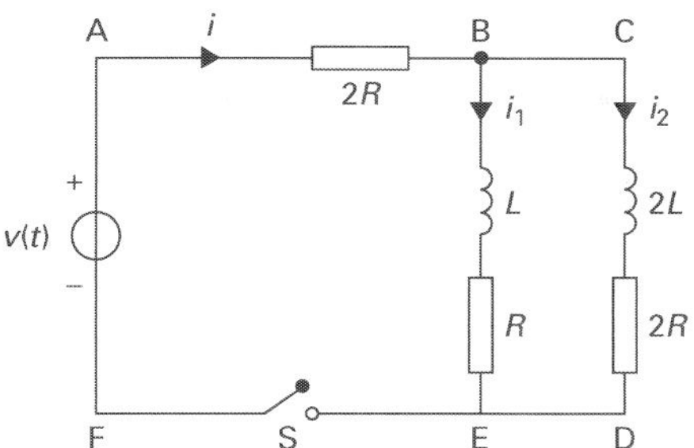

FIGURE 14.11
RL-network from exercise 14.26.

14.27 A string at rest is attached at $x = 0$ and $x = 2$. Let $u(x, t)$ be the position of the string at point x and at time t. At $t = 0$ the string is given an initial velocity $u_t(x, 0) = 2 \sin \pi x$. The equation of motion of the string is given by $u_{tt} = 4u_{xx}$ $(0 < x < 2, t > 0)$. The initial and boundary conditions are given by

$$\begin{cases} u(0, t) = u(2, t) = 0 & \text{for } t \geq 0, \\ u(x, 0) = 0 & \text{for } 0 \leq x \leq 2, \\ u_t(x, 0) = 2 \sin \pi x & \text{for } 0 < x < 2. \end{cases}$$

Solve this problem using the Laplace transform.

Part 5
Discrete transforms

INTRODUCTION TO PART 5

In the previous chapters we have seen how the Fourier transform of a continuous-time signal can be calculated using tables and properties. However, it is not always possible to apply these methods. The reason could be that we only know the continuous-time signal for a limited number of moments in time, or simply that the Fourier integral cannot be determined analytically. Starting from a limited set of data, one then usually has to rely on numerical methods in order to determine Fourier transforms or spectra. To turn a numerical method into a manageable tool for a user, it is first transformed into an algorithm, which can then be processed by a digital computer. The user then has a program at his/her disposal to calculate spectra or Fourier transforms. Calculating the spectrum of a continuous-time signal using a computer program can be considered as signal processing. When an algorithm for such a program is studied in more detail, then one notices that almost all calculations are implemented in terms of numbers, or sequences of numbers. In fact, the continuous-time signal is first transformed into a sequence of numbers (we will call this a *discrete-time signal*) representing the function values, and subsequently this sequence is processed by the algorithm. One then calls this *digital signal processing*. It is clear that because of the finite computing time available, and the limited memory capacity of a digital computer, the spectrum can only be determined for a finite number of frequencies, and is seldom exact. Moreover, when transforming the continuous-time signal into a discrete-time signal, there will in general be loss of information. However, because of the rapid development of the digital computer, especially the capacity of memory chips and the processor speed, capabilities have increased enormously and disadvantages have diminished. This is one of the reasons why the field of digital signal processing has aroused increasing interest (as is witnessed by the growing literature, e.g. *Digitale signaalbewerking* by A.W.M. van den Enden and N.A.M. Verhoeckx (in Dutch)).

Applications of digital signal processing can nowadays be found in several fields. For the consumer we have, for example, the compact disc and high definition tv. In industry digital computers are used to control industrial processess (robotics), photos originating from satellites are being analysed using so-called digital imaging techniques. In the medical sciences, digital processing is used in cardiology, for instance.

An important subject in the field of digital signal processing is the discrete transform of signals, that is, transforms defined for discrete-time signals. They have many similarities with the Fourier and Laplace transforms of continuous-time signals, treated in the previous parts.

Chapter 15, the first chapter of this part, focusses mainly on the problem of the transformation of a continuous-time signal into a discrete-time signal. One would like to minimize the loss of information which is always introduced by this transformation in various kinds of applications.

"

In chapter 16 the so-called discrete Fourier transform is treated. This transform is often used when analysing spectra of signals. The fast algorithms to calculate the discrete Fourier transform are collectively known as the Fast Fourier Transform, FFT for short. In chapter 17 we treat a frequently used example of an FFT algorithm.

The so-called z-transform, the subject of chapter 18, is a valuable tool used to describe discrete-time *systems*. For example, the frequency response of a discrete-time system (see chapter 1) can easily be obtained from the z-transform of the impulse response.

Finally, chapter 19 treats in particular the applications of discrete-time transforms to discrete-time systems. In chapter 1 these discrete-time systems have already been discussed in general terms. As such, one can consider chapter 19 as the closing chapter of this final part on discrete transforms.

Contents of Chapter 15

Sampling of continuous-time signals

Introduction *340*

Summary *352*

Selftest *353*

CHAPTER 15

Sampling of continuous-time signals

INTRODUCTION

In chapter 1 signals were divided into continuous-time and discrete-time signals. Ever since, we have almost exclusively discussed continuous-time signals. This chapter, being the first chapter of part 5, can be considered as sort of a transition from the continuous-time to the discrete-time signals. In section 15.1 we first introduce a number of important discrete-time signals, which are very similar to well-known continuous-time signals like the unit pulse or delta function. Subsequently, we pay special attention in section 15.2 to the transformation of a continuous-time signal into a discrete-time signal (sampling) and vice versa (reconstruction), leading to the formulation and the proof of the so-called *sampling theorem* in section 15.3. The sampling theorem gives a lower bound (the so-called Nyquist frequency) for the sampling frequency such that a given continuous-time signal can be transformed into a discrete-time signal *without loss of information*. We close with the treatment of the so-called aliasing problem in section 15.4. This problem arises when a continuous-time signal is transformed into a discrete-time signal using a sampling frequency which is too low.

LEARNING OBJECTIVES
After studying this chapter it is expected that you
- can describe discrete-time signals using unit pulses
- can describe periodic discrete-time signals using periodic unit pulses
- can explain the meaning of the terms sampling, sampling period and sampling
  frequency
- can explain the sampling theorem and can apply it
- can understand the reconstruction formula for band-limited signals
- can describe the consequences of the aliasing problem.

15.1   Discrete-time signals and sampling

In chapter 1 a discrete-time signal was defined as a complex-valued function having as domain the set $\mathbb{Z}$ of integers (see section 1.2.1). To make a clear distinction with continuous-time signals, we will denote discrete-time signals with square brackets enclosing the argument, so $f[n]$, $g[n]$, etc. By the way, the term discrete-*time* in signal theory is somewhat misleading. It suggests, as for continuous-time signals, that this refers to time as being the independent variable. This is not necessarily always the case.

In practice the signals will only be given on a limited part of $\mathbb{Z}$. In fact, the signal then consists of a *finite* sequence of numbers. The domain can then be extended to all integers by adding zeros to the sequence or by extending it periodically.

340

We will often omit the adjective 'discrete-time' whenever it is clear from the context that we are dealing with discrete-time signals.

EXAMPLE 15.1

For a given periodic continuous-time signal with period T and line spectrum c_n we can define a discrete-time signal by

$$f[n] = c_n \qquad \text{for } n \in \mathbb{Z}.$$

Here it is clear that the term discrete-time is misleading. Rather one should call this a discrete-frequency signal. ◄

EXAMPLE 15.2

For a given continuous-time signal $f(t)$ and a positive number T we define a discrete-time signal by

$$f[n] = f(nT) \qquad \text{for } n \in \mathbb{Z}.$$

◄

Sampling

Sampling period

Samples

Sampling frequency

The signal $f[n]$ from example 15.2 is called a *sampling* of the signal $f(t)$ at the points $0, \pm T, \pm 2T, \ldots$. The positive number T is called the *sampling period*, $1/T$ the sampling frequency, and the values $f(nT)$ the *samples* of the sampling. In this book, however, we mainly use angular frequencies and so here the *sampling frequency* ω_s is defined as

$$\omega_s = \frac{2\pi}{T}. \tag{15.1}$$

EXAMPLE 15.3

Sampling of the signal $f(t) = \sin(\omega_0 t)$, where ω_0 is a given positive frequency, with the following sampling frequencies leads to the corresponding discrete-time signals:

frequency   *signal*

$\omega_s = \omega_0 \qquad f[n] = \sin(2\pi n) = 0$

$\omega_s = 8\omega_0 \qquad f[n] = \sin(n\pi/4)$

$\omega_s = \pi\omega_0 \qquad f[n] = \sin(2n).$

Note that sampling with $\omega_s = 8\omega_0$ results in a periodic discrete-time signal $f[n] = f(n\pi/4\omega_0)$ having period $N = 8$ (see figure 15.1). For $\omega_s = \pi\omega_0$ however, we see that despite the periodicity of $f(t)$, the sampling is no longer periodic. ◄

An elementary signal is the so-called *discrete unit pulse* $\delta[n]$, also called the Kronecker delta function, which is defined as follows.

DEFINITION 15.1
Discrete unit pulse

The discrete unit pulse $\delta[n]$ *is defined by*

$$\delta[n] = \begin{cases} 1 & \text{for } n = 0, \\ 0 & \text{for } n \neq 0. \end{cases}$$

The discrete unit pulse has a nice and simple property, which is formulated in our following theorem.

THEOREM 15.1

For an arbitrary discrete-time signal $f[n]$ *one has*

$$f[n] = \sum_{k=-\infty}^{\infty} f[k]\delta[n-k] \qquad \text{for } n \in \mathbb{Z}. \tag{15.2}$$

Proof
Note that in the infinite sum in the right-hand side all terms are zero, except for the term with $k = n$. The latter equals $f[n]$. The sum is thus equal to $f[n]$ as well. ∎

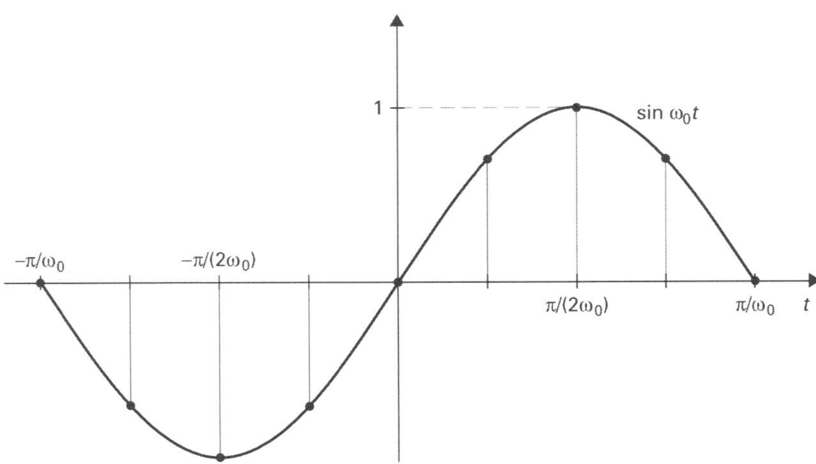

FIGURE 15.1
Sampling of $\sin(\omega_0 t)$ with frequency $8\omega_0$.

Put into words, the theorem above states that an arbitrary discrete-time signal can be written as a *superposition* of shifted unit pulses.

EXAMPLE 15.4

The signal $f[n]$ defined by

$$f[n] = \begin{cases} 1 & \text{for } n = 0, \\ -1 & \text{for } n = 1, \\ 2 & \text{for } n = -1, \\ 0 & \text{elsewhere,} \end{cases}$$

can also be written as follows:

$$f[n] = 2\delta[n+1] + \delta[n] - \delta[n-1].$$

In figure 15.2 this signal $f[n]$ is represented by a graph.   ◄

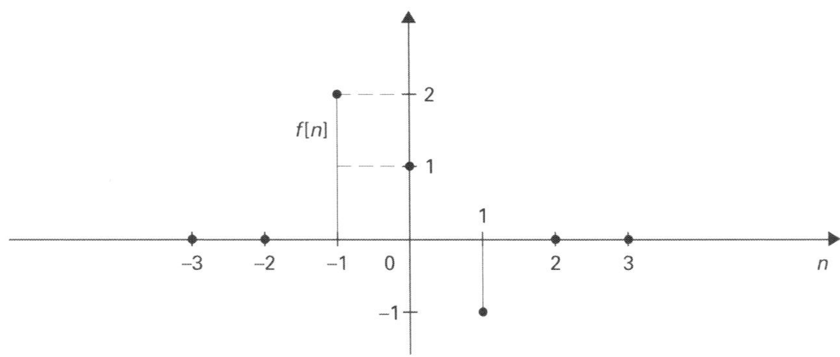

FIGURE 15.2
The signal $f[n] = 2\delta[n+1] + \delta[n] - \delta[n-1]$.

Besides in the form of formula (15.2), a *periodic* discrete-time signal $f[n]$ with period N, that is, $f[n + N] = f[n]$ for all $n \in \mathbb{Z}$ (see section 1.2.2), can also be written in another way as a superposition. To this end we first introduce the so-called periodic train of discrete unit pulses.

DEFINITION 15.2
Periodic train of discrete unit pulses

Let N be a positive integer. The periodic train of discrete unit pulses $\delta_N[n]$ with period N is given by

$$\delta_N[n] = \begin{cases} 1 & \textit{if } n \textit{ is an integer multiple of } N, \\ 0 & \textit{otherwise.} \end{cases}$$

In figure 15.3 a graph of $\delta_3[n]$ is drawn. It is easy to express an arbitrary periodic signal with period N as shifted trains of discrete unit pulses.

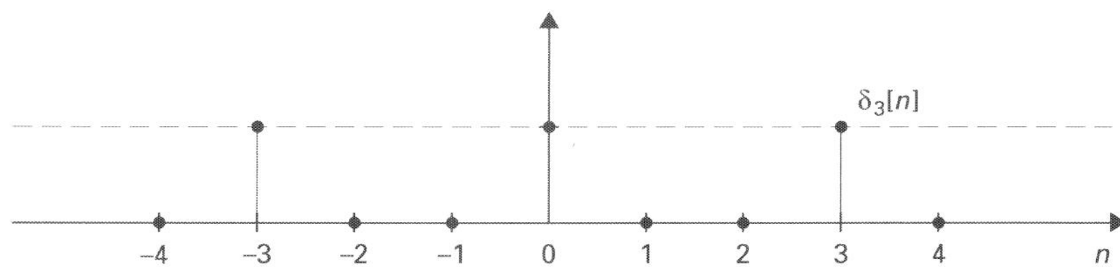

FIGURE 15.3
Periodic train of discrete unit pulses.

THEOREM 15.2

Let $f[n]$ be a periodic discrete-time signal with period N. Then

$$f[n] = \sum_{k=0}^{N-1} f[k]\delta_N[n-k] \qquad \textit{for } n \in \mathbb{Z}.$$

Proof
In the right-hand side of the sum above, each of the terms represents a periodic signal with period N. It is then easy to see that the sum itself also represents a periodic signal with period N. We thus only need to prove the equality for the values $n = 0, 1, 2, \ldots, N - 1$. But this follows immediately from the definition of $\delta_N[n]$. ∎

EXAMPLE 15.5

The signal $f[n]$ given by

$$f[0] = 1, \, f[1] = 0, \, f[2] = -1, \, f[3] = 0,$$

and furthermore periodic with period 4, can also be written as

$$f[n] = \delta_4[n] - \delta_4[n - 2].$$

◀

We close this section with the introduction of the discrete version of the unit step function $\epsilon(t)$.

DEFINITION 15.3
Discrete unit step function

The discrete unit step function $\epsilon[n]$ is defined by

$$\epsilon[n] = \begin{cases} 1 & \textit{for } n \geq 0, \\ 0 & \textit{for } n < 0. \end{cases}$$

The graph of $\epsilon[n]$ is drawn in figure 15.4. The signal $\epsilon[n]$ is thus an example of a causal signal. If $f[n]$ is an arbitrary discrete-time signal, then the signal $f[n]\epsilon[n]$ is a causal signal coinciding with the signal $f[n]$ for $n \geq 0$.

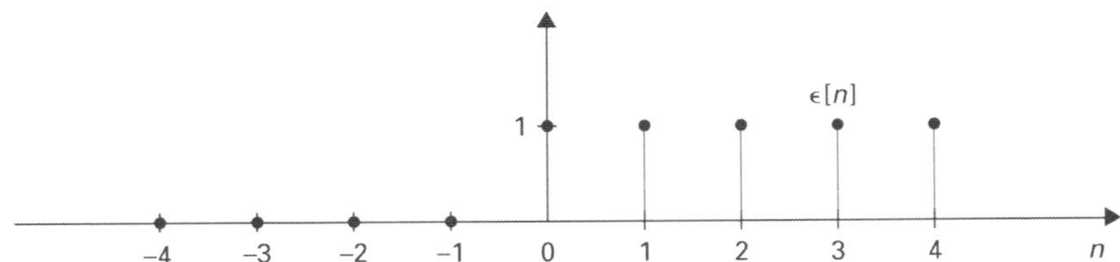

FIGURE 15.4
The discrete unit step function.

EXERCISES

15.1 A discrete-time signal is given by $f[n] = \epsilon[n-4] - \epsilon[n+1]$. Describe the signal as a superposition of discrete unit pulses.

15.2 A periodic discrete-time signal with period 5 is given by $f[-2] = 1$, $f[-1] = 0$, $f[0] = 1$, $f[1] = 0$, $f[2] = 1$. Describe the signal as a superposition of periodic trains of unit pulses.

15.3 Let the complex number $z = \frac{1}{2}(\sqrt{3} + i)$ be given. Consider the discrete-time signal $f[n] = z^n$. Show that $f[n]$ is periodic.

15.4 Show that $\epsilon[n] = \displaystyle\sum_{k=0}^{\infty} \delta[n-k]$.

15.5 Sketch the graph of the signal $f[n] = 2\delta_4[n-1] + \epsilon[n+1]$.

15.2   Reconstruction of continuous-time signals

Sampling turns a continuous-time signal into a discrete-time signal. Of course it is clear that this may result in loss of information. The loss will be reduced to zero if, for a given sampling, we are able to get a complete reconstruction of the continuous-time signal. Hence, the discrete signal must be converted back into a *Reconstruction* continuous-time signal. This process is called *reconstruction*. Several reconstruction methods exist. A simple method, based on linear interpolation, will be treated in this section. However, it will turn out that this method has some disadvantages. In section 15.3 we will treat a better method which, under certain conditions, will recover the original continuous-time signal.

In a *linear interpolation* one constructs from a given sampling $f[n] = f(nT)$ a continuous-time signal $f_r(t)$ having the following properties (see figure 15.5):

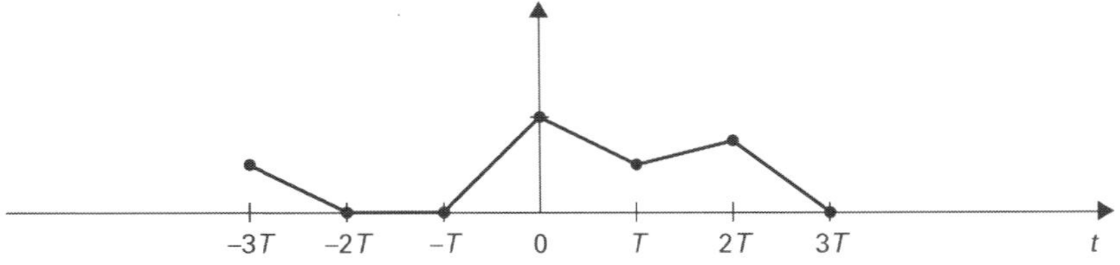

FIGURE 15.5
Linear interpolation.

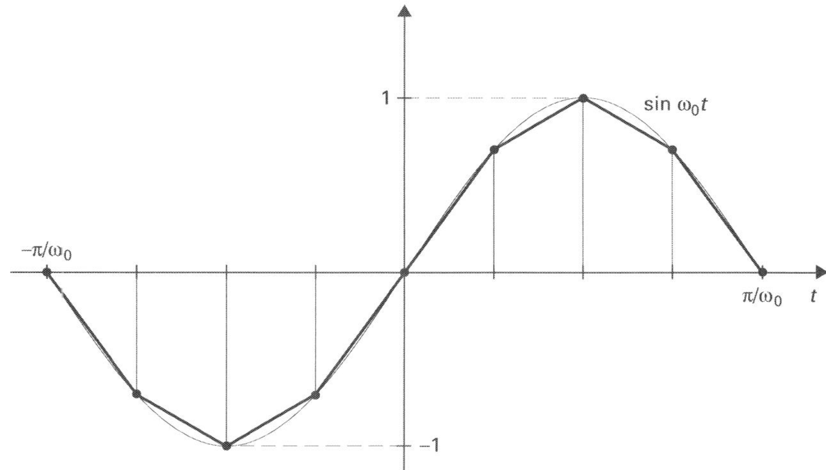

FIGURE 15.6
Reconstruction of $\sin(\omega_0 t)$ after sampling.

a $f(nT) = f[n]$ for all integer n;
b   between the two consecutive points nT and $(n+1)T$ the graph of $f_r(t)$ consists of a straight line.

Linear interpolation

We now say that for $t \in (nT, (n+1)T)$ the values $f_r(t)$ are obtained from the values $f[n]$ and $f[n+1]$ by *linear interpolation*. Reconstruction of the periodic signal $\sin(\omega_0 t)$ for sampling frequency $\omega_s = 8\omega_0$ leads to the signal consisting of the line elements as drawn in figure 15.6. A signal $f_r(t)$ obtained by linear interpolation from a sampling $f[n]$ can be written in a very nice way as a superposition of triangular pulses. To this end we will use the triangular pulse function $q_T(t)$ (see (6.12), where we called it the triangle function). This signal also arises as the reconstruction by linear interpolation of the discrete unit pulse $\delta[n]$ with sampling period T. The shifted signal $q_T(t - kT)$ can be considered as a reconstruction of $\delta[n - k]$ (see figure 15.7). Since according to theorem 15.1 we have

$$f[n] = \sum_{k=-\infty}^{\infty} f[k]\delta[n-k],$$

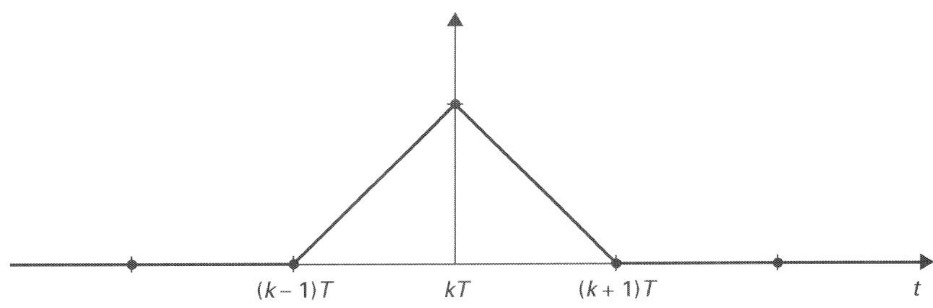

FIGURE 15.7
The triangular pulse function as linear interpolation of the discrete unit pulse.

one may thus expect that reconstructing an arbitrary signal $f(t)$ by linear interpolation will result in the following representation of the signal $f_r(t)$:

$$f_r(t) = \sum_{k=-\infty}^{\infty} f[k]q_T(t - kT).$$

That this is indeed the case is expressed by the following theorem.

THEOREM 15.3

Let $f[n]$ be the sampling of $f(t)$ with sampling period T and let $f_r(t)$ be the continuous-time signal obtained from $f[n]$ by linear interpolation. Then

$$f_r(t) = \sum_{k=-\infty}^{\infty} f[k]q_T(t - kT).$$

Proof
We have to show that on an interval I of the form $[nT, (n + 1)T]$ the graph of $f_r(t)$ consists of the line element connecting the points $(nT, f(nT))$ and $((n + 1)T, f((n + 1)T))$. Note that on the interval I only the triangular pulses $q_T(t - kT)$ with $k = n$ and $k = n + 1$ are unequal to 0. Hence one has for $t \in I$ that

$$f_r(t) = f[n]q_T(t - nT) + f[n + 1]q_T(t - (n + 1)T).$$

The graph of $f_r(t)$ on the interval I is thus a straight line and we have, moreover, that

$$f_r(nT) = f[n]q_T(0) + f[n + 1]q_T(-T) = f[n],$$
$$f_r((n + 1)T) = f[n]q_T(T) + f[n + 1]q_T(0) = f[n + 1].$$

This proves the theorem. ∎

EXAMPLE 15.6

Consider the signal $f(t) = \dfrac{\sin(\alpha t)}{t}$ with $\alpha > 0$. When sampled with sampling frequency $\omega_s = 2\alpha$, and then reconstructed by linear interpolation, it will lead to the following signal $f_r(t)$:

$$f[n] = \frac{\alpha \sin(n\pi)}{n\pi} = 0 \qquad \text{for } n \neq 0,$$
$$f[0] = \alpha,$$
$$f_r(t) = \alpha \, q_T(t).$$

◀

From figure 15.6 we can see that linear interpolation can result in a reasonable approximation of $f(t)$ if the sampling frequency is sufficiently large. A disadvantage is the fact that the graph of the signal $f_r(t)$ may have sharp turns at $t = nT$, which will lead to non-negligible contributions to the high-frequency components in the spectrum of $f_r(t)$ (see section 4.2 in chapter 4). These high-frequency components may not even occur at all in the spectrum of the original signal $f(t)$. Even if the reconstruction in the time domain is satisfactory at high sampling frequencies, there may still be considerable deviations in the frequency domain. Especially for signals with spectrum $F(\omega) = 0$ for $|\omega|$ greater than some specific value, linear interpolation will usually give a bad approximation in the frequency domain. In the next section we will derive that for these signals, the so-called *band-limited signals* (see definition 15.4), there exists a reconstruction method which, for a sufficiently high sampling frequency, yields exactly the original signal, meaning that $f_r(t) = f(t)$

for all $t \in \mathbb{R}$. Here, the reconstruction formula again has the form

$$f_{\mathrm{r}}(t) = \sum_{k=-\infty}^{\infty} f[k]\Psi(t - kT).$$

However, the signal $\Psi(t)$ is no longer the triangular pulse function $q_T(t)$, but a signal that is band-limited as well.

EXERCISE

15.6

Let $p_T(t)$ be the rectangular pulse function defined in (6.10) and let $f(t)$ be a continuous-time signal. Consider the reconstruction formula

$$f_{\mathrm{r}}(t) = \sum_{k=-\infty}^{\infty} f[k]p_T(t - kT).$$

Sketch the graph of $f_{\mathrm{r}}(t)$. Note: this reconstruction formula is sometimes called *zero-order interpolation*.

15.3   The sampling theorem

In this section we will derive a reconstruction formula which, for a sufficiently high sampling frequency, gives an exact reconstruction for band-limited signals. Band-limited signals are defined as follows.

DEFINITION 15.4
Band-limited signal

A signal $f(t)$ with spectrum $F(\omega)$ is called band-limited *if there exists an ω_{c} such that $F(\omega) = 0$ for $|\omega| > \omega_{\mathrm{c}}$.*

EXAMPLE 15.7

The signal $f(t) = \dfrac{\sin(\alpha t)}{t}$ with $\alpha > 0$ has spectrum $F(\omega) = \pi p_{2\alpha}(\omega)$ (see table 3). This signal is band-limited: $F(\omega) = 0$ for $|\omega| > \alpha$. This band-limited signal $\sin(\alpha t)/t$ will be used repeatedly in the reconstruction of signals, just as we have used the triangular pulse function $q_T(t)$ in the previous section.   ◀

When a continuous-time signal $f(t)$ does not contain frequencies greater than ω_{c}, that is, if $F(\omega) = 0$ for $|\omega| > \omega_{\mathrm{c}}$, then it will turn out that in a sampling with sampling frequency $\omega_{\mathrm{s}} > 2\omega_{\mathrm{c}}$ there is *no loss of information*. This then means that $f(t)$ can be recovered completely from the values $f[n]$, in other words, that the signal $f(t)$ is uniquely determined by the sampling $f[n]$.

If ω_{c} is the smallest non-negative frequency such that $F(\omega) = 0$ for $|\omega| > \omega_{\mathrm{c}}$, then $2\omega_{\mathrm{c}}$ is called the *Nyquist frequency*. The condition $\omega_{\mathrm{s}} > 2\omega_{\mathrm{c}}$ is called the *sampling condition*. The limiting value $\omega_{\mathrm{s}} = 2\omega_{\mathrm{c}}$ has the following plausible explanation. For the signal $f(t) = \sin(\omega_{\mathrm{c}} t)$ the spectrum $F(\omega)$ is (see table 5)

Nyquist frequency

Sampling condition

$$F(\omega) = \frac{\pi}{i}\left(\delta(\omega - \omega_{\mathrm{c}}) - \delta(\omega + \omega_{\mathrm{c}})\right).$$

The signal is band-limited: $F(\omega) = 0$ for $|\omega| > \omega_{\mathrm{c}}$. When we sample this signal with sampling frequency $\omega_{\mathrm{s}} = 2\omega_{\mathrm{c}}$, then $f[n] = \sin(n\omega_{\mathrm{c}}T) = \sin(2n\pi\omega_{\mathrm{c}}/\omega_{\mathrm{s}}) = \sin(n\pi) = 0$ for all integer n. Hence, the sampling $f[n]$ is equal to the sampling of the null signal. However, the null signal is also band-limited. This means that $f(t)$ is *not* uniquely determined by the samples if we use sampling frequency $2\omega_{\mathrm{c}}$, and so we have to choose a higher sampling frequency.

Loosely formulated, the Nyquist frequency equals twice the highest frequency occurring in a signal. If the sampling frequency is higher than the Nyquist frequency, then $f(t)$ can be reconstructed completely from the sampling $f[n]$. For the reconstruction one uses the function $\sin(\alpha t)/t$. This is formulated in the following theorem, which is called the *sampling theorem* or *Shannon's theorem*.

THEOREM 15.4
Sampling theorem

Let $f(t)$ be a band-limited signal with Nyquist frequency $2\omega_c$ and let $f[n]$ be the sampling of $f(t)$ at sampling frequency ω_s and sampling period $T = 2\pi/\omega_s$. If the sampling frequency ω_s satisfies the sampling condition $\omega_s > 2\omega_c$, then

$$f(t) = \sum_{n=-\infty}^{\infty} f[n]\frac{2\sin(\omega_s(t - nT)/2)}{\omega_s(t - nT)} \qquad for\ t \in \mathbb{R}. \tag{15.3}$$

*Proof**
The proof is divided into a number of steps. The proofs of any of these steps can be omitted without loss of continuity. One then only gets a general idea of the proof of the sampling theorem.

Step 1
     Introduce the function

$$F_s(\omega) = \sum_{k=-\infty}^{\infty} F(\omega - k\omega_s).$$

The function $F_s(\omega)$ is a periodic function of ω with period ω_s and fundamental frequency $2\pi/\omega_s = T$ (see section 7.3). As an illustration we have drawn the graphs of $F_s(\omega)$, for a certain given $F(\omega)$, in figure 15.8. In figure 15.8a we have $\omega_s > 2\omega_c$, while this is not the case in figure 15.8b. For the case $\omega_s > 2\omega_c$ we see

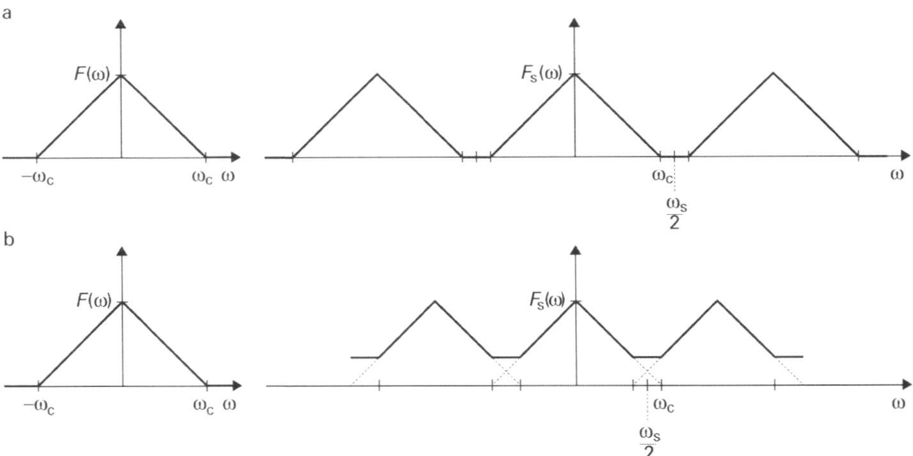

FIGURE 15.8
The spectrum $F_s(\omega)$ for $\omega_s > 2\omega_c$ (a) and $\omega_s < 2\omega_c$ (b).

that $F_s(\omega)$ is nothing else but the periodic extension of $F(\omega)$ with period ω_s to the entire ω-axis, and so we have $F_s(\omega) = F(\omega)$ for $-\omega_s/2 < \omega < \omega_s/2$. In all cases we have on the interval $[-\omega_s/2, \omega_s/2]$ that $F_s(\omega)$ is a sum of finitely many shifted copies of $F(\omega)$.

Step 2
     Prove that

$$F_s(\omega) = T \sum_{n=-\infty}^{\infty} f(nT)e^{-inT\omega}. \tag{15.4}$$

This result can be obtained by applying Poisson's summation formula (7.22) to the function $F_s(\omega)$:

$$F_s(\omega) = T \sum_{n=-\infty}^{\infty} f(-nT)e^{inT\omega} = T \sum_{n=-\infty}^{\infty} f(nT)e^{-inT\omega}.$$

The identity in step 2 is valid for all $\omega_s > 0$ and for any band-limited signal whose spectrum is piecewise smooth.

Step 3
    Our next step is to multiply the function $F_s(\omega)$ by the rectangular pulse $p_{\omega_s}(\omega)$. This gives rise to the function $F_r(\omega) = F_s(\omega)p_{\omega_s}(\omega)$.

Step 4
    Prove that $f_r(t) \leftrightarrow F_r(\omega)$, where

$$f_r(t) = \sum_{n=-\infty}^{\infty} f[n]\Psi(t - nT) \qquad \text{with} \qquad \Psi(t) = \frac{2\sin(\omega_s t/2)}{\omega_s t}.$$

By applying some properties, transforming $F_r(\omega)$ back to the time domain results in the following signal $f_r(t)$:

$$\frac{\sin(\omega_s t/2)}{\pi t} \quad \leftrightarrow \quad p_{\omega_s}(\omega),$$

$$\frac{\sin(\omega_s(t - nT)/2)}{\pi(t - nT)} \quad \leftrightarrow \quad e^{-in\omega T} p_{\omega_s}(\omega),$$

$$\sum_{n=-\infty}^{\infty} f(nT)\frac{2\sin(\omega_s(t - nT)/2)}{\omega_s(t - nT)} \quad \leftrightarrow \quad T \sum_{n=-\infty}^{\infty} f(nT)e^{-in\omega T} p_{\omega_s}(\omega),$$

$$f_r(t) \quad \leftrightarrow \quad F_r(\omega).$$

Step 5
    If $\omega_s > 2\omega_c$, then $F_r(\omega) = F(\omega)$ and so $f_r(t) = f(t)$.
In this last step we have to show that $f_r(t) = f(t)$ for all $t \in \mathbb{R}$. To show this, we will finally use the sampling condition $\omega_s > 2\omega_c$. Under this condition we have that $F_s(\omega) = F(\omega)$ on the interval $[-\omega_s/2, \omega_s/2]$ (see figure 15.8a). Hence, $F(\omega) = F_s(\omega)p_{\omega_s}(\omega)$ for all ω. From the uniqueness of the Fourier transform (theorem 7.4) it then follows immediately from an inverse transform that $f_r(t) = f(t)$.

In this proof we have assumed that $F(\omega)$ is a piecewise smooth function containing no delta components. The sampling theorem also holds under less restrictive conditions. For example, $F(\omega)$ may contain a number of delta components. We will no go into this any further. ∎

EXAMPLE 15.8
    The signal $f(t)$ defined by $f(t) = \dfrac{\sin^2(\pi t)}{\pi t^2}$ has spectrum $F(\omega) = \pi q_{2\pi}(\omega)$, where $q_{2\pi}(\omega)$ is the triangular pulse function being 0 outside the interval $[-2\pi, 2\pi]$. The Nyquist frequency is thus equal to 4π. Hence, using a sampling with sampling frequency $\omega_s > 4\pi$ it is possible to give an exact reconstruction of the signal. For the sampling period T one then has $2\pi/T > 4\pi$, so $T < 1/2$. This is satisfied by $T = 1/4$. For this value of T the reconstruction formula reads as follows:

$$f(t) = 16 \sum_{n=-\infty}^{\infty} \frac{\sin^2(n\pi/4)}{\pi n^2} \frac{\sin(4\pi(t - n/4))}{4\pi(t - n/4)}$$

$$= \frac{16}{\pi^2} \sum_{n=-\infty}^{\infty} \frac{(-1)^n \sin^2(n\pi/4) \sin(4\pi t)}{n^2(4t - n)}.$$

◀

EXAMPLE 15.9

The continuous-time signal $f(t) = \cos(\omega_0 t + \phi_0)$ is sampled with sampling frequency ω_s. Next, the signal is recovered from the samples using the reconstruction formula from the sampling theorem. The spectrum of $f(t)$ can easily be obtained using some properties. The result is: $F(\omega) = \pi [e^{i\phi_0} \delta(\omega - \omega_0) + e^{-i\phi_0} \delta(\omega + \omega_0)]$. The signal $f(t)$ is thus band-limited with Nyquist frequency $2\omega_0$. From the sampling theorem it then follows that for $\omega_s > 2\omega_0$ the reconstruction will give us the original signal back again.

◀

EXAMPLE 15.10

Human hearing will only register sound signals having frequencies that in general do not exceed the 20 kHz limit. This means that audible sound signals can approximately be considered as band-limited signals with a Nyquist frequency $80\,000\pi$. In order to sample audio signals without loss of information, one should use at least $40\,000$ samples per second. A compact disc, which essentially contains a sampling of an audio signal, will have to satisfy this condition.

◀

In principle, all the information in a band-limited signal should be contained in a sampling, provided that the sampling frequency is sufficiently large. A nice illustration of this fact is the energy-content. Let us assume that a given band-limited signal is an energy-signal. One should then be able to express the energy-content in terms of $f[n]$. This can be done as follows. Parseval's identity for continuous-time signals states that (see (7.19))

$$\int_{-\infty}^{\infty} | f(t) |^2 \, dt = \frac{1}{2\pi} \int_{-\infty}^{\infty} | F(\omega) |^2 \, d\omega.$$

The signal is band-limited and $\omega_s > 2\omega_c$, and using the functions introduced in the proof of the sampling theorem we can thus write down the following identity for the energy-content E:

$$E = \frac{1}{2\pi} \int_{-\frac{1}{2}\omega_s}^{\frac{1}{2}\omega_s} | F(\omega) |^2 \, d\omega = \frac{1}{T\omega_s} \int_{-\frac{1}{2}\omega_s}^{\frac{1}{2}\omega_s} | F_s(\omega) |^2 \, d\omega.$$

The function $F_s(\omega)$ is periodic with period ω_s and so we can again apply Parseval's theorem, but now for periodic functions, which results in:

$$E = \frac{1}{T} \sum_{n=-\infty}^{\infty} | c_n |^2.$$

In step 2 of the proof of the sampling theorem we saw that $c_n = Tf(-nT)$. From this we then obtain the desired expression:

$$E = T \sum_{n=-\infty}^{\infty} | f(-nT) |^2 = T \sum_{n=-\infty}^{\infty} | f[n] |^2.$$

Up to the factor T this is the energy-content of the discrete-time signal $f[n]$.

EXERCISES

15.7

Let a band-limited signal $\phi(t)$ be given. Show that for any signal $f(t)$ the convolution product $(\phi * f)(t)$ is also band-limited.

15.8*

The spectrum $F(\omega)$ of a signal is given by $F(\omega) = q_\pi(\omega)$. Give a sketch of $F_s(\omega)$ for $\omega_s = 3\pi/2$.

15.9   The spectrum $F(\omega)$ of a signal is given by $F(\omega) = P(\omega) p_{\omega_s}(\omega)$. Here $P(\omega)$ is a periodic function with period π and Fourier coefficients c_n given by

$$c_n = \begin{cases} 1 & \text{for } |n| = 1, \\ 0 & \text{otherwise.} \end{cases}$$

Calculate the inverse Fourier transform of $F(\omega)$.

15.10   Given is the signal $f(t)$ with spectrum $F(\omega) = (p_\pi * p_{2\pi})(\omega)$. For which values of the sampling period T can one reconstruct the signal from the sampling without loss of information?

15.4   The aliasing problem*

In this section we will look at the consequences of a sampling with a sampling frequency which is too low. This is a situation occurring often in practice. We will start with a simple example.

EXAMPLE

Using a tv-camera we record a rotating wheel having angular velocity ω_0, that is to say, having a speed of $f_0 = \omega_0/2\pi$ revolutions per second. We assume that the camera has frequency 25 Hz. This means that the camera produces 25 images per second of the wheel. If we also have $f_0 = 25$ Hz, then the camera always produces the same image. We do not see the rotation of the wheel; the angular frequency of the observed wheel is then equal to 0. If the angular velocity ω_0 of the wheel changes, then the direction of the observed rotation may even be *opposite* to the actual rotation of the wheel. The phenomenon occurring in this simple example can be considered as a consequence of a sampling with a sampling frequency which is too low. In order to understand this, we associate the continuous-time signal $e^{i\omega_0 t}$ with the rotating wheel. The sequence of images produced by the camera can be considered as a sampling of this signal with sampling frequency $\omega_0 = 50\pi$. This is the camera frequency converted to an angular frequency. Finally, we consider the observed rotation as a reconstruction of this sampling. Apparently, the reconstruction leads to a signal having a different angular frequency.   ◄

In this example the sampling condition is not satisfied. This is because the Nyquist frequency of $e^{i\omega_0 t}$ is $2\omega_0$ and in this example we have $\omega_s < 2\omega_0$. We can determine which frequency *is* observed by consulting, for example, the proof of the sampling theorem. In step 4 it says that the spectrum of the reconstructed signal $f_r(t)$ is given by

$$f_r(t) \leftrightarrow F_r(\omega) = F_s(\omega) p_{\omega_s}(\omega) \qquad \text{where} \qquad F_s(\omega) = \sum_{k=-\infty}^{\infty} F(\omega - k\omega_s).$$

Due to the construction of $F_s(\omega)$, high-frequency components in $F(\omega)$ may end up in the interval $I = [-\omega_s/2, \omega_s/2]$ by the shifts over a distance $k\omega_s$. The problem thus arising is called the *aliasing problem*. The function $F_s(\omega)$ of our example is drawn in figure 15.9. By shifting over a distance $k\omega_s$, the frequency component with frequency ω_0 ends up in the interval I at position ω_r. (We will not consider the situation where components end up in the endpoints of the interval.) The reconstructed signal is a signal containing only the frequency ω_r. The frequency ω_r can be interpreted as the angular frequency of the observed wheel.

Aliasing problem

In practice, when sampling audio for example, one will first lead the signal through a low-pass filter with a cut-off frequency, that is, the highest frequency that will pass through the filter, equal to half the sampling frequency. This avoids

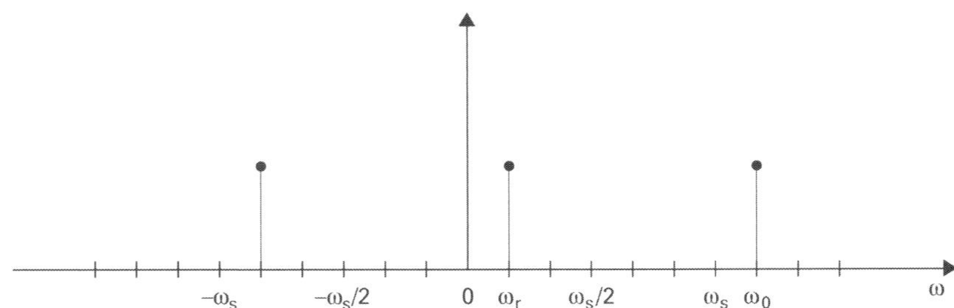

FIGURE 15.9
Reconstruction of $e^{i\omega_0 t}$.

the annoying aspect of the 'seeping through' of high-frequency components that are present in the original signal. Here one should take in mind, for example, a music recording in a room where a bat is present, but remaining unnoticed. Without the low-pass filter this could lead to a hum in the recording.

EXERCISES

15.11*   The periodic signal $f(t) = \cos(\omega_0 t + \phi_0)$ is sampled with sampling frequency $\omega_s = 3\omega_0/2$ and then reconstructed with the reconstruction formula from the sampling theorem. Determine the reconstruction.

15.12*   Given is a signal $f(t)$ with spectrum $F(\omega) = p_\pi(\omega)$. The signal is sampled with different sampling periods T and then reconstructed using the reconstruction formula from the sampling theorem. Determine the reconstruction $f_r(t)$ for each of the following sampling periods:
a $T = 4/3$,
b $T = 2$,
c $T = 8/3$.

SUMMARY

Discrete-time signals can in general be described as a superposition of shifted discrete unit pulses. In particular, the periodic discrete-time signals can be described as a superposition of periodic trains of discrete unit pulses. Of importance are the discrete-time signals arising from a sampling of a continuous-time signal. Such a discrete-time signal is given by $f[n] = f(nT)$, where T is the sampling period and $\omega_s = 2\pi/T$ is the sampling frequency.

One of the important problems in signal theory is choosing the sampling frequency in such a way that the continuous-time signal $f(t)$ can be reconstructed from the samples $f[n]$ without loss of information. That this is possible for band-limited signals is expressed by the so-called sampling theorem or Shannon's theorem. It states that if the sampling frequency ω_s satisfies the sampling condition $\omega_s > 2\omega_c$, where ω_c is the highest frequency occurring in the signal, then

$$f(t) = \sum_{n=-\infty}^{\infty} f[n] \frac{2\sin(\omega_s(t - nT)/2)}{\omega_s(t - nT)}.$$

The frequency $2\omega_c$ is called the Nyquist frequency. According to the sampling condition the Nyquist frequency should be a lower bound for the sampling frequency.

In practice the sampling condition is often not satisfied. Problems arising from this are called aliasing problems. High-frequency components in the original signal can end up as low-frequency components in the reconstruction. It is therefore important to know how the reconstruction proceeds in the frequency domain. By first leading the continuous-time signal through a low-pass filter, one can avoid the aliasing problem.

SELFTEST

15.13   Given is a linear time-invariant system with frequency response $H(\omega) = q_\pi(\omega)$.
a   Determine the impulse response.
b   To the system we apply a band-limited signal $u(t)$ with Nyquist frequency π. Show that the response $y(t)$ is also band-limited. In which band lie the frequencies of the output signal? Justify your answer.
c   For an input a sampling $u[n]$ with sampling period $T = 1$ is available, given by

$$u[n] = \begin{cases} 1 & \text{for } n = 0, \, |n| = 1, \\ 0 & \text{otherwise.} \end{cases}$$

Calculate the response $y(t)$.

15.14*   The spectrum $F(\omega)$ of a signal $f(t)$ is represented by the graph of figure 15.10.
a   Calculate $f(t)$.
b   The signal is sampled with sampling frequency ω_s. For which values of ω_s can one reconstruct the signal without loss of information? Justify your answer.

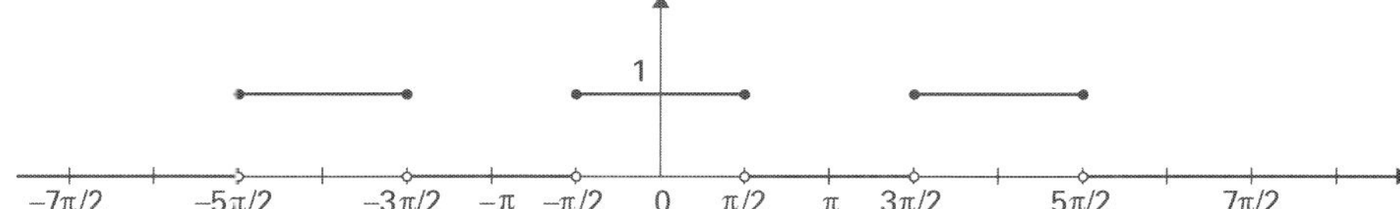

FIGURE 15.10
The spectrum $F(\omega)$ of exercise 15.14.

c   Let $T = 1$ be the sampling period and $f[n] = f(nT)$. Calculate the reconstruction $f_r(t)$ given by

$$f_r(t) = \sum_{n=-\infty}^{\infty} f[n]\Psi(t - nT) \qquad \text{with} \qquad \Psi(t) = \frac{2\sin(\omega_s t/2)}{\omega_s t}.$$

15.15   A periodic signal $f(t)$ with period 2π is sampled with sampling period $T = \pi/2$.
a   Show that

$$f[n] = \sum_{k=0}^{3} f[k]\delta_4[n - k].$$

b   The given periodic signal is band-limited with Nyquist frequency 3. Moreover, it is given that $f(-\pi/2) = f(\pi/2) = 1$, $f(0) = f(\pi) = 0$. Show that $f(t)$ is completely determined by these values.
c   Show that the line spectrum c_n of $f(t)$ satisfies $c_n = 0$ for $|n| \geq 2$.
d   Calculate c_0.

15.16   The spectrum $F(\omega)$ of the continuous-time signal $f(t)$ is given by $F(\omega) = p_{2\pi}(\omega)\cos\omega$. The signal is sampled with sampling period $T = 2/3$.

a   Is it possible to reconstruct $f(t)$ completely using the sampling $f[n] = f(nT)$? Justify your answer.

b   Determine the sampling $f[n]$ for $n = 0, \pm 1, \pm 2, \ldots$.

c   Show that

$$\int_{-\infty}^{\infty} f(t)\,dt = T \sum_{n=-\infty}^{\infty} f[n].$$

d   Calculate the energy-content of $f(t)$.

Contents of Chapter 16

The discrete Fourier transform

The discrete Fourier transform

INTRODUCTION

From parts 2 and 3 it is obvious that Fourier series and Fourier integrals play an important role in the analysis of continuous-time signals. In many cases we are forced to calculate the Fourier coefficients or the Fourier integral on the basis of a given sampling of the signal. We are therefore interested in a transformation that will transform a discrete-time signal, in this case a sampling, directly into the frequency domain. In general, such transformations are called *discrete* transforms. A particularly important discrete transform is the so-called *discrete Fourier transform*, abbreviated as DFT, and it will be the central theme of the present chapter. It arises naturally if one approximates the Fourier coefficients of a periodic continuous-time signal numerically using the trapezoidal rule. This is the subject of the first section of this chapter, which also introduces the DFT as a transform defined for periodic discrete-time signals. In the next section we introduce the inverse DFT in the so-called fundamental theorem of the discrete Fourier transform, which very much resembles the fundamental theorem of Fourier series. In the remaining sections, all kinds of properties of the DFT are treated, and again we will encounter many similarities with Fourier series and Fourier integrals. For example, one can formulate a Parseval theorem for periodic discrete-time signals. Following the introduction of the so-called *cyclical convolution* we can derive convolution theorems, which look very similar to the ones derived for continuous-time signals. Further applications of the DFT will not be considered until the next chapter, where we pay special attention to an efficient algorithm for the calculation of the DFT.

LEARNING OBJECTIVES
After studying this chapter it is expected that you
- know the definition of the discrete Fourier transform (DFT)
- can relate the DFT to the Fourier coefficients
- can formulate and prove the fundamental theorem of the discrete Fourier transform
- can apply the most frequently occurring properties of the DFT
- can calculate the DFT for some simple signals
- know the definition of the cyclical convolution
- can formulate, prove and apply the convolution theorems in the n-domain and the k-domain
- know Parseval's theorem for periodic discrete-time signals and can apply it to calculate the power of a periodic discrete-time signal.

16.1   Introduction and definition of the discrete Fourier transform

In this section we introduce a discrete transform for periodic discrete-time signals. Periodic discrete-time signals can for instance be obtained by sampling a periodic continuous-time signal with period T using a sampling frequency ω_s equal to an integer multiple of the fundamental frequency $\omega_0 = 2\pi/T$ of $f(t)$, so $\omega_s = N\omega_0$ for some integer $N > 0$. There are then N samples in the interval $[0, T)$ and the sampling $f[n] = f(nT/N)$ is, moreover, periodic with period N. Note that the sampling period is then equal to T/N. Using the sampling $f[n]$ we would now like to give an approximation for the Fourier coefficients c_k of the signal $f(t)$. For this we use the trapezoidal rule for the numerical approximation of integrals.

16.1.1 *Trapezoidal rule for periodic functions*

Let $g(t)$ be a periodic function with period T whose integral over the interval $[0, T]$ needs to be calculated. We start by dividing the interval of integration $[0, T]$ into the subintervals I_n of equal length T/N:

$$I_n = \left[\frac{(n-1)T}{N}, \frac{nT}{N}\right] \qquad \text{for } n = 1, 2, \ldots, N.$$

On each subinterval I_n we replace $g(t)$ by the linear interpolation by the linear function $l_n(t)$ given by

$$l_n(t) = \frac{N}{T}\left(g[n-1]\left(\frac{nT}{N} - t\right) + g[n]\left(t - \frac{(n-1)T}{N}\right)\right).$$

Note that in the case when $g(t)$ is real-valued, the graph of $l_n(t)$ consists of the line element connecting the points $((n-1)T/N, g[n-1])$ and $(nT/N, g[n])$ (see

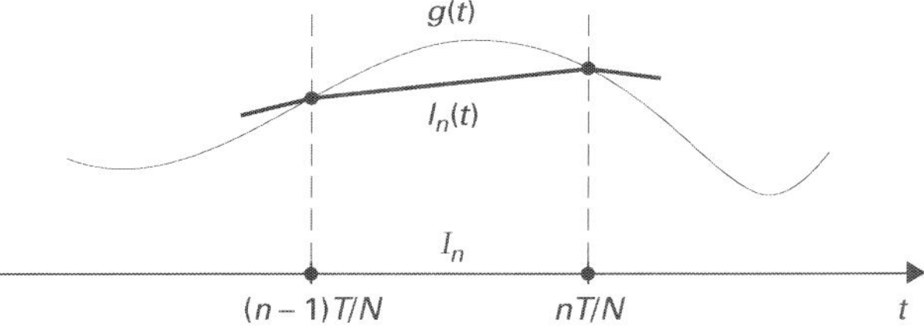

FIGURE 16.1
Approximation of a function by linear interpolation.

figure 16.1). Next we approximate the integral of $g(t)$ over the interval I_n by

$$\int_{(n-1)T/N}^{nT/N} g(t)\,dt \approx \int_{(n-1)T/N}^{nT/N} l_n(t)\,dt = \frac{T}{2N}(g[n-1] + g[n]).$$

Trapezoidal rule

By summing all of these approximations over the interval $[0, T]$ we obtain the so-called *trapezoidal rule*:

$$\int_0^T g(t)\,dt = \sum_{n=1}^{N}\int_{(n-1)T/N}^{nT/N} g(t)\,dt \approx \sum_{n=1}^{N}\frac{T}{2N}(g[n-1] + g[n])$$

$$= \frac{T}{2N}(g[0] + 2g[1] + \cdots + 2g[N-1] + g[N]).$$

Now it is given that $g(t)$ is periodic, so $g[0] = g[N]$, and hence

$$\int_0^T g(t)\,dt \approx \frac{T}{N}(g[0] + g[1] + \cdots + g[N-1]). \tag{16.1}$$

Observe that in (16.1) we have an elementary Riemann sum in which all function values have been taken in the left endpoint of the subintervals. We know that for Riemann integrable functions the Riemann sums will converge to the value of the Riemann integral if we let the length of the subintervals decrease. But piecewise smooth functions are Riemann integrable. For piecewise smooth signals we thus have

$$\frac{T}{N}(g[0] + g[1] + \cdots + g[N-1]) \rightarrow \int_0^T g(t)\,dt \qquad \text{when } N \rightarrow \infty. \tag{16.2}$$

Hence, (16.1) gives an approximation of the integral over one period, starting from a sampling $g[n]$ of the function $g(t)$.

16.1.2 An approximation of the Fourier coefficients

The Fourier coefficients of a periodic function $f(t)$ can be calculated from (see section 3.2.2)

$$c_k = \frac{1}{T} \int_0^T f(t)e^{-ik\omega_0 t}\,dt.$$

Applying the trapezoidal rule (16.1) with the integrand being the periodic function $g(t) = f(t)e^{-ik\omega_0 t}$ with $\omega_0 = 2\pi/T$, we obtain the following approximation for the Fourier coefficient c_k:

$$c_k \approx \frac{1}{T}\frac{T}{N} \sum_{n=0}^{N-1} f[n]e^{-ik\omega_0 nT/N} = \frac{1}{N} \sum_{n=0}^{N-1} f[n]e^{-2\pi ink/N}.$$

The sum is written as $F[k]$, so

$$F[k] = \sum_{n=0}^{N-1} f[n]e^{-2\pi ink/N},$$

and hence it follows that

$$c_k \approx \frac{1}{N}F[k]. \tag{16.3}$$

From (16.2) we know that $F[k]/N \rightarrow c_k$ if $N \rightarrow \infty$. For a given value of k the quotient $F[k]/N$ will be a good approximation of c_k for sufficiently large N. However, how large N should be in order to get a specific degree of accuracy for the approximation of c_k will in general not only depend on $f(t)$ but also on the value of k. We do not intend to go into this matter any further in this book. It is immediately clear however, that after a value of N is chosen, $F[k]/N$ cannot be a good approximation of c_k for *all* k. This is because the sequence $F[k]$ is periodic with period N, while for piecewise smooth signals we have $c_k \rightarrow 0$ for $|k| \rightarrow \infty$.

The periodicity of the sequence $F[k]$ can be verified by a substitution:

$$F[k+N] = \sum_{n=0}^{N-1} f[n]e^{-2\pi in(k+N)/N} = \sum_{n=0}^{N-1} f[n]e^{-2\pi ink/N} = F[k].$$

Here we used the relation

$$e^{-2\pi in(k+N)/N} = e^{-2\pi ink/N}e^{-2\pi in} = e^{-2\pi ink/N}.$$

In the following example we approximate the Fourier coefficients of a periodic function by applying (16.3).

EXAMPLE 16.1

Consider the periodic continuous-time signal $f(t)$ with period 2π, given on the interval $[0, 2\pi]$ by

$$f(t) = 1 - \frac{|t - \pi|}{\pi}.$$

It is easy to verify that this is an even signal. For this signal the Fourier coefficients can be calculated explicitly (perform these calculations yourself). They have the following values:

$$c_k = \begin{cases} -\dfrac{2}{(\pi k)^2} & \text{for } k \text{ odd,} \\[2ex] \dfrac{1}{2} & \text{for } k = 0, \\[2ex] 0 & \text{for } k \neq 0 \text{ and even.} \end{cases}$$

We want to compare these Fourier coefficients with the numbers $F[k]/N$ corresponding to the sampling at the points $2\pi n/N$ with $N = 128$ and calculated according to (16.3). The results are listed in table 16.1.

TABLE 16.1   The Fourier coefficients c_k and the approximation $F[k]/N$.

k	c_k	$F[k]/N$
00	$+0.500\,00$	$+0.500\,00$
01	$-0.202\,64$	$-0.202\,64$
03	$-0.022\,51$	$-0.022\,56$
05	$-0.008\,11$	$-0.008\,15$
$\cdots$	$\cdots$	$\cdots$
$\cdots$	$\cdots$	$\cdots$
$\cdots$	$\cdots$	$\cdots$
59	$-0.000\,06$	$-0.000\,12$
61	$-0.000\,05$	$-0.000\,12$
63	$-0.000\,05$	$-0.000\,12$

The values given here were rounded off after the fifth decimal place. The Fourier coefficients with an even index unequal to 0 have been omitted since these are all 0. After all, the signal is even. Here we also have $F[k] = F[-k]$ for all $k \in \mathbb{Z}$, as is the case for the Fourier coefficients in this example. In section 16.3 we will see that this follows from a symmetry property. Note that for small values of k we get a good approximation. ◄

In general, the numbers $F[k]/N$ with $|k| \leq N/2$ are used as an approximation for c_k. It is important, though, that at the points where $f(t)$ has a jump the sampling is given the value $(f(t+) + f(t-))/2$. For example, $f[0] = (f(0+) + f(T-))/2$.

16.1.3   Definition of the discrete Fourier transform

We have seen that the sequence $F[k]$ plays an important role in the approximation of the Fourier coefficients of a periodic signal $f(t)$ from a sampling $f[n]$. But its importance is not restricted to the determination of the Fourier coefficients. Elsewhere in signal theory, one will often encounter this sequence as well. There is thus ample reason to study this sequence in more detail and also to give it a name, as will be done in the following definition.

DEFINITION 16.1
N-point discrete Fourier transform

Let $f[n]$ be a periodic discrete-time signal with period N. The sequence $F[k]$ defined by

$$F[k] = \sum_{n=0}^{N-1} f[n]e^{-2\pi i nk/N} \qquad \text{for } k \in \mathbb{Z} \tag{16.4}$$

is called the N-point discrete Fourier transform of $f[n]$.

N-point DFT

The transform assigning the N-point discrete Fourier transform $F[k]$ to the signal $f[n]$ is called the N-point discrete Fourier transform, which is abbreviated as DFT. If it is clear from the context what the number of points in the DFT should be, then we will usually omit the additional 'N-point'.

We have seen that the sequence $F[k]$ is periodic with period N. Hence, considered as a discrete signal, $F[k]$ is again a periodic discrete signal. Hence, the DFT converts a periodic discrete signal into a periodic discrete signal again having the same period.

In the next section it will be shown that the signal $f[n]$ can be recovered from $F[k]$ by means of the inverse DFT. Hence, if $F[k]$ is known, then in principle $f[n]$

k-domain
n-domain

is also known. We will call $F[k]$ the description of the signal in the so-called k-*domain*, and $f[n]$ the description in the n-*domain* or time domain. Because of the close relationship of $F[k]$ with the Fourier coefficients, the k-domain is also called

Frequency domain
Discrete spectrum

the *frequency domain*, and $F[k]$ the *discrete spectrum* of $f[n]$. Finally, as for the continuous-time signals, we denote the transform pair $f[n]$, $F[k]$ by

$$f[n] \leftrightarrow F[k].$$

It is easy to show that the DFT is a linear transformation. Hence, if

$$f_1[n] \leftrightarrow F_1[k] \qquad \text{and} \qquad f_2[n] \leftrightarrow F_2[k],$$

then one has for arbitrary complex a and b that

$$af_1[n] + bf_2[n] \leftrightarrow aF_1[k] + bF_2[k]. \tag{16.5}$$

We close this section with a property of periodic discrete-time signals that may sometimes be useful when calculating the $F[k]$. The property implies that if the values of the signal over one full period are added together, then the outcome is the same regardless of the starting point of this summation over one full period. This property is formulated in the following lemma.

LEMMA 16.1

Let $g[n]$ be a periodic discrete signal with period N. Then one has for any integer j that

$$\sum_{n=j}^{j+N-1} g[n] = \sum_{n=0}^{N-1} g[n].$$

Proof
Let the integers l and m be given by the relationship $j = mN + l$, where $0 \le l \le N - 1$. Note that this determines l and m uniquely. Because of the periodicity of $g[n]$ one then has

$$\sum_{n=j}^{j+N-1} g[n] = g[j] + g[j+1] + \cdots + g[j+N-1]$$

$$= g[l] + \cdots + g[N-1] + g[N] + \cdots + g[N+l-1]$$

$$= g[l] + \cdots + g[N-1] + g[0] + \cdots + g[l-1]$$

$$= g[0] + g[1] + \cdots + g[N-1] = \sum_{n=0}^{N-1} g[n].$$

∎

Lemma 16.1 can immediately be applied to the DFT since the general term in the representation of $F[k]$ as a sum is periodic, as a function of n, with period N. Among other things it implies that the DFT for $N = 2M + 1$ can also be given by

$$F[k] = \sum_{n=-M}^{M} f[n]e^{-2\pi i n k/N}.$$

EXAMPLE 16.2

A periodic discrete-time signal $f[n]$ with period 5 is given by $f[-2] = -1$, $f[-1] = -2$, $f[0] = 0$, $f[1] = 2$, $f[2] = 1$. The 5-point DFT can be calculated as follows:

$$F[k] = -e^{4\pi i k/5} - 2e^{2\pi i k/5} + 2e^{-2\pi i k/5} + e^{-4\pi i k/5}$$

$$= -2i\sin(4\pi k/5) - 4i\sin(2\pi k/5).$$

◄

Note that the DFT in example 16.2 is purely imaginary. This has to do with the fact that $f[n]$ is an odd and real signal. Just as for the Fourier coefficients, the DFT has similar symmetry properties, as will be shown in section 16.3.

EXERCISES

16.1

Show that for the 2-point DFT of a periodic discrete-time signal $f[n]$ with period 2 one has $F[k] = f[0] + (-1)^k f[1]$.

16.2

Calculate the 2-point and 4-point DFT of the discrete-time signal $f[n] = (-1)^n$ for $n \in \mathbb{Z}$.

16.3

The periodic continuous-time signal $f(t)$ with period T is given on the interval $[0, T]$ by the graph of figure 16.2. Calculate the Fourier coefficients of $f(t)$ and compare c_0 with the value $F[0]$ of the N-point DFT of $f[n]$.

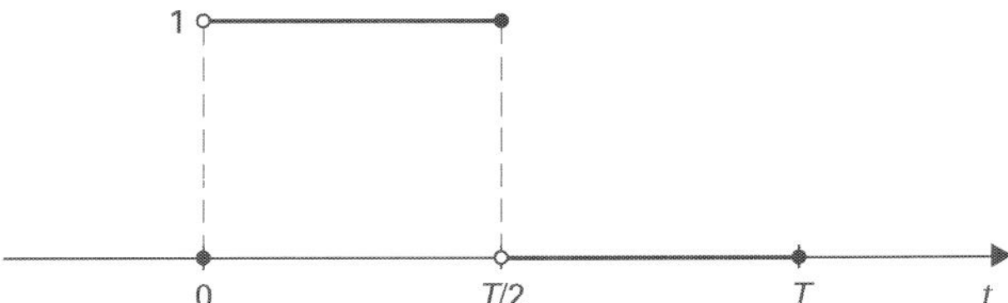

FIGURE 16.2
Graph of the periodic function from exercise 16.3.

16.4

The periodic continuous-time signal $f(t)$ with period T is given by the graph from figure 16.3. Calculate the Fourier coefficients of $f(t)$ and compare c_0 with the value $F[0]$ of the N-point DFT of $f[n]$.

16.5

In the previous two exercises 16.3 and 16.4, the continuous-time signals only differ at the jump discontinuity. Which of the values at these jumps do you prefer? Justify your answer.

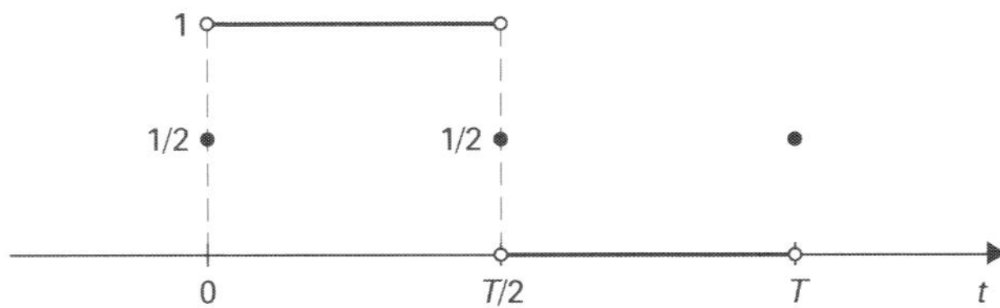

FIGURE 16.3
Graph of the periodic function from exercise 16.4.

16.6   A periodic discrete-time signal $f[n]$ with period 4 is given by $f[-2] = 1$, $f[-1] = 0$, $f[0] = 2$, $f[1] = 0$. Calculate the 4-point DFT of $f[n]$.

16.2   Fundamental theorem of the discrete Fourier transform

In the previous section the notation $f[n] \leftrightarrow F[k]$ already alluded to the fact that a periodic discrete-time signal $f[n]$ can be reconstructed completely from its discrete spectrum $F[k]$ or, formulated differently, that an *inverse* DFT exists. This is indeed the case, as will be shown in the present section. For this we will use an elegant property of the Nth roots of unity. These have been defined in section 2.1.2 as the roots in the complex plane of the equation $z^N = 1$. If we now put $w = e^{2\pi i/N}$, then the N distinct roots z_j $(j = 0, 1, \ldots, N - 1)$ of this equation are given by

$$z_j = e^{2\pi i j/N} = w^j \qquad \text{where } j = 0, 1, \ldots, N - 1.$$

For arbitrary integer n one then has

$$\frac{1}{N} \sum_{k=0}^{N-1} e^{2\pi ink/N} = \frac{1}{N} \sum_{k=0}^{N-1} w^{nk}$$

$$= \begin{cases} 1 & \text{if } n \text{ is an integer multiple of } N, \\ \dfrac{1}{N}\dfrac{1 - w^{nN}}{1 - w^n} = 0 & \text{otherwise.} \end{cases}$$

Note that because of the definition of the periodic train of unit pulses (see definition 15.2), this property can also be written as

$$\delta_N[n] = \frac{1}{N} \sum_{k=0}^{N-1} e^{2\pi ink/N} \qquad \text{for } n \in \mathbb{Z}. \tag{16.6}$$

We use this property to prove the following theorem. In this theorem we introduce the inverse DFT.

THEOREM 16.1
Fundamental theorem of the discrete Fourier transform

Let $f[n]$ be a periodic discrete-time signal with period N and DFT $F[k]$ given by $F[k] = \sum_{n=0}^{N-1} f[n]e^{-2\pi ink/N}$. Then one has for all integer n that

$$f[n] = \frac{1}{N} \sum_{k=0}^{N-1} F[k]e^{2\pi ink/N}. \tag{16.7}$$

Proof
Substitution of $F[k]$ into the sum in the right-hand side of (16.7) gives

$$\frac{1}{N}\sum_{k=0}^{N-1} F[k]e^{2\pi ink/N} = \frac{1}{N}\sum_{k=0}^{N-1}\sum_{l=0}^{N-1} f[l]e^{2\pi ik(n-l)/N}$$

$$= \frac{1}{N}\sum_{l=0}^{N-1} f[l]\sum_{k=0}^{N-1} e^{2\pi ik(n-l)/N}$$

$$= \sum_{l=0}^{N-1} f[l]\delta_N[n-l] = f(n).$$

Here we have used (16.6) following the change in the order of summation. This completes the proof. ∎

Inverse DFT

Identity (16.7) shows how the signal $f[n]$ can be recovered from $F[k]$. The transformation assigning the signal $f[n]$ to $F[k]$ is called the *inverse discrete Fourier transform*. From identity (16.7) it also follows that an arbitrary periodic discrete-time signal $f[n]$ with period N can be written as a linear combination of the time-harmonic signals $g_k[n] = e^{2\pi ink/N}$, where $k = 0, 1, \ldots, N - 1$, which are themselves periodic with period N.

The fundamental theorem of the DFT is very similar to the fundamental theorem of Fourier series. In the latter theorem, a periodic continuous-time signal with period T and fundamental frequency $\omega_0 = 2\pi/T$ is written as a superposition of the time-harmonic continuous-time signals $e^{ik\omega_0 t}$, where $k \in \mathbb{Z}$. For periodic discrete-time signals it is now quite appropriate to call the quantity $2\pi/N$ the *fundamental frequency*. The discrete signal $e^{2\pi ink/N}$, considered as function of n, then has frequency $2\pi k/N$, which is an integer multiple of $2\pi/N$. Since

Fundamental frequency

$$e^{2\pi in(k+N)/N} = e^{2\pi ink/N} \qquad \text{for all } n \in \mathbb{Z},$$

only finitely (N) many frequencies can be distinguished in the discrete-time situation. Hence, the sum occurring in the fundamental theorem of the DFT only contains a finite number of terms, which is in contrast to the Fourier series of a periodic continuous-time signal.

EXAMPLE 16.3

As a first example we consider the periodic train of unit pulses $\delta_N[n]$. The DFT of $\delta_N[n]$ follows immediately from the definition of $\delta_N[n]$ (verify this), and results in

$$\delta_N[n] \leftrightarrow F[k] = 1. \tag{16.8}$$

We see here that the DFT of the periodic train of discrete unit pulses is equal to the constant discrete signal 1. The Fourier coefficients of the periodic train of delta functions with period 2π are also mutually equal (verify this). Again we note the similarity between the DFT and the Fourier coefficients. Applying theorem 16.1 gives

$$\delta_N[n] = \frac{1}{N}\sum_{k=0}^{N-1} e^{2\pi ink/N}.$$

Here we see (16.6) re-appearing. ◀

EXAMPLE 16.4

Calculating the DFT of the periodic discrete-time signal $f_l[n] = e^{2\pi inl/N}$ with period N is an immediate application of (16.6). The result is

$$f_l[n] = e^{2\pi inl/N} \leftrightarrow N\delta_N[l - k].$$

Applying theorem 16.1 then gives:

$$f_l[n] = \sum_{k=0}^{N-1} \delta_N[l-k] e^{2\pi i n k/N} = e^{2\pi i n l/N}.$$

This indeed recovers $f_l[n]$.   ◄

Amplitude spectrum

Phase spectrum

In general, $F[k]$ will of course be complex. As for continuous-time signals we call the modulus $|F[k]|$ of the spectrum $F[k]$ the *amplitude spectrum* and arg $F[k]$ the *phase spectrum* of $f[n]$. The phase spectrum is determined up to a multiple of 2π.

In the next section we consider some properties of the DFT. We will show, for example, that a shift in the n-domain does not change the amplitude spectrum of a discrete-time signal.

EXERCISES

16.7   Calculate the 4-point inverse DFT of the discrete signal $F[k]$ with period 4 given by $F[0] = 1$, $F[1] = 0$, $F[2] = 0$, $F[3] = 1$.

16.8   For a periodic discrete-time signal $f[n]$ with period 4 the amplitude and phase spectrum are given by $|F[k]| = 2$ and arg $F[k] = \pi k/2$ respectively. Calculate $f[n]$ for $n = 1, 2, 3, 4$.

16.9   Given is the complex number z with $z \neq 0$. The periodic discrete-time signal $f[n]$ with period N is given by $f[n] = z^n$ for $n = 0, 1, \ldots, N-1$. Calculate the N-point DFT of $f[n]$.

16.3   Properties of the discrete Fourier transform

A large number of properties that we have encountered in the theory of Fourier series and integrals will return here. They will be treated in the present section. Moreover, they are summarized in table 12 at the back of the book.

16.3.1   Linearity

In the previous section we have already noted that the DFT is a linear transformation (see (16.5)).

16.3.2   Reciprocity

As in the case of the Fourier transform of continuous-time signals, the formulas for the DFT and the inverse DFT show a great similarity. As a consequence we can again formulate a reciprocity rule for the DFT. Let $F[k]$ be the DFT of $f[n]$. We will now calculate the DFT of $F[n]$. The expression for the inverse DFT reads as follows:

$$f[n] = \frac{1}{N} \sum_{k=0}^{N-1} F[k] e^{2\pi i n k/N}.$$

From this it follows, by interchanging the variables n and k, that

$$f[k] = \frac{1}{N} \sum_{n=0}^{N-1} F[n] e^{2\pi i n k/N}.$$

Replacing k by $-k$ in this identity, we obtain the expression

$$\sum_{n=0}^{N-1} F[n]e^{-2\pi ink/N} = Nf[-k].$$

Reciprocity

Apparently, the DFT of $F[n]$ is equal to $Nf[-k]$. We formulate this as the *reciprocity rule* for the DFT:

$$F[n] \leftrightarrow Nf[-k]. \tag{16.9}$$

EXAMPLE 16.5

The DFT of the signal $\delta_N[n]$ is the constant signal 1 (see (16.8)), so $\delta_N[n] \leftrightarrow 1$. From the reciprocity rule it follows that $1 \leftrightarrow N\delta_N[-k]$. Since $\delta_N[n]$ is an even signal (verify this yourself), we obtain that $1 \leftrightarrow N\delta_N[k]$. Written out in full this gives the expression

$$\sum_{n=0}^{N-1} e^{-2\pi ink/N} = N\delta_N[k]$$

and, after interchanging the variables, (16.6) again re-appears. ◀

16.3.3 Time reversal

Time reversal

By *time reversal* we will mean the operation in the n-domain which replaces n by $-n$, which implies that a reversal in time takes place. The result in the k-domain is easy: there will also be a reversal in frequency, which is summarized in

$$f[-n] \leftrightarrow F[-k]. \tag{16.10}$$

In fact, the DFT of the signal $f[-n]$ equals

$$\sum_{n=0}^{N-1} f[-n]e^{-2\pi ink/N} = \sum_{n=0}^{N-1} f[N-n]e^{2\pi i(N-n)k/N} = \sum_{n=1}^{N} f[n]e^{2\pi ink/N}$$
$$= F[-k].$$

Note that we used lemma 16.1 for the final equality in the calculation above. As a consequence we have for an even or odd periodic discrete-time signal that the DFT is also even or, respectively, odd. When, for example, $f[n]$ is even, then $f[-n] = f[n]$ for all integer n and so we have $F[-k] = F[k]$ for all integer k.

16.3.4 Conjugation

The discrete spectrum of the complex conjugate $\overline{f[n]}$ of a signal $f[n]$ can be found by a direct calculation:

$$\sum_{n=0}^{N-1} \overline{f[n]}e^{-2\pi ink/N} = \overline{\sum_{n=0}^{N-1} f[n]e^{2\pi ink/N}} = \overline{F[-k]}.$$

Conjugation

Conjugation in the n-domain thus implies a conjugation in the k-domain and a reversal of frequency, that is to say,

$$\overline{f[n]} \leftrightarrow \overline{F[-k]}. \tag{16.11}$$

For real signals this rule implies that the amplitude spectrum is an even function of k and that the phase spectrum is an odd function of k. In fact, if $f[n]$ is real, then $\overline{f[n]} = f[n]$ and so $\overline{F[-k]} = F[k]$, which means that

$$|F[-k]| = \left|\overline{F[k]}\right| = |F[k]|$$

and

$$\arg F[-k] = -\arg \overline{F[-k]} = -\arg F[k].$$

Moreover, real signals have the advantage that the amount of computation necessary to determine the DFT can be halved. Since $F[k]$ is periodic with period N, one has for real signals $f[n]$ that $F[N-k] = F[-k] = \overline{F[k]}$. This means that it suffices to calculate $F[k]$ for $0 \leq k \leq N/2$. The values of $F[k]$ for $N/2 < k < N$ then follow by conjugation.

EXAMPLE 16.6

In this example we return to section 16.1. In example 16.1 the Fourier coefficients of a periodic continuous-time signal have been determined using a 128-point DFT $F[k]$ (see table 16.1). Since the signal $f(t)$ in example 16.1 is real and even, these properties also hold for the sampling $f[n]$ and this implies that $F[-k] = \overline{F[k]}$ and $F[-k] = F[k]$. We conclude that $F[k]$ is also real and even. ◀

16.3.5 Shift in the *n*-domain

A discrete-time signal has been introduced as a function defined on the integers. Therefore, we can only allow a shift in the n-domain over an *integer*, say l. This is because if $f[n]$ is a discrete-time signal, then for integer l the signal $f[n-l]$ is again a discrete-time signal. A *shift in the n-domain* has the following consequence in the k-domain:

Shift in the n-domain

$$f[n-l] \leftrightarrow e^{-2\pi ilk/N} F[k]. \tag{16.12}$$

One can prove this property as follows:

$$\sum_{n=0}^{N-1} f[n-l]e^{-2\pi ink/N} = \sum_{n=-l}^{N-1-l} f[n]e^{-2\pi i(n+l)k/N}$$

$$= e^{-2\pi ilk/N} \sum_{n=0}^{N-1} f[n]e^{-2\pi ink/N} = e^{-2\pi ilk/N} F[k].$$

In this calculation we have applied lemma 16.1.

EXAMPLE 16.7

In this example we consider a sum of periodic trains of discrete unit pulses given by

$$f[n] = \sum_{l=-m}^{m} \delta_N[n-l],$$

where we assume that $2m < N$. A graph of $f[n]$ is shown in figure 16.4. To calculate the DFT of $f[n]$, we first use the linearity property (16.5). This means that for each term occurring in the description of $f[n]$ we take the DFT. By adding these, we obtain the DFT of $f[n]$. The DFT of $\delta_N[n-l]$ immediately follows from (16.8) and the shift property (16.12) and results in

$$\delta_N[n-l] \leftrightarrow e^{-2\pi ilk/N}.$$

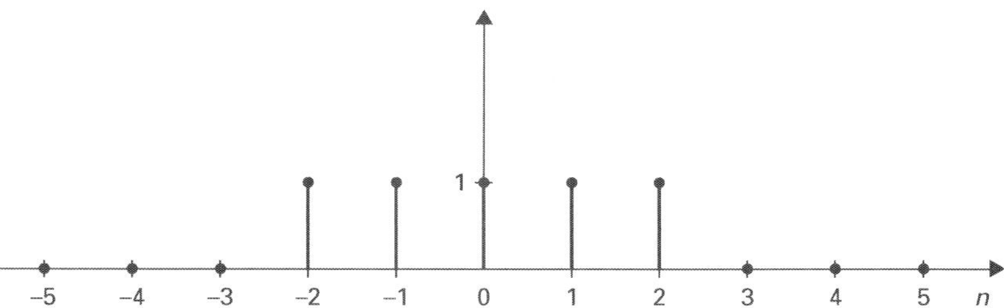

FIGURE 16.4
Sum of periodic trains of discrete unit pulses for $m = 2$ and $N = 10$.

Hence,

$$f[n] \leftrightarrow F[k] = \sum_{l=-m}^{m} e^{-2\pi ilk/N}.$$

This sum can be calculated explicitly. If, for convenience, we put $z = e^{-2\pi ik/N}$, then $F[k]$ is the sum of a finite geometric series with initial term z^{-m}, ratio z and final term z^{m}. Hence, the sum $F[k]$ equals

$$F[k] = \begin{cases} 2m + 1 & \text{if } k \text{ is a multiple of } N, \\ \dfrac{e^{2\pi imk/N} - e^{-2\pi i(m+1)k/N}}{1 - e^{-2\pi ik/N}} & \text{otherwise.} \end{cases}$$

We rewrite the second expression as follows:

$$\frac{e^{2\pi imk/N} - e^{-2\pi i(m+1)k/N}}{1 - e^{-2\pi ik/N}} = \frac{e^{\pi i(2m+1)k/N} - e^{-\pi i(2m+1)k/N}}{e^{\pi ik/N} - e^{-\pi ik/N}}$$
$$= \frac{\sin(\pi(2m + 1)k/N)}{\sin(\pi k/N)}.$$

The final result for $F[k]$ is then

$$F[k] = \begin{cases} 2m + 1 & \text{if } k \text{ is a multiple of } N, \\ \dfrac{\sin(\pi(2m + 1)k/N)}{\sin(\pi k/N)} & \text{otherwise.} \end{cases}$$

◀

From the shift property (16.12) it follows that the amplitude spectrum does not change under a shift in the n-domain, since

$$\left| e^{-2\pi ilk/N} F[k] \right| = |F[k]|.$$

The phase spectrum changes in a simple way:

$$\arg(e^{-2\pi ilk/N} F[k]) = \arg(F[k]) - 2\pi lk/N.$$

The change $-2\pi lk/N$ is linear in k. We then say that a shift in the n-domain causes a *linear phase shift*.

Linear phase shift

16.3.6   Shift in the *k*-domain

Shift in the k-domain   Similar to the *n*-domain, a *shift in the k-domain* will result in a multiplication by a complex exponential in the *n*-domain. Derive for yourself the following rule:

$$e^{2\pi inl/N} f[n] \leftrightarrow F[k - l]. \tag{16.13}$$

EXERCISES

16.10   The periodic continuous-time signal $f(t)$ with period 2π from example 16.1 satisfies the relation $f(t) + f(t + \pi) = 1$ for all $t \in \mathbb{R}$.
a   Prove this relation.
b   Show that for the given sampling with $N = 128$ one has $f[n] + f[n + N/2] = 1$ $(n \in \mathbb{Z})$.
c   Prove that the previous relation implies that $F[k] = 0$ for even k, except when k is an integer multiple of 128. What is $F[k]$ when k is an integer multiple of 128?

16.11   The periodic discrete-time signal $f[n]$ is given by

$$f[n] = \begin{cases} 2m + 1 & \text{if } n \text{ is a multiple of } N, \\[2ex] \dfrac{\sin(\pi(2m + 1)n/N)}{\sin(\pi n/N)} & \text{otherwise.} \end{cases}$$

Determine the DFT of $f[n]$.

16.12   Given is a real periodic discrete-time signal $f[n]$ with period 4. For the 4-point DFT $F[k]$ of $f[n]$ it is known that $F[0] = 1$, $F[1] = i$, $F[2] = 0$. Calculate $f[n]$ for $n = 0, 1, 2, 3$.

16.4   Cyclical convolution

For both the Fourier series and the Fourier integrals we have encountered the convolution product, or convolution for short. In both cases the corresponding convolution theorems showed that the convolution operation in the time domain is related to a multiplication in the frequency domain. For the DFT we will again come across a convolution.

Let the periodic discrete-time signals $f[n]$ and $g[n]$ with period N and DFT $F[k]$ and, respectively, $G[k]$ be given. We are then looking for a periodic discrete-time signal in the *n*-domain whose spectrum equals the product $F[k]G[k]$. This signal can be found as follows. Using definition 16.1 for the DFT we see that

$$F[k]G[k] = \sum_{l=0}^{N-1} f[l]G[k]e^{-2\pi ilk/N}.$$

The product $F[k]G[k]$ is written here as a linear combination of the signals $f[l]G[k]e^{-2\pi ilk/N}$. We subsequently apply the shift property (16.12), which results in

$$g[n - l] \leftrightarrow G[k]e^{-2\pi ilk/N}.$$

The linearity property then gives

$$\sum_{l=0}^{N-1} f[l]g[n - l] \leftrightarrow F[k]G[k].$$

It is now clear how to define the convolution for periodic discrete-time signals. Since the convolution product is in the first instance meant for periodic signals, this product will be called the *cyclical convolution product* or *cyclical convolution* for short.

DEFINITION 16.2
Cyclical convolution

The cyclical convolution product *of two periodic discrete-time signals $f[n]$ and $g[n]$ with period N is the discrete-time signal $(f * g)[n]$ defined by*

$$(f * g)[n] = \sum_{l=0}^{N-1} f[l]g[n-l]. \tag{16.14}$$

The corresponding convolution theorem, which has already been proven, can be formulated as follows.

THEOREM 16.2
Convolution in the n-domain

Let $f[n]$ and $g[n]$ be periodic discrete-time signals with period N. Let $F[k]$ be the DFT of $f[n]$ and $G[k]$ the DFT of $g[n]$. Then one has for the cyclical convolution product $(f * g)[n]$ that

$$(f * g)[n] \leftrightarrow F[k]G[k]. \tag{16.15}$$

From theorem 16.2 it follows that cyclical convolution is commutative, that is to say,

$$(f * g)[n] = (g * f)[n] \qquad \text{for } n \in \mathbb{Z}. \tag{16.16}$$

In fact, the ordinary product in (16.14) is commutative, and so the left-hand and right-hand sides of (16.16) have the same DFT.

EXAMPLE 16.8

In theorem 15.2 of chapter 15 we have seen that for a periodic discrete-time signal with period N one has

$$f[n] = \sum_{l=0}^{N-1} f[l]\delta_N[n-l] \qquad \text{for } n \in \mathbb{Z}.$$

This result also follows easily from the convolution theorem. In the right-hand side we have the cyclical convolution of the signals $\delta_N[n]$ and $f[n]$. Since $\delta_N[n] \leftrightarrow 1$, we have, according to theorem 16.2,

$$\sum_{l=0}^{N-1} f[l]\delta_N[n-l] \leftrightarrow F[k] \cdot 1 = F[k].$$

Theorem 15.2 now follows from the inverse transform. ◄

Because of the close relationship between the DFT and the inverse DFT, we may expect that the DFT of a product $f[n]g[n]$ in the n-domain will result in a convolution product in the k-domain. This is indeed the case, as is stated in the following theorem.

THEOREM 16.3
Convolution in the k-domain

Let $f[n]$ and $g[n]$ be periodic discrete-time signals with period N. Let $F[k]$ be the DFT of $f[n]$ and $G[k]$ the DFT of $g[n]$. Then

$$f[n]g[n] \leftrightarrow \frac{1}{N}(F * G)[k]. \tag{16.17}$$

Proof
We apply the inverse DFT to $F[k]$, which results in

$$f[n]g[n] = \frac{1}{N} \sum_{l=0}^{N-1} F[l]e^{2\pi iln/N} g[n].$$

We have thus written the product $f[n]g[n]$ in the n-domain as a linear combination of the signals $F[l]e^{2\pi iln/N}g[n]$. For these signals one can obtain the DFT by means of the shift property. We then find that

$$F[l]e^{2\pi iln/N}g[n] \leftrightarrow F[l]G[k-l].$$

Applying the linearity property, (16.17) follows, proving the theorem. $\blacksquare$

EXAMPLE 16.9

Let $f[n] = e^{2\pi in/N}$ and let $g[n]$ be an arbitrary periodic discrete-time signal with period N. Assume that $g[n] \leftrightarrow G[k]$. Since $e^{2\pi in/N} \leftrightarrow N\delta_N[k-1]$, it follows from the convolution theorem in the k-domain that

$$e^{2\pi in/N}g[n] \leftrightarrow \frac{1}{N}\sum_{l=0}^{N-1} N\delta_N[l-1]G[k-l] = G[k-1].$$

As a matter of fact, this result can be obtained more quickly by applying the shift property (16.13) in the k-domain. $\blacktriangleleft$

Power

As the final subject of this chapter we consider the *power* of a periodic discrete-time signal. This has been introduced in chapter 1 as the quantity P satisfying

$$P = \frac{1}{N}\sum_{n=0}^{N-1} |f[n]|^2.$$

From the inverse DFT we know that a periodic discrete-time signal $f[n]$ can be written as a linear combination of the signals $e^{2\pi ink/N}$ (see (16.7)):

$$f[n] = \frac{1}{N}\sum_{k=0}^{N-1} F[k]e^{2\pi ink/N}.$$

The power of the signal $F[k]e^{2\pi ink/N}/N$ in the n-domain equals $|F[k]|^2/N^2$ (check this yourself). An important consequence of our next theorem will be that the power of $f[n]$ is equal to the *sum* of the powers of the individual terms in the above expression for $f[n]$ (see (16.19)).

THEOREM 16.4
Parseval's theorem

Let $f[n]$ and $g[n]$ be periodic discrete-time signals with period N. Let $F[k]$ be the DFT of $f[n]$ and $G[k]$ the DFT of $g[n]$. Then

$$\sum_{n=0}^{N-1} f[n]\overline{g[n]} = \frac{1}{N}\sum_{k=0}^{N-1} F[k]\overline{G[k]}. \tag{16.18}$$

Proof
In the left-hand side of (16.18) we substitute for $f[n]$ the inverse DFT of $F[k]$ and we then change the order of summation, resulting in:

$$\sum_{n=0}^{N-1} f[n]\overline{g[n]} = \frac{1}{N}\sum_{n=0}^{N-1} \overline{g[n]}\sum_{k=0}^{N-1} F[k]e^{2\pi ink/N}$$

$$= \frac{1}{N}\sum_{k=0}^{N-1} F[k]\sum_{n=0}^{N-1} \overline{g[n]}e^{2\pi ink/N}$$

$$= \frac{1}{N}\sum_{k=0}^{N-1} F[k]\overline{\sum_{n=0}^{N-1} g[n]e^{-2\pi ink/N}}$$

$$= \frac{1}{N}\sum_{k=0}^{N-1} F[k]\overline{G[k]}.$$

This proves the theorem. ■

Replacing $g[n]$ by $f[n]$ we obtain from the theorem above that

$$\sum_{n=0}^{N-1} |f[n]|^2 = \frac{1}{N} \sum_{k=0}^{N-1} |F[k]|^2. \tag{16.19}$$

EXAMPLE 16.10

Let $f[n]$ be the periodic discrete-time signal with period N given by $f[n] = \sin(2\pi n/N)$. By changing to complex exponentials it follows that $f[n] = (e^{2\pi i n/N} - e^{-2\pi i n/N})/2i$. From this representation we immediately obtain the DFT of $f[n]$:

$$F[-1] = -N/2i,$$

$$F[1] = N/2i,$$

$$F[0] = F[2] = F[3] = \ldots = F[N-2] = 0.$$

According to (16.19) the power of $f[n]$ is then equal to

$$P = \frac{1}{N^2}\left(\frac{N^2}{4} + \frac{N^2}{4}\right) = \frac{1}{2}.$$

◀

EXERCISES

16.13   Let the periodic discrete-time signal $f[n]$ be given by $f[n] = \delta_N[n] + \delta_N[n-1]$. Calculate the cyclical convolution product of $f[n]$ with $f[n]$.

16.14   For the periodic discrete-time signal $f[n]$ with period N the N-point DFT is given by $F[k] = \cos(2\pi k/N)\sin(4\pi k/N)$. Determine $f[n]$ using the convolution theorem 16.2.

16.15   Let $f[n]$ and $g[n]$ be periodic discrete-time signals with period N and with $f[n] \leftrightarrow F[k]$ and $g[n] \leftrightarrow G[k]$. Prove the following *duality property*:

$$\sum_{n=0}^{N-1} f[n]G[n] = \sum_{k=0}^{N-1} F[k]g[k].$$

16.16   Determine the power of the periodic discrete-time signal $f[n]$ whose DFT is given by $F[k] = \cos^2(\pi k/N)$.

16.17   Determine the power of the periodic discrete-time signal $f[n]$ with period N given by

$$f[n] = \begin{cases} 2m+1 & \text{if } n \text{ is a multiple of } N, \\[2mm] \dfrac{\sin(\pi(2m+1)n/N)}{\sin(\pi n/N)} & \text{otherwise.} \end{cases}$$

Here m is a positive integer with $m < N/2$.

SUMMARY

The discrete Fourier transform (DFT) assigns to a periodic discrete-time signal $f[n]$ with period N, given in the time domain or n-domain, a periodic discrete signal

$F[k]$ in the frequency domain or k-domain:

$$F[k] = \sum_{n=0}^{N-1} f[n]e^{-2\pi ink/N}.$$

The DFT, also called the (discrete) spectrum of $f[n]$, arises in a natural way if we pose the problem of approximating the Fourier coefficients of a periodic continuous-time signal using the trapezoidal rule.

Through the inverse DFT one can recover a periodic discrete-time signal $f[n]$ from its spectrum $F[k]$:

$$f[n] = \frac{1}{N} \sum_{k=0}^{N-1} F[k]e^{2\pi ink/N}.$$

When deriving the inverse DFT, an important role is played by the fundamental property

$$\delta_N[n] = \frac{1}{N} \sum_{k=0}^{N-1} e^{2\pi ink/N}.$$

The DFT has similar properties to the Fourier transform of continuous-time signals. An overview of these properties can be found in table 12 at the back of this book. Another similarity with the Fourier theory of continuous-time signals is the convolution product, which is called cyclical convolution in the case of periodic discrete-time signals, and which is defined by

$$(f * g)[n] = \sum_{l=0}^{N-1} f[l]g[n-l].$$

The corresponding convolution theorem (the cyclical convolution theorem in the n-domain) states that the DFT of the cyclical convolution product is equal to the ordinary product of the DFTs of $f[n]$ and $g[n]$. Apart from a convolution theorem in the n-domain, there is also a cyclical convolution theorem in the k-domain, which states that the DFT transforms the ordinary product $f[n]g[n]$ in the n-domain into the cyclical convolution in the k-domain, up to a factor $1/N$.

Finally, one can also formulate the following Parseval identity for periodic discrete-time signals:

$$\sum_{n=0}^{N-1} f[n]\overline{g[n]} = \frac{1}{N} \sum_{k=0}^{N-1} F[k]\overline{G[k]},$$

where $F[k]$ and $G[k]$ are the DFTs of, respectively, $f[n]$ and $g[n]$. From this identity one can then obtain the power of a periodic discrete-time signal in the k-domain. One has the following result:

$$\sum_{n=0}^{N-1} |f[n]|^2 = \frac{1}{N} \sum_{k=0}^{N-1} |F[k]|^2.$$

SELFTEST

16.18   For a periodic continuous-time signal $f(t)$ with period T and fundamental frequency ω_0, the Fourier coefficients are given by $c_{-2} = c_2 = 1$, $c_{-1} = c_1 = 2$, $c_0 = 1$. The other Fourier coefficients are equal to zero. Let $g[n]$ be the periodic discrete-time signal with period 5 such that $g[-2] = c_{-2}, g[-1] = c_{-1}, g[0] = c_0,$ $g[1] = c_1, g[2] = c_2.$

a Calculate the 5-point DFT $G[k]$ of $g[n]$.

b Show that $f(2\pi m/5\omega_0) = G[-m]$ for $m \in \mathbb{Z}$.

c Prove that

$$\frac{1}{T} \int_0^T |f(t)|^2 \, dt = \frac{1}{5} \sum_{k=0}^{4} |G[k]|^2 .$$

16.19 The periodic discrete-time signals $f[n]$ and $g[n]$ with period N are given, for all integer n, by

$$f[n] = \delta_N[n+1] - \delta_N[n] + \delta_N[n-1],$$
$$g[n] = \cos(4\pi n/N).$$

a Determine the N-point DFT of $f[n]$.

b Calculate the cyclical convolution product of $f[n]$ and $g[n]$.

c Calculate the power of $g[n]$.

16.20 Given is a periodic discrete-time signal $f[n]$ with period T. This discrete-time signal is a sampling of the periodic continuous-time signal $f(t)$ with period T. The sampling period is equal to $T/5$. The signal $f(t)$ is even and real. Let $F[k]$ be the 5-point DFT of $f[n]$.

a Show that $F[k]$ is also even and real.

b It is given that $F[0] = 1, F[1] = 2, F[2] = 1$. Determine $f[n]$ for all integer n.

c Let furthermore be given that $f(t)$ is band-limited, with bandwidth $10\pi/T$. For which values of k do we have $c_k = F[k]/5$? Here the numbers c_k are the Fourier coefficients of $f(t)$. Justify your answer.

Contents of Chapter 17

The Fast Fourier Transform

The Fast Fourier Transform

INTRODUCTION

The N-point discrete Fourier transform (DFT) was defined in chapter 16 as a transformation converting a periodic discrete-time signal $f[n]$ in the n-domain into a periodic discrete signal $F[k]$ in the k-domain according to the relation

$$
\begin{aligned}
F[k] &= \sum_{n=0}^{N-1} f[n]e^{-2\pi ink/N} \\
&= f[0] + f[1]e^{-2\pi ik/N} + \cdots + f[N-1]e^{-2\pi ik(N-1)/N}.
\end{aligned}
$$

If we now want to compute $F[k]$ for a certain value of k, then in general one will need $N-1$ (complex) multiplications and $N-1$ (complex) additions to do so. If we call a multiplication or an addition in the complex plane an *elementary* operation, then in general we will thus need $2N-2$ elementary operations to compute $F[k]$ for any given k. If we want to determine $F[k]$ for all $k \in \{0, 1, \ldots, N-1\}$, then in first instance we expect that this would require $2N^2 - 2N$ elementary operations. We then say that in a direct calculation of an N-point DFT the number of elementary operations is of order N^2, for large N. In applications we will often have to compute DFTs based on a large number of points. This will then result in many elementary operations, and thus in a large number of round-off errors and a considerable computing time. Fortunately, one has come up with algorithms that reduce the number of elementary operations substantially. Most often, these algorithms are based on the factorization of N into prime numbers, and are known collectively as *Fast Fourier Transform*, abbreviated as FFT. In the most popular versions, N is a power of 2.

If N can be written as a product of integers, say $N = N_1 N_2$, then it will be derived in section 17.1 that the computation of the N-point DFT can be reduced to the computation of a number of N_1-point DFTs and N_2-point DFTs. We get a much better overall picture of this reduction if we interpret the $N_1 N_2$-point DFT as an operation on *matrices*. When N_1 and N_2 can be factorized as well, then we can again reduce the number of points in the DFTs to be calculated. In this way an FFT algorithm arises. A Fast Fourier Transform is thus not a transform, but in fact an *efficient numerical implementation* of the DFT.

In section 17.2 special attention is paid to an FFT where N is a power of 2, so $N = 2^m$. An FFT algorithm then boils down to a repeated computation of 2-point DFTs.

In the final section we will treat some applications of the FFT. These are, of course, applications of the DFT. We will examine the calculation of the *spectrum* of a continuous-time signal, the calculation of the *cyclical convolution* product, and the calculation of the so-called *cross-correlation* of two discrete-time signals.

LEARNING OBJECTIVES

After studying this chapter it is expected that you
- know the relation between an N-point DFT and the Nth roots of unity
- can interpret an $N_1 N_2$-point DFT as an operation on matrices
- can describe the importance and the advantage of an FFT
- can give a global description of an FFT algorithm based on $N = 2^m$ points
- know the meaning of the term butterfly in an FFT algorithm
- can describe how an FFT can be used to calculate the spectrum of a continuous-time signal
- can indicate the advantage of the FFT for the calculation of a cyclical convolution
- know the concepts of cross-correlation, autocorrelation and power spectrum
- know the relationship between autocorrelation and power spectrum
- can indicate the advantage of the FFT for the calculation of a cross-correlation or an autocorrelation.

17.1   The DFT as an operation on matrices

In this section we will examine how an N-point DFT, where N can be written as a product of positive integers (so $N = N_1 N_2$ with $N_1 \geq 2$, $N_2 \geq 2$), can be interpreted as an operation on an $N_1 \times N_2$-matrix. Such an interpretation is important in order to get a good grasp of an FFT algorithm.

At the basis of an FFT algorithm lies the relationship between an N-point DFT and the so-called Nth roots of unity, which have already been introduced in chapter 2 as the N distinct roots in the complex plane of the equation $z^N = 1$. To begin with, we introduce the Nth root of unity w_N as follows:

$$w_N = e^{2\pi i/N}. \tag{17.1}$$

In chapter 2 we have seen that the Nth roots of unity consist of integer powers of w_N. Since $w_N^{-nk} = e^{-2\pi i nk/N}$ for integer n and k, it follows from definition 16.1 that the DFT of a discrete-time signal $f[n]$ can also be written as

$$F[k] = \sum_{n=0}^{N-1} f[n]w_N^{-nk}. \tag{17.2}$$

We can subsequently write this in a very compact way by introducing the following polynomials $P_N(z)$:

$$P_N(z) = \sum_{n=0}^{N-1} f[n]z^n. \tag{17.3}$$

From (17.2) it now immediately follows that

$$F[k] = P_N(w_N^{-k}). \tag{17.4}$$

Computing an N-point DFT can apparently be viewed as the calculation of the values of a complex polynomial, where for z we take the Nth roots of unity.

EXAMPLE 17.1

Let $N = 3$, $f[0] = 1$, $f[1] = 0$ and $f[2] = 1$. From (17.1) it follows that $w_3 = e^{2\pi i/3} = (-1 + i\sqrt{3})/2$ and then the third roots of unity 1, w_3^{-1}, w_3^{-2} are given by 1, $-(1 + i\sqrt{3})/2$, $-(1 - i\sqrt{3})/2$ respectively. Because of (17.3), the polynomial $P_3(z)$ has the form $P_3(z) = 1 + z^2$. Hence, it follows from (17.4) that

$$F[0] = P_3(1) = 2,$$
$$F[1] = P_3(-\tfrac{1}{2}(1 + i\sqrt{3})) = \tfrac{1}{2}(1 + i\sqrt{3}),$$
$$F[2] = P_3(-\tfrac{1}{2}(1 - i\sqrt{3})) = \tfrac{1}{2}(1 - i\sqrt{3}).$$

$\blacktriangleleft$

If N can be written as a product of two integers, so $N = N_1 N_2$ with N_1 and N_2 integer and greater than or equal to 2, then there are simple relations between the roots of unity w_{N_1}, w_{N_2} and w_N. Verify for yourself that the following relations hold:

$$w_N^{N_2} = w_{N_1} \qquad \text{and} \qquad w_N^{N_1} = w_{N_2}. \tag{17.5}$$

One can use these relations cleverly in order to calculate the $N_1 N_2$-point DFT. To this end we assign in a unique way to each $k \in \{0, 1, \ldots, N - 1\}$ the integers μ_1 and μ_2 according to

$$k = \mu_1 N_2 + \mu_2 \qquad \text{where } 0 \le \mu_2 < N_2. \tag{17.6}$$

Hence, the number μ_1 is the result of the so-called integer division (division with remainder) of k by N_2, and μ_2 is the remainder in this division. In the same way we assign to each integer $n \in \{0, 1, \ldots, N - 1\}$ the integers ν_1 and ν_2 according to

$$n = \nu_2 N_1 + \nu_1 \qquad \text{where } 0 \le \nu_1 < N_1. \tag{17.7}$$

The numbers ν_1 and ν_2 thus arise from the integer division of n by N_1. Since both k and n lie between 0 and $N - 1$, we have

$$0 \le \mu_1, \nu_1 < N_1 \qquad \text{and} \qquad 0 \le \mu_2, \nu_2 < N_2.$$

Now consider (17.2) for the N-point DFT and substitute here for k and n the expressions in, respectively, (17.6) and (17.7). First note that the sum $\sum_{n=0}^{N-1} a_n$ is equal to the repeated sum $\sum_{\nu_1=0}^{N_1-1} \sum_{\nu_2=0}^{N_2-1} a_{\nu_2 N_1 + \nu_1}$. In addition we will use relations (17.5) and the identity $w_N^{N_1 N_2} = w_N^N = 1$. The calculation of the DFT then reads as follows:

$$F[k] = F[\mu_1 N_2 + \mu_2] = \sum_{\nu_1=0}^{N_1-1} \sum_{\nu_2=0}^{N_2-1} f[\nu_2 N_1 + \nu_1] w_N^{-(\mu_1 N_2 + \mu_2)(\nu_2 N_1 + \nu_1)}$$

$$= \sum_{\nu_1=0}^{N_1-1} w_{N_1}^{-\mu_1 \nu_1} w_N^{-\mu_2 \nu_1} \sum_{\nu_2=0}^{N_2-1} f[\nu_2 N_1 + \nu_1] w_{N_2}^{-\mu_2 \nu_2}$$

$$= \sum_{\nu_1=0}^{N_1-1} w_N^{-\mu_2 \nu_1} c[\nu_1, \mu_2] w_{N_1}^{-\mu_1 \nu_1},$$

where

$$c[\nu_1, \mu_2] = \sum_{\nu_2=0}^{N_2-1} f[\nu_2 N_1 + \nu_1] w_{N_2}^{-\mu_2 \nu_2}. \tag{17.8}$$

We have now found that

$$F[\mu_1 N_2 + \mu_2] = \sum_{\nu_1=0}^{N_1-1} w_N^{-\mu_2 \nu_1} c[\nu_1, \mu_2] w_{N_1}^{-\mu_1 \nu_1}. \tag{17.9}$$

At first, it may seem very complicated to calculate a DFT using (17.8) and (17.9). Still, it will be precisely these formulas that will give us a first clue how to determine a DFT efficiently. In order to get a good insight into the operations occurring in these formulas, we introduce a number of matrices consisting of N_1 rows and N_2 columns:

$$
M_f = \begin{pmatrix}
f[0] & f[N_1] & \cdots & f[N-2N_1] & f[N-N_1] \\
f[1] & f[N_1+1] & \cdots & f[N-2N_1+1] & f[N-N_1+1] \\
\vdots & \vdots & \ddots & \vdots & \vdots \\
f[N_1-2] & f[2N_1-2] & \cdots & f[N-N_1-2] & f[N-2] \\
f[N_1-1] & f[2N_1-1] & \cdots & f[N-N_1-1] & f[N-1]
\end{pmatrix}
$$

$$\tag{17.10a}$$

$$
C = \begin{pmatrix}
c[0,0] & c[0,1] & \cdots & c[0,N_2-2] & c[0,N_2-1] \\
c[1,0] & c[1,1] & \cdots & c[1,N_2-2] & c[1,N_2-1] \\
\vdots & \vdots & \ddots & \vdots & \vdots \\
c[N_1-2,0] & c[N_1-2,1] & \cdots & c[N_1-2,N_2-2] & c[N_1-2,N_2-1] \\
c[N_1-1,0] & c[N_1-1,1] & \cdots & c[N_1-1,N_2-2] & c[N_1-1,N_2-1]
\end{pmatrix}
$$

$$\tag{17.10b}$$

$$
M_F = \begin{pmatrix}
F[0] & F[1] & \cdots & F[N_2-2] & F[N_2-1] \\
F[N_2] & F[N_2+1] & \cdots & F[2N_2-2] & F[2N_2-1] \\
\vdots & \vdots & \ddots & \vdots & \vdots \\
F[N-2N_2] & F[N-2N_2+1] & \cdots & F[N-N_2-2] & F[N-N_2-1] \\
F[N-N_2] & F[N-N_2+1] & \cdots & F[N-2] & F[N-1]
\end{pmatrix}
$$

$$\tag{17.10c}$$

The element $M_f(\mu,\nu)$ in row μ ($\mu = 0,1,\ldots,N_1-1$) and column ν ($\nu = 0,1,\ldots,N_2-1$) is thus given by

$$M_f(\mu,\nu) = f[\nu N_1 + \mu]. \tag{17.11}$$

The element $M_F(\mu,\nu)$ in row μ ($\mu = 0,1,\ldots,N_1-1$) and column ν ($\nu = 0,1,\ldots,N_2-1$) is given by

$$M_F(\mu,\nu) = F[\mu N_2 + \nu]. \tag{17.12}$$

Having introduced these matrices, we return to (17.8) and (17.9). On the basis of (17.8) we note that for fixed ν_1 the row $c[\nu_1,0], \ldots, c[\nu_1, N_2-1]$ is the N_2-point DFT of the row $f[\nu_1], f[N_1+\nu_1], \ldots, f[(N_2-1)N_1+\nu_1]$. For fixed μ_2 we can interpret (17.9) as the N_1-point DFT of the column

$$
\begin{pmatrix}
c[0,\mu_2] \\
w_N^{-\mu_2} c[1,\mu_2] \\
\vdots \\
w_N^{-(N_1-2)\mu_2} c[N_1-2,\mu_2] \\
w_N^{-(N_1-1)\mu_2} c[N_1-1,\mu_2]
\end{pmatrix}.
$$

This is column μ_2 of matrix C, however, with its elements multiplied by the factors $w_N^{-\mu_2\nu_1}$. These factors are called *twiddle factors*. Formula (17.9) and the definition of the matrix M_F also show us that the N_1-point DFT of this column can be found in column μ_2 of matrix M_F.

Summarizing, we have now derived a method for the calculation of the $N_1 N_2$-point DFT of a discrete-time signal $f[n]$, which consists of the following steps:

Algorithm for an $N_1 N_2$-point DFT

- Construct the $N_1 \times N_2$ matrix M_f from the discrete signal $f[n]$.
- Calculate the N_2-point DFT of row μ ($\mu = 0, 1, \ldots, N_1 - 1$) of matrix M_f and put this in row μ of a new matrix C.
- Multiply the elements of matrix C by the twiddle factors, that is to say, multiply $c[\mu, v]$ by $w_N^{-\mu v}$; call the new matrix Ct.
- Calculate the N_1-point DFT of column v ($v = 0, 1, \ldots, N_2 - 1$) of matrix Ct and put this in column v of matrix M_F; then finally matrix M_F represents the $N_1 N_2$-point DFT of $f[n]$.

EXAMPLE 17.2

As an illustration of this method we calculate the 6-point DFT of the discrete-time signal $f[n]$ with $f[0] = 1$, $f[1] = 1$, $f[2] = 0$, $f[3] = 0$, $f[4] = 1$, $f[5] = 1$. We take $N_1 = 3$ and $N_2 = 2$. The matrix M_f then looks as follows:

$$M_f = \begin{pmatrix} f[0] & f[3] \\ f[1] & f[4] \\ f[2] & f[5] \end{pmatrix} = \begin{pmatrix} 1 & 0 \\ 1 & 1 \\ 0 & 1 \end{pmatrix}.$$

Next we have to take the 2-point DFT of the rows of this matrix. We are then dealing with the root of unity $w_2 = -1$ of $z^2 = 1$. Applying (17.4) in this situation is easy. For example, to calculate the 2-point DFT of the first row, we have to use the polynomial $P_2(z) = f[0] + f[3]z$. We subsequently substitute the values $z = 1$ and $z = -1$, resulting in the row $(f[0] + f[3], f[0] - f[3])$ as the 2-point DFT of the row $(f[0], f[3])$. Hence, determining a 2-point DFT is simply a calculation of the sum and the difference of the values of the signal. In this example the matrix C then looks as follows:

$$C = \begin{pmatrix} c[0, 0] & c[0, 1] \\ c[1, 0] & c[1, 1] \\ c[2, 0] & c[2, 1] \end{pmatrix} = \begin{pmatrix} f[0] + f[3] & f[0] - f[3] \\ f[1] + f[4] & f[1] - f[4] \\ f[2] + f[5] & f[2] - f[5] \end{pmatrix} = \begin{pmatrix} 1 & 1 \\ 2 & 0 \\ 1 & -1 \end{pmatrix}.$$

We multiply the elements of this matrix by the twiddle factors. This means that the element in row μ and column v is multiplied by $w_6^{-\mu v}$. In the row with $\mu = 0$ and the column with $v = 0$ nothing changes, since $w_6^{-\mu v} = w_6^0 = 1$ in these cases. In this example the element $c[1, 1]$ also remains unchanged since it was already equal to 0. The element $c[2, 1]$ is multiplied by $w_6^{-2 \cdot 1} = (e^{2\pi i/6})^{-2} = e^{-2\pi i/3} = -(1 + i\sqrt{3})/2$. The matrix Ct arising in this way looks as follows:

$$Ct = \begin{pmatrix} 1 & 1 \\ 2 & 0 \\ 1 & \frac{1}{2}(1 + i\sqrt{3}) \end{pmatrix}$$

We finally have to take the 3-point DFT of the columns of this matrix (as in example 17.1). We then obtain the 6-point DFT in the following matrix:

$$\begin{pmatrix} F[0] & F[1] \\ F[2] & F[3] \\ F[4] & F[5] \end{pmatrix} = \begin{pmatrix} 4 & \frac{1}{2}(3 + i\sqrt{3}) \\ -\frac{1}{2}(1 + i\sqrt{3}) & 0 \\ -\frac{1}{2}(1 - i\sqrt{3}) & \frac{1}{2}(3 - i\sqrt{3}) \end{pmatrix}.$$

◀

Calculating the DFT using this method will only lead to an advantage for large values of N. We will verify this by calculating the number of elementary operations. In doing so we will assume that the N_1-point DFT and the N_2-point DFT are determined by a direct calculation.

Number of operations of
$N_1 N_2$-point DFT

Determining the matrix C means that we have to calculate a total of N_1 DFTs based on N_2 points. For this we need $2N_1(N_2^2 - N_2)$ elementary operations (see the introduction to this chapter). After this, the elements of C have to be multiplied by the twiddle factors. Since the twiddle factors are 1 for the first row and the first columns, we need another $(N_1 - 1)(N_2 - 1)$ elementary operations to calculate the matrix Ct. Finally, to determine the matrix M_F we have to calculate N_2 DFTs based on N_1 points. This requires yet another $2N_2(N_1^2 - N_1)$ elementary operations. We conclude that the described method requires a total of

$$2N_1 N_2(N_1 + N_2) - 3N_1 N_2 - N_1 - N_2 + 1$$

elementary operations. For large N this is of order $N_1 N_2(N_1 + N_2)$. In the introduction to this chapter we have already noted that for large N a direct calculation of the N-point DFT requires a number of elementary operations of the order N^2. With the method described above we thus benefit by a factor $N_1 N_2/(N_1 + N_2)$.

When, moreover, N_1 or N_2 can be factorized even further, then we can again obtain a reduction when calculating the N_1-point DFT or the N_2-point DFT. We will then benefit even more.

We have already seen that the calculation of a 2-point DFT is extremely simple. Now if N is a power of 2, then the above implies that the N-point DFT can be calculated by a repeated application of 2-point DFTs. In the next section we will treat this case explicitly.

EXERCISES

17.1   Use (17.2) to show that the relation between $f[n]$ and the 5-point DFT $F[k]$ can be represented as follows (here $w = e^{2\pi i/5}$):

$$\begin{pmatrix} 1 & 1 & 1 & 1 & 1 \\ 1 & w^{-1} & w^{-2} & w^{-3} & w^{-4} \\ 1 & w^{-2} & w^{-4} & w^{-1} & w^{-3} \\ 1 & w^{-3} & w^{-1} & w^{-4} & w^{-2} \\ 1 & w^{-4} & w^{-3} & w^{-2} & w^{-1} \end{pmatrix} \begin{pmatrix} f[0] \\ f[1] \\ f[2] \\ f[3] \\ f[4] \end{pmatrix} = \begin{pmatrix} F[0] \\ F[1] \\ F[2] \\ F[3] \\ F[4] \end{pmatrix}.$$

17.2   Determine the inverse of the matrix from exercise 17.1 by applying the inverse DFT.

17.3   As in example 17.2, determine the 6-point DFT of the discrete signal given in example 17.2. However, now take $N_1 = 2$ and $N_2 = 3$.

17.4   Use a repeated 2-point DFT to calculate the 4-point DFT of the periodic discrete-time signal $f[n]$ with period 4 given by $f[-1] = 1$, $f[0] = 2$, $f[1] = 0$, $f[2] = 2$.

17.2   The N-point DFT with $N = 2^m$

In this section we pay special attention to the N-point DFT where N is a power of 2, so $N = 2^m$ for some integer $m \geq 1$. In the previous section we discussed a method to reduce the calculation of an N-point DFT to a calculation of DFTs on a smaller number of points. We will study that method again for the special case $N = 2^m$. Put $N_1 = 2$ and $N_2 = 2^{m-1}$. The $2 \times 2^{m-1}$ matrix M_f (see (17.10a)) then looks as follows:

$$M_f = \begin{pmatrix} f[0] & f[2] & \cdots & f[N-2] \\ f[1] & f[3] & \cdots & f[N-1] \end{pmatrix}.$$

To each row of this matrix we first have to apply the $N/2$-point DFT in order to calculate the matrix C. This means that we have to calculate the $N/2$-point DFT of the discrete-time signals $f[2n]$ and $f[2n+1]$, which are both periodic with period $N_2 = N/2$. Let us assume that in the k-domain this leads to the signals $A[k]$ and $B[k]$ respectively, so

$$f[2n] \leftrightarrow A[k] \qquad \text{and} \qquad f[2n+1] \leftrightarrow B[k]. \tag{17.13}$$

The matrix C then looks as follows:

$$C = \begin{pmatrix} A[0] & A[1] & \ldots & A[\nu] & \ldots & A[N_2 - 1] \\ B[0] & B[1] & \ldots & B[\nu] & \ldots & B[N_2 - 1] \end{pmatrix}.$$

The next step is to multiply the elements of this matrix by the twiddle factors. The element in row μ and column ν is multiplied by the twiddle factor $w_N^{-\mu\nu}$. The first row ($\mu = 0$) and the first column ($\nu = 0$) do not change. Hence, the matrix Ct looks like this:

$$Ct = \begin{pmatrix} A[0] & A[1] & \ldots & A[\nu] & \ldots & A[N_2 - 1] \\ B[0] & w_N^{-1} B[1] & \ldots & w_N^{-\nu} B[\nu] & \ldots & w_N^{-(N_2-1)} B[N_2 - 1] \end{pmatrix}.$$

Finally, we have to determine the 2-point DFT of the columns of this matrix. We have already seen that the calculation of a 2-point DFT is nothing else but taking sums and differences. The 2-point DFT of column ν of matrix Ct gives column ν of matrix M_F, which finally gives us the N-point DFT of the signal $f[n]$:

$$M_F = \begin{pmatrix} F[0] & F[1] & \ldots & F[\nu] & \ldots & F[N_2 - 1] \\ F[N_2] & F[N_2 + 1] & \ldots & F[N_2 + \nu] & \ldots & F[N - 1] \end{pmatrix},$$

with

$$\begin{pmatrix} F[\nu] \\ F[N_2 + \nu] \end{pmatrix} = \begin{pmatrix} A[\nu] + w_N^{-\nu} B[\nu] \\ A[\nu] - w_N^{-\nu} B[\nu] \end{pmatrix} \qquad \text{for } \nu = 0, 1, \ldots N_2 - 1. \tag{17.14}$$

To determine $A[k]$ and $B[k]$ we have to calculate the $N/2$-point DFT of the signals $f[2n]$ and $f[2n+1]$ respectively. For this one can use the same method, by changing to $N/4$-point DFTs. One can repeat this process until only 2-point DFTs remain. This results in an FFT algorithm for which we now want to determine the number of elementary operations.

Let $\phi(m)$ be the number of elementary operations necessary to calculate a 2^m-point DFT according to the method described above. For the calculation of $A[k]$ and $B[k]$ one thus needs $2\phi(m-1)$ elementary operations. In order to determine $F[k]$ from this, using (17.14), we see that two elementary operations are necessary to calculate $F[0]$ and $F[N_2]$, and that three elementary operations are necessary to calculate $F[\nu]$ and $F[N_2 + \nu]$ for $\nu = 1, 2, \ldots, N_2 - 1$. This gives a total of $2 + 3(N_2 - 1) = 3 \cdot 2^{m-1} - 1$ elementary operations. From this we obtain a recurrence relation for $\phi(m)$:

$$\phi(m) = 2\phi(m-1) + 3 \cdot 2^{m-1} - 1.$$

Since the calculation of one 2-point DFT requires one addition and one subtraction, we know that $\phi(1) = 2$. By induction one can then show that

$$\phi(m) = (3m - 2)2^{m-1} + 1.$$

Hence, for the FFT executed in the way described earlier, the number of elementary operations is of order $m \cdot 2^m = N(^2\log N)$. Compared to N^2 this is a considerable reduction for large values of N.

The calculation of $F[\nu]$ and $F[N_2 + \nu]$ from $A[\nu]$ and $B[\nu]$ according to (17.14)

and using the twiddle factors $w_N^{-\nu}$ is sometimes called a *butterfly* and can be represented by the scheme in figure 17.1. This scheme should be interpreted as follows.

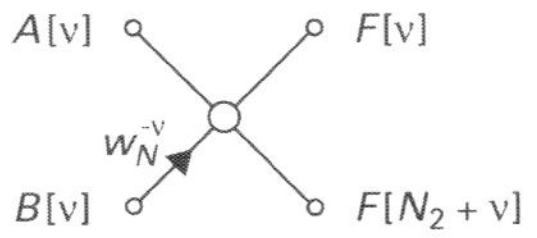

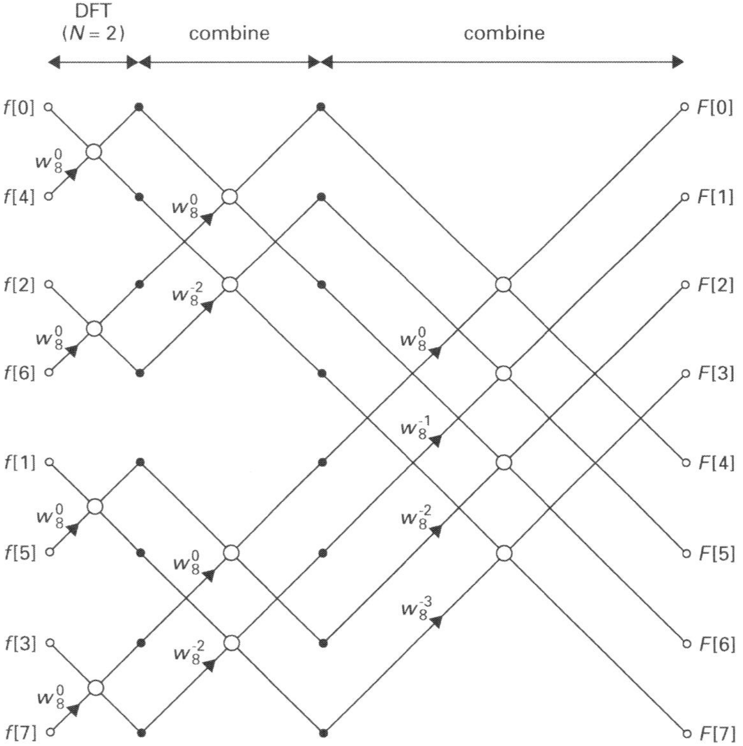

FIGURE 17.1
An FFT butterfly.

The term $B[v]$ is multiplied by w_N^{-v}. Next, the sum and difference of $A[v]$ and $B[v]w_N^{-v}$ give, respectively, $F[v]$ and $F[N_2 + v]$. One uses the term 'butterfly' because the scheme has the shape of a butterfly. An FFT algorithm for the calculation of a 2^m-point DFT thus consists of a repeated application of butterflies. In figure 17.2 the butterfly-scheme is given for $N = 8$. In this scheme we recognize

FIGURE 17.2
The 8-point FFT butterfly.

that one uses the first coefficients of the two 4-point DFTs to determine $F[0]$ and $F[4]$. To determine the latter two, we in turn use the first coefficients of the 2-point DFTs, etc. In figure 17.2 we see that at the start the sequence $f[0], f[1], \ldots, f[7]$ is mixed up. If we represent the numbers $0, 1, \ldots, 7$ as 3-digit binary numbers, then changing the order from $0, 1, \ldots, 7$ into $0, 4, 2, 6, 1, 5, 3, 7$ can be represented in the binary system as follows:

decimal number	binary representation	bit reversal	decimal representation
0	000	000	0
1	001	100	4
2	010	010	2
3	011	110	6
4	100	001	1
5	101	101	5
6	110	011	3
7	111	111	7

Bit reversal

Note how the new ordering arises by reversing the bitrow in the binary representation. We call this *bit reversal*. One can show for general $N = 2^m$ that the starting sequence of the butterfly scheme can be obtained by representing the numbers $0, 1, \ldots, N - 1$ as binary numbers with m digits and then applying bit reversal. We will not go into this any further.

EXERCISES

17.5   Give the butterfly scheme for the calculation of a 4-point DFT and use this to determine the 4-point DFT of the periodic discrete-time signal with period 4 given by $f[-1] = 2$, $f[0] = i$, $f[1] = 1$, $f[2] = i$.

17.6   Given is a periodic discrete-time signal $f[n]$ with period N and $f[2n + 1] = 1$ for all integer n. Moreover, the $N/2$-point DFT $A[k]$ of the signal $f[2n]$ is given. Calculate the N-point DFT of $f[n]$.

17.7   Given is a periodic discrete-time signal $f[n]$ with period $4N$. For the signals $f[4n]$, $f[4n + 1]$, $f[4n + 2]$, $f[4n + 3]$ the N-point DFTs are given by, respectively, $A[k]$, $B[k]$, $C[k]$, $D[k]$. Calculate the $4N$-point DFT of $f[n]$.

17.3   Applications

In the introduction we have already noted that an application of the FFT is in fact an application of the DFT. In chapter 16 we have seen an application of the DFT, namely, the calculation of Fourier coefficients. Applying the DFT is now becoming much more attractive since we have fast algorithms available to compute it. For this we prefer DFTs where the number of points is a power of 2. This implies that for the calculation of the Fourier coefficients of a periodic signal $f(t)$ with period T, the number of samples in $[0, T)$ should also be a power of 2.

Apart from this, it is quite natural to look at possible applications of the DFT to the calculation of the Fourier integral, or the spectrum, of a non-periodic continuous-time signal.

17.3.1   *Calculation of Fourier integrals*

The Fourier transform or spectrum $F(\omega)$ of a continuous-time signal $f(t)$, or, put differently, of a function defined on $\mathbb{R}$, has been defined in chapter 6 as the improper

Riemann integral

$$F(\omega) = \int_{-\infty}^{\infty} f(t)e^{-i\omega t}\, dt.$$

In order to calculate this integral, we are first of all confronted with the problem of *improper* integration, in this case integration over $\mathbb{R}$. If the signal is absolutely integrable, then

$$\left| F(\omega) - \int_{-T}^{T} f(t)e^{-i\omega t}\, dt \right| = \left| \int_{|t|\geq T} f(t)e^{-i\omega t}\, dt \right|$$

$$\leq \int_{|t|\geq T} |f(t)|\, dt \ \to\ 0 \qquad \text{for } T \to \infty. \tag{17.15}$$

Hence, for sufficiently large T and arbitrary ω, the integral

$$F_T(\omega) = \int_{-T}^{T} f(t)e^{-i\omega t}\, dt \tag{17.16}$$

will be a good approximation of $F(\omega)$. We can also write (17.16) in the form

$$\int_{-T}^{T} f(t)e^{-i\omega t}\, dt = \int_{-\infty}^{\infty} f(t)p_{2T}(t)e^{-i\omega t}\, dt.$$

The integral in (17.16) represents the spectrum of $f(t)p_{2T}(t)$, where $p_{2T}(t)$ is the rectangular pulse function with duration $2T$ (see (6.10)). We now say that $f(t)$ is *Time window* multiplied by a rectangular *time window* with width $2T$. In the frequency domain this corresponds to a convolution of $F(\omega)$ with

$$\frac{2\sin T\omega}{\omega}.$$

Multiplication by a time window introduces a first error when we determine the spectrum $F(\omega)$ in this way. By (17.15), this error is small if the time window is chosen sufficiently wide.

　　Next we want to calculate the integral in (17.16), after a suitable choice of T, using a sampling of $f(t)$ with $2N$ samples in the interval $[-T, T]$. The sampling period is thus equal to T/N. We put $t_n = nT/N$. Applying the trapezoidal rule (see section 16.1) to the integral in (17.16) leads to

$$F_T(\omega) \approx \frac{T}{2N}\left(f(-T)e^{i\omega T} + f(T)e^{-i\omega T} + 2\sum_{n=1-N}^{N-1} f(t_n)e^{-i\omega t_n} \right). \tag{17.17}$$

In order to use the DFT based on $2N$ points, we now define a periodic discrete-time signal $f_p[n]$ with period $2N$ as follows:

$$f_p[-N] = \tfrac{1}{2}(f(-T) + f(T)),$$

$$f_p[n] = f(t_n) \qquad \text{for } n = -N+1, -N+2, \ldots, N-1. \tag{17.18}$$

Substitution of $\omega = k\pi/T$ in (17.17) gives

$$F_T\left(\frac{k\pi}{T}\right) \approx \frac{T}{N}\sum_{n=-N}^{N-1} f_p[n]e^{-2\pi ikn/2N}. \tag{17.19}$$

Up to the factor T/N, the right-hand side equals the $2N$-point DFT of $f_p[n]$; see definition 16.1. Since $F_T(\omega)$ is an approximation of the spectrum $F(\omega)$, (17.19) shows us how a DFT can be used to approximate the spectrum of a non-periodic continuous-time signal. However, as in the case of the Fourier coefficients, this will not result in a reliable approximation for *all* k. In general one will get good results for $|k| < N/2$, with an error of the order of $(T/N)^2$. We will not go into this any further. We do, however, want to make the following remark. The choice of the width $2T$ of the time window apparently results in a distance π/T between two consecutive frequencies for which the spectrum is being determined. The number of frequencies for which the spectrum can be determined with a specific degree of accuracy is thus related to the number of samples in the chosen time window.

EXAMPLE 17.3

As an example of the use of a DFT to approximate the spectrum of a continuous-time signal, we consider the causal signal $f(t)$ given by $f(t) = e^{-2t}\epsilon(t)$. The spectrum $F(\omega)$ of this signal is known (see table 3):

$$F(\omega) = \frac{1}{2 + i\omega}.$$

We choose a time window with width $2T = 10$. This introduces a first error in the calculation of the spectrum, which can easily be estimated as follows:

$$|F(\omega) - F_5(\omega)| \leq \int_{|t| \geq 5} |f(t)|\, dt = \int_5^\infty e^{-2t}\, dt \approx 2.3 \cdot 10^{-5}.$$

Next we determine the discrete-time signal $f_p[n]$ according to (17.18). Special attention has to be paid to the discontinuity of $f(t)$ at $t = 0$. For $f_p[0]$ we have to choose (see section 16.1.2): $f_p[0] = \frac{1}{2}(f(0+) + f(0-)) = \frac{1}{2}$. We take $N = 128$ and determine the $2N$-point DFT of the signal $f_p[n]$. Multiplying this DFT by T/N then gives an approximation of the spectrum at the frequencies $k\pi/T$. In table 17.1 these approximations are given, together with the exact values accurate to 4 decimal digits.

TABLE 17.1   Approximation using a 256-point DFT and the exact values of the spectrum of a non-periodic signal.

k	*approximation*	*exact*
00	0.5002	0.5000
01	$0.4554 - 0.1429i$	$0.4551 - 0.1430i$
02	$0.3587 - 0.2252i$	$0.3585 - 0.2251i$
03	$0.2651 - 0.2493i$	$0.2648 - 0.2496i$
$\vdots$	$\vdots$	$\vdots$
63	$0.0016 - 0.0200i$	$0.0013 - 0.0252i$
64	$0.0015 - 0.0195i$	$0.0012 - 0.0248i$
65	$0.0015 - 0.0195i$	$0.0012 - 0.0244i$
$\vdots$	$\vdots$	$\vdots$
126	$0.0008 - 0.0002i$	$0.0003 - 0.0126i$
127	$0.0008 - 0.0005i$	$0.0003 - 0.0125i$
128	$0.0008 - 0.0007i$	$0.0003 - 0.0124i$

It is useless to tabulate $F[k]$ for $k > 128$. The signal $f(t)$ is real, so $\overline{F[k]} = F[-k] = F[256 - k]$. As far as the *quality* of the approximation is concerned, we note that the approximation is only satisfactory for relatively small values of k. In practice this is usually sufficient. ◀

The signal $f(t)$ in example 17.3 is causal since $f(t) = 0$ for $t < 0$. Hence,

$$F_T(\omega) = \int_{-T}^{T} f(t)e^{-i\omega t}\, dt = \int_0^{T} f(t)e^{-i\omega t}\, dt.$$

Applying the trapezoidal rule to the integral in the right-hand side and substituting $\omega = 2k\pi/T$ gives

$$F_T\left(\frac{2k\pi}{T}\right) \approx \frac{T}{N}\left(\frac{f(0+) + f(T-)}{2} + \sum_{n=1}^{N-1} f(t_n)e^{-2\pi ink/N}\right). \tag{17.20}$$

From this we conclude that the spectrum at $2k\pi/T$ can be calculated with an N-point DFT. However, the values thus calculated correspond exactly with the values calculated with the $2N$-point DFT using (17.19) with k replaced by $2k$:

$$\begin{aligned}
F_T\left(\frac{2k\pi}{T}\right) &\approx \frac{T}{N}\sum_{n=-N}^{N-1} f_p[n]e^{-2\pi ikn/N} \\
&= \frac{T}{N}\left(f_p[-N] + \sum_{n=0}^{N-1} f_p[n]e^{-2\pi ikn/N}\right).
\end{aligned}$$

Now according to (17.18) we have $f_p[-N] = f(T)/2$ and $f_p[0] = f(0+)/2$ (since $f_p[0] = (f(0+) + f(0-))/2$ and f is causal) and if we substitute this in the preceding formula, then we obtain precisely (17.20).

In this section we used the DFT to calculate the spectrum of a signal $f(t)$, after applying a time window, for a number of specific frequencies. By utilizing an FFT algorithm to compute the DFT, a spectrum can thus be approximated quite efficiently.

17.3.2 *Fast convolution*

According to definition 16.2, the cyclical convolution product of two periodic discrete-time signals $f[n]$ and $g[n]$ with period N is given by

$$(f * g)[n] = \sum_{l=0}^{N-1} f[l]g[n-l] \qquad \text{for } n \in \mathbb{Z}.$$

In order to calculate the convolution product for a specific value of n, it seems that we have to perform $N - 1$ additions and N multiplications; that is, $2N - 1$ elementary operations in total. A straightforward calculation of the convolution product for $n = 0, 1, \ldots, N-1$ would then require $2N^2 - N$ elementary operations. For large N the number of operations is thus of the order N^2. Again, this number of elementary operations can be reduced considerably by using an FFT algorithm. Here the convolution theorem plays an important role. First we calculate the N-point DFT of $f[n]$ and $g[n]$. Let us denote these by $F[k]$ and $G[k]$ respectively. By the convolution theorem 16.2, the N-point DFT of the convolution product is simply the product of $F[k]$ and $G[k]$. By an inverse DFT of $F[k]G[k]$ to the n-domain, we obtain the convolution product. Calculating the convolution product in this way requires fewer elementary operations than the direct method. For if we assume that the calculation of an N-point DFT using an FFT algorithm requires $N(^2\log N)$ elementary operations, then the total number of elementary operations to calculate the convolution product equals $2N(^2\log N) + N + N(^2\log N) = 3N(^2\log N) + N$. Here we assumed that the inverse DFT is also calculated using an FFT algorithm.

Compared to the number of the order N^2, this is still a considerable reduction for large values of N. Calculating a convolution product using the FFT is also called *fast convolution*.

Fast convolution

17.3.3 Fast correlation

An operation on signals that is related to convolution is the so-called *cross-correlation*.

DEFINITION 17.1
Cross-correlation

The cross-correlation *of two periodic discrete-time signals* $f_1[n]$ *and* $f_2[n]$ *with period N is defined by*

$$\rho_{1,2}[n] = \sum_{l=0}^{N-1} \overline{f_1[l]} f_2[n+l].$$ (17.21)

Contrary to the convolution product, the cross-correlation is *not* commutative (verify this yourself).

For the calculation of the cross-correlation $\rho_{1,2}[n]$ we will again use the DFT. Let $f_1[n] \leftrightarrow F_1[k]$ and $f_2[n] \leftrightarrow F_2[k]$. According to the shift property we then have $f_2[n+l] \leftrightarrow e^{2\pi i l k/N} F_2[k]$. Next we use the linearity of the DFT to obtain that

$$\rho_{1,2}[n] \leftrightarrow \sum_{l=0}^{N-1} \overline{f_1[l]} e^{2\pi i l k/N} F_2[k] = \overline{F_1[k]} F_2[k].$$ (17.22)

In order to calculate $\rho_{1,2}[n]$ in practice for large values of N, we see that, just as for the convolution, transforming it to the k-domain can result in a considerable advantage. Check for yourself that the number of elementary operations needed for the calculation of the cross-correlation by means of an FFT algorithm is equal to the number of elementary operations needed for the calculation of the fast convolution.

When $f_1[n] = f_2[n] = f[n]$ in formula (17.21), then we call $\rho_{1,2}[n]$ the *autocorrelation* of the signal $f[n]$. From (17.22) it follows that the N-point DFT of the autocorrelation is given by

Autocorrelation

$$\rho[n] \leftrightarrow \overline{F[k]} F[k] = |F[k]|^2 .$$

Power spectrum

The spectrum of the autocorrelation is called the *power spectrum* of the periodic discrete-time signal $f[n]$.

EXERCISES

17.8   Given is the causal continuous-time signal $f(t) = e^{-2t}\epsilon(t)$. In example 17.3 we used a $2N$-point DFT to approximate the spectrum at the frequencies $k\pi/T$. Afterwards we noted that the spectrum at the frequencies $2k\pi/T$ could be approximated using an N-point DFT. Find out whether the spectrum at the frequencies $(2k+1)\pi/T$ can also be approximated using an N-point DFT.

17.9   Given is a continuous-time signal $f(t)$ with the properties: $f(t) = 0$ for $|t| > T$ for some $T > 0$, $f(t)$ is continuous and odd. For $f(t)$ a sampling $f[n]$ is available with sampling period T/N. Give an efficient way to approximate the spectrum of $f(t)$, assuming that you can use efficient algorithms to calculate DFTs with an arbitrary number of points.

17.10 Calculate the autocorrelation of the periodic discrete-time signal with period N given by $f[n] = \cos(2\pi n/N)$, and use this to determine the power spectrum of $f[n]$.

17.11   Calculate the cross-correlation $\rho_{1,2}$ of $f_1[n] = \delta_N[n] + \delta_N[n-1]$ and $f_2[n] = \delta_N[n] + \delta_N[n+1]$ using the DFT.

17.12   Let the two sequences of complex numbers $a_1, a_2, \ldots, a_N$ and $b_1, b_2, \ldots, b_N$ be given. Indicate how the sum $\sum_{n=1}^{N} a_n b_n$ can be calculated using the DFT.

SUMMARY

An N-point DFT can be computed efficiently if we use a factorization of N into integers. Calculating an $N_1 N_2$-point DFT is then reduced to a calculation of N_1-point and N_2-point DFTs. This can be accomplished using properties of the Nth roots of unity $w_N = e^{2\pi i/N}$. This is because a DFT can be interpreted as the determination of a complex polynomial at the Nth roots of unity. The reduction method of an $N_1 N_2$-point DFT to N_1-point and N_2-point DFTs can be represented in a convenient way by considering the $N_1 N_2$-point DFT as a matrix calculation together with a multiplication by certain factors, called twiddle factors.

Efficient algorithms for the calculation of the DFT are collectively known as Fast Fourier Transform, abbreviated as FFT. Popular versions have the number of points in the corresponding DFT equal to a power of 2. In this situation the algorithm is based on a repeated application of 2-point DFTs. In comparison to a direct calculation of an N-point DFT, which requires a number of complex multiplications and additions of the order N^2, an FFT algorithm with $N = 2^m$ requires a number of multiplications and additions of the order $N(^2 \log N)$. This is a considerable reduction.

Applications of the FFT are in fact applications of the DFT. The Fast Fourier Transform is used, for example, to calculate Fourier coefficients and Fourier integrals. Using the convolution theorem, one can also calculate the cyclical convolution product efficiently, for large values of N, by means of the FFT. Closely related to the cyclical convolution product is the cross-correlation of two periodic discrete-time signals. There is a simple expression of the DFT of a cross-correlation as a product of the DFTs of the signals involved (only a conjugation enters). As a consequence, one can also calculate a cross-correlation efficiently using an FFT. The cross-correlation of a signal with itself gives the autocorrelation. The autocorrelation has the special property that its DFT equals the power spectrum of the signal.

SELFTEST

17.13   Given is a periodic discrete-time signal $f[n]$ with period 3. Let $w = \frac{1}{2}(-1 - i\sqrt{3})$. Show that the three-point DFT $F[k]$ of $f[n]$ is given by $F[k] = f[0] + w^k(f[1] + w^k f[2])$.

17.14   Given is a periodic discrete-time signal $f[n]$ with period $3N$. The N-point DFTs of the signals $f[3n]$, $f[3n+1]$, $f[3n+2]$ are given by $A[k]$, $B[k]$, $C[k]$ respectively. Calculate the $3N$-point DFT of $f[n]$.

17.15   Describe a method to calculate an N-point DFT efficiently for $N = 3^m$.

17.16   Given are the causal discrete-time signals $f[n]$ and $g[n]$ with $f[n] = 0$ and $g[n] = 0$ for $n > N$ and some $N > 0$. Indicate how one can use the DFT to calculate the sums

$$\sum_{l=0}^{N} f[l]g[n-l]$$

for integer n. Is it useful, as far as the number of additions and multiplications is concerned, to use the FFT? Justify your answer.

17.17   Given is a periodic discrete-time signal $f[n]$ with period 4. Moreover, the signal is even and real. Show that for the 4-point DFT one has

$$F[0] = f[0] + 2f[1] + f[2],$$
$$F[1] = f[0] - f[2],$$
$$F[2] = f[0] - 2f[1] + f[2],$$
$$F[3] = F[1].$$

Contents of Chapter 18

The z-transform

Introduction *391*

The z-transform

INTRODUCTION

The discrete Fourier transform, in the first instance intended for *periodic* discrete-time signals, arose in chapter 16 in a natural way in the context of the calculation of the Fourier coefficients of *periodic* signals. Calculating the spectrum of a *non-periodic* signal introduces in a similar way a Fourier transform that can be applied to *non-periodic* discrete-time signals. This transform can also be considered as a special version of the so-called *z-transform*, which will be studied first in the present chapter.

The z-transform assigns a function $F(z)$ to a discrete signal $f[n]$ and is defined in the complex z-plane as the sum of the 'two-sided' series

$$F(z) = \sum_{n=-\infty}^{\infty} f[n]z^{-n}.$$

The z-transform plays an important role in the analysis of discrete systems, which will be examined in the next chapter.

The treatment of the z-transform will follow the same path as the treatment of any of the other transforms in this book. First we present the definition, with appropriate attention being given to convergence problems (section 18.1). We subsequently establish a number of important properties (section 18.2).

However, there is also a difference. We will *not* formulate a fundamental theorem for the z-transform. The reason is that for a proper treatment of such a theorem, showing how a signal can be recovered from its z-transform, we would have to use some advanced theorems from the theory of integration of functions defined in the complex plane. This theory is outside the scope of this book. Reconstructing a signal from a given z-transform will therefore almost always be limited to situations frequently occurring in practice, which means that the z-transform is given as a *rational function* of z. In section 18.3 we will see how the technique of the partial fraction expansion can lead us to the original signal then.

As for almost every other signal-transform, the z-transform gives rise to the definition of a convolution product in the n-domain, and also to a corresponding convolution theorem. This is the subject of section 18.4. We will *not* formulate a Parseval theorem for the z-transform. In relation to the z-transform it is less appropriate and it is better suited for the Fourier transform of discrete signals.

Finally, in section 18.5, the Fourier transform of non-periodic discrete-time signals is treated. This arises from the z-transform by substituting the value $e^{i\omega}$ for z. The variable ω can then be interpreted as a frequency. Treating this transform, we will see that we can use the theory of the Fourier series to our advantage.

LEARNING OBJECTIVES

After studying this chapter it is expected that you
- know the definition of the z-transform and can indicate the region of convergence
- can readily apply some of the most frequently occurring properties of the z-transform
- can calculate the z-transform of some elementary signals
- know the definition of the convolution product for the z-transform and can formulate the corresponding convolution theorem
- know the definition of an absolutely summable signal and of a signal with a finite switch-on time
- can determine an absolutely summable discrete-time signal, or a signal with a finite switch-on time, whose z-transform is a given rational function
- know the definition of the Fourier transform of discrete-time signals
- can relate the Fourier transform to the z-transform and to Fourier series
- can derive the most frequently occurring properties of the Fourier transform from the theory of the Fourier series, and can apply these
- know Parseval's theorem for the Fourier transform and can apply it to the energy-content of a discrete-time signal.

18.1   Definition and convergence of the z-transform

In chapter 16 a Fourier transform has been introduced for periodic discrete-time signals, the so-called discrete Fourier transform, abbreviated as DFT. In this chapter we consider discrete-time signals $f[n]$ that are *non-periodic*. A transformation having some similarity with the DFT is the discrete transform which assigns to the non-periodic signal $f[n]$ the Fourier series

$$\sum_{n=-\infty}^{\infty} f[n]e^{-in\omega} \qquad \text{for } \omega \in \mathbb{R}.$$

The foundation of this transform is the so-called z-transform, which will be treated first. Specifically, if we substitute $z = e^{i\omega}$ in the Fourier series above, then we obtain the series

$$\sum_{n=-\infty}^{\infty} f[n]z^{-n},$$

which can then be considered for general complex z. In definition 18.1 the sum of this series will be called the z-transform of $f[n]$. Of course, this series is only meaningful for those values of z for which the series converges.

DEFINITION 18.1
z-transform

Let $f[n]$ be a discrete-time signal. The z-transform of $f[n]$ is defined by

$$F(z) = \sum_{n=-\infty}^{\infty} f[n]z^{-n}, \qquad (18.1)$$

for those values of z for which the series converges.

We note that the z-transform can be considered as some sort of two-sided power series, having not only positive integer powers of z, but also negative integer powers of z. For such series one has similar convergence properties as for power series. In order to find these properties, we will split the z-transform into two parts, the

Causal part

so-called *causal part* given by the series

$$\sum_{n=0}^{\infty} f[n]z^{-n}, \tag{18.2}$$

Anti-causal part

and the *anti-causal part* given by the series

$$\sum_{n=-\infty}^{-1} f[n]z^{-n}. \tag{18.3}$$

Note that the anti-causal part can be rewritten as

$$\sum_{n=-\infty}^{-1} f[n]z^{-n} = \sum_{n=1}^{\infty} f[-n]z^{n},$$

which transforms it into a power series in z with coefficients $f[-n]$.

Convergence of the z-transform means that the series

$$\sum_{n=-M}^{N} f[n]z^{-n}$$

should converge for $M \to \infty$ and $N \to \infty$, independently from each other. Consequently, the z-transform converges if and only if both the causal and the anti-causal part converge. We denote the sum of the causal part by $F_+(z)$ and the sum of the anti-causal part by $F_-(z)$. Hence, in the case of convergence we have

$$F_+(z) = \sum_{n=0}^{\infty} f[n]z^{-n},$$

$$F_-(z) = \sum_{n=-\infty}^{-1} f[n]z^{-n}, \tag{18.4}$$

$$F(z) = \sum_{n=-\infty}^{\infty} f[n]z^{-n} = F_-(z) + F_+(z).$$

It is thus important to find out for which values of z *both* parts converge. The anti-causal part is a power series in z. If we put $w = 1/z$, then we see that the causal part is a power series in w. Complex power series $\sum_{n=0}^{\infty} a_n z^n$ have a radius of convergence R (see chapter 2) for which we have one of the following three possibilities:

a   when $R = 0$, the power series converges only for $z = 0$;
b   when $R = \infty$, the power series converges absolutely for all complex z;
c   when $R > 0$, the power series converges absolutely for $|z| < R$ and diverges for $|z| > R$.

Now let R_2 be the radius of convergence of the anti-causal part and R_1^{-1} the radius of convergence of the power series $\sum_{n=0}^{\infty} f[n]w^n$. Then the anti-causal part converges absolutely for $|z| < R_2$ and the causal part converges for $|z| > R_1$.

- If $R_1 < R_2$, then we may conclude that the z-transform converges in the ring $R_1 < |z| < R_2$ (see figure 18.1). For $|z| > R_2$ the causal part converges while the anti-causal part diverges, and so the z-transform is divergent. For $|z| < R_1$ the anti-causal part converges while the causal part diverges, and so again the z-transform is divergent. We will call the ring $R_1 < |z| < R_2$ the *region of convergence* of the z-transform. The region of convergence does not necessarily

Region of convergence

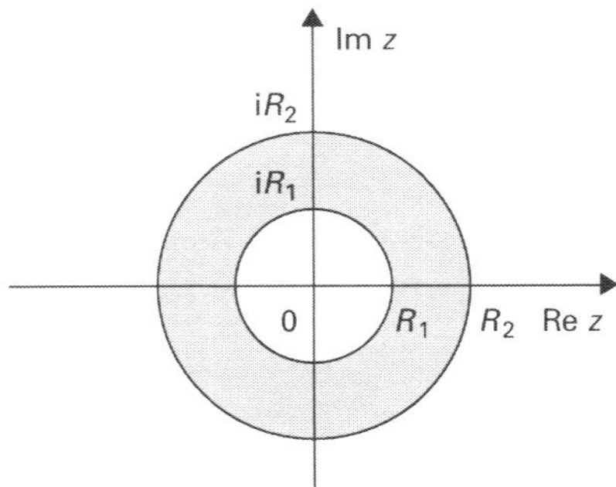

FIGURE 18.1
Region of convergence of the z-transform.

coincide with the set of all z for which the z-transform converges. Specifically, the z-transform may well be convergent at some of the points on the *boundary* of the ring. We will not go into this any further.

- If $R_1 > R_2$, then the z-transform diverges for every complex z and so the region of convergence is empty.
- If $R_1 = 0$ (which means that the causal part converges for every $z \neq 0$), then the region of convergence is the *interior* of the circle with radius R_2, with the exception of $z = 0$. If all terms of the causal part are zero, with the possible exception of the term with $n = 0$, then $z = 0$ also belongs to the region of convergence.
- If $R_2 = \infty$, then the region of convergence is the *exterior* of the circle with radius R_1.

Signals with a finite switch-on time

In practice we are usually dealing with signals that have been switched on at a certain moment in time. Such signals will be called *signals with a finite switch-on time*. For these signals there exists an N such that $f[n] = 0$ for all $n < N$. The anti-causal part then consists of only a finite number of non-zero terms and so this part converges for all z, implying $R_2 = \infty$. The region of convergence is then the exterior of a circle with radius R_1 (see figure 18.2). Examples of this are of course the *causal* signals. The mapping assigning the z-transform $F(z)$ to $f[n]$ is called

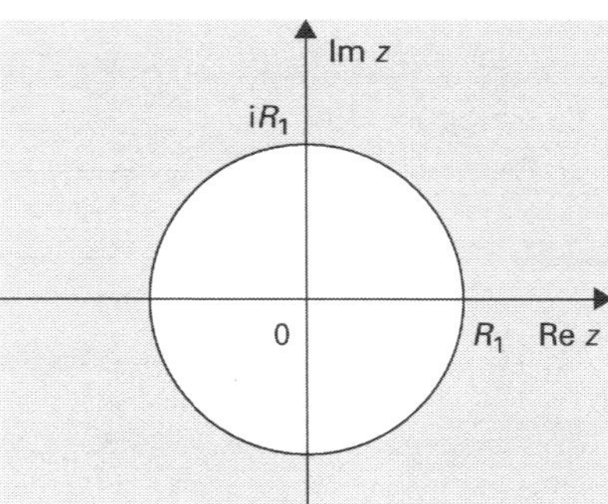

FIGURE 18.2
Region of convergence of a signal with a finite switch-on time.

z-transform

Transform pair

the *z-transform*, and when V is the region of convergence, then we denote this by the *transform pair*

$$f[n] \leftrightarrow F(z) \qquad \text{for } z \in V.$$

z-domain

We will also call $F(z)$ the description of the signal $f[n]$ in the so-called *z-domain*. The z-domain is the complex plane.

EXAMPLE 18.1

Let $f[n]$ be the signal given by

$$f[n] = \begin{cases} 1/n & \text{for } n > 0, \\ 0 & \text{for } n = 0, \\ 2^n & \text{for } n < 0. \end{cases}$$

The anti-causal part is the power series

$$\sum_{n=-\infty}^{-1} 2^n z^{-n} = \sum_{n=1}^{\infty} \left(\frac{z}{2}\right)^n.$$

This is a geometric series which is convergent for $|z| < 2$ and divergent for $|z| \geq 2$. The causal part is the series

$$0 + \sum_{n=1}^{\infty} \frac{z^{-n}}{n} = \sum_{n=1}^{\infty} \frac{1}{n} \left(\frac{1}{z}\right)^n.$$

This series can be considered as a power series in $1/z$, which converges for $|1/z| < 1$ (see chapter 2). Hence, the causal part converges for $|z| > 1$ and diverges for $|z| < 1$. The region of convergence of the z-transform of the signal $f[n]$ is thus the ring $1 < |z| < 2$. ◀

EXAMPLE 18.2

Let $f[n]$ be the causal signal given by $f[n] = a^n \epsilon[n]$, where $a \in \mathbb{C}$, $a \neq 0$. Since $f[n] = 0$ for $n < 0$, all terms of the anti-causal part of the z-transform are zero, and so this part converges for every complex z. Hence, $R_2 = \infty$. The causal part is a geometric series:

$$\sum_{n=0}^{\infty} a^n z^{-n} = \sum_{n=0}^{\infty} \left(\frac{a}{z}\right)^n.$$

This geometric series converges for $|z| > |a|$. The region of convergence is thus the exterior of the circle in the complex plane with radius $|a|$. The sum of the geometric series is equal to $z/(z - a)$. We have now found that

$$a^n \epsilon[n] \leftrightarrow \frac{z}{z - a} \qquad \text{for } |z| > |a|. \tag{18.5}$$

◀

EXAMPLE 18.3

Let $f[n]$ be the discrete-time signal defined by

$$f[n] = \begin{cases} 0 & \text{for } n \geq 0, \\ -a^n & \text{for } n < 0. \end{cases}$$

Here $a \in \mathbb{C}$, $a \neq 0$. All the terms in the causal part of the z-transform are zero, and so the causal part converges for every complex z and has sum 0. The anti-causal part is found as follows:

$$\sum_{n=-\infty}^{-1} \left(-a^n z^{-n}\right) = -\sum_{n=1}^{\infty} \left(\frac{z}{a}\right)^n.$$

Again this is a geometric series with ratio z/a, convergent for $|z| < |a|$ and with sum $z/(z - a)$. The region of convergence is thus the interior of the circle in the complex plane with radius $|a|$. We have now found that

$$f[n] \leftrightarrow \frac{z}{z - a} \qquad \text{for } |z| < |a|.$$

◄

EXAMPLE 18.4

The z-transform of the discrete unit pulse $\delta[n]$ is easy to calculate. A direct substitution gives

$$\sum_{n=-\infty}^{\infty} \delta[n]z^{-n} = 1,$$

and this series converges for every complex z. We thus have

$$\delta[n] \leftrightarrow 1 \qquad \text{for all } z \in \mathbb{C}. \tag{18.6}$$

Apparently, the region of convergence is the whole complex plane.   ◄

Examples 18.2 and 18.3 illustrate how important it is to indicate the region of convergence of the z-transform. We see that in both of the examples the z-transform equals $z/(z - a)$, while the signals $f[n]$ are different. Fortunately, the regions of convergence also differ.

From the above it should be clear that $f[n]$ is *not* uniquely determined when only $F(z)$ is given. In order to determine $f[n]$ uniquely one must also know the *region of convergence*. Determining an inverse z-transform thus requires us to be very careful, and in addition one will also need a mathematical tool that is outside the scope of this book. However, we have seen that for signals with a finite switch-on time, such as causal signals, the region of convergence is the exterior of a circle. In that case we can recover the signal $f[n]$ from the sum $F(z)$ only. We will return to this in section 18.3.

EXERCISES

18.1
Determine the region of convergence of the z-transform of a discrete-time signal $f[n]$ having only finitely many values $f[n]$ unequal to zero.

18.2
Sketch the region of convergence of the z-transform of the signal $f[n]$ given by $f[n] = (2^{-n} + 3^{-n})\epsilon[n]$.

18.3
Sketch the region of convergence of the z-transform of the following signals:
a $f[n] = (2^n + 3^n)\epsilon[-n]$,
b $f[n] = \cos(n\pi/2)\epsilon[n]$,
c $f[n] = \cos(n\pi/2)\epsilon[n] + (2^n + 3^n)\epsilon[-n]$.

18.2   Properties of the z-transform

The properties or rules that are treated here will use the transform pairs $f[n] \leftrightarrow F(z)$ and $g[n] \leftrightarrow G(z)$. An overview of the properties can be found in table 14 at the back of this book.

18.2.1   *Linearity*

Linearity

From definition 18.1 it follows immediately that the z-transform is a *linear* transformation. This means that for all complex a and b we have

$$af[n] + bg[n] \leftrightarrow aF(z) + bG(z). \tag{18.7}$$

18.2.2　Time reversal

Time reversal

By *time reversal* we mean the operation in the n-domain whereby n is replaced by $-n$. This has the following consequence for the z-transform:

$$f[-n] \leftrightarrow F\left(\frac{1}{z}\right). \tag{18.8}$$

Prove this property yourself.

18.2.3　Conjugation

The z-transform of $\overline{f[n]}$ can be found using the following direct calculation:

$$\sum_{n=-\infty}^{\infty} \overline{f[n]} z^{-n} = \overline{\sum_{n=-\infty}^{\infty} f[n](\overline{z})^{-n}} = \overline{F(\overline{z})}.$$

Conjugation

This implies the *conjugation* property

$$\overline{f[n]} \leftrightarrow \overline{F(\overline{z})}. \tag{18.9}$$

If the signal $f[n]$ is real, so $f[n] = \overline{f[n]}$, then it follows from (18.9) that $F(z) = \overline{F(\overline{z})}$. Conversely, if $F(z) = \overline{F(\overline{z})}$, then we can show that the signal $f[n]$ has to be real.

If $f[n]$ is a real signal and a is a zero of $F(z)$, then $\overline{a}$ is also a zero of $F(z)$. This follows from $\overline{F(\overline{a})} = F(a) = 0$, so $F(\overline{a}) = 0$. Hence, the zeros lie symmetrically with respect to the real axis (see figure 18.3).

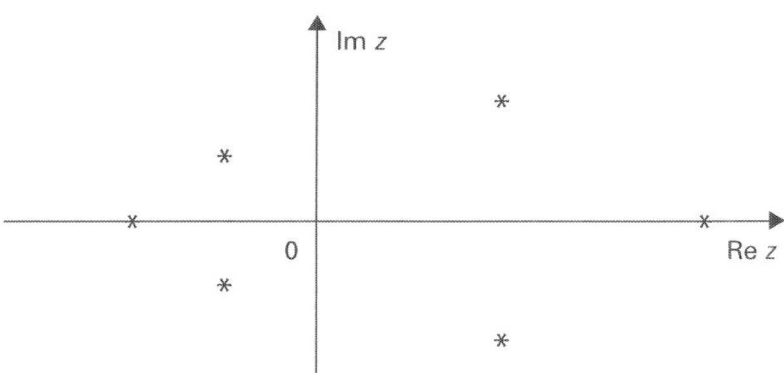

FIGURE 18.3
Zeros of the z-transform of a real signal.

18.2.4　Shift in the n-domain

In section 16.3.5 it was already noted that shifting in the n-domain is only allowed for an integer l. The corresponding operation in the z-domain can be derived as follows:

$$\sum_{n=-\infty}^{\infty} f[n-l] z^{-n} = \sum_{n=-\infty}^{\infty} f[n] z^{-(n+l)} = z^{-l} F(z).$$

Shift in the n-domain

From this we obtain the *shift* property of the z-transform:

$$f[n-l] \leftrightarrow z^{-l} F(z). \tag{18.10}$$

EXAMPLE 18.5

Using the discrete unit pulse, we can describe a discrete-time signal $f[n]$ having $f[n] \neq 0$ for only finitely many values of n by the finite sum

$$f[n] = \sum_{l=-N}^{N} f[l]\delta[n-l].$$

Here N is chosen such that $f[n] = 0$ for $|n| > N$. Since $\delta[n] \leftrightarrow 1$, applying the shift property (18.10) and the linearity (18.7) gives

$$F(z) = \sum_{l=-N}^{N} f[l]z^{-l}.$$

What shows up here is nothing else but definition 18.1 of the z-transform of the given signal.   ◄

18.2.5   Scaling in the z-domain

Instead of a shift in the z-domain, we now consider a scaling in the z-domain. This is because a shift in the z-domain will lead to a complicated and not very useful operation in the n-domain.

Scaling in the z-domain means that we replace z by z/a, where $a \in \mathbb{C}, a \neq 0$. We have

$$F\left(\frac{z}{a}\right) = \sum_{n=-\infty}^{\infty} f[n]\left(\frac{z}{a}\right)^{-n} = \sum_{n=-\infty}^{\infty} a^n f[n]z^{-n}.$$

Scaling in the z-domain

From this, the following rule for *scaling* in the z-domain immediately follows:

$$a^n f[n] \leftrightarrow F\left(\frac{z}{a}\right). \tag{18.11}$$

18.2.6   Differentiation in the z-domain

Complex power series have similar properties to real power series. One of these concerns the termwise differentiation of a power series (see chapter 2). We formulate this as follows. When the complex power series $\sum_{n=0}^{\infty} a_n z^n$ has radius of convergence R and when $S(z)$ is the sum for $|z| < R$, then $S(z)$ is differentiable for $|z| < R$ (see chapter 11 for differentiability of complex functions) and, moreover,

$$\frac{d}{dz}S(z) = \sum_{n=1}^{\infty} na_n z^{n-1} \qquad \text{for } |z| < R. \tag{18.12}$$

We will not prove this property. This property can be extended in an obvious way to the two-sided power series $\sum_{n=-\infty}^{\infty} a_n z^n$ that we encounter in the z-transform. Termwise differentiation is then allowed in the region of convergence, which is mostly a ring in the complex plane. Term-by-term differentiation gives

$$\frac{d}{dz}\sum_{n=-\infty}^{\infty} f[n]z^{-n} = -\sum_{n=-\infty}^{\infty} nf[n]z^{-n-1}.$$

Differentiation in the z-domain

From this we obtain the *differentiation* rule for the z-transform:

$$nf[n] \leftrightarrow -z\frac{d}{dz}F(z). \tag{18.13}$$

Calculating the z-transform of a signal will in general consist of applying one or more of the above properties, and using a number of frequently occurring transform pairs, some of which have been included in table 13. In the following example we will derive two such important transform pairs.

EXAMPLE 18.6

The z-transform of the discrete step signal $\epsilon[n]$ equals $F(z) = z/(z-1)$, with the convergence region being $|z| > 1$ (see example 18.2 with $a = 1$). From the differentiation rule it follows that

$$n\epsilon[n] \leftrightarrow -z\frac{d}{dz}\frac{z}{z-1} = \frac{z}{(z-1)^2} \qquad \text{for } |z| > 1.$$

Applying the shift property gives $(n-1)\epsilon[n-1] \leftrightarrow 1/(z-1)^2$. Again applying the differentiation rule leads to

$$n(n-1)\epsilon[n-1] \leftrightarrow -z\frac{d}{dz}\frac{1}{(z-1)^2} = \frac{2z}{(z-1)^3} \qquad \text{for } |z| > 1.$$

Note that $n(n-1)\epsilon[n-1] = n(n-1)\epsilon[n]$. This process can be repeated over and over again, say k times, which eventually results in the following transform pair:

$$n(n-1)\cdot\ldots\cdot(n-k+1)\epsilon[n] \leftrightarrow k!\frac{z}{(z-1)^{k+1}} \qquad \text{for } |z| > 1.$$

Using the binomial coefficients we can also express this transform pair as follows:

$$\binom{n}{k}\epsilon[n] \leftrightarrow \frac{z}{(z-1)^{k+1}} \qquad \text{for } |z| > 1.$$

The binomial coefficients are defined, as usual, by

$$\binom{n}{0} = 1, \qquad \binom{n}{k} = \frac{n(n-1)\cdot\ldots\cdot(n-k+1)}{k!} \qquad \text{for } k = 1, 2, \ldots.$$

Note that $\binom{n}{k} = 0$ for $k > n$. Applying the scaling property (18.11) we finally obtain for $a \neq 0$ and $k = 0, 1, 2, \ldots$ the transform pair

$$\binom{n}{k}a^n\epsilon[n] \leftrightarrow \frac{a^k z}{(z-a)^{k+1}} \qquad \text{for } |z| > |a|. \tag{18.14}$$

We can subsequently use the time reversal rule (18.8) to derive for $a \neq 0$ and $k = 0, 1, 2, \ldots$ that

$$\binom{n}{k}a^n\epsilon[-n-1] \leftrightarrow \frac{-a^k z}{(z-a)^{k+1}} \qquad \text{for } |z| < |a|. \tag{18.15}$$

In exercise 18.6 you are asked to derive this result. ◀

EXERCISES

18.4

Determine the z-transform $F(z)$ for each of the signals $f[n]$ given below, and determine the region of convergence as well:

a $f[n] = 2n\epsilon[n-2]$,
b $f[n] = 2n\epsilon[-n-2]$,
c $f[n] = (-1)^n\epsilon[-n]$,
d $f[n] = \epsilon[4-n]$,
e $f[n] = (n^2 + n4^n)\epsilon[n]$.

18.5   Determine the z-transform $F(z)$ for each of the following signals:
a $f[n] = \cos(n\pi/2)\epsilon[n]$,
b $f[n] = \sin(n\pi/2)\epsilon[n]$,
c $f[n] = e^{in\phi}\epsilon[n] + 2^n e^{in\phi}\epsilon[-n]$, for a given ϕ.

18.6   Derive formula (18.15).

18.3 The inverse z-transform of rational functions

In the introduction we remarked that we will not present an inverse of the z-transform. This does not mean that an inverse z-transform does not exist. In its most general form, the inverse z-transform requires a rather advanced mathematical tool. However, in practice one often encounters z-transforms that are *rational functions* of z. We will therefore confine ourselves to these functions when determining an inverse. A rational function $F(z)$ can be written in the form

$$F(z) = \frac{a_n z^n + a_{n-1}z^{n-1} + \cdots + a_1 z + a_0}{b_m z^m + b_{m-1}z^{m-1} + \cdots + b_1 z + b_0}. \tag{18.16}$$

The numerator and the denominator contain polynomials in the complex variable z. Denote the numerator by $P(z)$ and the denominator by $Q(z)$. We assume that $a_n \neq 0$ and $b_m \neq 0$. Hence, the degree of $P(z)$ is n and the degree of $Q(z)$ is m. Moreover, we will assume that numerator and denominator have no common factors, which means that there is no polynomial $D(z)$ with a degree greater than or equal to 1 which is a divisor of both $P(z)$ and of $Q(z)$. Now the problem is to find a discrete-time signal $f[n]$ with $f[n] \leftrightarrow F(z)$. We know that if the region of convergence is not indicated, the signal $f[n]$ is not uniquely determined. The distinct regions of convergence playing a role here can be derived from the location
Pole   of the zeros of $Q(z)$, which are called the *poles* of $F(z)$. In chapter 2 we noted that every polynomial in the complex plane can be completely factorized into linear factors, so

$$Q(z) = c(z - z_1)^{\nu_1}(z - z_2)^{\nu_2} \cdot \ldots \cdot (z - z_l)^{\nu_l}, \tag{18.17}$$

where $c \in \mathbb{C}$, $\nu_1, \nu_2, \ldots, \nu_l$ are integers greater than or equal to 1, and $z_1, z_2, \ldots,$
Order of a pole z_l are the distinct zeros of $Q(z)$. The point z_j is then called a pole of $F(z)$ of *order* ν_j. We now state without proof that the distinct regions of convergence that can play a role in the inverse transform of $F(z)$ are the rings bounded by the circles in the complex plane having radius $|z_j|$ with $j = 1, 2, \ldots, l$ (see figure 18.4), including the exterior of the largest circle and the interior of the smallest circle. For each ring another inverse transform can be obtained. This means that a signal $f[n]$ can be found whose z-transform is equal to the given $F(z)$ and whose region of convergence is equal to the given ring.

EXAMPLE 18.7   Let $F(z) = z/(z-1)$. There is only one pole at $z = 1$ of order 1. One can distinguish two regions of convergence, namely $|z| > 1$ and $|z| < 1$. From example 18.2 it follows that the signal $f[n] = \epsilon[n]$ corresponds to the region of convergence $|z| > 1$; from example 18.3 it follows that the signal $f[n] = -\epsilon[-n - 1]$ corresponds to the region of convergence $|z| < 1$.   ◀

In practice one usually works with signals having a finite switch-on time. For such signals the region of convergence of the z-transform is always the exterior of a circle in the complex plane, as we have already noted in section 18.1. In the theory of discrete systems we will encounter signals $h[n]$ whose z-transform is again a rational function and, moreover, having the property that the circle $|z| = 1$ belongs to the

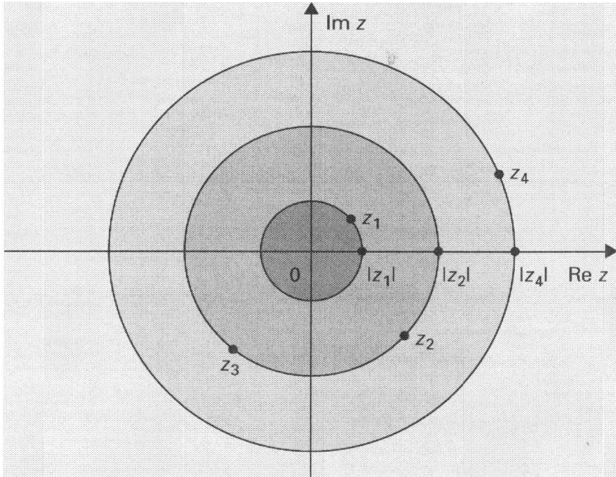

FIGURE 18.4
Regions of convergence of the z-transform.

Absolutely summable

region of convergence. This implies that the series $\sum_{n=-\infty}^{\infty} h[n]z^{-n}$ is absolutely convergent for $|z| = 1$. We will call signals $h[n]$ having this property *absolutely summable*. Hence, a signal $f[n]$ is absolutely summable if

$$\sum_{n=-\infty}^{\infty} |f[n]| < \infty. \tag{18.18}$$

Determining the inverse z-transform will be limited to absolutely summable signals and to signals with a finite switch-on time. We will use some examples to show how these signals can be obtained by using a partial fraction expansion in combination with a frequent application of (18.14) and (18.15).

EXAMPLE 18.8

Let the function $F(z)$ be given by

$$F(z) = \frac{z^3}{z^2 - 1}.$$

Motivated by (18.14) and (18.15), both containing a factor z in the numerator, we do *not* expand $F(z)$ in partial fractions, but instead $F(z)/z$:

$$\frac{F(z)}{z} = \frac{z^2}{z^2 - 1} = \frac{z^2}{(z - 1)(z + 1)}.$$

We cannot start with a partial fraction expansion immediately, since the degree of the numerator (2 in this case) is not smaller than the degree of the denominator (again 2 in this case). However, it is easy to see that $z^2 = 1 \cdot (z^2 - 1) + 1$. Hence,

$$\frac{z^2}{z^2 - 1} = 1 + \frac{1}{z^2 - 1}.$$

A partial fraction expansion of $1/(z^2 - 1)$ gives

$$\frac{1}{z^2 - 1} = \frac{1}{2}\left(\frac{1}{z - 1} - \frac{1}{z + 1}\right).$$

We thus obtain that

$$\frac{F(z)}{z} = 1 + \frac{1}{2}\left(\frac{1}{z-1} - \frac{1}{z+1}\right), \quad \text{or} \quad F(z) = z + \frac{1}{2}\left(\frac{z}{z-1} - \frac{z}{z+1}\right).$$

The poles of $F(z)$ are located at $z = 1$ and $z = -1$. There is thus no region of convergence containing the unit circle. The inverse z-transform of $F(z)$ corresponding to the region of convergence $|z| > 1$ results in the signal $f[n]$ with a finite switch-on time. We can now determine the signal $f[n]$ by applying the inverse transform to each of the terms in the above expression for $F(z)$, taking into account the region of convergence. The term z is easy: it arises from the discrete signal $\delta[n + 1]$; use example 18.4 and apply a shift in the n-domain. Since the region of convergence is $|z| > 1$, we use formula (18.14) for the remaining terms. This results in $f[n] = \delta[n + 1] + \frac{1}{2}\epsilon[n] - \frac{1}{2}(-1)^n\epsilon[n].$ ◀

EXAMPLE 18.9

Let the function $F(z)$ be given by

$$F(z) = \frac{z}{(z - \frac{1}{2})(z - 2)}.$$

A partial fraction expansion of $F(z)/z$ leads to

$$F(z) = \frac{2}{3}\left(\frac{z}{z-2} - \frac{z}{z - \frac{1}{2}}\right).$$

The rational function $F(z)$ has poles at $z = 2$ and $z = \frac{1}{2}$. First, we are interested in the absolutely summable signal $f[n]$ having z-transform $F(z)$. This means that the region of convergence has to contain the unit circle. We should keep this in mind when determining the inverse transform of the terms of $F(z)$. We therefore use (18.14) for the inverse transform of the term $z/(z - \frac{1}{2})$ and (18.15) for the term $z/(z - 2)$. This guarantees that the unit circle belongs to the region of convergence. For we then have

$$2^{-n}\epsilon[n] \leftrightarrow \frac{z}{z - \frac{1}{2}} \qquad \text{for } |z| > \frac{1}{2},$$

$$2^n\epsilon[-n - 1] \leftrightarrow -\frac{z}{z - 2} \qquad \text{for } |z| < 2.$$

The absolutely summable signal is therefore given by

$$f[n] = \frac{1}{3}(-2^{-n+1}\epsilon[n] - 2^{n+1}\epsilon[-n - 1]).$$

When, on the other hand, we are interested in the signal with a finite switch-on time having z-transform $F(z)$, then the region of convergence is the exterior of the circle $|z| = 2$. In this case the inverse transform of the term $z/(z-2)$ has to be determined using (18.14) as well. It then follows that

$$2^n\epsilon[n] \leftrightarrow \frac{z}{z - 2} \qquad \text{for } |z| > 2.$$

Hence, we now obtain that

$$f[n] = \frac{1}{3}(2^{n+1} - 2^{-n+1})\epsilon[n].$$

◀

EXAMPLE 18.10

Let the function $F(z)$ be given by

$$F(z) = \frac{1}{z(z - 2)^2}.$$

We calculate the *causal* signal $f[n]$ whose z-transform is the given function $F(z)$. The poles of $F(z)$ are at $z = 0$ (order 1) and at $z = 2$ (order 2). A partial fraction expansion of $F(z)/z$ gives

$$F(z) = \frac{1}{4}\left(1 + \frac{1}{z} - \frac{z}{z-2} + \frac{z}{(z-2)^2}\right).$$

Since we are looking for a causal signal, we use (18.14) to determine the inverse transform of the terms in the above expression for $F(z)$, resulting in

$$f[n] = \tfrac{1}{4}(\delta[n] + \delta[n-1] - 2^n \epsilon[n] + n2^{n-1}\epsilon[n]).$$

◄

We close this section with an example of a denominator having complex zeros.

EXAMPLE 18.11

Let the function $F(z)$ be given by

$$F(z) = \frac{z^4 + z^3 - 2z + 1}{z^2 + 2z + 2}.$$

The following steps will lead to the signal $f[n]$ with a finite switch-on time whose z-transform equals the given $F(z)$. Motivated by (18.14) and (18.15), both having a factor z in the numerator, we divide $F(z)$ by z, as in examples 18.8 – 18.10:

$$\frac{F(z)}{z} = \frac{z^4 + z^3 - 2z + 1}{z(z^2 + 2z + 2)}.$$

Since the degree of the numerator is greater than or equal to the degree of the denominator, we use a division to write the numerator in the form $z^4 + z^3 - 2z + 1 = (z-1)(z^3 + 2z^2 + 2z) + 1$ and hence

$$\frac{F(z)}{z} = z - 1 + \frac{1}{z(z^2 + 2z + 2)}.$$

Factorizing the denominator gives the poles of $F(z)$: $z(z^2 + 2z + 2) = z(z + 1 + i)(z + 1 - i)$. Next we apply a partial fraction expansion:

$$\frac{1}{z(z^2 + 2z + 2)} = \frac{A}{z} + \frac{B}{z+1+i} + \frac{C}{z+1-i}.$$

In order to determine the constants A, B and C we note that

$$1 = A(z^2 + 2z + 2) + Bz(z + 1 - i) + Cz(z + 1 + i).$$

Successive substitution of the poles $z = 0$, $z = -1 - i$, $z = -1 + i$ gives $A = 1/2$, $B = -(1 + i)/4$, $C = -(1 - i)/4$. Hence,

$$F(z) = z^2 - z + \frac{1}{2} - \frac{(1+i)z/4}{z+1+i} - \frac{(1-i)z/4}{z+1-i}.$$

The signal $f[n]$ with a finite switch-on time can now be obtained using (18.14) again, which results in

$$\begin{aligned}
f[n] &= \delta[n+2] - \delta[n+1] + \tfrac{1}{2}\delta[n] \\
&\quad - \tfrac{1}{4}(-1)^n(1+i)^{n+1}\epsilon[n] - \tfrac{1}{4}(-1)^n(1-i)^{n+1}\epsilon[n] \\
&= \delta[n+2] - \delta[n+1] + \tfrac{1}{2}\delta[n] - \tfrac{1}{4}(-1)^n((1+i)^{n+1} + (1-i)^{n+1})\epsilon[n].
\end{aligned}$$

Note that the rational function $F(z)$ has <u>real</u> coefficients. From this it follows that $F(z)$ satisfies the relation $F(z) = \overline{F(\bar{z})}$ (verify this yourself). In section 18.2.3 we have noted that $f[n]$ has to be a real signal then, although this is not immediately clear from the previous expression for $f[n]$. However, if we write $1 \pm i = \sqrt{2}e^{\pm \pi i/4}$, then it follows from

$$
\begin{aligned}
(1+i)^{n+1} + (1-i)^{n+1} &= \left(\sqrt{2}\right)^{n+1} \left(e^{i(n+1)\pi/4} + e^{-i(n+1)\pi/4}\right) \\
&= \left(\sqrt{2}\right)^{n+3} \cos \tfrac{1}{4}(n+1)\pi
\end{aligned}
$$

that

$$
f[n] = \delta[n+2] - \delta[n+1] + \tfrac{1}{2}\delta[n] + \left(\left(-\sqrt{2}\right)^{n-1} \cos \tfrac{1}{4}(n+1)\pi\right) \epsilon[n].
$$

◀

EXERCISES

18.7   Determine the real and absolutely summable signal $f[n]$ whose z-transform is given in example 18.10.

18.8   Determine the signal $f[n]$ with a finite switch-on time whose z-transform $F(z)$ is given by $F(z) = 1/(z^2 + 4)$.

18.9   Determine the absolutely summable signal $f[n]$ whose z-transform is given by $F(z) = 1/(z^2 + 4)$.

18.10   Determine the signal $f[n]$ with a finite switch-on time whose z-transform is given by $F(z) = z^3/((z + \tfrac{1}{2})(z + 3))$.

18.11   Determine the absolutely summable signal $f[n]$ whose z-transform is given by $F(z) = z^3/((z + \tfrac{1}{2})(z + 3))$.

18.4   Convolution

For each transform considered up to now, we were able to define, for two signals in the time domain, a convolution product whose transform was equal to the ordinary product of the transforms of the signals involved. This can also be done for the z-transform. Let us assume that for two discrete signals $f[n]$ and $g[n]$ the z-transforms are given by, respectively, $F(z)$ and $G(z)$. We then want to find a signal $h[n]$ in the n-domain whose z-transform is equal to the ordinary product $F(z)G(z)$. We are thus trying to find a discrete-time signal $h[n]$ such that

$$
\sum_{n=-\infty}^{\infty} h[n]z^{-n} = \sum_{l=-\infty}^{\infty} f[l]z^{-l} \sum_{k=-\infty}^{\infty} g[k]z^{-k}.
$$

In the right-hand side of this expression we have a product of two infinite series. In order to find an expression for $h[n]$, we will expand this product in a formal way and then sort the terms with respect to equal powers of z. By 'formal' we mean that we will not worry about possible convergence issues at the moment. The coefficient $h[n]$ of z^{-n} is then equal to an infinite sum:

$$
h[n] = \sum_{l=-\infty}^{\infty} f[l]g[n-l].
$$

The series in the right-hand side arises by multiplying the coefficient of z^{-l} in the series for $F(z)$ by the coefficient of z^{-n+l} in the series for $G(z)$ and then summing this over all l. It is now clear how we should define the convolution product of two discrete-time signals for the z-transform.

DEFINITION 18.2
*Convolution product of
discrete-time signals*

For two discrete-time signals $f[n]$ and $g[n]$ the convolution product $(f * g)[n]$ *is
defined by*

$$(f * g)[n] = \sum_{l=-\infty}^{\infty} f[l]g[n - l]. \tag{18.19}$$

In exercise 18.16 you will be asked to show that, similar to any of the other convolution products, the convolution product of discrete-time signals is commutative.

The expression in (18.19) is an infinite sum and so special attention should be given to convergence. We will not go into this in much detail here, and confine ourselves to the situation where the regions of convergence of the z-transforms of $f[n]$ and $g[n]$ are both non-empty. The product $F(z)G(z)$ is then certainly defined on the intersection of these two regions of convergence, and therefore we will subsequently assume that the intersection is also non-empty. In this situation one can show that the series in (18.19) even converges absolutely. The proof of this, as well as the proof of the following convolution theorem, will be omitted.

THEOREM 18.1
Convolution theorem

Let $f[n]$ and $g[n]$ be discrete-time signals having z-transforms $F(z)$ and $G(z)$ respectively, and let V be the intersection of the regions of convergence of these z-transforms. Then

$$(f * g)[n] \leftrightarrow F(z)G(z) \qquad for\ z \in V. \tag{18.20}$$

EXAMPLE 18.12

Let the discrete-time signals $f[n]$ and $g[n]$ be given by $f[n] = 2^{-n}\epsilon[n]$ and $g[n] = \epsilon[-n]$. The z-transform of $f[n]$ follows immediately from (18.14): $F(z) = z/(z - \frac{1}{2})$ with region of convergence $|z| > \frac{1}{2}$. The z-transform of $g[n]$ follows from (18.15): $G(z) = 1/(1 - z)$ with region of convergence $|z| < 1$. The intersection of these two regions of convergence is the ring $\frac{1}{2} < |z| < 1$. This region is non-empty and the convolution product is thus well-defined:

$$(f * g)[n] = \sum_{l=-\infty}^{\infty} 2^{-l}\epsilon[l]\epsilon[l - n] = \sum_{l=0}^{\infty} 2^{-l}\epsilon[l - n].$$

In this case one can determine the convolution product in a direct way. For $n < 0$ it equals $\sum_{l=0}^{\infty} 2^{-l} = 2$ and for $n \geq 0$ it equals $\sum_{l=n}^{\infty} 2^{-l} = 2^{1-n}$. Using $\epsilon[n]$ this can be written as

$$(f * g)[n] = 2\epsilon[-1 - n] + 2^{1-n}\epsilon[n].$$

One can also determine the convolution product using the convolution theorem. According to the convolution theorem, the z-transform of $(f * g)[n]$ equals

$$F(z)G(z) = \frac{z}{(z - \frac{1}{2})(1 - z)}.$$

By transforming this back to the n-domain one can obtain the convolution product as well. A partial fraction expansion gives

$$\frac{z}{(z - \frac{1}{2})(1 - z)} = \frac{2z}{z - \frac{1}{2}} + \frac{2z}{1 - z}.$$

In order to write the two terms in the right-hand side as sums of power series, we have to take into account that the ring $\frac{1}{2} < |z| < 1$ should be the region of convergence for these series. Both terms can then be written as a sum of a geometric series, converging in the indicated ring, as follows:

$$\frac{2z}{z - \frac{1}{2}} = \frac{2}{1 - (2z)^{-1}} = \sum_{n=0}^{\infty} 2^{1-n} z^{-n} \qquad \text{for } |z| > \frac{1}{2},$$

$$\frac{2z}{1 - z} = \sum_{n=0}^{\infty} 2z^{n+1} = \sum_{n=-\infty}^{-1} 2z^{-n} \qquad \text{for } |z| < 1.$$

Hence, for $\frac{1}{2} < |z| < 1$ we have

$$F(z)G(z) = \sum_{n=-\infty}^{-1} 2z^{-n} + \sum_{n=0}^{\infty} 2^{1-n} z^{-n}.$$

From this one immediately obtains the previously found expression for the convolution product. ◄

In practice we are often dealing with signals in the n-domain having a finite switch-on time. In the previous section we have seen that the region of convergence of the z-transform of such signals is the exterior of a circle. The convolution product of these kinds of signals is again such a signal, and so the region of convergence of the z-transform is then again the exterior of a circle. This can be shown as follows. When $f[n]$ and $g[n]$ are discrete signals such that $f[n] = g[n] = 0$ for $n < N$ and some N (N may also be negative), then the convolution product can be written as

$$(f * g)[n] = \sum_{l=-\infty}^{\infty} f[l]g[n - l] = \sum_{l=N}^{\infty} f[l]g[n - l].$$

If $n < 2N$, then $n - l < 2N - N = N$ for $l \geq N$ and so all terms in the above series are 0. The convolution product is thus a signal which is zero for $n < 2N$. Moreover, the series in the convolution product contains only a finite number of terms unequal to zero. This is because if $l > n - N$ we have $g[n - l] = 0$. We may thus conclude that

$$(f * g)[n] = \left(\sum_{l=N}^{n-N} f[l]g[n - l] \right) \epsilon[n - 2N].$$

In particular, if $f[n]$ and $g[n]$ are causal signals (which corresponds to $N = 0$), then the convolution product is causal as well and

$$(f * g)[n] = \left(\sum_{l=0}^{n} f[l]g[n - l] \right) \epsilon[n].$$

EXAMPLE 18.13

Let the causal signals $f[n]$ and $g[n]$ be given by $f[n] = 2^{-n}\epsilon[n]$ and $g[n] = 3^{-n}\epsilon[n]$, then

$$(f * g)[n] = \left(\sum_{l=0}^{n} 2^{-l} 3^{-(n-l)} \right) \epsilon[n] = 3^{-n} \left(\sum_{l=0}^{n} \left(\frac{3}{2} \right)^{l} \right) \epsilon[n].$$

Hence,

$$(f * g)[n] = (3 \cdot 2^{-n} - 2 \cdot 3^{-n})\epsilon[n].$$

The convolution product can also be found in a different way. By the convolution theorem, the z-transform of the convolution product equals $F(z)G(z)$, where

$F(z) = z/(z - \frac{1}{2})$ and $G(z) = z/(z - \frac{1}{3})$. Hence (after a partial fraction expansion),

$$(f * g)[n] \leftrightarrow \frac{3z}{z - \frac{1}{2}} - \frac{2z}{z - \frac{1}{3}}.$$

An inverse transform gives us the same convolution product as before. ◀

EXERCISES

18.12   We call N the switch-on time of a signal $f[n]$ if $f[n] = 0$ for $n < N$ and $f[N] \neq 0$, and we call M the switch-off time of $f[n]$ if $f[n] = 0$ for $n > M$ and $f[M] \neq 0$. Let $f[n]$ and $g[n]$ be discrete-time signals with switch-on times N_1 and N_2 respectively, and switch-off times M_1 and M_2 respectively. Give the switch-on and switch-off time of the convolution product of $f[n]$ and $g[n]$.

18.13   **a**   Calculate the causal signal $f[n]$ whose z-transform is given by $F(z) = z/(z^2 + 1)$.
b   Use the convolution theorem to calculate the causal signal $h[n]$ whose z-transform is given by $H(z) = z^2/(z^2 + 1)^2$.

18.14   Show that $(f * \epsilon)[n] = \sum_{l=-\infty}^{n} f[l]$.

18.15   Use the convolution theorem to show that $(f * \delta)[n] = f[n]$.

18.16   Show that the convolution product (18.19) is commutative.

18.5   Fourier transform of non-periodic discrete-time signals

In this section we will meet a Fourier transform for non-periodic discrete-time signals. In section 18.1 we have already noted that for a discrete-time signal $f[n]$ the infinite series

$$\sum_{n=-\infty}^{\infty} f[n]e^{-in\omega} \tag{18.21}$$

could serve as the Fourier transform of a non-periodic discrete-time signal $f[n]$. In this section we will see that this is indeed the case. To start with, we note that if $F(z)$ is the z-transform of $f[n]$, then the above series arises by substituting for z the complex number $z = e^{i\omega}$, which lies on the complex unit circle, into the series for the z-transform. Hence,

$$F(e^{i\omega}) = \sum_{n=-\infty}^{\infty} f[n]e^{-in\omega}.$$

An expression like this is well-known to us. In the right-hand side we have a complex Fourier series whose sum equals the function $F(e^{i\omega})$. This function is periodic with period 2π. It is then possible to recover $f[n]$ from $F(e^{i\omega})$ using the well-known formula (3.9) for the Fourier coefficients. Applying this formula gives (note the sign!)

$$f[n] = \frac{1}{2\pi} \int_{-\pi}^{\pi} F(e^{i\omega})e^{in\omega} \, d\omega. \tag{18.22}$$

We can now also interpret (18.22) by comparing it with the inverse Fourier transform of a continuous-time signal $f(t)$:

$$f(t) = \frac{1}{2\pi} \int_{-\infty}^{\infty} F_a(\omega)e^{i\omega t} \, d\omega.$$

Here $F_a(\omega)$ is the spectrum of $f(t)$. The signal $f(t)$ is written here as a (continuous) superposition of the continuous-time harmonic signals $e^{i\omega t}$ with frequency ω. Formula (18.22) will now be interpreted in the same way. The *discrete* signal $f[n]$ is written here as a (continuous) superposition of the *discrete-time* harmonic signals $e^{in\omega}$ with frequency ω. Analogous to the continuous-time situation, it is now quite natural to call $F(e^{i\omega})$ the *Fourier transform* or *spectrum* of the signal $f[n]$.

DEFINITION 18.3
Fourier transform or spectrum of a discrete-time signal

Let $f[n]$ be a discrete-time signal. The Fourier transform *or* spectrum *of $f[n]$ is the function $F(e^{i\omega})$ defined by*

$$F(e^{i\omega}) = \sum_{n=-\infty}^{\infty} f[n]e^{-in\omega} \qquad for\ \omega \in \mathbb{R}. \tag{18.23}$$

The transform assigning the spectrum $F(e^{i\omega})$ to the function $f[n]$ is called the *Fourier transform* for discrete signals. This should be distinguished from the DFT, which in first instance was created for *periodic* discrete-time signals.

Because of the strong similarity with the Fourier series and the z-transform, the Fourier transform for discrete-time signals will have a number of analogous properties. Of course, there is still the convergence issue. However, we will confine ourselves to discrete signals $f[n]$ that are absolutely summable, that is, for which $\sum_{n=-\infty}^{\infty} |f[n]| < \infty$. Then the z-transform of $f[n]$ converges absolutely on the unit circle in the complex plane.

We will now list a number of properties of the Fourier transform of discrete signals. We omit the proofs since these properties are directly related to similar properties of Fourier series or of the z-transform. In fact, these properties are simply reformulations of what we already know. In this listing of the properties, we use the transform pairs $f[n] \leftrightarrow F(e^{i\omega})$ and $g[n] \leftrightarrow G(e^{i\omega})$.

Linearity

$$af[n] + bg[n] \leftrightarrow aF(e^{i\omega}) + bG(e^{i\omega}), \tag{18.24}$$

Conjugation

$$\overline{f[n]} \leftrightarrow \overline{F(e^{-i\omega})}, \tag{18.25}$$

Shift in the n-domain

$$f[n-k] \leftrightarrow e^{-ik\omega} F(e^{i\omega}), \tag{18.26}$$

Shift in the ω-domain

$$e^{in\psi} f[n] \leftrightarrow F(e^{i(\omega-\psi)}), \tag{18.27}$$

Convolution in the n-domain

$$(f * g)[n] = \sum_{l=-\infty}^{\infty} f[l]g[n-l] \leftrightarrow F(e^{i\omega})G(e^{i\omega}), \tag{18.28}$$

Convolution in the ω-domain

$$f[n]g[n] \leftrightarrow \frac{1}{2\pi} \int_{-\pi}^{\pi} F(e^{i(\omega-u)})G(e^{iu})\,du, \tag{18.29}$$

Parseval's identity

$$\sum_{n=-\infty}^{\infty} f[n]\overline{g[n]} = \frac{1}{2\pi} \int_{-\pi}^{\pi} F(e^{i\omega})\overline{G(e^{i\omega})}\,d\omega. \tag{18.30}$$

From Parseval's identity it immediately follows that the energy-content of a discrete-time signal (see section 1.2.3) is given by

Energy-content

$$\sum_{n=-\infty}^{\infty} |f[n]|^2 = \frac{1}{2\pi} \int_{-\pi}^{\pi} \left| F(e^{i\omega}) \right|^2 d\omega. \tag{18.31}$$

EXAMPLE 18.14

Let $f[n]$ be the causal signal given by $f[n] = 2^{-n}\epsilon[n]$. The z-transform $F(z)$ equals $F(z) = z/(z - \frac{1}{2})$. The Fourier transform equals $F(e^{i\omega})$. According to the left-hand side of (18.31), the energy-content of $f[n]$ equals

$$\sum_{n=-\infty}^{\infty} |f[n]|^2 = \sum_{n=0}^{\infty} 4^{-n} = \frac{4}{3},$$

and according to the right-hand side it equals

$$\frac{1}{2\pi} \int_{-\pi}^{\pi} \left| F(e^{i\omega}) \right|^2 d\omega = \frac{1}{2\pi} \int_{-\pi}^{\pi} \frac{e^{i\omega}}{(e^{i\omega} - \frac{1}{2})} \frac{e^{-i\omega}}{(e^{-i\omega} - \frac{1}{2})} d\omega$$
$$= \frac{1}{2\pi} \int_{-\pi}^{\pi} \frac{1}{5/4 - \cos\omega} d\omega.$$

We have thus found that

$$\int_{-\pi}^{\pi} \frac{1}{5/4 - \cos\omega} d\omega = \frac{8\pi}{3}.$$

◄

EXERCISES

18.17

Given is a discrete-time signal $f[n]$ with spectrum $F(e^{i\omega})$. Show that

$$\sum_{l=-\infty}^{n} 2^{l-n} f[l] = \frac{1}{\pi} \int_{-\pi}^{\pi} \frac{F(e^{i\omega})}{2e^{i\omega} - 1} e^{in\omega} d\omega.$$

18.18

Show that the spectrum $F(e^{i\omega})$ of a real signal $f[n]$ satisfies $\overline{F(e^{i\omega})} = F(e^{-i\omega})$. Is the converse statement also true?

18.19

Calculate the energy-content of the signal $f[n]$ whose spectrum is given by $F(e^{i\omega}) = \cos\omega$.

18.20

The *autocorrelation* $\rho[n]$ of a discrete-time signal $f[n]$ is defined by

$$\rho[n] = \sum_{l=-\infty}^{\infty} \overline{f[l]} f[n + l].$$

Calculate the spectrum of $\rho[n]$.

18.21

Show that $\delta[n] = \frac{1}{2\pi} \int_{-\pi}^{\pi} e^{in\omega} d\omega$.

SUMMARY

In this chapter the z-transform $F(z) = \sum_{n=-\infty}^{\infty} f[n]z^{-n}$ has been introduced as an operation on non-periodic discrete-time signals. In general, the region of convergence of the z-transform is a ring in the complex plane, whose boundary is determined by the radius of convergence of the anti-causal part and of the causal part.

The properties of the z-transform are very similar to the other transforms that have been treated earlier. A table of properties is included at the back of this book (see table 14). For the reconstruction of a signal from a given z-transform it is important that the region of convergence is known. In practice, mainly signals occur whose z-transform is a rational function of z. If we then know that the signal is absolutely summable, or that it has a finite switch-on time, then we can recover the original signal by means of a partial fraction expansion. The absolutely summable signals will play a role in the study of stable discrete-time signals in chapter 19.

From the z-transform $F(z)$ of a signal $f[n]$ one obtains the Fourier transform of $f[n]$ by substituting $z = e^{i\omega}$. The quantity ω is then called the frequency, and the Fourier transform $F(e^{i\omega})$ the spectrum of $f[n]$. Properties of the Fourier transform follow immediately from the theory of the Fourier series, since the Fourier transform of $f[n]$ represents a Fourier series. Using Parseval's identity for periodic functions one can then easily obtain the energy-content of the signal $f[n]$.

SELFTEST

18.22   A discrete-time signal $f[n]$ is given by $f[n] = 2^{-n}\epsilon[n + 2]$.
    a   Determine the region of convergence of the z-transform $F(z)$ of $f[n]$ and calculate $F(z)$.
    b   Calculate the signal $g[n]$ given by $g[n] = \sum_{l=-\infty}^{n} f[l]$.
    c   Determine the Fourier transform of $f[n]$.

18.23   For a discrete-time signal $f[n]$ the z-transform $F(z)$ is given by

$$F(z) = \frac{z^2}{z + 2}.$$

    a   Calculate $f[n]$ if in addition it is given that $f[n]$ is absolutely summable.
    b   Determine the Fourier transform of $f[n]$.
    c   Calculate the spectrum of $2^n f[n]$.

18.24   For a causal discrete-time signal $f[n]$ the z-transform $F(z)$ is given by

$$F(z) = \frac{1}{z(4z^2 + 1)}.$$

    a   Show that $f[n]$ is absolutely summable and determine the spectrum of $f[n]$.
    b   Calculate $f[n]$.
    c   Is the signal $f[n]$ real? Justify your answer.

18.25   For a discrete-time signal $f[n]$ it is given that

$$f[n] = \sum_{l=0}^{n} g[l]g[n - l].$$

Here $g[n]$ is a real and causal signal whose spectrum is given by

$$G(e^{i\omega}) = \frac{1}{4 + \cos(2\omega)}.$$

Determine the z-transform of $f[n]$.

Contents of Chapter 19

Applications of discrete transforms

Introduction *412*

19.1 The impulse response *413*
19.2 The transfer function and the frequency response *419*
19.3 LTD-systems described by difference equations *424*

Summary *427*

Selftest *428*

Applications of discrete transforms

INTRODUCTION

Applications of discrete transforms can mainly be found in the processing of discrete signals in discrete-time systems. In chapter 1 we have already discussed such systems in general terms. Since we now have certain discrete transforms available, we are able to get a better understanding of the discrete systems. Hence, in the present chapter we we will focus on a further analysis of the discrete-time systems.

In systems theory we distinguish inputs and the corresponding outputs or responses of the system. For discrete-time systems these signals are discrete-time signals. A system can be described by giving the relation that exists between the inputs and the outputs. This can be done in several ways. For example, by describing the relation in the n-domain, or in the z-domain, or, just as important, by describing it in the frequency or ω-domain. In the latter case we have the relationship between the *spectra* of the input and outputs in mind.

Discrete transforms play a special role in linear time-invariant discrete systems, similar to the role played by the Fourier integral in continuous-time systems (see chapter 10). Linear time-invariant systems have already been introduced in chapter 1 (see section 1.3.2). *Discrete-time* systems that are linear and time-invariant will henceforth be called LTD-systems for short. In section 19.1 we will see that for an LTD-system the relationship between an input and the corresponding output can be described in the n-domain by means of a convolution product. The response to the discrete unit pulse, the so-called *impulse response*, is of fundamental importance here. In fact, an LTD-system is completely determined by the impulse response. Hence, all kinds of properties of a system can be derived from the impulse response. Some of these properties, such as stability and causality of systems, will be examined in section 19.1.

From the convolution theorem of the z-transform it will follow that for LTD-systems the relationship in the z-domain between the input and the output is simply a multiplication by the so-called *transfer function*. The behaviour of the system is completely determined by this transfer function. Hence, all kinds of properties of a discrete-time system can then be derived from the transfer function as well. Restricting the transfer function to the unit circle in the complex plane leads for stable systems to the well-known frequency response of a system, which has already been introduced in chapter 1. It shows precisely how the spectrum, or the Fourier transform, of a discrete-time signal is effected by a linear time-invariant system. The transfer function and the frequency response will be treated in section 19.2.

Next we consider in section 19.3 LTD-systems described by *difference equations*. These LTD-systems are of practical importance because precisely these systems can be realized by means of computers. These systems have a rational transfer function. The stability of an LTD-system follows from the location of the poles of this function.

LEARNING OBJECTIVES

After studying this chapter it is expected that you
- know the concepts of impulse response and step response of an LTD-system and can determine these in simple situations
- know and can apply the relationship between an input and the corresponding output using the impulse response
- can use the impulse response to verify the stability and the causality of an LTD-system
- know the importance of the concepts of transfer function and frequency response of an LTD-system
- can use the transfer function to verify the stability and the causality, in particular for a rational transfer function
- know the importance of the transfer function in relation to the response of an LTD-system to the input z^n
- can analyse causal LTD-systems described by a difference equation.

19.1   The impulse response

In this section we introduce the impulse response of an LTD-system and we show its importance in relation to LTD-systems. Therefore we will start with the definition of the impulse response.

DEFINITION 19.1
Impulse response

The impulse response $h[n]$ *of an LTD-system is the response to the discrete unit pulse* $\delta[n]$. *This is denoted by*

$$\delta[n] \mapsto h[n]. \tag{19.1}$$

As a first illustration of the importance of the impulse response, we consider a simple LTD-system, which calculates a weighted average of an input $u[n]$ in the following way:

$$y[n] = \frac{u[n] + 2u[n-1] + u[n-2]}{3} \qquad \text{for } n \in \mathbb{Z}. \tag{19.2}$$

At time n the output $y[n]$ is equal to a weighted average of the previous three values of the input $u[n]$ (a *running average*). It is not difficult to show that this defines an LTD-system (see exercise 19.1). For convenience we repeat the definitions of a linear and time-invariant system here (see definitions 1.1 and 1.2).

LTD-system

A discrete-time system L is called *linear* if for all complex a and b and all inputs $u[n]$ and $v[n]$ one has

$$au[n] + bv[n] \mapsto a(\mathrm{L}u)[n] + b(\mathrm{L}v)[n] \qquad \text{for } n \in \mathbb{Z}.$$

A discrete-time system L is called *time-invariant* if for all integer l and all inputs $u[n]$ one has

$$u[n - l] \mapsto (\mathrm{L}u)[n - l].$$

Time-invariant systems have the property that a shift in time in the input results in the same shift in the output. Apparently, the system consists of components whose operation does not vary in time. Keep in mind, for example, an electric network where, among other things, the capacity of a capacitor, or the inductance of an inductor, does not depend on time.

We now determine the impulse response $h[n]$ of the system defined by (19.2). Substitution of $\delta[n] = u[n]$ in (19.2) gives

$$h[n] = \frac{\delta[n] + 2\delta[n-1] + \delta[n-2]}{3}.$$

We conclude from this that

$$h[0] = \tfrac{1}{3}, \qquad h[1] = \tfrac{2}{3}, \qquad h[2] = \tfrac{1}{3}, \qquad h[n] = 0 \quad \text{for } n \neq 0, 1, 2.$$

Therefore, relation (19.2) can also be written as

$$y[n] = h[0]u[n] + h[1]u[n-1] + h[2]u[n-2] = \sum_{l=-\infty}^{\infty} h[l]u[n-l].$$

Apparently, the output $y[n]$ is nothing else but the convolution product of the signal $h[n]$ and the input $u[n]$. We will now show that this property holds in general for all LTD-systems. In doing so, we tacitly assume, as for LTC-systems (see chapter 10), that LTD-systems satisfy the superposition rule (see section 10.1), that is to say, the linearity property is not only valid for finite series of inputs, but also for convergent infinite series.

Let $u[n]$ be an arbitrary input of an LTD-system. Using the representation (see chapter 15, (15.2))

$$u[n] = \sum_{l=-\infty}^{\infty} u[l]\delta[n-l],$$

the input $u[n]$ has been written as a superposition of shifted discrete unit pulses. The system is time-invariant, so the response to $\delta[n-l]$ is $h[n-l]$. Applying the superposition rule, we find the following relation for the response $y[n]$ to the input $u[n]$:

$$y[n] = \sum_{l=-\infty}^{\infty} u[l]h[n-l].$$

This establishes the following important property of LTD-systems.

For an LTD-system with impulse response $h[n]$, the response $y[n]$ to an input $u[n]$ is given by

$$y[n] = (h * u)[n]. \tag{19.3}$$

If we know the impulse response of an LTD-system, then the above implies that the response to any arbitrary input can be determined by calculating a convolution product. Apparently, an LTD-system is completely characterized by its impulse response.

EXAMPLE 19.1

An LTD-system for which the relationship between an input $u[n]$ and the response $y[n]$ is given by

$$y[n] = u[n-1]$$

Time-delay unit

is called a *time-delay unit*. The impulse response of the time-delay unit thus equals $h[n] = \delta[n-1]$. By (19.3), the relation between $y[n]$ and $u[n]$ can then also be written as

$$y[n] = \sum_{l=-\infty}^{\infty} u[l]\delta[n-l-1].$$

◀

EXAMPLE 19.2

Let $a \in \mathbb{C}$. The LTD-system for which the relationship between an input $u[n]$ and the response $y[n]$ is described by

$$y[n] = au[n]$$

Multiplier

is called a *multiplier*. The impulse response equals $h[n] = a\delta[n]$. By (19.3), $y[n]$ can also be written as

$$y[n] = a \sum_{l=-\infty}^{\infty} u[l]\delta[n - l].$$

◄

The time-delay unit and the multiplier are important components in LTD-systems that can be realized by digital computers. We will not go into this any further in this book. Relation (19.3) between an input $u[n]$ and the response $y[n]$ of an LTD-system is one of the important applications of the convolution product. Here we state without proof that the condition for the existence of the convolution product is satisfied if we confine ourselves to bounded inputs and, moreover, assume that the impulse response is absolutely summable, that is to say (see (18.18)),

$$\sum_{n=-\infty}^{\infty} |h[n]| < \infty.$$

In this situation one can show that the series arising when we expand the convolution product in (19.3) is also absolutely convergent. Moreover, an LTD-system with an absolutely summable impulse response is *stable* (see theorem 19.1 below). Stability of systems has been defined in chapter 1 (definition 1.3) and means that bounded inputs result in bounded outputs. For LTD-systems this property can be verified using the impulse response, as is shown in the following theorem.

THEOREM 19.1
Stability

An LTD-system with impulse response $h[n]$ is stable if and only if

$$\sum_{n=-\infty}^{\infty} |h[n]| < \infty. \tag{19.4}$$

Proof
First assume that $h[n]$ is absolutely summable and that $\sum_{n=-\infty}^{\infty} |h[n]| = I < \infty$. If $u[n]$ is now a bounded input with $|u[n]| \leq L$ for all integer n and some $L > 0$, then

$$|y[n]| = \left| \sum_{l=-\infty}^{\infty} h[l]u[n - l] \right| \leq L \sum_{l=-\infty}^{\infty} |h[l]| = L \cdot I.$$

Hence, the output is bounded as well and it has upper bound $L \cdot I$. Therefore the system is stable.

Next let us assume that the given LTD-system is stable. A response to a bounded input is then bounded. The discrete unit pulse is a bounded signal. The response to this, the impulse response $h[n]$, is thus a bounded signal. Now consider the bounded input $u[n]$ given by

$$u[n] = \exp(-i \arg h[-n]).$$

By the stability, the corresponding output $y[n]$ is then also bounded and $y[0]$ has the value

$$y[0] = \sum_{l=-\infty}^{\infty} \exp(-i \arg h[-l])h[-l] = \sum_{l=-\infty}^{\infty} |h[l]|.$$

Apparently, the series in the right-hand side is convergent and has sum $y[0]$. This proves the theorem. ∎

We note here that the choice $u[n] = \exp(-i \arg h[-n])$ in the proof above is a consequence of the assumption that the impulse response may be complex-valued. Since $h[-l] = |h[-l]| \exp(i \arg h[-l])$, one has that

$$\exp(-i \arg h[-l])h[-l] = \exp(-i \arg h[-l])|h[-l]|\exp(i \arg h[-l])$$
$$= |h[-l]|.$$

EXAMPLE 19.3

The time-delay unit is a stable system with $h[n] = \delta[n-1]$. For, $\delta[n] \mapsto \delta[n-1] = h[n]$ and

$$\sum_{n=-\infty}^{\infty} |h[n]| = \sum_{n=-\infty}^{\infty} |\delta[n-1]| = \delta[0] = 1.$$

Verify for yourself that the stability of the time-delay unit can easily be checked in a direct way using definition 1.3. ◀

EXAMPLE 19.4

The multiplier is a stable system with $h[n] = a\delta[n]$. For, $\delta[n] \mapsto a\delta[n] = h[n]$ and

$$\sum_{n=-\infty}^{\infty} |h[n]| = \sum_{n=-\infty}^{\infty} |a\delta[n]| = |a|.$$

The stability of the multiplier is also easy to verify using the definition. ◀

EXAMPLE 19.5

Consider the system $u[n] \mapsto y[n]$ defined by

$$y[n] = \sum_{l=-\infty}^{n} u[l].$$

The impulse response $h[n]$ equals

$$h[n] = \sum_{l=-\infty}^{n} \delta[l] = \epsilon[n].$$

This system can therefore be described by

$$y[n] = \sum_{l=-\infty}^{n} u[l] = \sum_{l=-\infty}^{\infty} u[l]\epsilon[n-l] = (\epsilon * u)[n].$$

The system is an LTD-system, but it is *not* stable since

$$\sum_{n=-\infty}^{\infty} |\epsilon[n]| = \infty.$$

Also prove the instability using the definition. ◀

We have seen that the stability of an LTD-system is determined by the absolute summability of the impulse response. This is not very surprising. Property (19.3) states that an LTD-system is determined by the impulse response. This means that one should somehow be able to derive all kinds of properties of an LTD-system from the impulse response. For example, the *causality* of an LTD-system can also easily be checked using the impulse response, as our next theorem will show. Causality of a system was defined in chapter 1 (definition 1.4). For LTD-systems causality means that the response to a causal input is again causal. Recall that a signal $f[n]$ is causal if $f[n] = 0$ for $n < 0$.

THEOREM 19.2
Causality

An LTD-system is causal *if and only if the impulse response $h[n]$ is a causal signal.*

Proof
We will be using theorem 1.2 from chapter 1. From that theorem it follows that an LTD-system is causal if and only if the response to an arbitrary causal input is again causal.

First assume that the LTD-system is causal. Since $\delta[n] = 0$ for $n < 0$, the discrete unit pulse is a causal signal and so the response $h[n]$ is causal as well.

Let us now assume that $h[n]$ is a causal signal and let $u[n]$ be a causal input. Then the causality of $h[n]$ and (19.3) imply that

$$y[n] = \sum_{l=-\infty}^{\infty} u[l]h[n-l] = \sum_{l=-\infty}^{n} u[l]h[n-l].$$

Since $u[l] = 0$ for all $l < 0$, all terms in the above series are zero for $n < 0$. Hence, $y[n] = 0$ for $n < 0$, which means that $y[n]$ is a causal signal. This finishes the proof. ∎

Both the time-delay unit and the multiplier are examples of causal LTD-systems. Useful systems are usually required to be stable and causal.

Besides causality and stability one can also require a system to be real. A discrete-time system is called a *real* system if the response to any real input is again real.

Real discrete-time system

Show for yourself that an LTD-system is real if and only if the impulse response is real. The time-delay unit is an example of a real system, but the multiplier is a real system only if the number a is real.

In the above we have seen that one can obtain all kinds of properties of an LTD-system from the impulse response. Another special response that one may encounter in LTD-systems is the response to the discrete unit step function $\epsilon[n]$. This response

Step response

is called the *step response* and is denoted by $a[n]$, so

$$\epsilon[n] \mapsto a[n]. \tag{19.5}$$

If we know the step response of an LTD-system, then the impulse response is obtained in the following way. Since $\delta[n] = \epsilon[n] - \epsilon[n-1]$, the linearity and the time-invariance of the system imply that

$$h[n] = a[n] - a[n-1].$$

Hence, once the step response is known, one can use the impulse response and (19.3) to determine the response to an arbitrary input. We close this section with an example of an LTD-system that is known as a detection filter, which allows us to detect signals.

EXAMPLE 19.6
Detection filter

Let $x[n]$ be a given discrete-time signal with $x[n] = 0$ for $|n| > N$ and some $N > 0$. Consider the discrete-time system L given by

$$u[n] \mapsto y[n] = \sum_{m=-\infty}^{\infty} u[m]\overline{x[m-n]}.$$

The linearity of this system is easy to check. The time-invariance follows from

$$u[n-l] \mapsto \sum_{m=-\infty}^{\infty} u[m-l]\overline{x[m-n]} = \sum_{m=-\infty}^{\infty} u[m]\overline{x[m-(n-l)]}.$$

The impulse response $h[n]$ of the system is equal to $h[n] = \overline{x[-n]}$. Since only a finite number of terms of $x[n]$ are unequal to zero, it follows from theorem 19.1 that the system is stable. Moreover, theorem 19.2 implies that the system is causal if and only if $x[n] = 0$ for $n > 0$. Finally, we also note that the system is real if the signal $x[n]$ is real. We call this LTD-system L a *detection filter* because of the following. From the Cauchy–Schwarz inequality (see exercise 19.7) it follows that

$$
|y[0]| = \left| \sum_{m=-N}^{N} u[m]\overline{x[m]} \right| \leq \sum_{m=-N}^{N} |u[m]| \, |\overline{x[m]}| = \sum_{m=-N}^{N} |u[m]| \, |x[m]|
$$

$$
\leq \sqrt{ \sum_{m=-N}^{N} |u[m]|^2 \sum_{m=-N}^{N} |x[m]|^2 }.
$$

In addition, we know that in this inequality there is an equality if and only if for some $\alpha \in \mathbb{C}$

$$
u[n] = \alpha x[n] \qquad \text{for } n \in \mathbb{Z},
$$

independent of n. Hence, when a number of signals is applied to the system, including the signal $x[n]$ itself, then one can detect whether the input agrees with the signal $x[n]$ by measuring $y[0]$.										◀

EXERCISES

19.1 Show that the system defined by (19.2) is an LTD-system.

19.2 Determine the impulse response of the LTD-systems given below and verify property (19.3):
a $y[n] = \sum_{l=-\infty}^{n-1} 2^{l-n} u[l]$,
b $y[n] = (u[n-1] + u[n+1])/2$,
c $y[n] = \sum_{l=n}^{\infty} 2^{l-n} u[l]$.

19.3 Which of the discrete-time systems given in exercise 19.2 are causal and which of them are stable? Justify your answer.

19.4 For an LTD-system the impulse response is given by $h[n] = \delta[n] - 2\delta[n-1] + \delta[n-2]$.
a Determine the step response.
b Determine the response to an arbitrary input $u[n]$.

19.5 For an LTD-system the step response is given by $a[n] = 2^{-n}\epsilon[n] - 3^{-n}\epsilon[n-1]$. Determine the response to the input $u[n] = 4^{-n}\epsilon[n]$.

19.6 For a system L the response $y[n]$ to an input $u[n]$ is given by $y[n] = a_0 u[n] + a_1 u[n-1] + \cdots + a_N u[n-N]$.
a Show that the system L is an LTD-system.
b Show that the system L is causal and stable.
c Under which conditions is the system L real? Justify your answer.
d Show that property (19.3) holds.

19.7* In this exercise the Cauchy–Schwarz inequality is derived for sequences of complex numbers. Given are the (finite) sequences of complex numbers $a_1, a_2, \ldots, a_m$ and $b_1, b_2, \ldots, b_m$. The Cauchy–Schwarz inequality reads as follows:

$$
\left| \sum_{n=1}^{m} a_n \overline{b_n} \right| \leq \sqrt{ \sum_{n=1}^{m} |a_n|^2 } \cdot \sqrt{ \sum_{n=1}^{m} |b_n|^2 }.
$$

Here we have an equality-sign if and only if there exists an $\alpha \in \mathbb{C}$ such that

$$
a_n = \alpha b_n \qquad \text{for } n = 1, 2, \ldots, m.
$$

a   Show that for all complex λ one has

$$\sum_{n=1}^{m} \mid b_n - \lambda a_n \mid^2 = \left(\sum_{n=1}^{m} \left| a_n^2 \right| \right) \mid \lambda \mid^2 - 2\mathrm{Re}\left(\lambda \sum_{n=1}^{m} a_n \overline{b_n} \right) + \sum_{n=1}^{m} \left| b_n^2 \right| \geq 0.$$

b   Let $\phi = \arg(\sum_{n=1}^{m} a_n \overline{b_n})$ and $\lambda = \beta e^{-i\phi}$ with β arbitrary real. Show that for all real β one has

$$\sum_{n=1}^{m} \left| a_n^2 \right| \beta^2 - 2\beta \left| \sum_{n=1}^{m} a_n \overline{b_n} \right| + \sum_{n=1}^{m} \left| b_n^2 \right| \geq 0.$$

c   Show that part b implies that

$$\left| \sum_{n=1}^{m} a_n \overline{b_n} \right| \leq \sqrt{\sum_{n=1}^{m} \mid a_n \mid^2} \cdot \sqrt{\sum_{n=1}^{m} \mid b_n \mid^2}.$$

d   Show that if we have an equality-sign in the above inequality, then a $\beta \geq 0$ exists such that

$$\sum_{n=1}^{m} \left| a_n^2 \right| \beta^2 - 2\beta \left| \sum_{n=1}^{m} a_n \overline{b_n} \right| + \sum_{n=1}^{m} \left| b_n^2 \right| = 0.$$

e   Show that for the value of β found in part d one has

$$\sum_{n=1}^{m} \left| b_n - \beta e^{i\phi} a_n \right|^2 = 0.$$

From this it follows that $b_n = \beta e^{i\phi} a_n$ for all n. Hence, in the case of an equality the sequence b_n is a (complex) multiple of the sequence a_n.

19.2   The transfer function and the frequency response

In the previous section we have seen that property (19.3) gives an important description of an LTD-system in the n-domain. In this section we study the relation between the input and the output in the z-domain, which will lead to the introduction of the so-called transfer function. Moreover, we will study how an LTD-system effects the spectrum of an input. Here the frequency response of an LTD-system, introduced in chapter 1 ((1.8)), comes into play.

It is quite obvious to apply the convolution theorem to (19.3). When $U(z)$ and $H(z)$ denote the z-transforms of the signals $u[n]$ and $h[n]$ respectively, then it follows from the convolution theorem that the z-transform $Y(z)$ of the output $y[n]$ is given by

$$Y(z) = H(z)U(z). \tag{19.6}$$

Apparently, the relation between the input and the output has a simple description in the z-domain. It consists of an ordinary multiplication by the z-transform $H(z)$ of the impulse response $h[n]$, which we will call the transfer function of the system.

DEFINITION 19.2
Transfer function

The transfer function $H(z)$ *of a system is defined by*

$$H(z) = \sum_{n=-\infty}^{\infty} h[n]z^{-n}, \tag{19.7}$$

for those values of $z \in \mathbb{C}$ for which the series converges.

EXAMPLE 19.7

The transfer function $H(z)$ of the time-delay unit, as given in example 19.1, is equal to the z-transform of the impulse response $\delta[n-1]$, and so $H(z) = 1/z$ for $z \neq 0$. ◄

EXAMPLE 19.8

The transfer function $H(z)$ of the multiplier, as given in example 19.2, is equal to the z-transform of the impulse response $h[n] = a\delta[n]$, and so $H(z) = a$ for all $z \in \mathbb{C}$.   ◄

EXAMPLE 19.9

The transfer function $H(z)$ of the detection filter, as given in example 19.6, is equal to the z-transform of the impulse response $h[n] = \overline{x[-n]}$, where $x[n]$ is a discrete-time signal with $x[n] = 0$ for $|n| > N$. When $X(z)$ denotes the z-transform of $x[n]$, then it follows from properties (18.8) and (18.9) that $H(z) = \overline{X(1/\overline{z})}$. ◄

Since an LTD-system is completely determined by the transfer function $H(z)$, one should be able to derive all kinds of properties of the system from the transfer function. The first property we consider is the *causality* of an LTD-system. An LTD-system is causal if and only if $h[n]$ is causal. Then the z-transform $H(z)$ consists of only the causal part

$$H(z) = \sum_{n=0}^{\infty} h[n]z^{-n},$$

and we know that the region of convergence is then the exterior of a circle in the complex plane. Moreover, from the z-transform it can be derived that

$$\lim_{|z|\to\infty} H(z) = h[0]. \tag{19.8}$$

One can show that the converse is also true. This means that we have the following theorem.

THEOREM 19.3

Let an LTD-system with transfer function $H(z)$ be given. The LTD-system is causal if and only if:
a   *the region of convergence of $H(z)$ is the exterior of a circle;*
b $\lim_{|z|\to\infty} H(z)$ *exists.*

Theorem 19.3 has some special consequences for LTD-systems whose transfer function is a *rational function* of z, so $H(z) = P(z)/Q(z)$, where $P(z)$ and $Q(z)$ are polynomials in z without common zeros. The region of convergence is the exterior of a circle. For a rational function $H(z)$ this is the circle passing through a pole of $H(z)$ with maximal distance to the origin. Since $H(z)$ has a limit for $|z| \to \infty$, this implies that the degree of the numerator $P(z)$ cannot be greater than the degree of the denominator $Q(z)$.

Next we consider the *stability* of a system. From theorem 19.1 we know that a system is stable if and only if $\sum_{n=-\infty}^{\infty} |h[n]| < \infty$. But since

$$\sum_{n=-\infty}^{\infty} |h[n]| = \sum_{n=-\infty}^{\infty} |h[n]z^{-n}| \qquad \text{for } |z| = 1,$$

we may conclude that an LTD-system is stable if and only if the transfer function is absolutely convergent on the unit circle $|z| = 1$ in the complex plane. As a consequence we have the following theorem for causal LTD-systems having a rational transfer function.

THEOREM 19.4

Let L be a causal LTD-system with a rational transfer function $H(z) = P(z)/Q(z)$, where $P(z)$ and $Q(z)$ are polynomials of degree n and m respectively, and without

common zeros. Then the following statements hold:
a $n \le m$;
b *the system is stable if and only if the poles of $H(z)$ lie inside the unit circle in the complex plane.*

Proof
Statement a follows from theorem 19.3, as we have already noted. We now prove statement b.

First assume that L is stable. The region of convergence of $H(z)$ must then contain the unit circle. Theorem 19.3 then implies that the region of convergence is the exterior of a circle. For a rational function $H(z)$ this is the circle passing through a pole of $H(z)$ with maximal distance to the origin. Hence, all poles of $H(z)$ lie inside the unit circle.

Now assume that the poles lie inside the unit circle. By the causality, the region of convergence is the exterior of a circle. This must contain the unit circle. Therefore the system is stable. ∎

EXAMPLE 19.10

For a causal LTD-system L the transfer function $H(z)$ is given by

$$H(z) = \frac{z}{4z^2 + 1}.$$

Because of the causality of the system, the impulse response is a causal signal. The region of convergence of $H(z)$ is thus the exterior of a circle in the complex plane. Since $H(z)$ is a rational function with poles at $z = \pm i/2$, we conclude that the region of convergence is $|z| > 1/2$. This contains the unit circle and so the causal system is stable. The impulse response can be found by an inverse transform of $H(z)$, resulting in

$$h[n] = 2^{-n-1} \sin(n\pi/2)\epsilon[n].$$

◀

The transfer function plays an important role in the response to the special input

$$u[n] = z^n \qquad \text{with } z \in \mathbb{C}.$$

If z belongs to the region of convergence of $H(z)$, then

$$z^n \mapsto H(z)z^n. \tag{19.9}$$

This can immediately be derived from (19.3) as follows:

$$y[n] = \sum_{l=-\infty}^{\infty} h[l]z^{n-l} = \left(\sum_{l=-\infty}^{\infty} h[l]z^{-l} \right) z^n.$$

Eigenfunction
Eigenvalue

Because of (19.9), the signal z^n is sometimes called an *eigenfunction* of the LTD-system, and $H(z)$ the corresponding *eigenvalue*.

In chapter 1 the frequency response $H(e^{i\omega})$ of an LTD-system has been defined using the response of an LTD-system to the input $e^{in\omega}$:

$$e^{in\omega} \mapsto H(e^{i\omega})e^{in\omega}. \tag{19.10}$$

We now see that this rule follows immediately from (19.9) by substituting $z = e^{i\omega}$. This then justifies the notation $H(e^{i\omega})$ for the frequency response of an LTD-system. Apparently one has the following important property:

Frequency response

The frequency response is equal to the spectrum of the impulse response:

$$H(e^{i\omega}) = \sum_{n=-\infty}^{\infty} h[n]e^{-in\omega}. \tag{19.11}$$

Now the spectrum of $h[n]$ is only defined for those frequencies for which the series in the right-hand side of (19.11) converges. However, if the system is *stable*, then $h[n]$ is absolutely summable and the above series is therefore absolutely convergent for all real ω since $|h[n]e^{-in\omega}| = |h[n]|$. Hence, for stable systems the frequency response is defined for *all* frequencies ω. One can even show that the function $H(e^{i\omega})$ is a continuous function of ω.

Since $H(e^{i\omega})$ is complex, we can write

$$H(e^{i\omega}) = \left| H(e^{i\omega}) \right| e^{i\Phi(\omega)}. \tag{19.12}$$

Amplitude spectrum

Phase spectrum

As usual we call $|H(e^{i\omega})|$ and $\Phi(\omega)$, respectively, the *amplitude* and *phase spectrum*. The phase spectrum is determined up to a multiple of 2π. The significance of the amplitude and phase spectrum for LTD-systems will now be demonstrated for a real and stable system by looking at the response to the following sinusoidal signal $u[n]$ for a positive frequency ω:

$$u[n] = A \cos(n\omega + \phi).$$

Here A, ϕ and ω are real constants and $\omega > 0$ is the frequency of the signal. By changing to complex exponentials, we can write the input as

$$u[n] = \tfrac{1}{2} A(e^{i\phi} e^{in\omega} + e^{-i\phi} e^{-in\omega}).$$

Applying the linearity of the system and property (19.10) leads to the following expression for the output $y[n]$:

$$y[n] = \tfrac{1}{2} A(e^{i\phi} H(e^{i\omega}) e^{in\omega} + e^{-i\phi} H(e^{-i\omega}) e^{-in\omega}).$$

Now check for yourself that for real systems one has

$$H(e^{-i\omega}) = \overline{H(e^{i\omega})},$$

and then note that this implies that $y[n] = A\,\mathrm{Re}(H(e^{i\omega})e^{i\phi}e^{in\omega})$. Finally we use (19.12) to write $y[n]$ as follows:

$$y[n] = A \left| H(e^{i\omega}) \right| \cos(n\omega + \phi + \Phi(\omega)). \tag{19.13}$$

The output has the same frequency as the input, only the amplitude is multiplied by a factor $|H(e^{i\omega})|$ and there is a phase shift $\Phi(\omega)$.

Often, LTD-systems are described by indicating what the frequency response is like. When $U(e^{i\omega})$ is the spectrum of $u[n]$ and $Y(e^{i\omega})$ is the spectrum of the corresponding output $y[n]$, then it follows from (19.6) that

$$Y(e^{i\omega}) = H(e^{i\omega})U(e^{i\omega}). \tag{19.14}$$

This property shows us how the LTD-system effects the spectrum of an input. If we know the frequency response, then by (18.22) the impulse response $h[n]$ is

$$h[n] = \frac{1}{2\pi} \int_{-\pi}^{\pi} H(e^{i\omega}) e^{in\omega} \, d\omega, \tag{19.15}$$

and so the LTD-system is again completely determined. The output can then be found using (19.14) and again (18.22):

$$y[n] = \frac{1}{2\pi} \int_{-\pi}^{\pi} H(e^{i\omega})U(e^{i\omega})e^{in\omega}\,d\omega. \tag{19.16}$$

From (19.16) we conclude that if $u[n]$ contains the components $U(e^{i\omega})e^{in\omega}$, then $y[n]$ contains the components $H(e^{i\omega})U(e^{i\omega})e^{in\omega}$. The component having frequency ω gets a phase shift $\Phi(\omega)$ and is amplified by a factor $|H(e^{i\omega})|$.

EXAMPLE 19.11
Ideal low-pass filter

Consider an LTD-system whose frequency response on the interval $(-\pi, \pi)$ is given by

$$H(e^{i\omega}) = \begin{cases} 1 & \text{for } |\omega| < \omega_c < \pi, \\ 0 & \text{for } \omega_c < |\omega| < \pi. \end{cases} \tag{19.17}$$

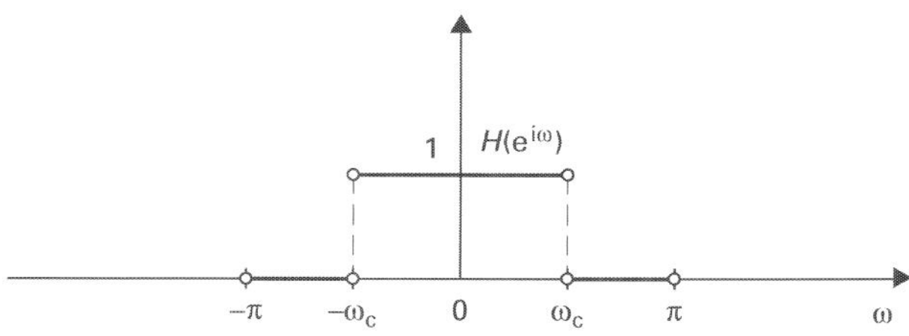

FIGURE 19.1
Frequency response of an ideal low-pass filter.

A graph of the frequency response is drawn in figure 19.1. The LTD-system defined by (19.17) is called an *ideal low-pass filter* since the components of an input $u[n]$ with frequencies ω in the *pass-band* $(-\omega_c, \omega_c)$ can pass undisturbed, while the other components are blocked completely. The impulse response follows from (19.15) and in our example it is equal to

$$h[n] = \frac{1}{2\pi} \int_{-\omega_c}^{\omega_c} e^{in\omega}\,d\omega = \frac{\sin(n\omega_c)}{n\pi}.$$

◀

The problem with the ideal low-pass filter is that the frequency response cannot be written as a rational function of $e^{i\omega}$, in other words, that the transfer function is not a rational function of z. In practice this implies that an ideal low-pass filter can only be approximated. In this book we will not go into this any further and so we refer to the relevant literature (e.g. *Digitale signaalbewerking* by A.W.M. van den Enden and N.A.M. Verhoeckx (in Dutch)).

The LTD-systems with a rational transfer function are important, because they can be realized using time-delay units, multipliers and adders and, moreover, can be described by means of so-called difference equations. In the next section we will study these LTD-systems described by difference equations.

EXERCISES

19.8

For an LTD-system the impulse response is given by $h[n] = 2^{-n}\cos(n\phi)\epsilon[n]$. Here ϕ is a real constant. Calculate the transfer function of the system and verify whether the system is stable.

19.9 For a stable LTD-system the transfer function is given by

$$H(z) = \frac{z^2}{9z^2 + 6z + 1}.$$

a Determine the impulse response.
b Calculate the response to the input $u[n] = \sin(n\pi/2)$.

19.10 For a causal LTD-system the transfer function is given by

$$H(z) = \frac{z+1}{(z-1)(2z+1)}.$$

a Determine the impulse response $h[n]$.
b Is the system stable? Justify your answer.

19.11 For a stable LTD-system the frequency response is given by $H(e^{i\omega}) = 1 + 2\cos(2\omega)$.
a Determine the response to the input $u[n] = \delta[n-2]$.
b Is the system causal? Justify your answer.

19.12 Show that the amplitude and phase spectrum of a real LTD-system are, respectively, an even and an odd function of the frequency ω.

19.13 For a discrete band-pass filter the frequency response is given by the graph from figure 19.2. Determine the impulse response of the filter.

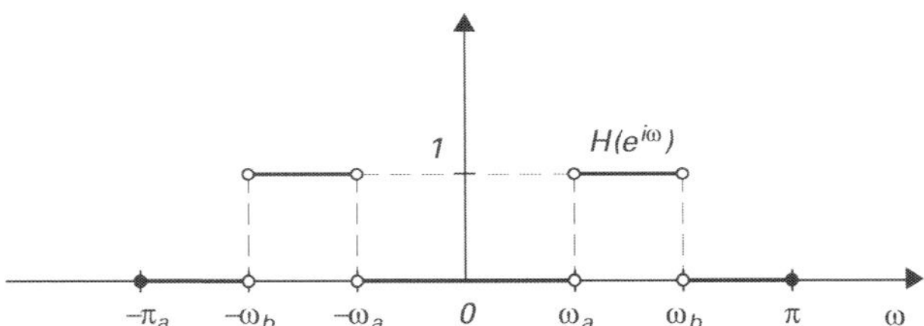

FIGURE 19.2
Frequency response of the ideal band-pass filter from exercise 19.13.

19.14 Let L be a stable LTD-system with frequency response $H(e^{i\omega})$ and $u[n]$ an input with finite energy-content and spectrum $U(e^{i\omega})$. Show that the energy-content of the output $y[n]$ is given by

$$\frac{1}{2\pi} \int_{-\pi}^{\pi} \left| H(e^{i\omega}) U(e^{i\omega}) \right|^2 d\omega.$$

19.3 LTD-systems described by difference equations

An LTD-system can be specified by describing the relationship between an input $u[n]$ and the corresponding response $y[n]$. As we have seen, this can be done in several ways. For example, in the n-domain by using the impulse response $h[n]$, or in the z-domain by means of the transfer function $H(z)$. In this section we will examine an important class of LTD-systems, namely the LTD-systems that can be described by difference equations. We have seen that the time-delay unit, introduced in example 19.1, can be described by the equation

$$y[n] = u[n-1].$$

A somewhat less elementary example of an LTD-system is obtained by also adding $u[n]$ and other delays of $u[n]$ to such an equation. More generally, one can consider the following equation:

$$y[n] = a_0 u[n] + a_1 u[n-1] + \cdots + a_N u[n-N]. \qquad (19.18)$$

An example of this is the LTD-system that calculates a weighted average of an input according to (19.2). When the input $u[n]$ is known, then the output $y[n]$ can easily be calculated by substituting $u[n]$, $u[n-1]$, ..., $u[n-N]$ into (19.18). In exercise 19.6 you were asked to show that (19.18) does indeed define an LTD-system, being causal as well.

A slightly more complicated equation is obtained by allowing in addition delays of $y[n]$. The most general situation considered by us is described by

$$y[n] + b_1 y[n-1] + \cdots + b_M y[n-M] = a_0 u[n] + \cdots + a_N u[n-N]. \qquad (19.19)$$

Difference equation

We call (19.19) a *difference equation*. Here the signal $u[n]$ is considered as a given signal. A difference equation is thus an equation of type (19.19) for the unknown signal $y[n]$ and with $u[n]$ given. Equation (19.18) is a special case of (19.19): for $b_1 = b_2 = \ldots = b_M = 0$ one obtains (19.18). In order to determine $y[n]$ from (19.18) we only need the signal $u[n]$. This is in contrast to (19.19). In the latter equation we also need $y[n-1]$, $y[n-2]$, ..., $y[n-M]$ in order to determine $y[n]$.

Non-recursive system

Recursive system

A system described by (19.18) is called a *non-recursive* system. A system described by (19.19) is called a *recursive* system.

There is yet another complication when defining a system by means of (19.19). For a given $u[n]$ there will, in general, be *several* solutions $y[n]$ satisfying (19.19). If we know the values of the output $y[n]$ for $n = -1, -2, \ldots, -M$, as the result of a given input $u[n]$, then we see that substituting $n = 0$ in (19.19) will lead to the value $y[0]$. By subsequently substituting $n = 1$, we obtain the value $y[1]$. This process can be repeated, and so we see that eventually $y[n]$ will be uniquely determined for every $n \geq 0$. When, for example, the system is causal and $u[n]$ is a causal input, then the output $y[n]$ is causal and so $y[-1] = 0$, $y[-2] = 0$, ..., $y[-M] = 0$. Consequently, $y[n]$ is uniquely determined by the difference equation.

We will confine ourselves to *causal* LTD-systems described by (19.19). First we will show that a causal LTD-system is completely determined by (19.19). To this end we apply the z-transform to the left- and right-hand side of (19.19). If we put

$$u[n] \leftrightarrow U(z) \qquad \text{and} \qquad y[n] \leftrightarrow Y(z),$$

then it follows from the linearity and the shift property (18.10) for the z-transform that

$$(1 + b_1 z^{-1} + \cdots + b_M z^{-M})Y(z) = (a_0 + a_1 z^{-1} + \cdots + a_N z^{-N})U(z).$$

The transfer function $H(z)$ is thus equal to

$$H(z) = \frac{Y(z)}{U(z)} = \frac{a_0 + a_1 z^{-1} + \cdots + a_N z^{-N}}{1 + b_1 z^{-1} + \cdots + b_M z^{-M}}. \qquad (19.20)$$

This shows that the transfer function is a rational function:

$$H(z) = z^{M-N} \frac{a_N + a_{N-1}z + \cdots + a_0 z^N}{b_M + b_{M-1}z + \cdots + z^M}.$$

In order to give a complete description of the LTD-system, we also have to know the region of convergence of the transfer function. However, the system is causal and for a causal system with a rational transfer function we have by theorem 19.4 that the degree of the numerator of $H(z)$ is not greater than the degree of the denominator.

One can easily verify that this is satisfied for all M and N. Moreover, the region of convergence is the exterior of the circle in the complex plane passing through a pole of $H(z)$ with maximal distance to the origin. Therefore, the region of convergence is known. Hence, there is only one causal LTD-system described by (19.19).

EXAMPLE 19.12

A causal system is described by the difference equation

$$y[n] - 3y[n-1] + 2y[n-2] = u[n] + u[n-1].$$

The transfer function can immediately be read off from the difference equation:

$$H(z) = \frac{1 + z^{-1}}{1 - 3z^{-1} + 2z^{-2}} = \frac{z + z^2}{z^2 - 3z + 2} = \frac{z + z^2}{(z-1)(z-2)}.$$

This has poles at $z = 1$ and $z = 2$. Because of the causality of the system, the region of convergence is the exterior of the circle $|z| = 2$. As we have seen in section 18.3, the impulse response can now be found by a partial fraction expansion. The result is

$$h[n] = (3 \cdot 2^n - 2)\epsilon[n].$$

◀

As a matter of fact, the LTD-system in the example above is not stable. For stable LTD-systems with a rational transfer function, the unit circle must be contained in the region of convergence, according to theorem 19.4. In example 19.12 this is not the case.

EXAMPLE 19.13

A causal LTD-system is given by

$$y[n] - 3y[n-1] + 3y[n-2] - y[n-3] = u[n-2].$$

We determine the step response $a[n]$ as follows. The transfer function of the given LTD-system can easily be read off from the difference equation:

$$H(z) = \frac{z^{-2}}{1 - 3z^{-1} + 3z^{-2} - z^{-3}} = \frac{z}{(z-1)^3} \qquad \text{for } |z| > 1.$$

The step response $a[n]$ is the response to the discrete signal $\epsilon[n]$, whose z-transform equals $z/(z-1)$. By (19.6) we have

$$a[n] \leftrightarrow \frac{z}{(z-1)^3} \cdot \frac{z}{z-1} = \frac{z^2}{(z-1)^4} \qquad \text{for } |z| > 1.$$

The step response can now be found by an inverse transform, after first applying the partial fraction expansion technique. We then find that

$$a[n] \leftrightarrow \frac{z^2}{(z-1)^4} = \frac{z}{(z-1)^3} + \frac{z}{(z-1)^4} \qquad \text{for } |z| > 1.$$

An inverse transform then gives

$$a[n] = \left(\binom{n}{2} + \binom{n}{3} \right) \epsilon[n].$$

The system is not stable, though. The unit circle is not contained in the region of convergence. Also note that the step response is not bounded, while the input $\epsilon[n]$ is bounded.

◀

EXERCISES

19.15   A causal LTD-system is described by the following difference equation:
$y[n] + \frac{1}{2}y[n-1] = u[n]$.
  a Calculate the impulse response of the system.
  b Calculate the response to the input $u[n] = (\frac{1}{2})^n \epsilon[n]$.
  c Is the system stable? Justify your answer.

19.16   A causal LTD-system is described by the following difference equation:
$y[n] - y[n-2] = u[n-1]$.
  a Calculate the transfer function of the system.
  b Is the system real? Justify your answer.
  c Calculate the response to the input $u[n] = \cos(n\phi)$. Here ϕ is a real constant.

19.17   A causal system is described by the following difference equation:
$y[n] - \frac{1}{4}y[n-2] = u[n] + u[n-1]$.
  a Calculate the impulse response of the system.
  b Is the system stable? Justify your answer.
  c Calculate the step response.
  d Calculate the response to the input $u[n] = \epsilon[n] + \epsilon[n-2]$.

SUMMARY

The LTD-systems are very well suited for the application of the z-transform and the Fourier transform of discrete-time signals. An important role in the theory of LTD-systems is played by the impulse response, the transfer function, and the frequency response. In the time domain or n-domain the relation between an input $u[n]$ and the corresponding response $y[n]$ is described by the convolution product

$$y[n] = (h * u)[n].$$

Here $h[n]$ is the response to the discrete unit pulse $\delta[n]$. The response $h[n]$ is called the impulse response. Applying the convolution theorem for the z-transform to the convolution product above immediately leads to the relation

$$Y(z) = H(z)U(z),$$

which can be regarded as a description of the LTD-system in the z-domain. Here $H(z)$ is the z-transform of the impulse response $h[n]$; $H(z)$ is called the transfer function. An LTD-system is thus completely determined by the impulse response, or by the transfer function. One has, for example, that an LTD-system is stable or causal if and only if the impulse response is absolutely summable or, respectively, causal. Since the transfer function is the z-transform of the impulse response, the stability also follows from the region of convergence of $H(z)$. If the unit circle in the complex plane is contained in the region of convergence of $H(z)$, then the corresponding LTD-system is stable. For an LTD-system the transfer function $H(z)$ can also be interpreted as the eigenvalue with eigenfunction the signal z^n. In particular, we have for stable systems that the response to the input $e^{in\omega}$, containing the single frequency ω, is equal to $H(e^{i\omega})e^{in\omega}$. The function $H(e^{i\omega})$ is called the frequency response. The significance of the frequency response for an LTD-system can especially be seen in the spectra of the input and the output. This is because

$$Y(e^{i\omega}) = H(e^{i\omega})U(e^{i\omega}).$$

In practice, an important category of LTD-systems is those systems that can be described by difference equations. The transfer function is then a rational function of z. If its poles lie inside the unit circle, then the system is stable.

SELFTEST

19.18   For a causal LTD-system the transfer function $H(z)$ is given by

$$H(z) = \frac{z + 1}{4z^2 + 4z + 1}.$$

a   Calculate the impulse response $h[n]$.
b   Calculate the response to the input $u[n] = (-1)^n$, where $n \in \mathbb{Z}$.
c   Is the system stable? Justify your answer.
d   Is the system real? Justify your answer.
e   Let $u[n]$ be an input with spectrum $U(e^{i\omega}) = \cos 2\omega$. Calculate the response $y[n]$ to $u[n]$.

19.19   Given is a stable LTD-system. We apply a periodic input $u[n]$ with period N to the LTD-system.
a   Show that the response $y[n]$ to $u[n]$ is also periodic with period N.
b   Let $U[k]$ and $Y[k]$ be the N-point DFT of $u[n]$ and $y[n]$ respectively, and let $H(z)$ be the transfer function of the system. Prove that $Y[k] = H(e^{2\pi ik/N})U[k]$.

19.20   For an LTD-system the frequency response is given by $H(e^{i\omega}) = \cos 2\omega$.
a   Calculate the impulse response $h[n]$.
b   To the LTD-system a periodic discrete-time signal $u[n]$ with period 4 is applied, whose 4-point DFT $F[k]$ is given by $F[0] = 1$, $F[1] = -1$, $F[2] = 0$, $F[3] = 1$. Calculate the response $y[n]$ to the input $u[n]$.

19.21   For a stable LTD-system the step response is given by $a[n] = n^2(\frac{1}{2})^n \epsilon[n]$.
a   Calculate the transfer function and the impulse response of the system.
b   Is the system causal? Justify your answer.
c   Calculate the response to the input $u[n] = e^{in\omega}$ (ω a real constant).

19.22   A causal LTD-system is described by the following difference equation:
$y[n] - \frac{1}{2}y[n - 1] = u[n - 1] + u[n - 2]$.
a   Determine the step response $a[n]$ for $n \geq 0$.
b   Is the system stable? Justify your answer.

19.23   A causal LTD-system is described by the following difference equation:
$6y[n] - 5y[n - 1] + y[n - 2] = 6u[n] - 6u[n - 2]$.
a   Calculate the impulse response $h[n]$ of the system.
b   Determine the frequency response of the system.
c   Describe the inputs $u[n]$ that are completely blocked by the system, that is to say, whose response $y[n]$ is identically zero.

19.24   For a stable LTD-system the frequency response is given by $H(e^{i\omega}) = 1 + 2\cos\omega + \cos 2\omega$.
a   Calculate the impulse response $h[n]$ of the system.
b   To the system we apply the input $u[n]$ whose spectrum $U(e^{i\omega})$ is given by $U(e^{i\omega}) = 1 + \sin\omega + \sin 2\omega$. Let $y[n]$ be the response to the given input $u[n]$. Calculate the energy-content of the signal $y[n]$.

Literature

There is an overwhelming amount of literature available on Fourier and Laplace transforms. In it, one can roughly distinguish two main trends. On the one hand the theoretical literature for a mathematically oriented audience, on the other hand the literature where the applications play a central role. In much of the literature aimed at the applications, the results are presented without precise conditions or proofs. The mathematical literature, however, is mostly of a very theoretical nature and assumes quite a lot of prerequisites, such as the so-called Lebesgue integral. In the following survey one will find very few books from the latter category. A number of the books mentioned below do require a mathematical background which goes beyond what we have assumed for this book. Two rather elementary standard books on mathematical analysis, which could be consulted in order to obtain the required background, are for example (these books are available in several editions):

Apostol, T.M., *Mathematical analysis*. Reading, Addison-Wesley, 1957.
Kaplan, W., *Advanced calculus*. Boston, Addison-Wesley, 1984, 3rd ed.

We do hope that the books listed here offer a good opportunity for a more elaborate study of the many different aspects of both the theory and the applications of the Fourier and Laplace transforms.

Bracewell, R.N., *The Fourier transform and its applications*. New York, McGraw-Hill, 1986, 2nd ed., revised.

A real classic (the first edition is from 1965), and generally considered as one of the standard works in the field of the applications of the Fourier integral, especially in signal theory. More specifically, it contains applications to filters, sampling, convolution, imaging (antennas and television), and sound (noise). The discrete Fourier transform and the Fast Fourier Transform (FFT) are also treated extensively.

Brigham, E.O., *The fast Fourier transform*. Englewood Cliffs, Prentice-Hall, 1974.

After a rather sketchy treatment of the Fourier integral and convolution, the determination of the discrete Fourier transform using the FFT is then treated very thoroughly. This book is a standard work on the FFT.

Churchill, R.V. and J.W. Brown, *Fourier series and boundary value problems*. New York, McGraw-Hill, 1978, 3rd ed.

Applying Fourier analysis to boundary value problems (see sections 5.2 and 10.4) is the central issue. Although emphasis is put on Fourier series, more general series of orthogonal functions and the Fourier integral are treated as well.

Doetsch, G., *Einführung in Theorie und Anwendung der Laplace Transformation*. Basel, Birkhäuser Verlag, 1958 (in German).
Doetsch, G., *Guide to the applications of Laplace transforms*. London, Van Nostrand, 1961.

The book mentioned first gives a very thorough treatment of the Laplace transform. Although only the Riemann integral is used, this book does assume a solid mathematical background. The Laplace transform of distributions is also treated, but it is assumed that the reader is familiar with the theory of distributions. Applications to differential equations are treated extensively. The second book is in fact a very compact version of the first one. The most relevant results are given without proof. The book then concentrates on the applications to differential and difference equations.

Dym, H. and H.P. McKean, *Fourier series and integrals.* New York, Academic Press, 1972.

This book requires a good mathematical background and uses the so-called Lebesgue integral (the only book in this survey which does), although this theory is explained in the first chapter. We mention this book mainly because it is considered as one of the standard works as far as the theory of Fourier series and Fourier integrals is concerned. It also contains a whole range of applications of Fourier analysis in mathematics and physics.

Enden, A.W.M. van den and N.A.M. Verhoeckx, *Digitale signaalbewerking.* Overberg (gem. Amerongen), Delta Press, 1987 (in Dutch).

This book (in Dutch) treats in a well-organized manner discrete systems and several discrete transforms. The theory is then applied to the design of discrete filters. Moreover, problems are discussed that arise when processing signals of finite word length. These are discrete signals whose *values* are discrete as well. The book is clearly aimed at the applications. The style is clear and it points the reader at important aspects of several of the transforms. Mathematical rigour is comparable to what is normally seen in mathematics books for engineers. When studying part 5 of our book, van den Enden and Verhoeckx is recommended to those who are interested in the style of the people who apply this material.

Hanna, J. and J.H. Rowland, *Fourier series, transforms and boundary value problems.* New York, Wiley, 1990.

A clearly written introduction to partial differential equations, mainly aimed at students in the technical sciences and engineering. Clear physical motivations are given of the equations under consideration, as well as of the interpretations of the obtained solutions. One also discusses, for example, existence and uniqueness of the solutions of the heat equation and the wave equation at an elementary level.

Körner, T.W., *Fourier analysis.* Cambridge, Cambridge University Press, 1990.

A book covering a wide variety of subjects and consisting of some hundred short chapters ('essays'). Alternately the theories of Fourier series and the Fourier integral are developed and a large number of different applications is treated (a small sample: approximation, the age of the earth, the transatlantic cable, the heat equation).

Papoulis, A., *Circuits and systems.* New York, Holt, Rinehart & Winston, 1980.

Of the large number of books on linear systems we only mention here this well-known book by Papoulis. It is quite comprehensive and treats both analogue and digital systems, and all transforms treated by us occur in it (Laplace transform, z-transform, Fourier series, Fourier integral, discrete Fourier transform, and FFT). There is a strong emphasis on the practical applications in electrical circuits and networks.

Papoulis, A., *Signal analysis.* Singapore, McGraw-Hill, 1984.

Again we only mention the book by Papoulis as one of the standard works in the vast amount of literature on signal analysis. Again all transforms treated by us occur in it. Here there is a strong emphasis on the applications in signal analysis: filters, windowing and data smoothing are some of the main subjects.

Senior, T.B.A., *Mathematical methods in electric engineering.* Cambridge, Cambridge University Press, 1986.

In this book the emphasis is on the Laplace transform and the applications to systems theory. The theory of complex functions, necessary in order to use the fundamental theorem of the Laplace transform (see section 13.5), is treated extensively. Besides this, Fourier series and the Fourier integral are treated in some detail.

Spiegel, M.R., *Schaum's outline of theory and problems of Fourier analysis with applications to boundary value problems.* New York, McGraw-Hill, 1976.
Spiegel, M.R., *Schaum's outline of theory and problems of Laplace transforms.* New York, McGraw-Hill, 1965.

These are two books from the well-known 'Schaum's outline' series. After a short summary of the most well-known results from the theory and the applications, worked examples follow, together with a large number of exercises. In the first book the emphasis is on Fourier series, the Fourier integral and the applications to boundary value problems (as in the previously mentioned book by R.V. Churchill and J.W. Brown). In the second book the emphasis is on the Laplace transform and the application to differential equations. Complex function theory is also treated briefly. Part of the material on Fourier series and the Fourier integral from the first book can also be found in the second book. These books are very suitable as practising material.

Tolstov, G.P., *Fourier series.* New York, Dover, 1962.

This book is mainly devoted to a thorough treatment of the theory of the Fourier series. Still, it is of an elementary character, meaning that one can read it with a minimum of mathematical prerequisites. Besides Fourier series, the Fourier integral is treated briefly, as well as other series of orthogonal functions.

Walker, P.L., *The theory of Fourier series and integral.* Chicester, Wiley, 1986.

One of the few books on Fourier series and the Fourier integral in this survey with a theoretical character, but which does not use the so-called Lebesgue integral. Therefore, this book is also intended for engineering students.

Zemanian, A.H., *Distribution theory and transform analysis.* New York, Dover, 1987.

It is quite hard to find an elementary book on distribution theory. This book contains a very comprehensive treatment of the distribution theory that is still very accessible. Besides the general theory, both the Fourier transform and the Laplace transform of distributions are treated.

Tables of transforms and properties

TABLE 1   Fourier coefficients of periodic functions with period
$T = 2\pi/\omega_0$

no.	$f(t)$	$c_n = \dfrac{1}{T}\displaystyle\int_{-T/2}^{T/2} f(t)e^{-in\omega_0 t}\,dt$	condition(s)
1	$p_{a,T}(t)$	$\dfrac{2\sin(n\omega_0 a/2)}{n\omega_0 T},\ c_0 = \dfrac{a}{T}$	$0 \le a \le T$
2	$q_{a,T}(t)$	$\dfrac{4\sin^2(n\omega_0 a/2)}{n^2\omega_0^2 aT},\ c_0 = \dfrac{a}{T}$	$0 < a \le T/2$
3	$2t/T$	$\dfrac{i(-1)^n}{\pi n},\ c_0 = 0$	

TABLE 2   Properties of the Fourier transform of periodic functions

no.	$f(t), g(t)$	c_n, d_n	condition(s)
1	$af(t) + bg(t)$	$ac_n + bd_n$	$a, b \in \mathbb{C}$
2	$\overline{f(t)}$	$\overline{c_{-n}}$	
3	$f(t - t_0)$	$e^{-in\omega_0 t_0} c_n$	$t_0 \in \mathbb{R}$
4	$f(-t)$	c_{-n}	
5	$f(t)$ even and real	c_n even and real	
6	$f(t)$ odd and real	c_n odd and imaginary	
7	$f(t)g(t)$	$\displaystyle\sum_{k=-\infty}^{\infty} c_k d_{n-k}$	
8	$(f * g)(t)$	$c_n d_n$	
9	$\displaystyle\int_{-T/2}^{t} f(\tau)\, d\tau$	$\dfrac{c_n}{in\omega_0}$	$c_0 = 0$
10	$f'(t)$	$in\omega_0 c_n$	
11	$\dfrac{1}{T} \displaystyle\int_{-T/2}^{T/2} f(t)\overline{g(t)}\, dt = \sum_{n=-\infty}^{\infty} c_n \overline{d_n}$		
12	$\dfrac{1}{T} \displaystyle\int_{-T/2}^{T/2} \lvert f(t) \rvert^2\, dt = \sum_{n=-\infty}^{\infty} \lvert c_n \rvert^2$		

TABLE 3   Fourier transforms of non-periodic functions

no.	$f(t)$	$F(\omega) = \displaystyle\int_{-\infty}^{\infty} f(t)e^{-i\omega t}\,dt$	*condition(s)*
1	$p_a(t)$	$\dfrac{2\sin(a\omega/2)}{\omega}$	$a > 0$
2	$\dfrac{\sin at}{t}$	$\pi p_{2a}(\omega)$	$a > 0$
3	$q_a(t)$	$\dfrac{4\sin^2(a\omega/2)}{a\omega^2}$	$a > 0$
4	$\dfrac{\sin^2 at}{t^2}$	$\pi a q_{2a}(\omega)$	$a > 0$
5	$e^{-a\lvert t\rvert}$	$\dfrac{2a}{a^2 + \omega^2}$	$a > 0$
6	$\dfrac{1}{a^2 + t^2}$	$\dfrac{\pi}{a}e^{-a\lvert\omega\rvert}$	$a > 0$
7	$\epsilon(t)e^{-at}$	$\dfrac{1}{a + i\omega}$	$\operatorname{Re} a > 0$
8	$\epsilon(t)te^{-at}$	$\dfrac{1}{(a + i\omega)^2}$	$\operatorname{Re} a > 0$
9	$\epsilon(t)e^{-at}\sin bt$	$\dfrac{b}{(a + i\omega)^2 + b^2}$	$\operatorname{Re} a > 0, b \in \mathbb{R}$
10	$\epsilon(t)e^{-at}\cos bt$	$\dfrac{a + i\omega}{(a + i\omega)^2 + b^2}$	$\operatorname{Re} a > 0, b \in \mathbb{R}$
11	e^{-at^2}	$\sqrt{\dfrac{\pi}{a}}\,e^{-\omega^2/4a}$	$a > 0$

TABLE 4   Properties of the Fourier transform of non-periodic functions

no.	$f(t), g(t)$	$F(\omega), G(\omega)$	condition(s)
1	$af(t) + bg(t)$	$aF(\omega) + bG(\omega)$	$a, b \in \mathbb{C}$
2	$\overline{f(t)}$	$\overline{F(-\omega)}$	
3	$f(t - a)$	$e^{-i\omega a} F(\omega)$	$a \in \mathbb{R}$
4	$e^{iat} f(t)$	$F(\omega - a)$	$a \in \mathbb{R}$
5	$f(at)$	$\lvert a \rvert^{-1} F(a^{-1}\omega)$	$a \in \mathbb{R}, a \neq 0$
6	$f(t)$ even and real	$F(\omega)$ even and real	
7	$f(t)$ odd and real	$F(\omega)$ odd and imaginary	
8	$f^{(n)}(t)$	$(i\omega)^n F(\omega)$	
9	$(-it)^n f(t)$	$F^{(n)}(\omega)$	
10	$\displaystyle\int_{-\infty}^{t} f(\tau)\, d\tau$	$\dfrac{F(\omega)}{i\omega}$	$F(0) = 0$
11	$F(-t)$	$2\pi f(\omega)$	
12	$(f * g)(t)$	$F(\omega)G(\omega)$	
13	$f(t)g(t)$	$\dfrac{1}{2\pi}(F * G)(\omega)$	

no.		
14	$\displaystyle\int_{-\infty}^{\infty} f(t)\overline{g(t)}\, dt = \frac{1}{2\pi} \int_{-\infty}^{\infty} F(\omega)\overline{G(\omega)}\, d\omega$	
15	$\displaystyle\int_{-\infty}^{\infty} \lvert f(t) \rvert^2\, dt = \frac{1}{2\pi} \int_{-\infty}^{\infty} \lvert F(\omega) \rvert^2\, d\omega$	
16	$\displaystyle\sum_{n=-\infty}^{\infty} f(nT) = \frac{1}{T} \sum_{n=-\infty}^{\infty} F\left(\frac{2\pi n}{T}\right) \quad (T > 0)$	

TABLE 5   Fourier transforms of distributions

no.	$T(t)$	$U(\omega) = \langle U, \phi \rangle = \langle T, \mathcal{F}\phi \rangle$	condition(s)
1	$\delta(t)$	1	
2	1	$2\pi\delta(\omega)$	
3	$\delta(t-a)$	$e^{-ia\omega}$	$a \in \mathbb{R}$
4	e^{iat}	$2\pi\delta(\omega-a)$	$a \in \mathbb{R}$
5	$\cos at$	$\pi(\delta(\omega-a)+\delta(\omega+a))$	$a \in \mathbb{R}$
6	$\sin at$	$-\pi i(\delta(\omega-a)-\delta(\omega+a))$	$a \in \mathbb{R}$
7	$\displaystyle \mathcyr{sh}(t) = \sum_{k=-\infty}^{\infty} \delta(t-k)$	$\displaystyle \mathcyr{sh}\left(\frac{\omega}{2\pi}\right) = 2\pi \sum_{k=-\infty}^{\infty} \delta(\omega-2\pi k)$	
8	$\mathrm{pv}(1/t)$	$-\pi i \operatorname{sgn}\omega$	
9	$\operatorname{sgn} t$	$-2i\,\mathrm{pv}(1/\omega)$	
10	$\epsilon(t)$	$\pi\delta(\omega) - i\,\mathrm{pv}(1/\omega)$	
11	$\delta^{(n)}$	$(i\omega)^n$	
12	t^n	$2\pi i^n \delta^{(n)}(\omega)$	

TABLE 6   Properties of the Fourier transform of distributions

no.	$T(t), S(t)$	$U(\omega), V(\omega)$	condition(s)
1	$aT(t) + bS(t)$	$aU(\omega) + bV(\omega)$	$a, b \in \mathbb{C}$
2	$T(t - a)$	$e^{-ia\omega}U(\omega)$	$a \in \mathbb{R}$
3	$e^{iat}T(t)$	$U(\omega - a)$	$a \in \mathbb{R}$
4	$T(at)$	$\lvert a \rvert^{-1} U(a^{-1}\omega)$	$a \in \mathbb{R}, a \neq 0$
5a	$T(t)$ even	$U(\omega)$ even	
5b	$T(t)$ odd	$U(\omega)$ odd	
6	$T^{(n)}(t)$	$(i\omega)^n U(\omega)$	
7	$(-it)^n T(t)$	$U^{(n)}(\omega)$	
8	$U(-t)$	$2\pi T(\omega)$	
9	$(T * S)(t)$	$U(\omega)V(\omega)$	
10	$T(t)S(t)$	$\dfrac{1}{2\pi}(U * V)(\omega)$	

TABLE 7   Laplace transforms of causal functions

no.	$f(t)$	$F(s) = (\mathcal{L}f)(s)$ $= \int_0^\infty f(t)e^{-st}\,dt$	half-plane of convergence	condition(s)
1	1	$\dfrac{1}{s}$	$\operatorname{Re} s > 0$	
2	e^{at}	$\dfrac{1}{s-a}$	$\operatorname{Re} s > \operatorname{Re} a$	$a \in \mathbb{C}$
3	t^n	$\dfrac{n!}{s^{n+1}}$	$\operatorname{Re} s > 0$	
4	$\sin at$	$\dfrac{a}{s^2 + a^2}$	$\operatorname{Re} s > 0$	$a \in \mathbb{R}$
5	$\cos at$	$\dfrac{s}{s^2 + a^2}$	$\operatorname{Re} s > 0$	$a \in \mathbb{R}$
6	$\sinh at$	$\dfrac{a}{s^2 - a^2}$	$\operatorname{Re} s > a$	$a \in \mathbb{R}$
7	$\cosh at$	$\dfrac{s}{s^2 - a^2}$	$\operatorname{Re} s > a$	$a \in \mathbb{R}$
8	$\sin(at + b)$	$\dfrac{a \cos b + s \sin b}{s^2 + a^2}$	$\operatorname{Re} s > 0$	$a, b \in \mathbb{R}$
9	$\cos(at + b)$	$\dfrac{s \cos b - a \sin b}{s^2 + a^2}$	$\operatorname{Re} s > 0$	$a, b \in \mathbb{R}$
10	$t^n e^{-at}$	$\dfrac{n!}{(s + a)^{n+1}}$	$\operatorname{Re} s > \operatorname{Re}(-a)$	$a \in \mathbb{C}$
11	$\epsilon(t - a)$	$\dfrac{e^{-as}}{s}$	$\operatorname{Re} s > 0$	$a > 0$

TABLE 8   Properties of the Laplace transform of causal functions

no.	$f(t), g(t)$	$F(s), G(s)$	*condition(s)*
1	$af(t) + bg(t)$	$aF(s) + bG(s)$	$a, b \in \mathbb{C}$
2	$\epsilon(t-a)f(t-a)$	$e^{-as}F(s)$	$a \geq 0$
3	$e^{at}f(t)$	$F(s-a)$	$a \in \mathbb{C}$
4	$f(at)$	$a^{-1}F(a^{-1}s)$	$a > 0$
5	$f^{(n)}(t)$	$s^n F(s)$	
6	$(-1)^n t^n f(t)$	$F^{(n)}(s)$	
7	$\displaystyle\int_0^t f(\tau)\,d\tau$	$\dfrac{F(s)}{s}$	
8	$(f * g)(t)$	$F(s)G(s)$	
9	$f(t)$	$\dfrac{\int_0^T f(t)e^{-st}\,dt}{1 - e^{-sT}}$	$f(t+T) = f(t)$

10	$\displaystyle\lim_{s \to \infty} sF(s) = f(0+)$
11	$\displaystyle\lim_{t \to \infty} f(t) = \lim_{s \to 0} sF(s)$

TABLE 9　　　Laplace transforms of distributions

no.	$T(t)$	$U(s) = \langle T(t), e^{-st} \rangle$	condition(s)
1	$\delta(t)$	1	
2	$\delta^{(n)}(t)$	s^n	
3	$\delta(t - a)$	e^{-as}	$a > 0$
4	$\delta^{(n)}(t - a)$	$s^n e^{-as}$	$a > 0$

TABLE 10　　　Properties of the Laplace transform of distributions

no.	$T(t), S(t)$	$U(s), V(s)$	condition(s)
1	$aT(t) + bS(t)$	$aU(s) + bV(s)$	$a, b \in \mathbb{C}$
2	$T(t - a)$	$e^{-as} U(s)$	$a \geq 0$
3	$T^{(n)}(t)$	$s^n U(s)$	
4	$(T * S)(t)$	$U(s)V(s)$	

TABLE 11　　　Discrete Fourier transforms of periodic discrete-time signals with period N

no.	$f[n]$	$F[k] = \sum_{n=0}^{N-1} f[n]e^{-2\pi ink/N}$	condition(s)
1	$\delta_N[n]$	1	
2	$e^{2\pi inl/N}$	$N\delta_N[k - l]$	
3	$\displaystyle\sum_{l=-m}^{m} \delta_N[n - l]$	$\begin{cases} 2m + 1 & \text{if } k \text{ is a multiple of } N \\ \dfrac{\sin((2m + 1)k\pi/N)}{\sin(k\pi/N)} & \text{otherwise} \end{cases}$	$0 \leq m < \frac{N}{2}$

TABLE 12 Properties of the discrete Fourier transform of periodic discrete-time signals

no.	$f[n], g[n]$	$F[k], G[k]$	condition(s)
1	$af[n] + bg[n]$	$aF[k] + bG[k]$	$a, b \in \mathbb{C}$
2	$\overline{f[n]}$	$\overline{F[-k]}$	
3	$f[n - l]$	$e^{-2\pi i l k/N} F[k]$	$l \in \mathbb{Z}$
4	$e^{2\pi i l n/N} f[n]$	$F[k - l]$	$l \in \mathbb{Z}$
5	$f[-n]$	$F[-k]$	
6	$F[n]$	$Nf[-k]$	
7	$(f * g)[n]$	$F[k]G[k]$	
8	$f[n]g[n]$	$\dfrac{1}{N}(F * G)[k]$	

no.		
9	$\displaystyle\sum_{n=0}^{N-1} f[n]\overline{g[n]} = \frac{1}{N}\sum_{k=0}^{N-1} F[k]\overline{G[k]}$	
10	$\displaystyle\sum_{n=0}^{N-1} |f[n]|^2 = \frac{1}{N}\sum_{k=0}^{N-1} |F[k]|^2$	

TABLE 13 z-transforms of non-periodic discrete-time signals

no.	$f[n]$	$F(z) = \sum\limits_{n=-\infty}^{\infty} f[n]z^{-n}$	convergence region	condition(s)
1	$\delta[n]$	1	$z \in \mathbb{C}$	
2	$a^n \epsilon[n]$	$\dfrac{z}{z-a}$	$\lvert z \rvert > \lvert a \rvert$	$a \in \mathbb{C}$
3	$a^n \epsilon[-n-1]$	$\dfrac{-z}{z-a}$	$\lvert z \rvert < \lvert a \rvert$	$a \in \mathbb{C}$
4	$\dbinom{n}{k} a^n \epsilon[n]$	$\dfrac{a^k z}{(z-a)^{k+1}}$	$\lvert z \rvert > \lvert a \rvert$	$a \in \mathbb{C}, k \in \mathbb{N}$
5	$\dbinom{n}{k} a^n \epsilon[-n-1]$	$\dfrac{-a^k z}{(z-a)^{k+1}}$	$\lvert z \rvert < \lvert a \rvert$	$a \in \mathbb{C}, k \in \mathbb{N}$
6	$\cos(\omega_0 n + \phi_0)\epsilon[n]$	$\dfrac{z^2\cos\phi_0 - z\cos(\phi_0-\omega_0)}{z^2 - 2z\cos\omega_0 + 1}$	$\lvert z \rvert > 1$	$\omega_0, \phi_0 \in \mathbb{R}$

TABLE 14   Properties of the z-transform of non-periodic discrete-time signals

no.	$f[n], g[n]$	$F(z), G(z)$	condition(s)
1	$af[n] + bg[n]$	$aF(z) + bG(z)$	$a, b \in \mathbb{C}$
2	$f[-n]$	$F(1/z)$	
3	$\overline{f[n]}$	$\overline{F(\overline{z})}$	
4	$f[n-l]$	$z^{-l} F(z)$	$l \in \mathbb{Z}$
5	$a^n f[n]$	$F(z/a)$	$a \neq 0$
6	$nf[n]$	$-z\dfrac{d}{dz} F(z)$	
7	$(f * g)[n]$	$F(z)G(z)$	

TABLE 15   Properties of the Fourier transform of non-periodic discrete-time signals

no.	$f[n], g[n]$	$F(e^{i\omega}) = \sum\limits_{n=-\infty}^{\infty} f[n]e^{-in\omega},\ G(e^{i\omega})$	condition(s)
1	$af[n] + bg[n]$	$aF(e^{i\omega}) + bG(e^{i\omega})$	$a, b \in \mathbb{C}$
2	$\overline{f[n]}$	$\overline{F(e^{-i\omega})}$	
3	$f[n-k]$	$e^{-ik\omega}F(e^{i\omega})$	$k \in \mathbb{Z}$
4	$e^{-in\omega_0}f[n]$	$F(e^{i(\omega+\omega_0)})$	$\omega_0 \in \mathbb{R}$
5	$f[-n]$	$F(e^{-i\omega})$	
6	$(f * g)[n]$	$F(e^{i\omega})G(e^{i\omega})$	
7	$f[n]g[n]$	$\dfrac{1}{2\pi}\displaystyle\int_{-\pi}^{\pi} F(e^{i(\omega-u)})G(e^{iu})\,du$	

8	$\sum\limits_{n=-\infty}^{\infty} f[n]\overline{g[n]} = \dfrac{1}{2\pi}\displaystyle\int_{-\pi}^{\pi} F(e^{i\omega})\overline{G(e^{i\omega})}\,d\omega$
9	$\sum\limits_{n=-\infty}^{\infty} \lvert f[n]\rvert^2 = \dfrac{1}{2\pi}\displaystyle\int_{-\pi}^{\pi} \left\lvert F(e^{i\omega})\right\rvert^2 d\omega$

Index

Made in the USA
Monee, IL
07 July 2026

56550234R00251